To Vimla, Neeru, Neeraj, and Trevor

Ram P. Kanwal

Generalized Functions Theory and Technique

Second Edition

Birkhäuser
Boston • Basel • Berlin

Ram P. Kanwal
Department of Mathematics
The Pennsylvania State University
University Park, PA 16802
USA

Library of Congress Cataloging-in-Publication Data

Kanwal, Ram P.
Generalized functions : theory and technique / Ram P. Kanwal. --
2nd ed.
p. cm.
Includes bibliographical references (p. 449- 456) and index.
ISBN-13: 978-1-4684-0037-3 e-ISBN-13: 978-1-4684-0035-9
DOI: 10.1007/978-1-4684-0035-9
: acid-free paper)
1. Theory of distributions (Functional analysis) I. Title.
QA324.K36 1997
515'.782--dc21 97-13419
CIP

AMS Subject Classifications: 46F10-46F12, 42A05, 42B05, 35D05, 35E05
35I25, 34E05, 34E15, 45D05

Printed on acid-free paper
 Birkhäuser
Softcover reprint of the hardcover 2nd edition 1998

ISBN-13: 978-1-4684-0037-3

Reformatted from author's disk by TEXniques, Inc., Boston, MA.

9 8 7 6 5 4 3 2 1

Contents

Preface to the Second Edition

This second edition of Generalized Functions has been strengthened in many ways. The already extensive set of examples has been expanded. Since the publication of the first edition, there has been tremendous growth in the subject and I have attempted to incorporate some of these new concepts. Accordingly, almost all the chapters have been revised. The bibliography has been enlarged considerably.

Some of the material has been reorganized. For example, Chapters 12 and 13 of the first edition have been consolidated into Chapter 12 of this edition by a judicious process of elimination and addition of the subject matter. The new Chapter 13 explains the interplay between the theories of moments, asymptotics, and singular perturbations. Similarly, some sections of Chapter 15 have been revised and included in earlier chapters to improve the logical flow of ideas. However, two sections are retained. The section dealing with the application of the probability theory has been revised, and I am thankful to Professor Z.L. Crvenković for her help. The new material included in this chapter pertains to the modern topics of periodic distributions and microlocal theory.

I have demonstrated through various examples that familiarity with the generalized functions is very helpful for students in physical sciences and technology. For instance, the reader will realize from Chapter 6 how the generalized functions have revolutionized the Fourier analysis which is being used extensively in many fields of scientific activity. As is demonstrated in Chapter 10, one of the main areas of application is to the theory of partial differential equations. Following the motivation and the theme of the first edition, the additional material can be understood without the knowledge of advanced mathematics. Accordingly, the book remains accessible to first year graduate students. Indeed, I have taught a course on this subject at the The Pennsylvania State University for many years, and the students' response was very gratifying.

I have had the good fortune of doing research work with Professor Ricardo Estrada on various aspects of generalized functions. I take this opportunity to thank him for his two decades of collaboration. Finally, I am thankful to the editors and the staff of Birkhäuser for their helpful cooperation in the publication of this edition.

Ram P. Kanwal
The Pennsylvania State University

Preface to the First Edition

There has recently been a significant increase in the number of topics for which generalized functions have been found to be very effective tools. Familiarity with the basic concepts of this theory has become indispensable for students in applied mathematics, physics, and engineering, and it is becoming increasingly clear that methods based on generalized functions not only help us to solve unsolved problems but also enable us to recover known solutions in a very simple fashion.

This book contains both the theory and applications of generalized functions, with a significant feature being the quantity and variety of applications of this theory. I have attempted to furnish a wealth of applications from various physical and mathematical fields of current interest and have tried to make the presentation direct yet formal. Definitions and theorems are stated precisely, but rigor is minimized in favor of comprehension of techniques. Many examples are presented to illustrate the concepts, definitions, and theorems. Except for a few research topics, the mathematical background expected from a student is available in undergraduate courses in advanced calculus, ordinary and partial differential equations, and boundary value problems. Accordingly, most of the material is easily accessible to senior undergraduate and graduate students in mathematical, physical, and engineering sciences. The chapters that are suitable for a one-semester course are furnished with sets of exercises.

I hope that this book will encourage applied mathematicians, scientists, and engineers to make use of the powerful tools of generalized functions.

My thanks are due to many former students and my colleagues whose reactions and comments helped me in the preparation of this text. In particular, I thank A. Alawneh, R. Ayoub, R. Estrada, D.L. Jain, A. Krall, S. Obaid, R. Rostamian, B.K. Sachdeva, and I.M. Sheffer. A special word of gratitude goes to S. Obaid who also checked the manuscript. I am also grateful to the staff of the Academic Press for their cooperation.

CHAPTER 1

The Dirac Delta Function and Delta Sequences

1.1. The Heaviside function

The Heaviside function $H(x)$ is defined to be equal to zero for every negative value of x and to unity for every positive value of x, that is,

$$H(x) = \begin{cases} 0, & x < 0, \\ 1, & x > 0. \end{cases} \tag{1}$$

It has a jump discontinuity at $x = 0$ and is also called the *unit step function*. Its value at $x = 0$ is usually taken to be $\frac{1}{2}$. Sometimes it is taken to be a constant c, $0 < c < 1$, and then the function is written $H_c(x)$. If the jump in the Heaviside function is at a point $x = a$, then the function is written $H(x - a)$. Observe that

$$H(-x) = 1 - H(x), \qquad H(a - x) = 1 - H(x - a). \tag{2}$$

The functions $H(x)$, $H(x - a)$, and $H(a - x)$ are drawn in Figure 1.1.

We shall come across the Heaviside function with various arguments. For example, let us examine $H(ax + b)$. If $a > 0$, then this function is zero when $ax + b < 0$ or $x + b/a < 0$ and is unity when $x + b/a > 0$; that is, $H(ax + b) = H(x + b/a)$, $a > 0$. Similarly, when $a < 0$, we set $a = -A$, where $A > 0$, and find that $H(ax + b) = H(-x - b/a)$. Thus

$$H(ax + b) = H(x + b/a)H(a) + H(-x - b/a)H(-a). \tag{3}$$

The step function permits the annihilation of a part of the graph of a function $F(x)$. For instance, $y = H(x - a)F(x)$ is zero before $x = a$ and equal to $F(x)$ after $x = a$. Similarly, the function $H(x - a)F(x - a)$ translates the graph of $F(x)$, as shown in Figure 1.2.

The function $H(x)$ will prove very useful in the study of the generalized functions, especially in the discussion of functions with jump discontinuities. For instance, let $F(x)$ be a function that is continuous everywhere except for the point $x = \xi$, at which point it has a jump discontinuity

$$F(x) = \begin{cases} F_1(x), & x < \xi, \\ F_2(x), & x > \xi. \end{cases} \tag{4}$$

Then it can be written

$$F(x) = F_1(x)H(\xi - x) + F_2(x)H(x - \xi). \tag{5}$$

It is shown in Figure 1.3.

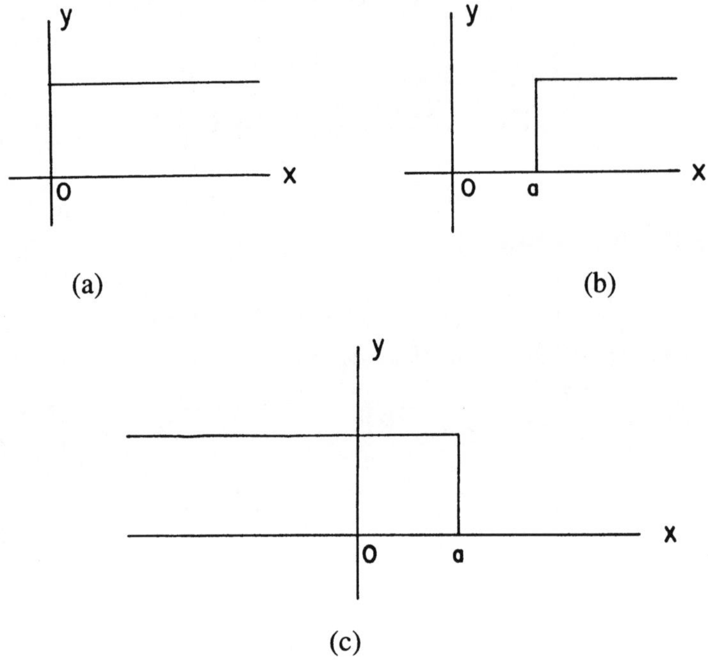

Figure 1.1. (a) $H(x)$; (b) $H(x-a)$; (c) $H(a-x)$.

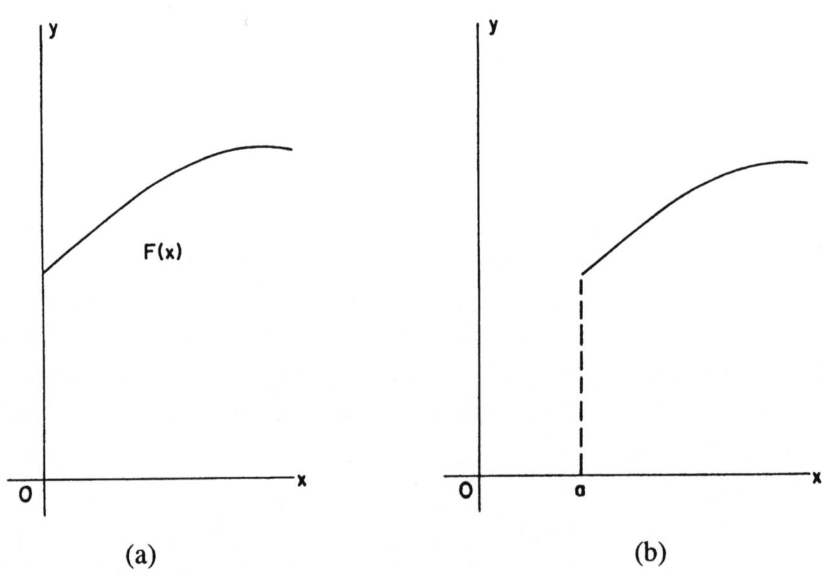

Figure 1.2. (a) $F(x)$; (b) $H(x-a)F(x-a)$.

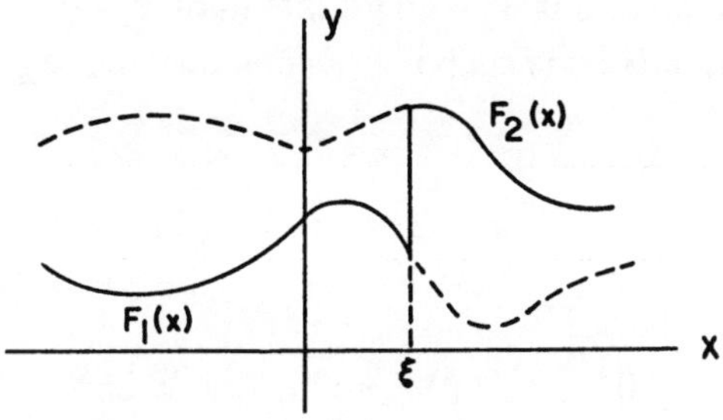

Figure 1.3. (a) $F(x) = F_1(x)H(\xi - x) + F_2(x)H(x - \xi)$.

This concept can be extended to enable us to write a function that has jump discontinuities at several points. For instance, we can write the function

$$F(x) = \begin{cases} x^2, & 0 < x < 1, \\ 3, & 1 < x < 2, \\ 0, & x > 2, \end{cases}$$

as

$$F(x) = x^2 H(x) - H(x-1)(-3 + x^2) - 3H(x-2).$$

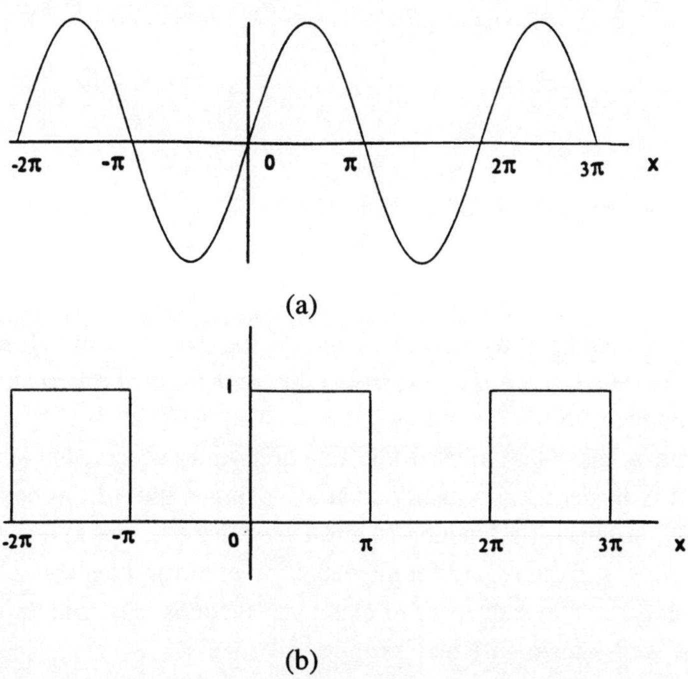

Figure 1.4. (a) $y = \sin x$; (b) $y = H(\sin x)$.

One can similarly write functions with an infinite number of jump discontinuities. For example, the periodic function $\sin x$ is such that its positive and negative values alternate;

$$\begin{aligned} \sin x &< 0 \quad \text{for} \quad (2n-1)\pi < x < 2n\pi \,, \\ \sin x &> 0 \quad \text{for} \quad 2n\pi < x < (2n+1)\pi \,. \end{aligned}$$

Accordingly,

$$H(\sin x) = \begin{cases} 0, & (2n-1)\pi < x < 2n\pi \,, \\ 1, & 2n\pi < x < (2n+1)\pi \,. \end{cases} \tag{6}$$

$$= \sum_{n=-\infty}^{\infty} [H(x - 2n\pi) - H(x - (2n+1)\pi)] \,. \tag{7}$$

The functions $\sin x$ and $H(\sin x)$ are shown in Figure 1.4.

1.2 The Dirac Delta Function

In physical problems one often encounters idealized concepts such as a force concentrated at a point ξ or an impulsive force that acts instantaneously. These forces are described by the Dirac delta function $\delta(x-\xi)$, which has several significant properties:

$$\delta(x-\xi) = 0 \,, \qquad x \neq \xi \,, \tag{1}$$

$$\int_a^b \delta(x-\xi)dx = \begin{cases} 0, & a, b < \xi \quad \text{or} \quad \xi < a, b \,, \\ 1, & a \le \xi \le b \,, \end{cases} \tag{2}$$

and

$$\int_{-\infty}^{\infty} \delta(x-\xi)dx = 1 \,. \tag{3}$$

Equation (3) is a special case of the general formula

$$\int_{-\infty}^{\infty} \delta(x-\xi)f(x)dx = f(\xi) \,, \tag{4}$$

where $f(x)$ is a sufficiently smooth function (this will become clear as we go along). Relation (4) is called the *sifting property* or the *reproducing property* of the delta function, and (3) is obtained from it by putting $f(x) = 1$.

Although scientists have used this function with success, the language of classical mathematics is inadequate to justify such a function. Indeed, properties (1) and (2) are contradictory, because if a function is zero everywhere except at one point, its integral is necessarily zero, without regard for the definition used for the integral.

Fortunately, certain sequences of classical functions exist and have property (4). For instance, the well-known Dirichlet formula

$$\lim_{m\to\infty} \int_{-\infty}^{\infty} f(x) \frac{\sin mx}{\pi x} dx = f(0) \tag{5}$$

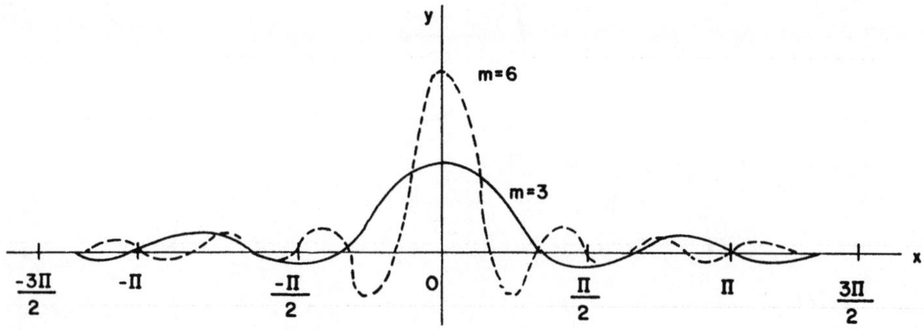

Figure 1.5. $s_m(x) = (\sin mx)/\pi x$.

satisfies the sifting property. This suggests that we may define the delta function as the limit of a sequence of suitable functions. The next section is devoted to this concept.

1.3. The Delta Sequences

Here we consider various sequences whose limit is the delta function. We have already mentioned sequence (5) which we now discuss in detail as our first example.
Example 1.

$$s_m(x) = \sin mx/\pi x\,, \qquad m = 1, 2, \ldots . \tag{1}$$

It is clear that for fixed m as $|x|$ becomes large $s_m(x)$ becomes small (see Figure 1.5). Recall the formula

$$\int_0^\infty \frac{\sin x}{x} dx = \frac{\pi}{2}\,. \tag{2}$$

Changing x to $-y$, we obtain

$$\int_{-\infty}^0 \frac{\sin x}{x} dx = \frac{\pi}{2}\,. \tag{3}$$

Adding (2) and (3) yields

$$\int_{-\infty}^\infty \frac{\sin x}{\pi x} dx = 1\,.$$

The result of changing x to mx in the latter formula is

$$\int_{-\infty}^\infty s_m(x)dx = \int_{-\infty}^\infty \frac{\sin mx}{\pi x} dx = 1\,. \tag{4}$$

This takes care of property (3) of the delta function.

Let us now attend to the sifting property and examine

$$\int_{-\infty}^{\infty} \frac{\sin mx}{\pi x} f(x)dx ,$$

where $f(x)$ is differentiable with $f'(x)$ continuous and bounded.

Observe that, for any $b > a > 0$,

$$\int_a^b s_m(x)dx = \frac{1}{\pi}\int_a^b \frac{\sin mx}{x}dx = \frac{1}{\pi}\int_{am}^{bm} \frac{\sin y}{y}dy ,$$

and

$$\lim_{m\to\infty} \frac{1}{\pi}\int_{am}^{bm} \frac{\sin y}{y}dy \to 0 .$$

Similarly, for $a < b < 0$,

$$\lim_{m\to\infty} \frac{1}{\pi}\int_{am}^{bm} \frac{\sin y}{y}dy \to 0 .$$

Thus

$$\begin{aligned}\lim_{m\to\infty}\int_{-\infty}^{\infty} s_m(x)f(x)dx &= \lim_{m\to\infty}\int_{-\varepsilon}^{\varepsilon} \frac{\sin mx}{\pi x} f(x)dx \\ &= f(0)\left[\lim_{m\to\infty}\int_{-\varepsilon}^{\varepsilon} \frac{\sin mx}{\pi x}dx\right] \\ &= f(0)\left[\lim_{m\to\infty}\int_{-\infty}^{\infty} \frac{\sin mx}{\pi x}dx\right] = f(0) ,\end{aligned}$$

where ε is any positive number, no matter how small, and we have used relation (4) as well as the mean value theorem of integral calculus. This proves the sifting property.

Example 2. A very important example of a delta sequence is

$$s_m(x) = \frac{1}{\pi}\frac{m}{1+m^2x^2} . \tag{5}$$

It is instructive to interpret (5) as a continuous charge distribution on a line (see Figure 1.6). It is clear that for $m \gg 1$, $s_m(x) \ll 1$, except for a peak of m/π at $x = 0$. The total charge $r_m(x)$ to the left of the point x is

$$r_m(x) = \int_{-\infty}^{x} s_m(u)du = \frac{1}{2} + \frac{1}{\pi}\tan^{-1} mx , \tag{6}$$

the *cumulative charge distribution* (see Figure 1.7). Since

$$\int_{-\infty}^{\infty} s_m(u)du = 1 \tag{7}$$

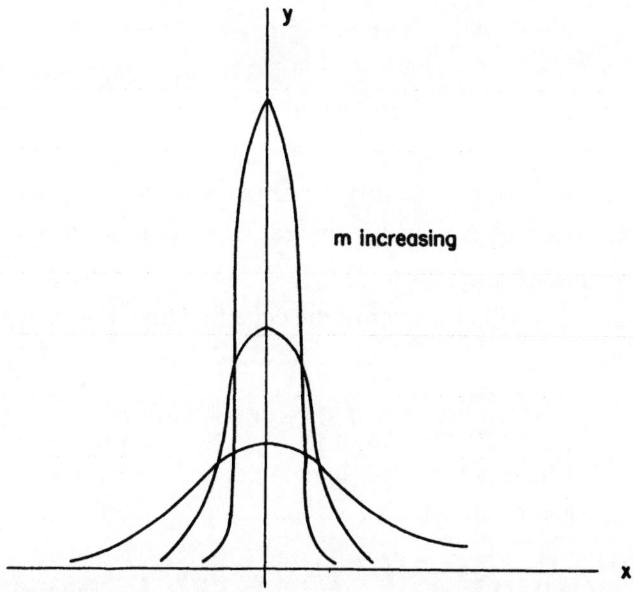

Figure 1.6. $s_m(x) = m/[\pi(1 + m^2x^2)]$.

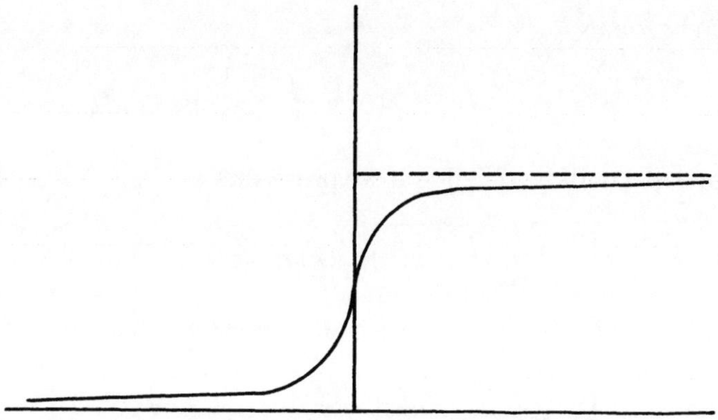

Figure 1.7. $r_m(x) = \frac{1}{2} + (1/\pi)(\tan^{-1} mx)$.

for every m, it follows that the total charge on the line is always equal to unity.

Furthermore,

$$\lim_{m\to\infty} s_m(x) = \begin{cases} 0, & x \neq 0, \\ \infty, & x = 0; \end{cases} \tag{8a}$$

$$\lim_{m\to\infty} r_m(x) = \begin{cases} 0\,, & x < 0\,, \\ \frac{1}{2}\,, & x = 0\,, \\ 1\,, & x > 0\,. \end{cases} \tag{8b}$$

As m increases, the charge is pushed toward the origin. Thus $\lim_{m\to\infty} s_m(x)$ describes the charge density due to a positive unit charge located at $x = 0$. It therefore resembles (we have still to prove the sifting property) the Dirac delta function and is not an ordinary function. The corresponding cumulative charge distribution, which arises from $\lim_{m\to\infty} r_m(x)$, is the Heaviside function $H(x)$.

The sequence $s_m(x)$ will characterize the delta function if we can prove that it satisfies the sifting property

$$\lim_{m\to\infty} \int_{-\infty}^{\infty} f(x) s_m(x) dx = f(0) \tag{9}$$

for a function $f(x)$ which is bounded, integrable, and continuous at $x = 0$. The proof follows by writing equation (9) as

$$\int_{-\infty}^{\infty} f(x) s_m(x) dx = \int_{-\infty}^{\infty} f(0) s_m(x) dx + \int_{-\infty}^{\infty} g(x) s_m(x) dx\,, \tag{10}$$

where

$$g(x) = f(x) - f(0).$$

In view of (7), (10) becomes

$$\int_{-\infty}^{\infty} f(x) s_m(x) dx = f(0) + \int_{-\infty}^{\infty} g(x) s_m(x) dx\,.$$

Thus, the sifting property will be satisfied if we prove that

$$\lim_{m\to\infty} \int_{-\infty}^{\infty} g(x) s_m(x) dx = 0\,. \tag{11}$$

We have, therefore, to establish that for any $\varepsilon > 0$ there exists an index N such that

$$\left| \int_{-\infty}^{\infty} g(x) s_m(x) dx \right| < \varepsilon\,, \qquad m > N\,.$$

For this purpose let A be a positive number (soon to be specified) that divides the interval $-\infty$ to ∞ into three parts, so that

$$\int_{-\infty}^{\infty} g s_m dx = \int_{-\infty}^{-A} g s_m dx + \int_{A}^{\infty} g s_m dx + \int_{-A}^{A} g s_m dx = I_1 + I_2 + I_3\,.$$

For the integral I_3, let the maximum of $|g|$ in $-A \le x \le A$ be denoted $M(A)$. Then

$$|I_3| \le \int_{-A}^{A} |g| s_m dx \le M(A) \int_{-A}^{A} \frac{m}{\pi(1 + m^2 x^2)} dx$$
$$= M(A) \left[\frac{2}{\pi} \tan^{-1} mA \right] \le M(A).$$

Since $g(0) = 0$ and $g(x)$ is continuous at $x = 0$, we have $\lim_{A\to 0} M(A) = 0$. Consequently, for any $\varepsilon > 0$, there exists a real number A sufficiently small that $|I_3| < \varepsilon/2$, and this holds independent of M.

With the number A so chosen, it remains to show that $|I_1 + I_2|$ is small for sufficiently large m. Since $f(x)$ is bounded and $|g(x)| < |f(x)| + |f(0)|$, it follows that $|g(x)|$ is bounded in $-\infty < x < \infty$, say $|g| < b$. Then

$$|I_1 + I_2| \le b\left[\int_{-\infty}^{-A} s_m dx + \int_A^{\infty} s_m dx\right] = b\left(1 - \frac{2}{\pi}\tan^{-1} mA\right).$$

With the number A fixed, $\lim_{m\to\infty} (2/\pi)\tan^{-1} mA = 1$. This means that we can find N such that

$$b[1 - (2/\pi)\tan^{-1} mA] < \varepsilon/2, \qquad m > N.$$

With this choice of N, we have

$$\left|\int_{-\infty}^{\infty} g(s)s_m(x)dx\right| \le |I_1 + I_2 + I_3| \le |I_1 + I_2| + |I_3| < \varepsilon$$

and relation (11) follows. This completes the proof.

Example 3. Various books in physics leave the reader with the impression that $\delta(x) = +\infty$ at $x = 0$. The following sequence illustrates that this is not always true:

$$s_m(x) = \begin{cases} -m, & |x| < 1/2m, \\ 2m, & 1/2m \le |x| \le 1/m, \\ 0 & \text{otherwise.} \end{cases} \tag{12}$$

This sequence is shown in Figure 1.8, which shows that $\lim_{m\to\infty} s_m(x) = -\infty$, although it yields the unit positive charge for large m; indeed, it is a delta sequence, as will now be proved.

The cumulative charge distribution $r_m(x)$ is

$$r_m(x) = \int_{-\infty}^{x} s_m(u)du = 0, \qquad x < -\frac{1}{m};$$

$$r_m(x) = \int_{-1/m}^{x} 2m\,du = 2m\left(x + \frac{1}{m}\right), \qquad -\frac{1}{m} \le x \le -\frac{1}{2m};$$

$$\begin{aligned} r_m(x) &= \int_{-1/m}^{-1/2m} 2m\,du + \int_{-1/2m}^{x} (-m)du \\ &= 1 - m\left(x + \frac{1}{2m}\right), \qquad -\frac{1}{2m} \le x \le \frac{1}{2m}; \end{aligned}$$

$$\begin{aligned} r_m(x) &= \int_{-1/m}^{-1/2m} 2m\,du + \int_{-1/2m}^{1/2m} (-m)du + \int_{1/2m}^{x} du \\ &= 2m(x - 1/2m), \qquad 1/2m \le x \le 1/m. \end{aligned}$$

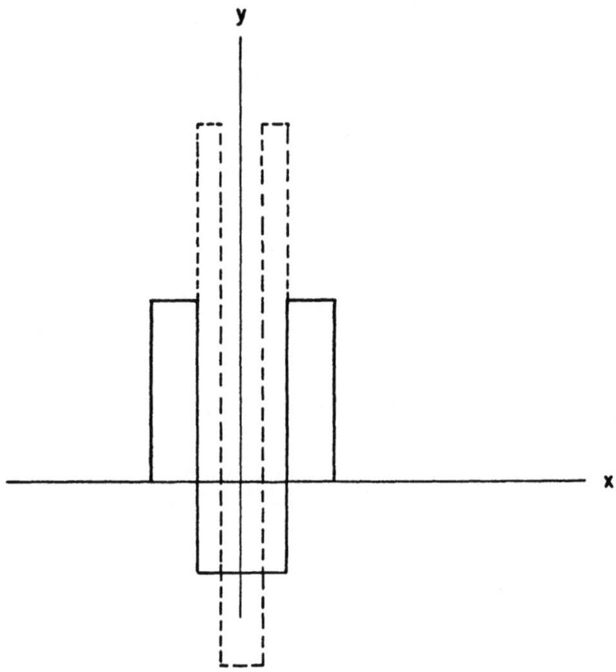

Figure 1.8. A delta sequence whose limit tends to $-\infty$.

and

$$r_m(x) = 1\,, \qquad x \geq 1/m\,.$$

In particular,

$$\int_{-\infty}^{\infty} s_m(x)dx = 1\,. \tag{13}$$

Let $f(x)$ be bounded, integrable, and continuous at $x = 0$. Then

$$\begin{aligned}\int_{-\infty}^{\infty} f(x)s_m(x)dx &= f(0)\int_{-\infty}^{\infty} s_m(x)dx + \int_{-\infty}^{\infty} g(x)s_m(x)dx \\ &= f(0) + \int_{-\infty}^{\infty} g(x)s_m(x)dx\,,\end{aligned}$$

where $g(x)$ is defined as in (10). From the definition of $s_m(x)$ we have

$$\begin{aligned}\int_{-\infty}^{\infty} g(x)s_m(x)dx =& 2m\int_{-1/m}^{-1/2m} g(x)dx - m\int_{-1/2m}^{1/2m} g(x)dx \\ &+ 2m\int_{1/2m}^{1/m} g(x)dx\,.\end{aligned} \tag{14}$$

Since $g(x)$ is continuous at $x = 0$ and $g(0) = 0$, it follows that for any given $\varepsilon > 0$ there exists an integer N such that

$$|g(x)| < \varepsilon/3\,, \qquad |x| < 1/N\,.$$

Thus from (14) we find that, for all $m > N$,

$$\begin{aligned}\int_{-\infty}^{\infty} g(x)s_m(x)dx &\leq 2m\int_{-1/m}^{-1/2m} |g(x)|dx + m\int_{-1/2m}^{1/2m} |g(x)|dx \\ &\quad + 2m\int_{1/2m}^{1/m} |g(x)|dx \\ &\leq \varepsilon/3\left[2m\left(-\frac{1}{2m}+\frac{1}{m}\right)+m\left(\frac{1}{2m}+\frac{1}{2m}\right)\right. \\ &\quad \left. + 2m\left(\frac{1}{m}-\frac{1}{2m}\right)\right], \\ &= \varepsilon .\end{aligned}$$

and the sifting property follows.

Example 4. Next we show that the Gaussian sequence from the theory of statistics,

$$(m/\pi)^{1/2}e^{-mx^2}, \tag{15}$$

defines a delta sequence. We prove that for a function $f(x)$ that is bounded, integrable, and continuous at $x = 0$

$$\lim_{m\to\infty}\int_{-\infty}^{\infty}\left(\frac{m}{\pi}\right)^{1/2} e^{-mx^2} f(x)dx = f(0)\,. \tag{16}$$

Indeed, proceeding as in the previous examples, we find that

$$\int_{-\infty}^{\infty}\left(\frac{m}{\pi}\right)^{1/2} e^{-mx^2} f(x)dx = f(0) + \int_{-\infty}^{\infty}\left(\frac{m}{x}\right)^{1/2} e^{-mx^2} g(x)dx\,. \tag{17}$$

Also

$$\left|\int_{-\infty}^{\infty}\left(\frac{m}{\pi}\right)^{1/2} e^{-mx^2} g(x)dx\right| = \left|\int_{-\infty}^{-A} + \int_{A}^{\infty} + \int_{-A}^{A}\left(\frac{m}{\pi}\right)^{1/2} e^{-mx^2} g(x)dx\right|, \tag{18}$$

where A is a finite real number. In view of the boundedness of $f(x)$,

$$|g(x)| = |f(x) - f(0)| < M\,,$$

and we have

$$\begin{aligned}\left|\int_{-\infty}^{-A}\left(\frac{m}{\pi}\right)^{1/2} e^{-mx^2} g(x)dx\right| &\leq M\left|\int_{-\infty}^{-A}\left(\frac{m}{\pi}\right)^{1/2} e^{-mx^2} dx\right| \\ &= \frac{M}{\sqrt{\pi}}\int_{-\infty}^{-A\sqrt{m}} e^{-y^2}dy \\ &\to 0 \quad \text{as} \quad m\to\infty\,.\end{aligned}$$

Similarly, the second integral on the right side tends to zero as $m \to \infty$.

Finally, we exploit the continuity of $f(x)$ at $x = 0$. Given $\varepsilon > 0$, there exists a $\delta > 0$ such that

$$|f(x) - f(0)| < \varepsilon \qquad \text{for} \quad |x| < \delta .$$

If we choose A such that $2A < \delta$, then we have

$$\begin{aligned}
\left| \int_{-A}^{-A} \left(\frac{m}{\pi}\right)^{1/2} e^{-mx^2} g(x) dx \right| &\le \int_{-A}^{-A} \left(\frac{m}{\pi}\right)^{1/2} e^{-mx^2} |g(x)| dx \\
&\le \varepsilon \int_{-A}^{A} \left(\frac{m}{\pi}\right)^{1/2} e^{-mx^2} dx \\
&= \frac{\varepsilon}{\sqrt{\pi}} \int_{A\sqrt{m}}^{-A\sqrt{m}} e^{-y^2} dy \\
&\to \varepsilon \quad \text{as} \quad m \to \infty .
\end{aligned} \tag{19}$$

Combining (17)–(19), we observe that the sifting property (16) has been established.

From these examples we find (intuitively) a sequence of functions each of which has its maximum value at $x = 0$ and as we move along the sequence, the maximum value increases while the graph of the function gets narrower so that it leads to the sifting property. The sequences of functions which lead to the delta function in this manner are called delta-convergent sequences. We, therefore, have the following definition:

Definition. A sequence $s_m(x)$ is called a *delta-convergent sequence* if

$$\lim_{m\to\infty} \int_{-\infty}^{\infty} s_m(x) f(x) dx = f(0) ,$$

for all functions $f(x)$ sufficiently smooth in $-\infty < x < \infty$. Thus we can say that for a delta-convergence sequence

$$\lim_{m\to\infty} s_m(x) = \delta(x) .$$

In these examples we have taken the unit charge located at $x = 0$. If it is located at $x = \xi$, then the preceding formulas become

$$\lim_{m\to\infty} \int_{-\infty}^{\infty} s_m(x - \xi) f(x) dx = f(\xi)$$

and

$$\lim_{m\to\infty} s_m(x - \xi) = \delta(x - \xi) .$$

For example,

$$\lim_{m\to\infty} \frac{1}{\pi} \left\{ \frac{m}{1 + m^2(x - \xi)^2} \right\} = \delta(x - \xi) . \tag{20}$$

The sifting property,

$$\int_{-\infty}^{\infty} f(x)\delta(x - \xi) dx = f(\xi) ,$$

is interpreted as the *action* of the generalized function $\delta(x-\xi)$ on $f(x)$; that is, when $\delta(x-\xi)$ acts on a suitably smooth function $f(x)$, it sifts out the value $f(\xi)$ at $x=\xi$.

1.4. A Unit Dipole

We have seen in previous examples how certain convergent sequences converge to a delta function and represent idealized concepts such as a unit charge. We shall now prove that the derivative with respect to x of a delta-convergent sequence gives a sequence that represents a unit dipole.

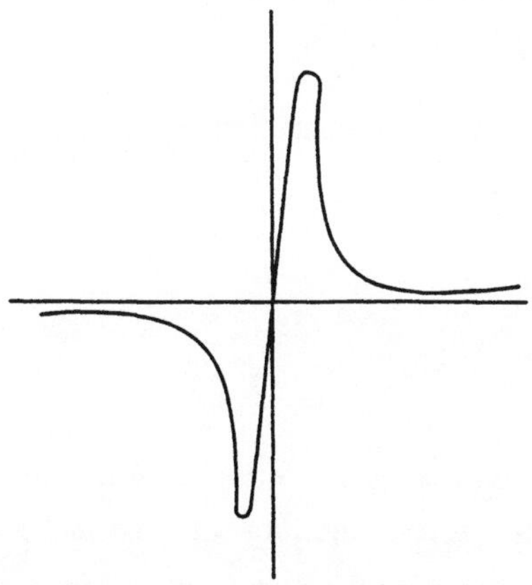

Figure 1.9. $t_m(x) = 2m^3x/\pi(1+m^2x^2)$.

Let charge $\pm m$ be located at $x=\pm\varepsilon$, respectively. The product $2m\varepsilon$ is known as the dipole moment of the charge configuration. When we let $\varepsilon \to 0$ and $m \to \infty$ in such a way that $2m\varepsilon = 1$, we get the dipole moment equal to 1, i.e., a unit dipole. Our contention is that it can be approximated by a continuous charge distribution that is the derivative of a corresponding distribution of a unit charge. For this purpose we shall take the sequence $s_m(x) = m/\pi(1+m^2x^2)$ and show that

$$t_m(x) = -\frac{ds_m}{dx} = \frac{2m^3x}{\pi(1+m^2x^2)^2},$$

describes a unit dipole. A sketch of $t_m(x)$ for large values of m is given in Figure 1.9. Furthermore,

$$\int_a^b t_m(x)dx = s_m(a) - s_m(b) = \frac{m}{\pi(1+m^2a^2)} - \frac{m}{\pi(1+m^2b^2)}. \tag{1}$$

If neither a nor b is zero, each term on the right side of (1) approaches zero as $m \to \infty$, and the total charge in any such interval goes to zero as $m \to \infty$. However, $\int_0^\varepsilon t_m(x)dx \to +\infty$ (like a positive point charge just to the right of the origin), and $\int_{-\varepsilon}^0 t_m(x)dx \to -\infty$ (like a negative point charge just to the left of the origin), whereas the first moment about the origin is $\int_{-\infty}^{\infty} x t_m(x)dx = 1$. Consequently, for large m, $t_m(x)$ approaches a unit dipole located at $x = 0$ and is thus a dipole sequence.

Finally, let us consider the action $A_m[f]$ of $t_m(x)$ on a suitably smooth function $f(x)$,

$$\begin{aligned}
A_m[f] &= \lim_{m\to\infty} \int_{-\infty}^{\infty} t_m(x) f(x)dx \\
&= \lim_{m\to\infty} \int_{-\infty}^{\infty} -\frac{ds_m}{dx} f(x)dx \\
&= \lim_{m\to\infty} [-s_m f]_{-\infty}^{\infty} + \lim_{m\to\infty} \int_{-\infty}^{\infty} s_m \frac{df}{dx} dx \\
&= (df/dx)(0)\,. \qquad (2)
\end{aligned}$$

In the next chapter we shall prove that a generalized function with property (2) is the derivative of the delta function.

1.5. The Heaviside Sequences

Finally, let us mention that we can also define a Heaviside function on the same lines. Indeed, we have already come across such a sequence, namely, (1.3.6). Another example is

$$h_m(x) = \begin{cases} 1\,, & x > 1/2m\,, \\ mx + \frac{1}{2}\,, & -1/2m \le x \le 1/2m\,, \\ 0\,, & x < -1/2m\,. \end{cases} \qquad (1)$$

To prove that it is a Heaviside sequence, observe that

$$\begin{aligned}
\int_{-\infty}^{\infty} h_m(x) f(x)dx &= \int_{-\infty}^{-1/2m} h_m(x) f(x)dx \\
&\quad + \int_{-1/2m}^{1/2m} h_m(x) f(x)dx + \int_{1/2m}^{\infty} f(x)dx\,.
\end{aligned}$$

As $m \to \infty$ we have

$$\lim_{m\to\infty} \int_{-\infty}^{\infty} h_m(x) f(x)dx = \int_{0}^{\infty} f(x)dx = \langle H(x), f(x)\rangle\,,$$

where the symbol $\langle \phi(x), \psi(x)\rangle$ stands for

$$\langle \phi(x), \psi(x)\rangle = \int_{-\infty}^{\infty} \phi(x)\psi(x)dx. \qquad (2)$$

Note that occasionally, when the range of integration is the entire space, we shall omit the limits on the integral sign.

Exercises

Show the following:

1. $H(t-|x|) = H(t+x)H(t-x) = H(t^2-x^2)H(t)$.

2. $H\{(x-a)(x-b)\} = \begin{cases} 0, & \min(a,b) < x < \max(a,b), \\ 1, & x < \min(a,b) \text{ and } x > \max(a,b). \end{cases}$

3. $H(e^t-\pi) = H(t-\ln\pi) = \begin{cases} 0, & t < \ln\pi, \\ 1, & t > \ln\pi. \end{cases}$

4. $\int_{-\infty}^{x} H(x-\xi)dx = (x-\xi)H(x-\xi)$.

5. Identify the Heaviside functions $H(\cos x)$, $H(\sinh x)$, and $H(\cosh x)$.

6. Consider the sequences of functions

$$(a) \quad s_m(x) = \begin{cases} 0, & |x| > 1/2m, \\ m, & |x| < 1/2m; \end{cases}$$

$$(b) \quad s_m(x) = \begin{cases} 0, & |x| > 1/2m, \\ 4m^2x + 2m, & -1/2m \le x \le 0, \\ -4m^2x + 2m, & 0 \le x \le 1/2m. \end{cases}$$

Sketch these functions and the corresponding cumulative distributions and prove that $\lim_{m\to\infty} s_m(x) = \delta(x)$.

7. Show that the sequences (a) $\frac{1}{2}me^{-m|x|}$, (b) $(1/\pi)m/(e^{mx}+e^{-mx})$, and (c) $\frac{m}{\pi}\frac{\sin^2 mx}{m^2x^2}$ are delta sequences.

8. Let $s_m(x)$ be a sequence of nonnegative functions such that

$$(a) \quad \int_{-\infty}^{\infty} s_m(x)dx = 1;$$

$$(b) \quad \lim_{m\to\infty} \int_a^b s_m(x)dx = \begin{cases} 0, & a,b > 0, \quad \text{or} \quad a,b < 0, \\ 1, & a < 0, \quad \text{and} \quad b > 0. \end{cases}$$

Show that $s_m(x)$ is a delta sequence.

9. By differentiating the sequence of Example 3 show that it yields a unit dipole and is thus a dipole sequence.

10. By differentiating the Heaviside sequence (1.5.1) show that $s_m(x) = dh_m/dx$ is a delta sequence.

11. Show that the sequence

$$s_m(x) = \begin{cases} m^2x + m\,, & -1/m \le x < 0\,; \\ m - m^2x\,, & 0 \le x < 1/m\,, \\ 0\,, & \text{all other values of } x\,; \end{cases}$$

is a delta sequence.

12. If $f(x)$ is a nonnegative function satisfying $\int_{-\infty}^{\infty} f(x)dx = 1$, show that $\{mf(mx)\}$ is a delta sequence.

13. Let $s_m(x)$ be a delta sequence consisting of even functions. Let $g(x)$ be an integrable function having a jump discontinuity at 0. Show that

$$\lim_{m\to\infty} \int_{-\infty}^{\infty} s_m(x)g(x)dx = \frac{1}{2}[g(0+) + g(0-)]\,,$$

where $g(0\pm)$ stands for the limits of $g(x)$ as $x \to 0$ from the right and from the left; i.e., $g(0\pm) = \lim_{\varepsilon\to 0} g(0 \pm i|\varepsilon|)$.

14. Show that the sequence

$$t_m(x) = \begin{cases} -m^2\,, & 0 \le x < 1/m\,, \\ m^2\,, & -1/m < x < 0\,, \\ 0\,, & |x| \ge m; \end{cases}$$

is a dipole sequence.

15. Show that each member of the sequences

$$s_m(x) = \frac{1}{2\pi}\left\{\frac{m}{1 + m^2(x+t)^2} + \frac{m}{1 + m^2(x-t)^2}\right\},$$

and

$$s_m(x) = \frac{1}{2\pi}\{\tan^{-1} m(x+t) - \tan^{-1} m(x-t)\}\,,$$

satisfies the wave equation $\partial^2 u/\partial t^2 - \partial^2 u/\partial x^2 = 0$. Deduce that $s(x) = \frac{1}{2}\{\delta(x+t) + \delta(x-t)\}$ and $s(x) = \frac{1}{2}\{H(x+t) - H(x-t)\}$ are also solutions of this equation.

16. (a) Prove that

$$s_m(x) = \begin{cases} c_m(1 - x^2)^m\,, & 0 \le |x| \le 1\,, \\ 0\,, & |x| > 1\,, \end{cases}$$

where

$$c_m = \frac{(2m+1)!}{2^{2m+1}(m!)^2},$$

is a delta sequence. Sketch $s_m(x)$ for $m = 1, 2, 3, 4$.

(b) Show that the relation

$$P_m(x) = \int_{-1}^{1} f(x+t)s_m(t)dt\,, \qquad 0 \le x \le 1\,,$$

yields a sequence of polynomials.

(c) With the help of relations (a) and (b), prove Weierstrass's approximation theorem: If a function $f(x)$ is continuous on the closed interval $[a, b]$, then there exists a sequence of polynomials $P_m(x)$ such that $\lim_{n\to\infty} P_m(x) = f(x)$.

Hints: (i) There is no loss of generality in taking the interval [0, 1] and in assuming that $f(x)$ vanishes at $x = 0$ and $x = 1$. (ii) The required polynomials are the ones given in (b).

CHAPTER 2

The Schwartz–Sobolev Theory of Distributions

2.1. Some Introductory Definitions

Let R_n be a real n-dimensional space in which we have a Cartesian system of coordinates such that a point P is denoted by $x = (x_1, x_2, \ldots, x_n)$ and the distance r, of P from the origin, is $r = |x| = (x_1^2 + x_2^2 + \cdots + x_n^2)^{1/2}$. Let k be an n-tuple of nonnegative integers, $k = (k_1, k_2, \ldots, k_n)$, the so-called *multiindex* of order n; then we define

$$|k| = k_1 + k_2 + \cdots + k_n\,, \qquad x^k = x_1^{k_1} x_2^{k_2} \cdots x_n^{k_n}\,,$$

$$k! = k_1! k_2! \cdots k_n!\,, \qquad \binom{k}{p} = \frac{k!}{k!(k-p)!}$$

and

$$D^k = \frac{\partial^{|k|}}{\partial x_1^{k_1} \partial x_2^{k_2} \cdots \partial x_n^{k_n}} = \frac{\partial^{k_1+k_2+\cdots+k_n}}{\partial x_1^{k_1} \partial x_2^{k_2} \cdots \partial x_n^{k_n}} = D_1^{k_1} D_2^{k_2} \cdots D_n^{k_n}\,, \tag{1}$$

where $D_j = \partial/\partial x_j$, $j = 1, 2, \ldots, n$. For the one-dimensional case D^k reduces to d/dx. Furthermore, if any component of k is zero, the differentiation with respect to the corresponding variable is omitted. For instance, in R_3, with $k = (3, 0, 4)$, we have

$$D^k = \partial^7/\partial x_1^3 \partial x_3^4 = D_1^3 D_3^4\,.$$

A differential operator L of order p is defined as

$$L = \sum_{|k| \le p} a_k(x) D^k\,, \tag{2}$$

where $a_k(x)$ are given functions and the sum is taken over all multiindices k of order n. For example, when $n = 1$, we have the ordinary differential operator

$$L = a_p(x) d^p/dx^p + a_{p-1}(x) d^{p-1}/dx^{p-1} + \cdots + a_0(x)\,. \tag{3}$$

As another example, the second-order partial differential operator in R_2 is

$$\begin{aligned} L = {} & a_{2,0}(x)\partial^2/\partial x_1^2 + a_{1,1}(x)\partial^2/\partial x_1 \partial x_2 + a_{0,2}(x)\partial^2/\partial x_2^2 \\ & + a_{1,0}(x)\partial/\partial x_1 + a_{0,1}(x)\partial/\partial x_2 + a_{0,0}(x)\,. \end{aligned} \tag{4}$$

We shall use the following notation for the integral (in the Lebesgue sense) of a function $f(x) = f(x_1, x_2, \ldots, x_n)$ over an n-dimensional region R:

$$\int \cdots \int_R f(x_1, x_2, \ldots, x_n) dx_1 dx_2 \cdots dx_n = \int_R f(x) dx\,, \tag{5}$$

where $dx = dx_1 dx_2 \cdots dx_n$. In the sequel we shall encounter similar integrals over hypersurfaces of dimensions $n-1$ in R_n, such as a two-dimensional surface S in R_3 and a curve C in R_2.

Definition. A function $f(x)$ is *locally integrable* in R_n if $\int_R |f(x)|dx$ exists for every bounded region R in R_n. A function $f(x)$ is locally integrable on a hypersurface in R_n if $\int_S |f(x)|dS$ exists for every bounded region S in R_{n-1}.

The class of locally integrable functions is rather wide. All piecewise continuous functions are locally integrable. Some functions that are infinite, such as $1/r^m$, $m < n$, are also locally integrable. However, $\delta(x)$ is not a locally integrable function for the following reason. Let $I_1, I_2, \ldots, I_n$ be a sequence of nested intervals i.e., each interval is included in the preceding one whose length tends to zero; then $\lim_{n\to\infty} \int_{I_n} f(x)dx = 0$, but $\delta(x)$ gives a finite value.

Another concept that plays a crucial role in the theory of distributions is the support of a function, defined as follows:

Definition. The *support* of a function $f(x)$ is the closure of the set of all points x such that $f(x) \neq 0$.

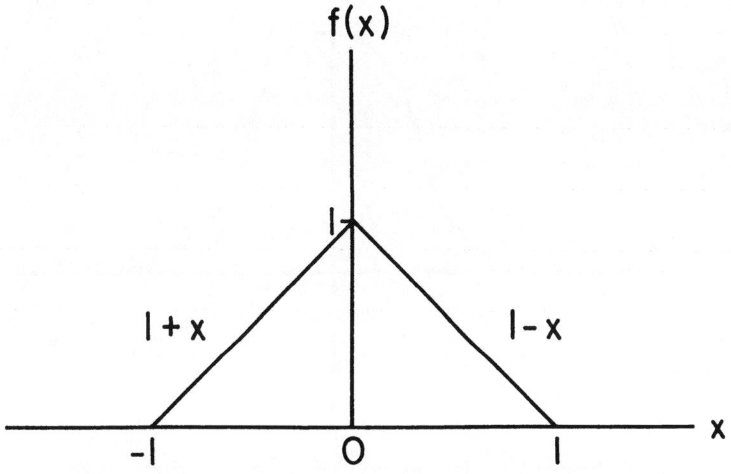

Figure 2.1. A function with compact support.

We shall denote the support of f by supp f. For example, for $f(x) = \sin x$, $x \in R_1$, the support of $f(x)$ consists of the whole real line, even though $\sin x$ vanishes at $x = n\pi$. If supp f is a bounded set, then f is said to have *compact support*. For instance, the support of the function (Figure 2.1)

$$f(x) = \begin{cases} 0, & -\infty < x \le -1, \\ x+1, & -1 < x < 0, \\ 1-x, & 0 \le x < 1, \\ 0, & 1 \le x < \infty, \end{cases}$$

is compact.

2.2. Test Functions

We have observed that an operational quantity such as $\delta(x)$ becomes meaningful if it is first multiplied by a sufficiently smooth auxiliary function and then integrated over the entire space. This point of view is also taken as the basis for the definition of an arbitrary generalized function. Accordingly, consider the space D consisting of real-valued functions $\phi(x) = \phi(x_1, x_2, \ldots, x_n)$, such that the following hold:

(1) $\phi(x)$ is an infinitely differentiable function defined at every point of R_n. This means that $D^k\phi$ exists for all multiindices k. Such a function is also called a C^∞ function.

(2) There exists a number A such that $\phi(x)$ vanishes for $r > A$. This means that $\phi(x)$ has compact support. Then $\phi(x)$ is called a test function.

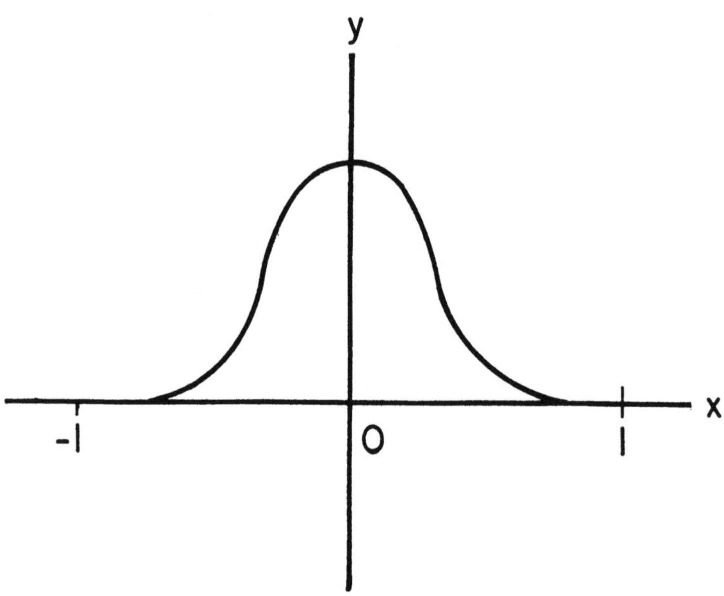

Figure 2.2. The test function defined by Eq. (2.2.1) for $a = 1$.

The prototype of a test function belonging to D is

$$\phi(x) = \begin{cases} \exp\left(-\dfrac{a^2}{a^2 - r^2}\right), & r < a, \\ 0, & r > a, \end{cases} \tag{1}$$

shown in Figure 2.2 (where $a = 1$). Its support is clearly $|r| \le a$. By taking $a = 1$ and multiplying $\phi(x) = \phi(x, 1)$ by a suitable normalizing factor we can construct the function

$$\phi(x, 1) = \begin{cases} c \exp\left(-\frac{1}{1-r^2}\right), & r < 1, \\ 0, & r > 1, \end{cases}$$

satisfying the conditions

$$\phi \in D\,, \quad \int_{R_n} \phi(x)dx = 1\,, \quad \operatorname{supp}\phi = A(\equiv |r| \le 1)\,, \quad x \in R_n\,, \tag{2}$$

so that

$$c^{-1} = \int_{|r|\le 1} \exp(-1/1 - r^2)dx\,.$$

This function, in turn, gives us the test function, the so called *cap* function

$$\phi_\varepsilon(x) = c_\varepsilon \phi(x/\varepsilon) = \begin{cases} c_\varepsilon \exp\left(-\frac{\varepsilon^2}{\varepsilon^2 - r^2}\right), & r < \varepsilon\,, \\ 0\,, & r \ge \varepsilon\,, \end{cases} \tag{3}$$

with the properties

$$\int_{R_n} \phi_\varepsilon(x)dx = 1\,, \qquad \operatorname{supp}\phi_\varepsilon = A(= |r| \le \varepsilon)\,,$$

$$c_\varepsilon^{-1} = \int_{|r|\le\varepsilon} \exp\left(-\frac{\varepsilon^2}{\varepsilon^2 - r^2}\right) dx\,.$$

We can also form a delta sequence out of this function. Indeed, the sequence

$$\phi_m(x) = \begin{cases} c_m \exp\left(-\frac{1}{1-m^2r^2}\right), & |r| < \frac{1}{m}\,, \\ 0\,, & |r| \ge \frac{1}{m}\,, \end{cases} \tag{5}$$

where

$$c_m^{-1} = \int_{r\le 1/m} \exp\left(-\frac{1}{1 - m^2r^2}\right) dx\,,$$

is a delta sequence (see Exercise 17). The proof follows by appealing to the generalization of Exercise 8 of Chapter 1. We shall show in Chapter 3 (see Example 2 of Section 3.3) that the sequence $\phi_\varepsilon(x)$, depending on a parameter ε, as defined by (3) also approaches $\delta(x) = \delta(x_1, \ldots, x_n)$ as $\varepsilon \to 0$.

The following properties of the test functions are evident.

(1) If ϕ_1 and ϕ_2 are in D, then so is $c_1\phi_1 + c_2\phi_2$, where c_1 and c_2 are real numbers. Thus D is a linear space.

(2) If $\phi \in D$, then so is $D^k\phi$.

(3) For a C^∞ function $f(x)$ and for a $\phi \in D$, $f\phi \in D$.

(4) If $\phi(x_1, x_2, \ldots, x_m)$ is an m-dimensional test function and $\psi(x_{m+1}, x_{m+2}, \ldots, x_n)$ is an $(n - m)$-dimensional test function, then $\phi\psi$ is an n-dimensional test function in the variables $x_1, x_2, \ldots, x_n$.

Note that the definition of D does not demand that all its elements have the same support. Take, for example, the functions $\phi(x)$ as defined by (1) in R_1, then $\phi(x)$ and $\phi(x - 3)$ are both members of D although their supports are, respectively, $(-1, 1)$ and $(2, 4)$.

It is more convenient to work with the basic notions of convergence than to introduce an inner product into this space. The kind of convergence we need is defined as follows:

Definition. A sequence $\{\phi_m\}$, $m = 1, 2, \ldots$, where $\phi_m \in D$, converges to ϕ_0 if the following two conditions are satisfied:

(1) All ϕ_m as well as ϕ_0 vanish outside a common region.
(2) $D^k\phi_m \to D^k\phi_0$ uniformly over R_n as $m \to \infty$ for all multiindices k.

It is not difficult to show that $\phi_0 \in D$ and hence that D is closed (or is complete) with respect to this definition of convergence. For the special case $\phi_0 = 0$, the sequence $\{\phi_m\}$ is called a null sequence.

Example 1. The sequence

$$\{(1/m)\phi(x, a)\}, \tag{6}$$

where $\phi(x, a)$ is defined by (1), is a null sequence. However, the sequence $\{(1/m)\phi(x/m, a)\}$ is not a convergent sequence, because the support of the function $\phi(x/m, a)$ is the sphere with radius ma, which is different for different m.

In addition to the space D of test functions, we shall use certain subspaces of D. For a region R in R_n, the space D_R contains those test functions whose support lies in R, that is,

$$D_R \equiv \{\phi : \phi \in D, \qquad \text{supp}\,\phi \subset R\}. \tag{7}$$

It is clearly a linear subspace of D. For example, D_x and D_y are two one-dimensional subspaces of test functions $\phi(x)$ and $\phi(y)$ and are contained in D_{xy}, which is the space of test functions $\phi(x, y)$ in R_2. The convergence in D_R is defined in the same manner as that in the space D.

2.3. Linear Functionals and the Schwartz–Sobolev Theory of Distributions

A linear functional t on the space D is an operation (or a rule) by which we assign to every test function $\phi(x)$ a real number denoted $\langle t, \phi\rangle$, such that

$$\langle t, c_1\phi_1 + c_2\phi_2\rangle = c_1\langle t, \phi_1\rangle + c_2\langle t, \phi_2\rangle \tag{1}$$

for arbitrary test functions ϕ_1 and ϕ_2 and real numbers c_1 and c_2. It follows that

$$\langle t, 0\rangle = 0, \tag{2}$$

and

$$\left\langle t, \sum_{j=1}^{m} c_j\phi_j \right\rangle = \sum_{j=1}^{m} c_j\langle t, \phi_j\rangle, \tag{3}$$

where c_j are arbitrary real numbers. The next concept is that of the continuity of the linear functionals. It is defined as follows:

Definition. A linear functional on D is *continuous* if and only if the sequence of numbers $\langle t, \phi_m \rangle$ converges to $\langle t, \phi \rangle$ when the sequence of test functions $\{\phi_m\}$ converges to the test function ϕ (in the sense of the convergence as defined in the previous section). Thus

$$\lim_{m\to\infty} \langle t, \phi_m \rangle = \langle t, \lim_{m\to\infty} \phi_m \rangle \,. \tag{4}$$

We now have all the tools for defining the concept of distributions.

Definition. A continuous linear functional on the space D of test functions is called a *distribution.*

Regular distributions

The set of distributions that are most useful are those generated by locally integrable functions. Indeed, every locally integrable function $f(x)$ generates a distribution through the formula

$$\langle f, \phi \rangle = \int_{R_n} f(x)\phi(x)dx \,. \tag{5}$$

Linearity of this functional is obvious. To prove its continuity we observe that

$$|\langle f, \phi \rangle| \leq \max_{x \in \operatorname{supp} \phi} |\phi(x)| \int_{\operatorname{supp} \phi} |f(x)| dx < \infty \,.$$

Thus, if the sequence $\{\phi_m\}$ converges to zero, then so does $\langle f, \phi_m \rangle$. Hence, it is continuous. Distributions defined by (5) are called *regular*. All other distributions are called *singular*. However, we may use formula (5) symbolically for a singular distribution also.

Distributions can be defined by partial differential operators as well. If $f(x)$ is locally integrable function, we can define a distribution as

$$\langle f, \phi \rangle = \int_{R_n} f(x) D^k \phi(x) dx \,, \qquad \phi \in D \,. \tag{6}$$

Remark 1. The constant c has three meanings in this book: (a) c as a number; (b) c as a constant point function; and (c) as a constant point function, c is locally integrable and generates the distribution

$$\langle c, \phi \rangle = \int_{R_n} c\phi(x)dx = c \int_{R_n} \phi(x)dx \,. \tag{7}$$

It will be clear from the text what c stands for whenever it occurs. The zero distribution on D has the property

$$\langle 0, \phi \rangle = 0 \,, \qquad \phi \in D \,. \tag{8}$$

Remark 2. The definition of a distribution can be extended to include complex-valued functions of a real variable. The arbitrary constants c_1 and c_2 occurring in the foregoing definition are then complex numbers. The space of test functions is then called $D^{(c)}$. Accordingly, we have the following definition:

Definition. A distribution t is a complex-valued functional on $D^{(c)}$ such that

(1) $\langle t, c_1\phi_1 + c_2\phi_2\rangle = c_1\langle t, \phi_1\rangle + c_2\langle t, \phi_2\rangle$,
(2) $\lim_{m\to\infty}\langle t, \phi_m\rangle = \langle t, \lim_{m\to\infty}\phi_m\rangle$,

where $\phi_m(x)$ are elements of $D^{(c)}$.

Remark 3. The distribution $\langle f, \phi\rangle$ will also be denoted f.

Space D′

The space of all distributions on D is denoted D'. The distributions t_1 and t_2 give rise to a new distribution $t = c_1t_1 + c_2t_2$ such that

$$\langle t, \phi\rangle = \langle c_2t_1 + c_2t_2, \phi\rangle = c_1\langle t_1, \phi\rangle + c_2\langle t_2, \phi\rangle . \tag{9}$$

It is easily verified that this satisfies the requirements of a distribution. Thus D' is itself a linear space. It is called the *dual space* of D and is a larger space than D. It forms a generalization of the class of locally integrable functions because it contains functions such as $\delta(x)$ (see Example 2 in Section 4) that are not locally integrable. For this reason distributions are also called *generalized* or *symbolic* functions. We shall use the terms "distribution" and "generalized function" interchangeably.

Another basic idea in the theory of distributions is embedded in the following theorem.

Theorem. *Two continuous functions that produce the same regular distributions are identical.*

Proof. The theorem follows from (5) as the only continuous function $f(x)$ for which $\langle f, \phi\rangle = 0$, for all $\phi \in D$, is $f(x) \equiv 0$, $x \in R_n$. This is because if there is a point x_0 such that $f(x_0) \neq 0$, then there exists a spherical neighborhood of radius ε about x_0 in which $f(x) \neq 0$ (say it is positive). Now the test function $\phi_\varepsilon[(x - x_0)/\varepsilon]$ defined by (2.2.3) is positive in the sphere $|x - x_0| < \varepsilon$ and vanishes outside it, for this test function $\langle f, \phi\rangle$ is greater than zero, which contradicts the hypothesis. The same is true for $f(x_0) < 0$. This proves the theorem.

For locally integrable functions this theorem does not hold, because we may alter the values of $f(x)$ on a set of measure zero without altering the regular distribution. Accordingly, we stipulate that if $f(x)$ and $g(x)$ are locally integrable functions and are equal almost everywhere, then they generate the same distribution, and we have

$$\langle f, \phi\rangle = \langle g, \phi\rangle , \qquad \phi \in D .$$

2.4. Examples

Example 1. The Heaviside distribution in R_n is

$$\langle H_R, \phi \rangle = \int_R \phi(x)dx\,, \qquad \text{where} \qquad H_R(x) = \begin{cases} 1\,, & x \in R\,, \\ 0\,, & x \notin R\,. \end{cases} \tag{1}$$

For R_1, (1) becomes

$$\langle H, \phi \rangle = \int_0^\infty \phi(x)dx\,. \tag{2}$$

Since $H(x)$ is a piecewise continuous function, this is a regular distribution.

Example 2. The Dirac delta distribution in R_n is

$$\langle \delta(x - \xi)\,, \phi(x) \rangle = \phi(\xi)\,, \tag{3}$$

for ξ is a fixed point in R_n. Linearity of this functional follows from the relation

$$\langle \delta, c_1\phi_1 + c_2\phi_2 \rangle = c_1\phi_1(\xi) + c_2\phi_2(\xi) = c_1\langle \delta, \phi_1 \rangle + c_2\langle \delta, \phi_2 \rangle\,,$$

where c_1 and c_2 are arbitrary real constants.

To prove continuity we observe that $\lim_{m\to\infty}\langle \delta, \phi_m \rangle = \lim_{m\to\infty} \phi_m(\xi)$. However, if $\phi_m(x) \to 0$, then $\phi_m(\xi) \to 0$, and we have continuity.

Thus the delta function is a distribution. We observed earlier that the delta function is not locally integrable. This distribution is therefore a singular distribution.

Example 3. Generalized functions related to the delta function. Various related generalized functions are useful in electrical and electronic engineering problems. For example, an infinite sequence of impulses (i.e., an infinite row of delta functions) is described by the Dirac comb:

$$III(x) = \sum_{n=-\infty}^{\infty} \delta(x - n)\,. \tag{4}$$

This is also called the *sampling* or *replicating* function because it gives the information about a function $f(x)$ at $x = n$:

$$III(x)f(x) = \langle III(x)\,,\ f(x) \rangle = \sum_{n=-\infty}^{\infty} f(n)\delta(x - n)\,,$$

as is clear in the Figure 2.3.

Other quantities of interest are the impulse pair functions, which are defined as (Figure 2.4)

$$II(x) = \frac{1}{2}\delta(x + \frac{1}{2}) + \frac{1}{2}\delta(x - \frac{1}{2})\,, \tag{5}$$

$$I_1(x) = \frac{1}{2}\delta(x + \frac{1}{2}) - \frac{1}{2}\delta(x - \frac{1}{2}) \tag{6}$$

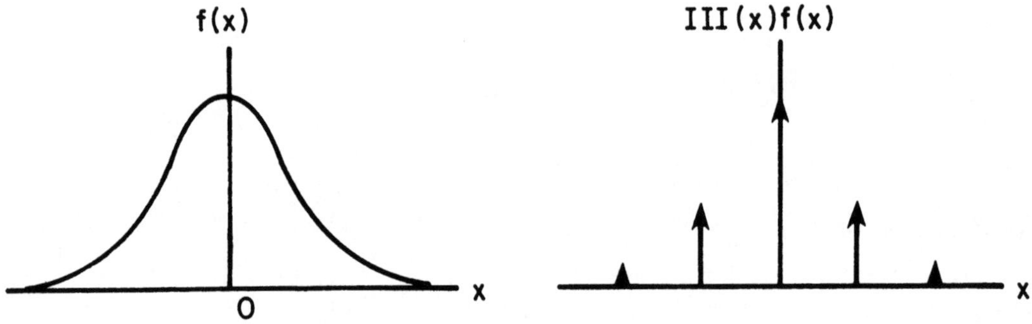

Figure. 2.3. The sampling function.

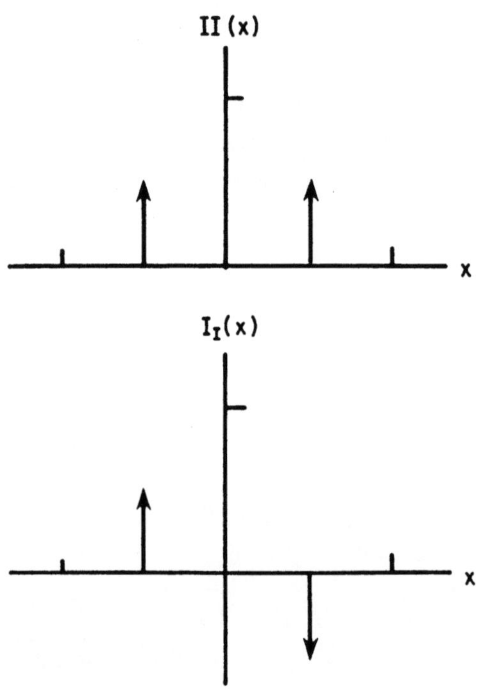

Figure 2.4. The impulse pair functions.

Example 4. The function $1/x$ does not define a regular distribution because $\int_{-\infty}^{\infty}[\phi(x)/x]dx$ is not convergent for all the test function (e.g., $\phi(x) = c$). However, we can employ the notion of the Cauchy principal value and define

$$\langle \mathcal{P}\left(\frac{1}{x}\right), \phi\rangle = \int_{-\infty}^{*\infty} \frac{\phi(x)}{x} dx = \lim_{\varepsilon\to 0} \int_{|x|\geq\varepsilon} \frac{\phi(x)}{x} dx$$

$$= \lim_{\varepsilon\to 0+} \left(\int_{-\infty}^{-\varepsilon} + \int_{\varepsilon}^{\infty}\right) \frac{\phi(x)}{x} dx\,, \qquad \phi \in D\,, \tag{7}$$

where the $*$ to the right of the integral sign indicates the principal value. This limit exists for the following reason. Since $\phi(x)$ is differentiable at $x = 0$, there is a function $\psi(x)$ continuous at $x = 0$ such that

$$\phi(x) = \phi(0) + x\psi(x)\,. \tag{8}$$

Let $[-A, A]$ be the support of the test function $\phi(x)$, so that for $\varepsilon > 0$,

$$\int_{|x|>\varepsilon} \frac{\phi(x)}{x} dx = \int_{A\geq|x|>\varepsilon} \left(\frac{\phi(0)}{x} + \psi(x)\right) dx$$

$$= \int_{A\geq|x|>\varepsilon} \psi(x) dx$$

$$\to \int_{-A}^{A} \psi(x) dx\,, \qquad \text{as} \quad \varepsilon \to 0\,. \tag{9}$$

The functional $\langle \mathcal{P}(1/x), \phi\rangle$ so defined is clearly linear. To prove its continuity we appeal to relation (9) and find that

$$\left\langle \mathcal{P}\left(\frac{1}{x}\right), \phi\right\rangle = \int_{-A}^{A} \psi(x) dx \leq 2A \max|\psi(x)|\,, \qquad -A \leq x \leq A\,,$$

by the mean value theorem. Thus $\mathcal{P}(1/x)$ is a distribution. This distribution is also called a *pseudofunction*; we shall write it as $Pf(1/x)$ in the sequel. Many more pseudofunctions are defined and analyzed in Chapter 4.

Example 5. From Example 4 it follows that the functions

$$\delta^{\pm}(x) = \frac{1}{2}\delta(x) \mp (1/2\pi i)Pf(1/x) \tag{10}$$

are also singular distributions on D and are called the *Heisenberg distributions*. We can further show that

$$\langle \delta^{\pm}(x), \phi(x)\rangle = \mp \lim_{\varepsilon\to 0} \frac{1}{2\pi i} \int_{-\infty}^{\infty} \frac{\phi(x)}{x \pm i\varepsilon} dx\,. \tag{11}$$

To prove this relation we proceed as follows. Let $\phi(x)$ be a test function, and let $\psi(x)$ be the continuous function satisfying (8) and where the support of $\phi(x)$ is $[-A, A]$. Then for each $\varepsilon > 0$,

$$\begin{aligned}\int_{-\infty}^{\infty} \frac{\phi(x)}{x+i\varepsilon}dx &= \int_{-A}^{A} \frac{x-i\varepsilon}{x^2+\varepsilon^2}\phi(x)dx \\ &= \phi(0)\int_{-A}^{A} \frac{x-i\varepsilon}{x^2+\varepsilon^2}dx + \int_{-A}^{A} \frac{x-i\varepsilon}{x^2+\varepsilon^2}x\psi(x)dx \\ &= -2i\phi(0)\tan^{-1}\frac{A}{\varepsilon} + \int_{-A}^{A} \frac{x-i\varepsilon}{x^2+\varepsilon^2}x\psi(x)dx\,.\end{aligned}$$

Thus,

$$\begin{aligned}\lim_{\varepsilon\to 0}\frac{1}{2\pi i}\int_{-\infty}^{\infty} \frac{\phi(x)}{x+i\varepsilon}dx &= -\frac{1}{2}\phi(0) + \frac{1}{2\pi i}\int_{-A}^{A} \psi(x)dx \\ &= -\frac{1}{2}\phi(0) + \frac{1}{2\pi i}\int_{-\infty}^{*\infty} \frac{\phi(x)}{x}dx\,, \qquad (12)\end{aligned}$$

where we have used results (7) and (9). Similarly,

$$\lim_{\varepsilon\to 0}\frac{1}{2\pi i}\int_{-\infty}^{\infty} \frac{\phi(x)}{x-i\varepsilon}dx = \frac{1}{2}\phi(0) + \frac{1}{2\pi i}\int_{-\infty}^{*\infty} \frac{\phi(x)}{x}dx\,. \qquad (13)$$

Relations (12) and (13) are called Plemelj formulas. Writing them as

$$\lim_{\varepsilon\to 0}\frac{1}{2\pi i}\int_{-\infty}^{\infty} \frac{\phi(x)}{x\pm i\varepsilon}dx = \mp\frac{1}{2}\langle\delta,\phi\rangle + \frac{1}{2\pi i}\left\langle Pf\left(\frac{1}{x}\right),\phi\right\rangle, \qquad (14)$$

we obtain the required formulas (11). Symbolically, we can write them as

$$\delta^{\pm}(x) = \mp\lim_{\varepsilon\to 0}\frac{1}{2\pi i}\frac{1}{x\pm i\varepsilon} = \mp\frac{1}{2\pi i}\frac{1}{x\pm i0}, \qquad (15)$$

so that

$$\delta(x) = \delta^{+}(x) + \delta^{-}(x) = \lim_{\varepsilon\to 0}\frac{1}{\pi}\frac{\varepsilon}{x^2+\varepsilon^2}, \qquad (16)$$

$$\frac{1}{\pi i}Pf\left(\frac{1}{x}\right) = \delta^{-}(x) - \delta^{+}(x) = \lim_{\varepsilon\to 0}\frac{1}{\pi i}\frac{x}{x^2+\varepsilon^2}. \qquad (17)$$

Note that by setting $\varepsilon = 1/m$ we get the delta sequence of Example 2 in Section 1.2.

Formulas (14) are sometimes written as

$$1/(x+i0) = -i\pi\delta(x) + Pf(1/x)\,, \qquad (18)$$

$$1/(x-i0) = i\pi\delta(x) + Pf(1/x)\,, \qquad (19)$$

and are called the Sokhotski-Plemelj equations.

From these formulas we derive the value of the jump of the function $1/(x+iy) = \frac{1}{z}$ across the x-axis as

$$\left[\frac{1}{x-i0} - \frac{1}{x+i0}\right] = 2\pi i\,\delta(x)\,.$$

Example 6. Let us find the solution of the equation

$$(x-\xi)t(x-\xi)=g(x)\,. \tag{20}$$

The homogeneous equation

$$(x-\xi)t(x-\xi)=0\,, \tag{21}$$

clearly has the solution $\delta(x-\xi)$, because $\int_{-\infty}^{\infty}(x-\xi)\delta(x-\xi)\phi(x)dx=0$. The generalized particular solution is

$$g(x)Pf\left(\frac{1}{x-\xi}\right), \tag{22}$$

as can be readily verified. Accordingly, the solution of equation (20) is

$$t(x-\xi)=\delta(x-\xi)+g(x)Pf\left(\frac{1}{x-\xi}\right). \tag{23}$$

This solution was first derived by Dirac and is very useful in transport theory.

Example 7. Single-layer distribution. Let us discuss a distribution in R_n that is zero outside a hypersurface S and that is not the zero distribution. Distributions of this kind are called surface distributions and have important applications. They are generalizations of the δ function.

Let $\sigma(x)$ be a locally integrable function defined on S. In the language of electrostatics, we have a charge density of strength $\sigma(x)dS$, spread over S; i.e., the surface element dS at $\xi\in S$ carries a concentrated source density of strength $\sigma(\xi)dS$. The corresponding volume source density in space is $\sigma(\xi)dS_\xi\delta(x-\xi)$. Accordingly, the total volume density is

$$t(x)=\int_S\sigma(\xi)\delta(x-\xi)dS_\xi\,.$$

The action of t on a test function ϕ is

$$\begin{aligned}\langle t,\phi\rangle=\left\langle\int_S\sigma(\xi)\delta(x-\xi)dS\,,\ \phi\right\rangle&=\int_{R_n}dx\phi(x)\int_S\sigma(\xi)\delta(x-\xi)dS\\&=\int_S\phi(\xi)\sigma(\xi)dS\,.\end{aligned} \tag{24}$$

Note that we can write this relation as $\langle\sigma(x)\delta(S),\phi(x)\rangle$, where $\delta(S)$ denotes the surface distribution of unit density over S, because

$$\int\phi(x)\sigma(x)\delta(S)dS=\int_S\phi(\xi)\sigma(\xi)dS\,. \tag{25}$$

From (24) it follows that we may call t the distribution corresponding to a surface layer of sources spread over S with density $\sigma(x)$. This layer is called a *single* (or *simple*) layer.

Example 8. Many physical quantities can be expressed by distributions in a natural way. In this example we consider a material system formed by a closed interval $[a, b]$ on which a mass is distributed with density $\rho(x)$. The total mass M of the system is

$$M = \int_a^b \rho(x)dx \,. \tag{26}$$

The abscissa X of the mass center is the first moment

$$X = \frac{1}{M}\int_a^b x\rho(x)dx \,. \tag{27}$$

Next, we extend the function $\rho(x)$ outside $[a, b]$ by zero so that we can write (26) and (27) as the distributions

$$M = \langle 1, \rho\rangle \,, \qquad X = (1/M)\langle x, \rho\rangle \,, \tag{28}$$

respectively.

As a special case, let us put $\rho = m\delta(x - A)$. Then (28) becomes

$$M = \langle 1, m\delta(x - A)\rangle = m \,, \qquad X = (1/m)\langle x, m\delta(x - A)\rangle = mA/m = A \,.$$

Thus, the distribution $m\delta(x - A)$ completely characterizes the material part of abscissa A and mass M. This concept can be readily generalized to include a system of material points as well as various other moments.

2.5. Algebraic Operations on Distributions

We have already defined taking the sum of two distributions and multiplying a distribution by a constant. We now define some other algebraic operations.

Linear change of variables

Let $\langle f, \phi\rangle$ be a regular distribution generated by a function $f(x)$ that is locally integrable in R_n. Let $x = Ay - a$, where A is an $n \times n$ matrix with $\det A \neq 0$ and a is a constant vector, be a nonsingular linear transformation of the space R_n onto itself. Then we have

$$\begin{aligned}\langle f(Ay - a)\,,\ \phi(y)\rangle &= \int_{R_n} f(Ay - a)\phi(y)dy \\ &= \frac{1}{|\det A|}\int_{R_n} f(x)\phi[A^{-1}(x + a)]dx \\ &= \frac{1}{|\det A|}\langle f(x), \phi[A^{-1}(x + a)]\rangle \,, \end{aligned} \tag{1}$$

where A^{-1} is the inverse of the matrix A.

Fortunately, the same definition applies to a singular distribution because $[A^{-1}(x+a)]$ is a valid operation on $\phi(x)$. Accordingly, we have

$$\langle t(Ay-a), \phi(y)\rangle = \langle t(x), \phi[A^{-1}(x+a)]/|\det A|\rangle . \tag{2}$$

Remark 1. For a simple translation, i.e., when $A = I$ the unit matrix, (2) yields

$$\langle t(y-a), \phi(y)\rangle = \langle t(x), \phi(x+a)\rangle . \tag{3}$$

For example

$$\langle \delta(y-a), \phi(y)\rangle = \langle \delta(x), \phi(x+a)\rangle = \phi(a) . \tag{4}$$

The distribution $\delta(y-a)$ is sometimes denoted $\delta_a(y)$.

A distribution t is said to be *invariant* with respect to translation over a distance a, or to be *periodic* with respect to a, if

$$\langle t(y-a), \phi(y)\rangle = \langle t(x), \phi(x+a)\rangle = \langle t, \phi\rangle . \tag{5}$$

For instance, the regular distribution (on $D^{(c)}$),

$$\langle e^{i\omega x}, \phi(x)\rangle = \int_{-\infty}^{\infty} e^{i\omega x}\phi(x)\,dx ,$$

which is generated by the locally integrable function $e^{i\omega x}$, is invariant with respect to translation over the distance $2n\pi$, $n = 0, 1, 2, \ldots$. That is,

$$\langle e^{i\omega(x\pm 2n\pi)}, \phi(x)\rangle = \langle e^{i\omega x}\phi(x)\rangle .$$

Remark 2. For a simple scale expansion, $A = cI$, $a = 0$, (2) becomes

$$\langle t(cy), \phi(y)\rangle = (1/|c^n|)\langle t, \phi(x/c)\rangle . \tag{6}$$

A distribution is called *homogeneous* of degree λ if

$$t(cx) = c^{\lambda} t(x) , \tag{7}$$

for $c > 0$. This means that the relation

$$\langle t(cy), \phi(y)\rangle = c^{-n}\langle t(x), \phi(x/c)\rangle = c^{\lambda}\langle t, \phi\rangle ,$$

or

$$\langle t(x), \phi(x/c)\rangle = c^{\lambda+n}\langle t, \phi\rangle , \tag{8}$$

holds.

For the special case of the delta function, (6) becomes

$$\langle \delta(cy), \phi(y)\rangle = (1/|c^n|)\langle \delta(x), \phi(x/c)\rangle = (1/|c^n|)\langle \delta(x), \phi(x)\rangle ,$$

or

$$\delta(cx) = (1/|c^n|)\delta(x) . \tag{9}$$

Thus, $\delta(x) = \delta(x_1, \ldots, x_n)$ is a homogeneous distribution of degree $-n$.

The reflection of a distribution in the origin means that we take $x = -y$, then (2) yields

$$\langle t(-y), \phi(y) \rangle = \langle t(x), \phi(-x) \rangle . \tag{10}$$

For delta distribution, (10) becomes

$$\delta(-y), \phi(y) \rangle = \langle \delta(x), \phi(-x) \rangle = \phi(0) = \langle \delta(x), \phi(x) \rangle ,$$

or

$$\delta(-x) = \delta(x) ;$$

that is, the delta function is an even function. This result can also be obtained with the help of the sifting property of the delta function. Indeed, we have

$$\int_{-\infty}^{\infty} \delta(x-a)\delta(x-b)dx = \delta(a-b) = \delta(b-a) , \qquad a \neq b ,$$

which proves that $\delta(a-b)$ is even. This integral formula also gives us the orthonormal property of the delta function.

A distribution that is invariant under reflection in the origin, so that

$$\langle t(-y), \phi(y) \rangle = \langle t(x), \phi(-x) \rangle = \langle t, \phi \rangle , \tag{11}$$

is called a *centrally symmetric* or *even distribution*. Similarly, a distribution with the property

$$\langle t(-x), \phi(x) \rangle = \langle t(x), \phi(-x) \rangle = -\langle t, \phi \rangle , \tag{12}$$

is called a *skew symmetric* or a *odd distribution*.

Remark 3. For a simple rotation, $a = 0$ and $A' = A^{-1}$, where A' and A^{-1} are, respectively, the transpose and the inverse of A; we then have

$$\langle t(Ay), \phi(y) \rangle = \langle t(x), \phi(A'x)/|\det A| \rangle . \tag{13}$$

Product of a distribution and a function

In general, it is difficult to define the product of two generalized functions. It may not be possible to do so even if they are locally integrable, as is clear from the example $f(x) = g(x) = 1/x^{1/2}$. However, we can always assign a meaning to the product of a generalized function $t(x)$ and an infinitely differentiable function $\psi(x)$ by setting

$$\langle \psi t, \phi \rangle = \langle t, \psi\phi \rangle , \tag{14}$$

because $\psi\phi$ is an element of D.

Example 1. For a constant c, we know that $\langle ct, \phi \rangle = c\langle t, \phi \rangle$. Because of relation (14) we have $\langle ct, \phi \rangle = \langle t, c\phi \rangle = c\langle t, \phi \rangle$ for a function c. Thus both results are consistent.

Example 2. $\langle\psi\delta, \phi\rangle = \langle\delta, \psi\phi\rangle = \psi(0)\phi(0) = \psi(0)\langle\delta, \phi\rangle = \langle\psi(0)\delta, \phi\rangle$ or

$$\psi(x)\delta(x) = \psi(0)\delta(x)\,. \tag{15}$$

It follows that in this special case it is sufficient for the function $\psi(x)$ to be continuous at the origin. More generally,

$$\psi(x)\delta(x-\xi) = \psi(\xi)\delta(x-\xi)\,. \tag{16}$$

Example 3. Let us show that

$$x^n \, Pf(1/x) = x^{n-1}\,, \qquad n \text{ a positive integer.} \tag{17}$$

We have

$$\left\langle x^n Pf\left(\frac{1}{x}\right), \phi(x)\right\rangle = \int_{-\infty}^{*\infty} \frac{x^n\phi(x)}{x}dx = \int_{-\infty}^{\infty} x^{n-1}\phi(x)dx = \langle x^{n-1}, \phi\rangle\,.$$

2.6. Analytic Operations on Distributions

In this section we present the definition of the operation of taking the derivative of a distribution. This definition should yield consistent results when we differentiate a distribution that is also a classical function. Therefore, let us first consider a regular distribution generated by a continuously differentiable function $f(x) = f(x_1, x_2, \ldots, x_n)$,

$$\langle f, \phi\rangle = \int_{R_n} f(x)\phi(x)dx\,.$$

This integral representation has a real and not a merely symbolic meaning. When we integrate the integral

$$\int_{R_n} \frac{\partial f}{\partial x_j}\phi(x)dx\,, \qquad j = 1, \ldots, n\,,$$

by parts, we get

$$\int_{R_n} \frac{\partial f}{\partial x_j}\phi(x)dx = -\int_{R_n} f(x)\frac{\partial\phi}{\partial x_j}dx\,, \tag{1}$$

where we have used the fact that $\phi(x)$ has compact support. Since $\partial\phi/\partial x_j$ is also a test function, we write (1)

$$\langle\partial f/\partial x_j, \phi\rangle = \langle f, -\partial\phi/\partial x_j\rangle = -\langle f, \partial\phi/\partial x_j\rangle\,.$$

This helps us in defining the distributional derivative $\partial t/\partial x_j$;

$$\langle\partial t/\partial x_j, \phi\rangle = -\langle t, \partial\phi/\partial x_j\rangle\,. \tag{2}$$

This result is remarkable, for we have transformed the difficulty of differentiation of a generalized function to the differentiation of a test function that has derivatives of all orders. This is precisely what happened to the algebraic operations in the previous section.

Let us show that the functional defined by (2) is a distribution. Linearity is obvious, and continuity follows from the fact that if $\{\phi_m\}$ is null sequence, then so is the sequence $\{\partial\phi_m/\partial x_j\}$ (recall the definition of convergence).

The partial derivatives of higher orders can be defined by repeating this process. This results in the general relation

$$\langle D^k t, \phi\rangle = \left\langle \frac{\partial^{|k|} t}{\partial x_1^{k_1} \cdots \partial x_n^{k_n}}, \phi \right\rangle = (-1)^{|k|}\langle t, D^k\phi\rangle\,, \tag{3}$$

where D^k is defined by (2.1.1).

In R_1 we have only ordinary derivatives, and relations (2) and (3) become

$$\langle t', \phi\rangle = -\langle t, \phi'\rangle\,, \tag{4}$$

and

$$\langle t^{(n)}, \phi\rangle = (-1)^n\langle t, \phi^{(n)}\rangle\,. \tag{5}$$

Properties of the generalized (or distributional) derivatives

(1) From (3) and (5) it follows that a generalized function is infinitely differentiable.

(2) The equality $t'(x) = 0$ holds if and only if the distribution $t(x)$ is a constant. From this relation it follows that if $t_1'(x) = t_2'(x)$, then $t_1(x)$ and $t_2(x)$ differ by a constant function.

(3) The generalized derivative agrees with the classical one whenever the latter exists.

(4) Because

$$\left\langle \frac{\partial}{\partial x_1}\left(\frac{\partial t}{\partial x_2}\right), \phi \right\rangle = \langle D^2 t, \phi\rangle = \left\langle t, \frac{\partial^2\phi}{\partial x_1 \partial x_2}\right\rangle = \left\langle t, \frac{\partial^2\phi}{\partial x_2 \partial x_1}\right\rangle = \left\langle \frac{\partial}{\partial x_2}\left(\frac{\partial t}{\partial x_1}\right), \phi \right\rangle,$$

it follows that

$$\frac{\partial^2 t}{\partial x_1 \partial x_2} = \frac{\partial^2 t}{\partial x_2 \partial x_1} = D^2 t\,. \tag{6}$$

In general,

$$D^{|k_1|+|k_2|} t = D^{|k_2|+|k_1|} t\,. \tag{7}$$

Thus, the result of differentiation does not depend on the order of differentiation.

(5) The derivative of the product of a distribution t and a function $f(x) \in C$ is

$$\frac{\partial}{\partial x_j}(ft) = \frac{\partial f}{\partial x_j} t + f\frac{\partial t}{\partial x_j}\,.$$

Indeed, for $\phi \in D$,

$$\left\langle \frac{\partial (ft)}{\partial x_j}, \phi \right\rangle = -\left\langle ft, \frac{\partial \phi}{\partial x_j} \right\rangle = -\left\langle t, f\frac{\partial \phi}{\partial x_j} \right\rangle = -\left\langle t, \frac{\partial (f\phi)}{\partial x_j} - \frac{\partial f}{\partial x_j}\phi \right\rangle$$

$$= -\left\langle t, \frac{\partial (f\phi)}{\partial x_j} \right\rangle + \left\langle t, \frac{\partial f}{\partial x_j}\phi \right\rangle$$

$$= \left\langle \frac{\partial t}{\partial x_j}, f\phi \right\rangle + \left\langle \frac{\partial f}{\partial x_j} t, \phi \right\rangle$$

$$= \left\langle f\frac{\partial t}{\partial x_j}, \phi \right\rangle + \left\langle \frac{\partial f}{\partial x_j} t, \phi \right\rangle$$

$$= \left\langle f\frac{\partial t}{\partial x_j} + \frac{\partial f}{\partial x_j} t, \phi \right\rangle,$$

as desired. The higher-order derivatives of the product can be defined in a similar fashion and we have

$$D^k(ft) = \sum_{m+n=k} (k!/m!n!)(D^m f)(D^n t)$$

$$= \sum_{p<k} \frac{k!}{p!(k-p)!}(D^{k-p} f)(D^p t), \tag{8}$$

where the summation is from $p = (0, 0, \ldots, 0)$ to $k = (k_1, k_2, \ldots, k_n)$. It can be readily proved by induction and forms Exercise 11.

A differential operator and its adjoint

With the differential operator as defined in Section 2.1 and the principles of differentiation as given in this section, we have

$$\langle Lt, \phi \rangle = \left\langle \sum_{|k|\le p} a_k D^k t, \phi \right\rangle = \left\langle t, \sum_{|k|\le p} (-1)^{|k|} D^k(a_k \phi) \right\rangle = \langle t, L^*\phi \rangle \tag{9}$$

where the pth-order differential operator L^*, given by

$$L^*\phi = \sum_{|k|\le p} (-1)^{|k|} D^k(a_k\phi), \tag{10}$$

is called the formal adjoint operator to L. If $L = L^*$, the operator L is self-adjoint. For example, the Laplacian operator ∇^2,

$$\nabla^2 = \frac{\partial^2}{\partial x_1^2} + \frac{\partial^2}{\partial x_2^2} + \cdots + \frac{\partial^2}{\partial x_n^2},$$

is self-adjoint, because

$$\langle \nabla^2 t, \phi \rangle = \langle t, \nabla^2 \phi \rangle, \qquad \phi \in D.$$

When L is an ordinary differential operator,

$$L = a_p(x)\frac{d^p}{dx^p} + a_{p-1}\frac{d^{p-1}}{dx^{p-1}} + \cdots + a_1\frac{d}{dx} + a_0\,, \tag{11}$$

(10) yields

$$L^*\phi = \sum_{k=0}^{p}(-1)^k\frac{d^k(a_k\phi)}{dx^k}\,. \tag{12}$$

Definition. A distribution E is said to be a *fundamental solution* for the differential operator L if

$$LE = \delta\,. \tag{13}$$

We end this section with two important theorems in which we use, among other concepts, Examples 1 and 2 of Section 2.7.

Theorem 1. *If $f(x)$ is a classical function and*

$$f(x) + a_0\delta(x-\xi) + \cdots + a_n\delta^{(n)}(x-\xi) = 0 \tag{14}$$

on the whole axis, $-\infty < x < \infty$, then $f(x) = 0$ and $a_0 = \cdots = a_n = 0$.

Proof. We proceed by induction. The case $n = 0$ is obvious because $\delta(x-\xi)$ is a generalized function and $f(x)$ is a classical function.

Suppose that the theorem holds for $n-1$. Then it follows that $f(x) = 0$ for $x \neq \xi$, so (14) integrates to

$$c + a_0H(x-\xi) + a_1\delta(x-\xi) + \cdots + a_n\delta^{(n-1)}(x-\xi) = 0\,.$$

However, we have assumed that the theorem holds for $n-1$, so

$$c + a_0H(x-\xi) = 0\,, \qquad a_1 = a_2 = \cdots = a_n = 0\,.$$

From the relation $c + a_0H(x-\xi) = 0$, we find that $a_0 = 0$, and we have proved the theorem.

Theorem 2. *Let a function $f(x)$ be n-times continuously differentiable; then*

$$\begin{aligned} f(x)\delta^{(n)}(x) =&(-1)^n f^{(n)}(0)\delta(x) + (-1)^{n-1}nf^{(n-1)}(0)\delta'(x) \\ &+ (-1)^{n-2}\frac{n(n-1)}{2!}f^{(n-2)}(0)\delta''(x) + \cdots + f(0)\delta^{(n)}(x)\,. \end{aligned} \tag{15}$$

Proof. As always, we have to show validity with the help of a test function. Accordingly, we examine

$$\begin{aligned}\int_{-\infty}^{\infty}\{f(x)\delta^{(n)}(x)\}\phi(x)dx &= \int_{-\infty}^{\infty}\{f(x)\phi(x)\}\delta^{(n)}(x)dx\\ &= [\{f(x)\phi(x)\}\delta^{(n-1)}(x)]_{-\infty}^{\infty}\\ &\quad -\int_{-\infty}^{\infty}\{f(x)\phi(x)\}'\delta^{(n-1)}(x)dx\\ &= -\int_{-\infty}^{\infty}\{f(x)\phi(x)\}'\delta^{(n-1)}(x)dx\,.\end{aligned}$$

After a succession of similar integrations by parts, we obtain

$$\int_{-\infty}^{\infty}\{f(x)\phi(x)\}\delta^{(n)}(x)dx = (-1)^n\int_{-\infty}^{\infty}\{f(x)\phi(x)\}^{(n)}\delta(x)dx\,.$$

Substitution of the formula

$$\begin{aligned}\{f(x)\phi(x)\}^{(n)} =& f^{(n)}(x)\phi(x) + nf^{(n-1)}(x)\phi'(x)\\ &+\frac{n(n-1)}{2!}f^{(n-2)}(x)\phi''(0) + \cdots + f(0)\phi^{(n)}(x)\,,\end{aligned} \tag{16}$$

in the preceding relation and the application of the sifting property yields

$$\begin{aligned}\int_{-\infty}^{\infty}\{f(x)\delta^{(n)}(x)\}\phi(x)dx =& (-1)^n\Big\{f^{(n)}(0)\phi(0) + nf^{(n-1)}(0)\phi'(0)\\ &+\frac{n(n-1)}{2!}f^{(n-2)}(0)\phi''(0) + \cdots + f(0)\phi^{(n)}(0)\Big\}\,.\end{aligned} \tag{17}$$

Because

$$\int_{-\infty}^{\infty}\phi(x)\delta^{(k)}(x)dx = (-1)^k\int_{-\infty}^{\infty}\phi^{(k)}(x)\delta(x)dx = (-1)^k\phi^{(k)}(0)\,,$$

relation (17) can be written

$$\begin{aligned}\langle f(x)\delta^{(n)}(x), \phi(x)\rangle =& (-1)^n\Big\{f^{(n)}(0)\langle\delta(x), \phi(x)\rangle\\ &+(-1)^{-1}nf^{(n-1)}(0)\langle\delta'(x), \phi(x)\rangle\\ &+(-1)^{-2}\frac{n(n-1)}{2!}f^{(n-2)}(0)\langle\delta''(x), \phi(x)\rangle\\ &+\cdots+(-1)^{-n}\langle\delta^{(n)}(x), \phi(x)\rangle\Big\}\,,\end{aligned}$$

which is equivalent to the symbolic formula (15).

For the special cases $n = 1, 2, 3$, (15) becomes

$$f(x)\delta'(x) = -f'(0)\delta(x) + f(0)\delta'(x)\,, \tag{18}$$

$$f(x)\delta''(x) = f''(0)\delta(x) - 2f'(0)\delta'(x) + f(0)\delta''(x)\,, \tag{19}$$

$$f(x)\delta'''(x) = f'''(0)\delta(x) + 3f''(0)\delta'(x) - 3f'(0)\delta''(x) + f(0)\delta'''(x)\,, \tag{20}$$

respectively. For instance, it follows from (18) that

$$\sin x\delta'(x) = -(\cos 0)f'(0)\delta(x) + (\sin 0)\delta'(x) = -\delta(x) \tag{21}$$

and

$$\cos x\delta'(x) = -(-\sin 0)\delta(x) + (\cos 0)\delta'(x) = \delta'(x)\,. \tag{22}$$

If we introduce the function $\delta(x-\xi)$ instead of $\delta(x)$ in formula (15) and use the binomial coefficient

$$\binom{n}{k} = \frac{n!}{k!(n-k)!}\,,$$

we have the general formula

$$f(x)\delta^{(n)}(x-\xi) = (-1)^n \sum_{k=0}^{n} (-1)^k \binom{n}{k} f^{(n-k)}(\xi)\delta^{(k)}(x-\xi)\,. \tag{23}$$

Corollary.

$$\begin{aligned}(f(x)H(x))^{(n)} = {} & f^{(n)}(x)H(x) + f^{(n-1)}(0)\delta(x) + f^{(n-2)}(0)\delta'(x) \\ & + \cdots + f(0)\delta^{(n-1)}(x)\,.\end{aligned} \tag{24}$$

Proof. Observe that

$$\begin{aligned}(f(x)H(x))^{(n)} = {} & f^{(n)}(x)H(x) + nf^{(n-1)}(x)H'(x) \\ & + \cdots + \frac{n(n-1)\cdots(n-r+1)}{r!} f^{(n-r)}(x)H^{(r)}(x) \\ & + \cdots + f(x)H^{(n)}(x)\,.\end{aligned}$$

Because $H^{(r)}(x) = \delta^{(r-1)}(x)$, we get (24) on using (15) and striking out the terms that cancel each other.

When we put $H(x-\xi)$ for $H(x)$ in relation (24), we have

$$\begin{aligned}(f(x)H(x-\xi))^{(n)} = {} & f^{(n)}(x)H(x-\xi) + f^{(n-1)}(\xi)\delta(x-\xi) \\ & + f^{(n-2)}(\xi)\delta'(x-\xi) + \cdots + f(\xi)\delta^{(n-1)}(x-\xi)\,.\end{aligned} \tag{25}$$

Another useful formula that follows from (15) is

$$x^n\delta^{(m)}(x) = \begin{cases} 0\,, & m < n\,, \\ (-1)^n n!\,\delta(x)\,, & m = n\,, \\ (-1)^n \frac{m!}{(m-n)!}\delta^{(m-n)}(x)\,, & m < n\,. \end{cases} \tag{26}$$

Its derivation is left as Exercise 12 at the end of the chapter.

A few simple illustrations of this result are

$$\begin{gathered} x\delta(x)=0\,, \qquad x\delta'(x)=-\delta(x)\,, \qquad x\delta^{(3)}(x)=-3\delta''(x)\,, \\ x^2\delta''(x)=-2\delta(x)\,, \qquad x^2\delta^{(4)}(x)=-12\delta''(x)\,. \end{gathered} \tag{27}$$

All these results can be extended to R_n. For instance, the formulas corresponding to (23) and (26) are (see Exercise 22)

$$f(x)D^k\delta(x-\xi)=(-1)^{|k|}\sum_{|p|<|k|}(-1)^{|p|}\frac{k!}{p!(k-p)!}(D^{k-p}f(\xi))(D^p\delta(x-\xi)) \tag{28}$$

and

$$x^n D^m\delta(x)=\begin{cases} 0\,, & |m|<|n|\,, \\ (-1)^{|n|}n!\,\delta(x)\,, & m=n\,, \\ (-1)^{|n|}\frac{m!}{(m-n)!}\delta^{|m-n|}(x)\,, & |m|>|n|\,. \end{cases} \tag{29}$$

As a simple example of (29) we find that the solution of the partial differential equation

$$r^2\nabla^2 u(x)=2n\delta(x)\,, \tag{30}$$

where $r=|x|$, is $u(x)=\delta(x)$.

2.7. Examples

Example 1. (a). Recall that the Heaviside distribution in R_1 is

$$\langle H,\phi\rangle=\int_{-\infty}^{\infty}H(x)\phi(x)dx=\int_0^{\infty}\phi(x)dx\,.$$

According to the foregoing definition of a derivative, we have

$$\langle H',\phi\rangle=-\langle H,\phi'\rangle=-\int_0^{\infty}\phi'(x)dx=\phi(0)=\langle\delta,\phi\rangle\,,$$

or

$$dH(x)/dx=\delta(x)\,, \tag{1}$$

i.e., the derivative of the Heaviside distribution is the Dirac distribution. Thus, $H(x)$ is the fundamental solution of the differential operator d/dx.

(b). Let us find the distributional derivative of the function

$$x_+=\begin{cases} x\,, & x>0\,, \\ 0\,, & x\le 0\,. \end{cases}$$

For this purpose we observe that for a test function $\phi(x)$ we have

$$\langle x'_+, \phi\rangle = -\langle x_+, \phi'\rangle = -\int x\phi'(x)dx = \int_0^\infty \phi(x)dx = \langle H(x), \phi(x)\rangle\,,$$

so that $(d/dx)(x_+) = H(x)$. With the help of this result we find that the function

$$E(x) = \frac{d^k}{dx^k}\left(\frac{x_+^{k-1}}{(k-1)!}\right)$$

is the fundamental solution of the differential operator d^k/dx^k. This result is generalized to R_n in Exercise 10.

Example 2. For the delta function we have, first,

$$\langle \partial\delta(x-\xi)/\partial x_j, \phi\rangle = -\langle\delta(x-\xi)\,, \partial\phi/\partial x_j\rangle = -\partial\phi/\partial x_j|_{x=\xi}\,. \tag{2}$$

In R_1, (2) becomes

$$\langle\delta'(x-\xi)\,, \phi(x)\rangle = -\phi'(\xi)\,. \tag{3}$$

Comparing (3) with (1.3.2), we find that $-\delta'(x)$ is the dipole distribution. This process can be continued to yield in R_n.

$$\langle D^k\delta(x-\xi), \phi\rangle = (-1)^{|k|}D^k\phi\big|_{x-\xi}\,, \tag{4}$$

and, in R_1,

$$\langle\delta_\xi^{(n)}, \phi\rangle = (-1)^n\left.\frac{d^n\phi}{dx^n}\right|_{x=\xi} = -\left\langle\delta_\xi\,, \frac{d^n\phi(x-\xi)}{d\xi}\right\rangle. \tag{5}$$

Second, for a function $f(x)\in C^\infty$, we have

$$\begin{aligned}\left\langle f\frac{\partial\delta_\xi}{\partial x_j}, \phi\right\rangle = \left\langle\frac{\partial\delta_\xi}{\partial x_j}, f\phi\right\rangle &= -\left\langle\delta_\xi, \frac{\partial(f\phi)}{\partial x_j}\right\rangle\\ &= -\left\langle\delta(x-\xi), f\frac{\partial\phi}{\partial x_j} + \frac{\partial f}{\partial x_j}\phi\right\rangle\\ &= -f(\xi)\left.\frac{d\phi}{\partial x_j}\right|_{x=\xi} - \left.\frac{\partial f}{\partial x_j}\right|_{x=\xi}\phi(\xi)\\ &= f(\xi)\left\langle\frac{\partial\delta_\xi}{\partial x_j}, \phi\right\rangle - \left.\frac{\partial f}{\partial x_j}\right|_{x=\xi}\langle\delta_\xi, \phi\rangle\,, \end{aligned} \tag{6}$$

or

$$f\frac{\partial\delta_\xi}{\partial x_j} = f(\xi)\frac{\partial\delta_\xi}{\partial x_j} - \left.\frac{\partial f}{\partial x_j}\right|_{x=\xi}\delta_\xi\,.$$

Note that this result is valid even when f is merely a C^1 function.

Example 3. Let us study the distribution $|x|$.

$$\langle |x|, \phi\rangle = \int_{-\infty}^\infty |x|\phi(x)dx = \int_0^\infty x\phi(x)dx - \int_{-\infty}^0 x\phi(x)dx\,. \tag{7}$$

Then

$$\langle |x|', \phi\rangle = -\langle |x|, \phi'\rangle = -\int_{-\infty}^{\infty} |x|\phi'(x)dx$$
$$= \int_{-\infty}^{0} x\phi'(x)dx - \int_{0}^{\infty} x\phi'(x)dx\,,$$

which, when integrated by parts, yields

$$\langle |x|', \phi\rangle = \int_{0}^{\infty} \phi(x)dx - \int_{-\infty}^{0} \phi(x)dx = \int_{-\infty}^{\infty} \text{sgn}\,(x)\phi(x)dx\,, \tag{8}$$

where sgn x, which denotes the *signum function*, is

$$\text{sgn}\,x = \begin{cases} 1\,, & x > 0\,, \\ -1\,, & x < 0\,. \end{cases} \tag{9}$$

Thus

$$\langle |x|', \phi\rangle = \langle \text{sgn}\,x, \phi\rangle\,, \qquad \text{or} \qquad |x|' = \text{sgn}\,x\,. \tag{10}$$

Note that

$$\text{sgn}\,x = H(x) - H(-x)\,. \tag{11}$$

Hence the second differentiation is readily obtained:

$$|x|'' = \text{sgn}\,'x = \delta(x) + \delta(-x) = 2\delta(x)\,. \tag{12}$$

These and their associated functions play a useful part in many fields of analysis. For example, we can evaluate

$$I = \int_{-1}^{1} |x| f''(x)dx\,,$$

by using the foregoing results:

$$I = [|x|f'(x)]_{-1}^{1} - \int_{-1}^{1} |x|' f'(x)dx$$
$$= f'(1) - f'(-1) - [|x|' f(x)]_{-1}^{1} + \int_{-1}^{1} |x|'' f(x)dx$$
$$= f'(1) - f'(-1) - [\text{sgn}\,(x) f(x)]_{-1}^{1} + 2\int_{-1}^{1} \delta(x) f(x)dx$$
$$= f'(1) - f'(-1) - f(1) - f(-1) + 2f(0)\,.$$

In passing we observe from (11) that

$$H(x) = \frac{1}{2}(1 + \text{sgn}\,(x))\,. \tag{13}$$

Moreover, if we define the reversed function $\phi^{\vee}(s)$ by

$$\phi^{\vee}(s) = \phi(-s)\,, \tag{14}$$

then we can write (11) as

$$\operatorname{sgn}(x) = H(x) - H^{\vee}(x). \tag{15}$$

Example 4. The Heisenberg distributions, explained in Example 5 of Section 2.4, can also be obtained by differentiating the regular distribution $\ln(x + i0)$, which is generated by the locally integrable function

$$\ln(x + i0) = \lim_{\varepsilon \to 0} \ln(x + i\varepsilon).$$

We may write

$$\ln(x + i\varepsilon) = \ln|x + i\varepsilon| + i \arg(x + i\varepsilon),$$

and

$$i \lim_{\varepsilon \to 0} \arg(x + i\varepsilon) = i \lim_{\varepsilon \to 0} \tan^{-1} \frac{\varepsilon}{x} = i\pi H(-x).$$

Thus

$$\lim_{\varepsilon \to 0} \ln(x + i\varepsilon) = \ln|x| + i\pi H(-x), \tag{16}$$

and we have

$$\begin{aligned} \left\langle \frac{d}{dx} \ln(x + i0), \phi \right\rangle &= \lim_{\varepsilon \to 0} \left\langle \frac{1}{x + i\varepsilon}, \phi \right\rangle \\ &= \left\langle Pf\left(\frac{1}{x}\right) - i\pi\delta(x), \phi \right\rangle, \qquad \phi \in D, \end{aligned}$$

which is equivalent to relation (2.4.18) except for a constant factor. Result (2.4.19) can be obtained in a similar manner.

Example 5. Double-layer distribution. Let S be a piecewise smooth double-sided hypersurface in R_n and let $\widehat{n}$ be the unit normal to S. On S we define a locally integrable function $\tau(x)$ that varies continuously on it. Consider the generalized function $-\tau(\xi)(\partial/\partial n)\delta[(x - \xi)]$, $\xi \in S$. This is clearly a generalization of the dipole distribution $-\delta'(x)$. In the language of electrostatics, we have normally oriented dipoles spread over S with surface density $\tau(\xi)$. Thus, the surface element dS_ξ at ξ carries a dipole field whose volume density is

$$-\tau(\xi)\frac{\partial}{\partial n_x}[\delta(x - \xi)]dS_\xi.$$

This gives the total volume density

$$-\int_S \tau(\xi)\frac{\partial}{\partial n_\xi}[\delta(x - \xi)]dS_\xi, \tag{17}$$

whose action on a test function ϕ is

$$\begin{aligned} \left\langle -\int_S \tau(\xi)\frac{\partial}{\partial n_\xi}\delta(x - \xi)dS_\xi, \phi \right\rangle &= -\int_{R_n} dx\, \phi(x) \int_S \tau(\xi)\frac{\partial}{\partial n_\xi}[\delta(x - \xi)]dS_\xi \\ &= \int_S \tau(\xi)\frac{\partial\phi(\xi)}{\partial n} dS_\xi. \end{aligned} \tag{18}$$

We can also write (18) as $\langle -(\partial/\partial n)(\tau\delta(S)), \phi\rangle$, because

$$-\left\langle \frac{\partial}{\partial n}(\tau\delta(S)), \phi\right\rangle = \left\langle \tau\delta(S), \frac{\partial\phi}{\partial n}\right\rangle = \int_S \tau(\xi)\frac{\partial\phi}{\partial n} dS_\xi .$$

Accordingly, $(\partial/\partial n)(\tau\delta(S))$ defines a double layer (or a dipole layer) over the surface S.

Example 6. Multipoles. Let us consider the case $n = 3$. Then the generalized function $\delta(x - x_0)$ represents a uniform surface charge along the plane $x = x_0$, while $\delta(x - x_0)\delta(y - y_0)$ represents a uniform line charge along the straight line $x = x_0$, $y = y_0$. Similarly, the function $\delta'(x - x_0)$ represents a plane layer of dipoles with axis along the x axis whereas $\delta'(x - x_0)\delta(y - y_0)$ represents a straight line $x = x_0$, $y = y_0$ of dipoles with axis along the x axis. Similarly, $\delta'(x - x_0)\delta(y - y_0)\delta(z - z_0)$ represents a dipole at the point (x_0, y_0, z_0) with axis along the x axis.

We can continue in this fashion and define multipoles. For example, a quadrupole is defined as the limit of positive and negative dipoles with axis along the x axis displaced an infinitesimal distance along that axis has a volume charge density $\delta''(x - x_0)\delta(z - z_0)$. Incidentally, all these charge densities have unit strength.

Various vector quantities such as electric current density J can be represented by the generalized functions. For instance, $\delta(x - x_0)e_y$, where e_y is a unit vector in the y direction, represents a uniform surface current of charge on the plane $x = x_0$ moving in the y direction. The function

$$\delta(x - x_0)\delta(y - y_0)e_z$$

represents a line current on the line $x = x_0$, $y = y_0$ moving in the z direction. Similarly, the function $\delta(x - x_0)\delta(y - y_0)\delta(z - z_0)e_z$ represents a point charge moving in the z direction.

Order of a distribution

From Example 1 we find that $\delta(x)$ is the derivative of a regular distribution $H(x)$. For this reason, the delta distribution is called a distribution of order 1. In general, a distribution is said to be of order n if it can be expressed as the nth derivative of a regular distribution but cannot be expressed as a derivative of lower order of a regular distribution. This means that the higher the order of a distribution, the more singular it is. A distribution of order zero is a regular distribution. To generalize and summarize these ideas we have: If t is a distribution of finite order, then there exists a locally integrable function f and a multiindex k such that $t = D^k f$.

2.8. The Support and Singular Support of a Distribution

The reader must have realized by now that we cannot talk about local properties of distributions. We can only consider their action on the class of test functions. Nevertheless, we can say that the distribution t vanishes in the neighborhood N of a point x if $\langle t, \phi \rangle = 0$ there for all $\phi \in D$. A distribution t vanishes in a region R of R_n if it vanishes in a neighborhood of every part of R. Accordingly, the distributions f and g are said to be equal in R if $f - g = 0$ for all $x \in R$.

A point x is called an *essential* point of a distribution t if there does not exist a neighborhood of x in which t vanishes. The closure of the set of all essential points is defined as support of t and denoted as supp t. Technically, we can describe the support as the least closed set outside of which t vanishes. If f is an ordinary function, then its support as a distribution coincides with its support as a function. Furthermore, it follows that if supp t and supp ϕ have no points in common, then $\langle t, \phi \rangle = 0$. It is interesting to compare this with the zero distribution; if $x \notin \text{supp}\, t$, then there is a neighborhood of x such that $t = 0$ in that neighborhood, and t also vanishes in any region consisting of such neighborhoods.

Example 1. Supp $H(x)$, where $H(x)$ is the Heaviside function, is the half-axis $x > 0$.

Example 2. The support of $\delta(x)$ and each of its derivatives is a single point $x = 0$. The converse is also true. A distribution t that has support of one point is a finite linear combination of $\delta(x)$ and its derivatives (see Exercise 25).

Example 3. If the distribution $t = 0$ for $x \in R$ then also $D^k t = 0$ for $x \in R$ so that supp $D^k t \subset \text{supp}\, t$.

The proof follows by observing that if $\phi \in D_R$, then also $D^k \phi \in D_R$. Accordingly,

$$\langle D^k t, \phi \rangle = (-1)^{|k|} \langle t, D^k \phi \rangle = 0 \,,$$

from which it follows that $D^k t = 0$ for $x \in R$.

Example 4.

$$\text{supp}\, a(x)t(x) \subset \text{supp}\, a \cap \text{supp}\, t \,.$$

This may be proved by noting that $x \in \text{supp}\, a(x)t(x)$ implies that $a(x)t(x) \neq 0$ for all neighborhoods of x. This means that $a(x) \neq 0$ and $t(x) \neq 0$ for all neighborhoods of x. Accordingly, $x \in \text{supp}\, a$ and $x \in \text{supp}\, t$; that is, $x \in \text{supp}\, a \cap \text{supp}\, t$, as required.

Example 5. The distribution $|t|^2 = t_1^2 + t_2^2 + \cdots + t_n^2$ has the entire space R_n as its support, although $|t|^2$ vanishes at $t = 0$. The reason is that it is the behavior of a distribution over the neighborhood of a point that determines whether the point is in the support of t.

Another useful concept is that of a *singular* support. This is defined as the closed set of points where $t(x)$ is not a smooth function. It is the least closed set outside of which t is equal to a function $f(x)$ possessing derivatives of all orders. For example, both $\delta(x)$ and $H(x)$ have as singular support the single point $x = 0$. Similarly, the distribution $|x|$ also has the point $x = 0$ as a singular support.

Example 6. As a final example, let us consider the functional

$$\int_{-1}^{1} (\sin 2x)\phi(x)dx .$$

By writing it as

$$\int_{-1}^{1} (\sin 2x)\phi(x)dx = \int_{-\infty}^{\infty} f(x)\phi(x)dx ,$$

where

$$f(x) = \begin{cases} \sin 2x , & -1 < x < 1 , \\ 0 & \text{otherwise} , \end{cases}$$

it becomes clear that the support of this functional is the closed interval $[-1, 1]$ while the singular support consists of the two points -1 and 1.

Exercises

1. Let a sequence of test functions $\{\phi_m(x)\}$, $m = 1, 2, \ldots$ converge in D to zero, and let $\psi(x)$ be an infinitely differentiable function. Show that $\{\psi\phi_m\}$ also converges to zero.

2. Compute the first four distributional derivatives of $\sin x$, $|\sin x|$, $x \cos x$, and $|x \cos x|$.

3. Find the first distributional derivatives of
   (a) $[1 - H(x)] \cos x$,
   (b) $|x| \sin x$,
   (c) $\{H(x+2) + H(x) - H(x-2)\}e^{-2x}$.

4. Find the second distributional derivatives of $e^{-|x|}$ and $\sin |x|$.

5. Show that

$$f(x) = \tanh(1/x) , \qquad x \neq 0,$$

(draw it!) has a jump discontinuity at $x = 0$. Find its distributional derivative.

6. Find the second distributional derivative of the function

$$f(x) = \begin{cases} x , & |x| \leq 1 , \\ 1 , & x > 1 , \\ -1 , & x < -1 . \end{cases}$$

7. Show that
   (a) $e^x \delta(x) = \delta(x)$,
   (b) $(\sin ax)\delta'(x) = -a\delta(x)$,
   (c) $(d^2/dx^2)[H(x) \sin ax] = a\delta(x) - a^2 H(x) \sin ax$,
   (d) $\frac{d^2}{dx^2} e^{ik|x|} = 2i\, k\delta(x) - k^2 e^{ikx}$.

8. Show that $[f'(t)]^\vee = -[f^\vee(t)]'$, where the symbol $^\vee$ is defined in Example 3 of Section 2.7.

9. Recall that $J_0(x)$ is a solution of the Bessel equation

$$\frac{d^2y}{dx^2} + \frac{1}{x}\frac{dy}{dx} + y(x) = 0 .$$

Show that $J_0(x)H(x)$ is also a solution.

10. (a). Consider the function

$$H(x) = H(x_1)H(x_2)\cdots H(x_n) , \qquad x \in R_n ,$$

and prove that

$$\frac{\partial^n H(x)}{\partial x_1 \partial x_2 \cdots \partial x_n} = \delta(x) .$$

(b). Show that the fundamental solution of the differential operator

$$D^k = \frac{\partial^{|k|}}{\partial x_1^{k_1} \cdots \partial x_n^k} ,$$

in R_n is

$$E(x) = \frac{(x_1)_+^{k-1} \cdots (x_n)_+^{k-1}}{[(k-1)!]^n} .$$

(c). Show that the function

$$E(x_1, x_2) = x_1 H(x_1) H(x_2) e^{x_2}$$

is the fundamental solution of the operator

$$\frac{\partial^2}{\partial x_1 \partial x_2}\left(\frac{\partial}{\partial x_2} - 1\right) .$$

11. Establish formula (2.6.8).

12. (a) Prove formula (2.6.26), namely,

$$x^n \delta^{(m)}(x) = \begin{cases} 0 , & m < n , \\ (-1)^n n! \delta(x) , & m = n , \\ (-1)^n \frac{m!}{(m-n)!} \delta^{(m-n)}(x) , & m < n . \end{cases}$$

(b) Deduce that

$$e^{-\lambda x} \delta^{(n)}(x) = \sum_{m=0}^{n} \binom{n}{m} \lambda^{n-m} \delta^{(m)}(x) .$$

13. Verify that

$$\frac{1}{x}\delta^{(n)} = -\frac{1}{n+1}\delta^{(n+1)} + c\delta ,$$

where c is a constant.

14. Show that the collections $\{(-1)^m \delta^{(m)}(x)\}_{m=0}^{\infty}$ and $\{x^n/n!\}_{n=0}^{\infty}$ form a biorthogonal set. That is

$$\left\langle (-1)^m \delta^{(m)}(x), \frac{x^n}{n!} \right\rangle = \begin{cases} 0 & \text{if} \quad n \neq m, \\ 1 & \text{if} \quad n = m. \end{cases}$$

15. Power distributions x_+^λ and x_-^λ are defined as

$$\langle x_+^\lambda, \phi \rangle = \int_0^\infty s^\lambda \phi(s) ds, \qquad \langle x_-^\lambda, \phi \rangle = \int_{-\infty}^0 |s|^\lambda \phi(s) ds,$$

respectively, where Re $\lambda > 0$. This means that the distribution x_+^λ is to be identified with the function

$$x_+^\lambda = \begin{cases} x^\lambda, & x > 0, \\ 0, & x \le 0, \end{cases}$$

and a similar definition holds for x_-^λ. Show that

$$\frac{d}{dx} x_+^\lambda = \lambda x_+^{\lambda-1}, \qquad \frac{d}{dx} x_-^\lambda = -\lambda x_-^{\lambda-1}.$$

These functions are studied in detail in Chapter 4.

16. Show that

$$\int_{-\infty}^{\infty} H(x-s) s_+^n ds = \frac{x_+^{n+1}}{n+1}, \qquad n = 1, 2, 3, \ldots.$$

17. Show that the sequence of test functions $\{\phi_m(x)\}$, where

$$\phi_m(x) = \begin{cases} c_m \exp\left(-\frac{1}{1-m^2r^2}\right), & r < 1/m. \\ 0, & r \ge 1/m, \end{cases}$$

and

$$c_m^{-1} = \int_{r \le m^{-1}} \exp\left(-\frac{1}{1-m^2r^2}\right) dx,$$

is a delta sequence.

18. Show that $2a\delta(at+x)\delta(at-x) = \delta(x)\delta(t)$. *Hint.* Examine the action of $2a\delta(at+x)\delta(at-x)$ and then use the transformation of coordinate system

$$s = at - x, \quad y = at + x.$$

19. Let the function $[x]$ be defined as the mean of the greatest integer less than x and the greatest integer less than or equal to x. Show that

(a) $[x]' = III(x)$,

(b) $(d/dx)\{[x]H(x)\} = III(x)H(x) - \frac{1}{2}\delta(x)$.

20. Show that $II(x) = \delta(2x^2 - \frac{1}{2})$. For more results of this nature see Section 3.1.

21. Show that

$$\delta(xy) = \frac{\delta(x) + \delta(y)}{(x^2 + y^2)^{1/2}} = \frac{\delta(x)}{y} + \frac{\delta(y)}{x} .$$

22. Derive the expansion of $f(x)D^k\delta(x)$ in R_2 and R_3 on the same lines as Theorem 2 of Section 2. Establish formula (2.6.28), namely,

$$f(x)D^k\delta(x - \xi) = (-1)^{|k|} \sum_{p \leq k} (-1)^{|p|} \frac{k!}{p!(k-p)!} (D^{k-p} f(\xi))(D^p \delta(x - \xi)) .$$

23. Show that the singular support of t is contained in the support of t.

24. Find the support and the singular support of the functionals

(a) $\int_0^\infty s^2 \phi(s) ds$,

(b) $\int_0^1 e^{-1/s} \phi(s) ds$,

(c) $\int_{-1}^1 e^{-1/(1-s^2)} \phi(s) ds$.

25. Show that every distribution $t(x)$ that has compact support is of finite order [2]. Then deduce that every distribution $t(x)$ whose support is the point $x = 0$ has the form $t(x) = \sum_{i=1}^n c_i \delta^{(i)}(x)$.

26. Show that the set $\delta, \delta', \ldots, \delta^{(m)}$ constitutes a linearly independent system.
Hint. Appeal to the function x^a, $a < m$ and examine the quantity $\langle \sum_{i=1}^m \alpha_i \delta^{(i)}(x), x^a \rangle$.

CHAPTER 3

Additional Properties of Distributions

3.1. Transformation Properties of the Delta Distribution

Some algebraic operations on the delta function were studied in the last chapter. In subsequent chapters we shall be required to transform this function to certain curvilinear coordinates. For this purpose we devote an entire section to this topic. Let us first study the meaning of the function $\delta[f(x)]$ and prove the result

$$\delta[f(x)] = \sum_{m=1}^{n} \frac{\delta(x - x_m)}{|f'(x_m)|}, \tag{1}$$

where x_m runs through the simple zeros of $f(x)$.

Suppose that $f(x)$ has a simple zero only at $x = x_1$ and that $f'(x_1) > 0$, so that the transformation is one-to-one in the neighborhood of x_1. Then the substitution $y = f(x)$ in the functional $\int_{-\infty}^{\infty} \delta[f(x)]\phi(x)dx$ gives

$$\int_{-\infty}^{\infty} \delta(y)\frac{\phi[x(y)]}{f'(x)}dy = \frac{\phi(x_1)}{f'(x_1)} = \int_{-\infty}^{\infty} \frac{\delta(x - x_1)\phi(x)dx}{f'(x_1)}$$

because only a neighborhood of $y = 0$ contributes to the integral. For the case $f'(x_1) < 0$, the direction of integration is reversed, and the right side of the above relation becomes $-\phi(x_1)/f'(x_1)$. Combining these two cases, we have

$$\delta[f(x)] = \delta(x - x_1)/|f'(x_1)|. \tag{2}$$

When the function $f(x)$ has simple zeros at $x_1, x_2, \ldots, x_n$, we have the required relation (1). If $f(x)$ has higher-order zeros, no significance is attached to $\delta[f(x)]$.

The following alternative approach is instructive and helpful in solving specific problems [3]. To fix the ideas, let us again assume that $f(x)$ has a simple zero only at x_1 (and thus $f(x_1) = 0$) but that $f'(x_1) \neq 0$, which we again assume to be positive. Furthermore, let us first consider a finite interval $\alpha < x_1 < \beta$. Thus $f(x)$ increases monotonically from $f(\alpha)$ to $f(\beta)$ as x goes from α to β, so that

$$H[f(x)] = H(x - x_1), \tag{3}$$

where H is the Heaviside function. Differentiating (3), we obtain

$$(d/dx)H[f(x)] = \delta(x - x_1),$$

or

$$\delta[f(x)] = [f'(x_1)]^{-1}\delta(x - x_1). \tag{4}$$

For a monotonically decreasing function

$$H[f(x)] = H(x_1 - x)\,, \tag{5}$$

whose differential is $\delta[f(x)] = -[f'(x_1)]^{-1}\delta(x - x_1)$. Combining this with (4) we rederive (2). The extension to more simple zeros and infinite interval is now trivial. Indeed, if there are n zeros of $f(x)$, then

$$\delta[f(x)] = \sum_{m=1}^{n} \frac{\delta(x - x_m)}{|f'(x_m)|}\,,$$

as desired.

Incidentally, it follows from (3) and (5) that

$$H'[f(x)] = \sum_{m=1}^{n} \frac{1}{2}\delta(x - x_m)\{\operatorname{sgn} f(x_m + \varepsilon_m) - \operatorname{sgn} f(x_m - \varepsilon_m)\}\,, \tag{6}$$

where ε_m is chosen so small that the interval $(x_m - \varepsilon_m, x_m + \varepsilon_m)$ contains no zero other than x_m.

We can go a step further and evaluate $\delta'[f(x)]$, where $f(x)$ is a twice continuously differentiable function that increases monotonically from $f(\alpha)$ to $f(\beta)$ as x increases from α to β. If $f(\xi) = 0$, where $\alpha < \xi < \beta$, then

$$\delta'[f(x)] = \frac{1}{|f'(\xi)|^2}\left\{\delta'(x - \xi) + \frac{f''(\xi)}{f'(\xi)}\delta(x - \xi)\right\}. \tag{7}$$

Indeed, for a test function $\phi(x)$ and for $t = f(x)$, we have

$$\begin{aligned}
\langle \phi(x), \delta'[f(x)]\rangle &= \int_{-\infty}^{\infty} \phi(x)\delta'[f(x)]dx = \int_{-\infty}^{\infty} \phi\{f^{-1}(t)\}\delta')t_{\{}f^{-1}(t)\}'dt \\
&= -\frac{d}{dt}[\phi\{f^{-1}(t)\}\{f^{-1}(t)\}']_{t=f(\xi)} = -\frac{d}{dt}\left[\phi(x)\frac{dx}{dt}\right]_{x=\xi} \\
&= -\left[\frac{d}{dx}\left\{\frac{\phi(x)}{f'(x)}\right\}\frac{1}{f'(x)}\right]_{x=\xi} \\
&= -\frac{1}{f'(\xi)}\left[\frac{\phi'(\xi)}{f'(\xi)} - \frac{\phi(\xi)f''(\xi)}{|f'(\xi)|^2}\right] \\
&= \left\langle \phi(x), \frac{1}{|f'(\xi)|^2}\left\{\delta'(x - \xi) + \frac{f''(\xi)}{f'(\xi)}\delta(x - \xi)\right\}\right\rangle,
\end{aligned}$$

which proves (7).

Alternatively, we can use the formula $\delta[f(x)] = [f'(\xi)]^{-1}\delta(x-\xi)$, and obtain

$$\begin{aligned}\delta'[f(x)] &= \frac{d\delta[f(x)]}{d[f(x)]} = \frac{1}{f'(x)}\frac{d}{dx}\delta[f(x)] \\ &= \frac{1}{f'(x)}\frac{d}{dx}\left[\frac{1}{f'(\xi)}\delta(x-\xi)\right] = \frac{1}{f'(x)}\left[\frac{\delta'(x-\xi)}{f'(\xi)}\right] \\ &= \frac{1}{f'(\xi)}\left[\frac{1}{f'(x)}\delta(x-\xi)\right] \\ &= \frac{1}{f'(\xi)}\left[\frac{\delta'(x-\xi)}{f'(\xi)} + \frac{f''(\xi)}{|f'(\xi)|^2}\delta(x-\xi)\right],\end{aligned}$$

where in the last step we have used the relation

$$g(x)\delta'(x-\xi) = g(\xi)\delta'(x-\xi) - g'(\xi)\delta(x-\xi),$$

from Section 2.6.

Formula (1) can be extended to higher dimensions. Let $x = (x_1, \dots, x_n) \in R_n$ and consider $\delta(f(x))$ where $f(x) = (f_1(x), \dots, f_m(x))$ is strictly monotonic. We confine ourselves to the case when $x_1 = (x_{11}, \dots, x_{1n})$ is the only root of $f(x) = 0$. Then we have the result

$$\delta(f(x)) = \frac{\delta(x-x_1)}{|J|}, \tag{8}$$

where $J = \det(\partial f_i/\partial x_j)$ stands for the Jacobian.

Example 1.

$$\delta(ax+b) = (1/|a|)\delta(x+b/a). \tag{9}$$

The proof follows by differentiating $H(ax+b)$.

Example 2.

$$\delta[(x-a)(x-b)] = [1/(b-a)][\delta(x-a)+\delta(x-b)], \qquad b > a. \tag{10}$$

The function $f(x)$ has two simple zeros, $x_1 = a$, $x_2 = b$. Appealing to (1), we obtain

$$\delta[(x-a)(x-b)] = \frac{\delta(x-a)}{|a-b|} + \frac{\delta(x-b)}{|b-a|} = \frac{1}{b-a}[\delta(x-a)+\delta(x-b)],$$

as required. In particular, if $b > 0$ and $a = -b$, then

$$\delta(x^2-b^2) = [\delta(x-b)+\delta(x+b)]/2b.$$

Example 3. $\delta(x^3+3x) = \frac{1}{3}\delta(x)$. In this case

$$f(x) = x^3+3x \qquad f(x) = 0 \quad \text{for} \quad x = 0. \qquad x^2+3 = 0.$$

Thus $x = 0$ is a simple zero and $f'(x) = 3x^2+3$, $f'(0) = 3$. so that

$$\delta(x^3+3x) = \frac{1}{3}\delta(x-0) = \frac{1}{3}\delta(x). \tag{11}$$

Example 4. $\delta(\tan x) = \delta(x)$, $-\frac{1}{2}\pi < x < \frac{1}{2}\pi$. Since $x = 0$ is a simple zero in this interval, we have

$$f(x) = \tan x\,, \qquad f'(x) = \sec^2 x\,, \qquad f'(0) = 1\,.$$

Thus

$$\delta(\tan x) = \delta(x)\,. \tag{12}$$

Example 5. $\delta(\cos x) = \delta(x - \pi/2)$. $0 < x < \pi$. Here $x = \pi/2$ is a simple zero and $f'(x) = -\sin x$, $f'(\pi/2) = -1$, so that

$$\delta(\cos x) = \delta(x - \pi/2)/|-1| = \delta(x - \pi/2)\,. \tag{13}$$

Example 6. With the help of Example 2 let us evaluate $\int_{-2\pi}^{2\pi} e^{\pi x}\delta(x^2 - \pi^2)dx$;

$$\int_{-2\pi}^{2\pi} e^{\pi x}\delta(x^2 - \pi^2)dx = \frac{1}{2\pi}\int_{-2\pi}^{2\pi} e^{\pi x}[\delta(x+\pi) + \delta(x-\pi)]dx = \frac{\cosh \pi^2}{\pi}\,.$$

Example 7. Result (1) remains true even for an infinite set of simple zeros. For instance, for $-\infty < x < \infty$, we have

$$\delta(\sin x) = \sum_{n=-\infty}^{\infty} \frac{\delta(x - n\pi)}{|(\cos n\pi)|} = \sum_{n=-\infty}^{\infty} \delta(x - n\pi)\,. \tag{14}$$

This can also be obtained by differentiating the identity (1.1.7).

Example 8. Let us evaluate $\delta'(\sinh 2x)$. Since $\sinh 2x = 0$ at $x = 0$, we appeal to (7):

$$\delta'(\sinh 2x) = \frac{1}{f'(0)}\left\{\frac{\delta'(x)}{f'(0)} + \frac{f''(0)}{|f'(0)|^2}\delta(x)\right\}\,.$$

Substituting in the values $f'(0) = 2\cosh 0 = 2$, $f''(0) = 4\sinh 0 = 0$. We derive the required value

$$\delta'(\sinh 2x) = \delta'(x)/4.$$

Example 9. Let us evaluate the integral

$$I = \int_{-\infty}^{\infty} (\cos x + \sin x)\delta'(x^3 + x^2 + x)dx\,. \tag{15}$$

Since $f(x) = x^3 + x^2 + x$ and $f(0) = 0$, we have from (7)

$$\begin{aligned} \delta'(x^3 + x^2 + x) &= \frac{1}{(3x^2 + 2x + 1)^2|_{x=0}}\left\{\delta'(x) + \frac{(6x+2)|_{x=0}}{(3x^2 + 2x + 1)|_{x=0}}\delta(x)\right\} \\ &= \delta'(x) + 2\delta(x)\,. \end{aligned} \tag{16}$$

Substituting (16) in (15), we get

$$\begin{aligned} I &= \int_{-\infty}^{\infty} (\cos x + \sin x)(\delta'(x) + 2\delta(x))dx \\ &= -[-\sin x + \cos x]_{x=0} + 2(\cos x + \sin x]_{x=0} = -1 + 2 = 1\,. \end{aligned}$$

Curvilinear coordinates

Let us now study the form of the delta function when we change from Cartesian to curvilinear coordinates. Since there is no basic difference in the analysis whether we study the general case of n-dimensional space or the special case $n = 2$, we first assume the latter for simplicity. We shall then state the result for the general case [4]. Accordingly, let us transform from Cartesian coordinates (x_1, x_2) to curvilinear coordinates (ξ_1, ξ_2) where

$$x_1 = u_1(\xi_1, \xi_2), \qquad x_2 = u_2(\xi_1, \xi_2), \tag{17}$$

and u_1, u_2 are single-valued continuously differentiable functions of their arguments. The transformation is assumed to be nonsingular, so the Jacobian

$$J = \det[\partial(u_1, u_2)/\partial(\xi_1, \xi_2)], \tag{18}$$

is nonzero.

Suppose that under this transformation the point (β_1, β_2) in the ξ system corresponds to the point (α_1, α_2) in the x system (see Figure 3.1). Then with the change of coordinates (17), the equation

$$\iint_{R_2} \phi(x_1, x_2)\delta(x_1 - \alpha_1)\delta(x_2 - \alpha_2)dx_1dx_2 = \phi(\alpha_1, \alpha_2),$$

becomes

$$\iint_{R_2} \phi(\xi_1, \xi_2)\delta[u_1(\xi_1, \xi_2) - \alpha_1]\delta[u_2(\xi_1, \xi_2) - \alpha_2]|J|d\xi_1d\xi_2 = \phi(\beta_1, \beta_2), \tag{19}$$

where $\phi(x_1, x_2) \in D$. From (19) it follows that

$$\delta[u_1(\xi_1, \xi_2) - \alpha_1]\delta[u_2(\xi_1, \xi_2) - \alpha_2]|J| = \delta(\xi_1 - \beta_1)\delta(\xi_2 - \beta_2),$$

or

$$\delta(x_1 - \alpha_1)\delta(x_2 - \alpha_2) = \delta(\xi_1 - \beta_1)\delta(\xi_2 - \beta_2)/|J|. \tag{20}$$

This analysis is readily extended to n dimensions and yields the following theorem:

Theorem. *Let curvilinear coordinates $\xi_1, \xi_2, \ldots, \xi_n$ be obtained from Cartesian coordinates $x_1, x_2, \ldots, x_n$ by the transformation law*

$$x_i = u_i(\xi_1, \ldots, \xi_n), \qquad i = 1, \ldots, n.$$

Furthermore, let the point P with coordinates $(\alpha_1, \ldots, \alpha_n)$ in the x coordinate system have the coordinates $(\beta_1, \ldots, \beta_n)$ in the ξ coordinate system. Then

$$\delta(x_1 - \alpha_1) \cdots \delta(x_n - \alpha_n) = |J|^{-1}\delta(\xi_1 - \beta_1) \cdots \delta(\xi_n - \beta_n), \tag{21}$$

where the Jacobian J of the transformation is nonzero.

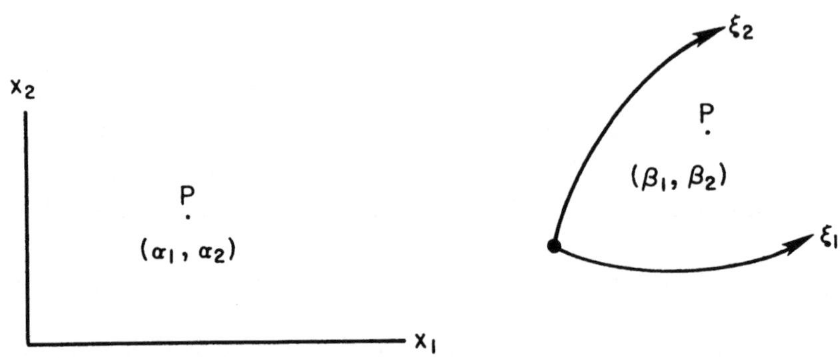

Figure 3.1. Transformation to curvilinear coordinates.

Example 10. The transformation from rectangular Cartesian coordinates (x, y) to plane polar coordinates (r, θ) is

$$x = r\cos\vartheta\,, \qquad y = r\sin\vartheta\,. \tag{22}$$

Thus, $J = r$, and (21) becomes

$$\delta(x - x_1)\delta(y - y_1) = \delta(r - r_1)\delta(\vartheta - \vartheta_1)/r\,. \tag{23}$$

The transformation from three-dimensional rectangular Cartesian coordinates (x, y, z) to spherical polar coordinates (r, ϑ, φ) is

$$x = r\sin\vartheta\cos\varphi\,, \qquad y = r\sin\vartheta\sin\varphi\,, \qquad z = r\cos\vartheta\,. \tag{24}$$

This yields $J = r^2\sin\vartheta$, and

$$\delta(x - x_1)\delta(y - y_1)\delta(z - z_1) = \delta(r - r_1)\delta(\vartheta - \vartheta_1)\delta(\varphi - \varphi_1)/(r^2\sin\vartheta)\,. \tag{25}$$

In the event that $J = 0$, so that at some point $\xi_i = \beta_i, i = 1, \ldots, n$, (9) fails to give us the required transformation, such a point is called a *singular point* of the transformation. In the foregoing example of plane polar coordinates $J = r = 0$ at the origin $(0, 0)$. At this point, $r = 0$ while ϑ may take on any value. A coordinate, such as ϑ in this example, which is either many valued or has no determinate value at a singular point of the transformation, is called an *ignorable point*. In the case of spherical polar coordinates, $J = r^2\sin\theta$ vanishes for all points on $\vartheta = 0$ (the z axis) and there ϕ may take on any value. Thus ϕ is an ignorable coordinate in this case. Also, $J = 0$ at the origin, where both ϑ and ϕ are ignorable.

Let us pursue the case $J = 0$ and restrict ourselves to $n = 2$. Suppose that (α_1, α_2) is a singular point of the transformation (17) while ξ_2 is an ignorable coordinate. Thus, at the singular point, $\xi_1 = \beta_1$ and ξ_2 is indeterminate, so that the test function $\phi(\xi_1, \xi_2)$ depends only on $\xi_1 = \beta_1$ there. Accordingly, we denote it $\phi(\beta_1)$. Then the generalized function

$$\delta(x_1 - \alpha_1)\delta(x_2 - \alpha_2)$$

also becomes a function of ξ_1 only, and we denote it $t(\xi_1)$. In this case, relation (19) should yield

$$\begin{aligned}\phi(\alpha_1, \alpha_2) &= \iint_{R_2} \delta(\xi_1 - \alpha_1)\delta(\xi_2 - \alpha_2)\phi(\xi_1, \xi_2)|J|d\xi_1 d\xi_2 \\ &= \iint_{R_2} t(\xi_1)\phi(\xi_1, \xi_2)|J|d\xi_1 d\xi_2 \\ &= \phi(\beta_1)\,. \end{aligned} \tag{26}$$

This equation is satisfied if we set $J_1 = \int |J|d\xi_2$, where the integration is along the entire domain of ξ_2, and take $t(\xi_1) = \delta(\xi_1 - \beta_1)/J_1$. Thus

$$\delta(x - \alpha_1)\delta(x - \alpha_2) = \delta(\xi_1 - \beta_1)/J_1\,. \tag{27}$$

For instance, in the case of the plane polar coordinates, the origin is the singular point and $J = r$. Thus $J_1 = \int_0^{2\pi} r\,d\theta = 2\pi r$, and (27) becomes

$$\delta(x)\delta(y) = \delta(r)/2\pi r\,. \tag{28}$$

In the case of spherical polar coordinates, φ is the ignorable coordinate. Thus $J_1 = \int_0^{2\pi} r^2 \sin\theta d\varphi = 2\pi r^2 \sin\theta$, and (27) becomes

$$\delta(x)\delta(y)\delta(z - z_1) = \delta(r - r_1)\delta(\vartheta)/(2\pi r^2 \sin\vartheta)\,. \tag{29}$$

However, at the origin both ϑ and φ are ignorable coordinates, so that

$$J_1 = \int_0^{\pi}\int_0^{2\pi} r^2 \sin\vartheta\, d\vartheta\, d\varphi = 4\pi r^2\,,$$

which yields

$$\delta(x)\delta(y)\delta(z) = \delta(r)/4\pi r^2\,. \tag{30}$$

3.2. Convergence of Distributions

A very important concept in the theory of distributions is that of convergence. A sequence of distributions $t_m \in D'$, $m = 1, 2, \ldots$, is said to *converge* to a distribution $t \in D'$ if

$$\lim_{m\to\infty} \langle t_m, \phi\rangle = \langle t, \phi\rangle \qquad \text{for all} \quad \phi \in D\,. \tag{1}$$

When the sequence of distributions consists of locally integrable functions $f_m(x)$, there is the following relationship between classical convergence and distributional convergence (also called weak convergence). If $\{f_m(x)\}$ converges uniformly to $f(x)$ over each finite interval, then $\langle f_m, \phi\rangle \to \langle f, \phi\rangle$. Indeed,

$$\lim_{m\to\infty}\langle f_m, \phi\rangle = \lim_{m\to\infty}\int_{-\infty}^{\infty} f_m(x)\phi(x)dx$$
$$= \int_{-\infty}^{\infty} f(x)\phi(x)dx\,, \qquad \phi \in D\,,$$

where we have used the uniform convergence of the sequence $\{f_m(x)\}$.

An important result in connection with the convergence of the sequence of distributions $\{t_m\}$ is given in the following theorem.

Theorem 1. *If the sequence $\{t_m\}$ of distributions converges to t, then the sequence $\{D^k t_m\}$ converges to $\{D^k t\}$.*

Proof.

$$\lim_{m\to\infty}\langle D^k t_m, \phi\rangle = \lim_{m\to\infty}(-1)^{|k|}\langle t_m, D^k\phi\rangle = (-1)^{|k|}\langle t, D^k\phi\rangle$$
$$= \langle D^k t, \phi\rangle\,, \qquad \phi \in D\,.$$

Definition. A series of distributions $\sum_{m=1}^{\infty} s_m$ converges to t if the sequence of partial sums $\{t_m\} = \left\{\sum_{m=1}^{k} s_m\right\}$ converges to t. We write $\sum_{m=1}^{\infty} s_m = t$. In view of Theorem 1, we have the important result that *every convergent sequence or series of distributions can be differentiated term by term as often as required.*

We have seen above that if a sequence of locally integrable functions converges uniformly to a function f, then it also converges to f in the sense of distributions. An important question in this connection arises regarding the sequence of functions that converge pointwise or fail to converge. Theorem 1 suggests the following procedure. Let $\{f_m(x)\}$ be the sequence in question. Suppose that an integer k can be found such that $\{f_m(x)\}$ is the kth derivative of a sequence $\{g_m(x)\}$ that converges uniformly to $g(x)$. Then it follows from the theorem that $\{f_m\} \to g^{(k)}$ where $g^{(k)}$ is the kth generalized derivative of g.

Let us extend the definition of distributional convergence to a sequence of distributions $t_\alpha(x)$ depending on a real parameter α. That is,

$$t_\alpha \to t \quad \text{as } \alpha \to \alpha_0 \qquad \text{if} \qquad \lim_{\alpha\to\alpha_0}\langle t_\alpha, \phi\rangle = \langle t, \phi\rangle\,, \qquad \phi \in D\,. \tag{2}$$

It can be readily proved that t is a distribution. The differentiation theorem in this case gives

$$\lim_{\alpha\to\alpha_0}\langle D^k t_\alpha, \phi\rangle = \lim_{\alpha\to\alpha_0}(-1)^{|k|}\langle t_\alpha, D^k\phi\rangle = (-1)^{|k|}\langle t, D^k\phi\rangle = \langle D^k t, \phi\rangle\,.$$

Inasmuch as the limits of the sequence $\{t_m\}$ and $\{t_\alpha\}$ are themselves distributions, we have the following important result.

Theorem 2. *The space D' is complete with respect to weak convergence.*

3.3. Delta Sequences with Parametric Dependence

In Section 1.2 we gave various delta convergent sequences. With the help of the parametric dependence, we can construct a sequence of functions that approach a delta function. We shall present this analysis in n dimensions, but we must first study the volume and surface area of the n-dimensional sphere of radius r.

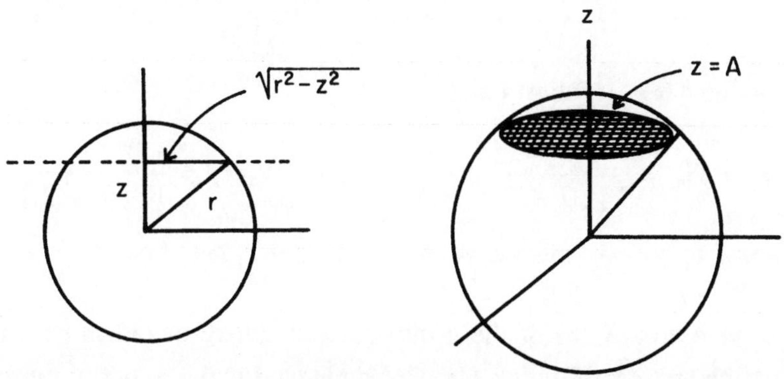

Figure 3.2.

Let $V_n(r)$ and $S_n(r)$ denote, respectively, the volume and the surface area of the n-dimensional sphere of radius r. To obtain a formula for $V_n(r)$ we first relate it to the volume of an $(n-1)$-dimensional sphere. Let z denote a coordinate on an axis through the origin. If A is a constant smaller than r, the intersection of the hypersurface $z = A$ and the n-dimensional sphere is an $(n-1)$-dimensional sphere of radius $(r^2 - z^2)^{1/2}$ (see Figure 3.2, drawn for $n = 2$ and 3). Then

$$V_n(r) = \int_{-r}^{r} V_{n-1}[(r^2 - z^2)^{1/2}]dz . \tag{1}$$

For example,

$$V_1(r) = \int_{-r}^{r} dz = 2r , \qquad n = 1 ,$$

$$V_2(r) = \int_{-r}^{r} 2(r^2 - z^2)^{1/2} dz = 4r^2 \int_0^1 (1 - u^2)^{1/2} du = \pi r^2 , \qquad n = 2 ,$$

$$V_3(r) = \int_{-r}^{r} \pi (r^2 - z^2) dz = 2\pi r^3 \int_0^1 (1 - u^2) du = \frac{4}{3}\pi r^3 , \qquad n = 3 .$$

Continuing in this manner, we have $V_n(r) = C_n r^n$, where C_n is defined by the recurrence formula

$$C_{n+1} = 2C_n \int_0^1 (1 - u^2)^{n/2} du . \tag{2}$$

This formula can be put in a simple form if we appeal to the beta function

$$\beta(p, q) = \int_0^1 t^{p-1}(1 - t)^{q-1} dt , \qquad p, q > 0 ,$$

as well as to the relation of $\beta(p,q)$ to the gamma function $\Gamma(p)$:

$$\beta(p,q) = \Gamma(p)\Gamma(q)/\Gamma(p+q), \qquad \text{where} \quad \Gamma(p) = \int_0^\infty t^{p-1}e^{-t}dt,$$

then it can be easily proved that

$$\int_0^1 (1-u^2)^{n/2}du = \frac{1}{2}\frac{\Gamma\left(\frac{1}{2}\right)\Gamma\left[\frac{1}{2}(n+2)\right]}{\Gamma\left[\frac{1}{2}(n+3)\right]}.$$

Thus, for an integer n, it follows that

$$C_n = \frac{\pi^{n/2}}{\Gamma[\frac{1}{2}(n+2)]}, \qquad \text{and} \qquad V_n(r) = \frac{\pi^{n/2}r^n}{\Gamma[\frac{1}{2}(n+2)]}. \tag{3}$$

It is interesting to observe that the volume of the unit sphere first increases with n and then decreases.

The surface area $S_n(r)$ of the n-dimensional sphere of radius r may be derived by considering the volume of a thin spherical shell. To the first order of thickness Δr, this is $S_n(r)\,\Delta r$,

$$\begin{aligned} S_n(r)\,\Delta r = V_n(r) - V_n(r-\Delta r) &= \frac{\pi^{n/2}}{\Gamma[\frac{1}{2}(n+2)]}[r^n - (r-\Delta r)^n] \\ &= \frac{\pi^{n/2}}{\Gamma[\frac{1}{2}(n+2)]} r^n \left(\frac{n\,\Delta r}{r}\right) + O(\Delta r)^2. \end{aligned}$$

Thus

$$S_n(r) = 2\pi^{n/2}r^{n-1}/\Gamma(n/2). \tag{4}$$

From these relations it follows that

$$S_n(r) = \frac{dV_n}{dr}, \qquad S_n(1) = \frac{2\pi^{n/2}}{\Gamma(n/2)}, \qquad S_n(r) = S_n(1)r^{n-1}. \tag{5}$$

Suppose we are dealing with a function $f(x)$ that depends only on r (say, $g(r)$); then

$$\int_{R_n} f(x)dx = \int_0^\infty g(r)\frac{dV_n}{dr}dr = \int_0^\infty g(r)S_n(r)dr = S_n(1)\int_0^\infty g(r)r^{n-1}dr. \tag{6}$$

The construction of a sequence of functions approaching the delta function is explained in the following theorem:

Theorem. *Let $f(x) = f(x_1, \ldots, x_n)$ be a nonnegative locally integrable function satisfying the condition*

$$\int_{R_n} f(x)dx = 1, \tag{7}$$

and define

$$f_\alpha(x) = (1/\alpha^n)f(x/\alpha) = (1/\alpha^n)f(x_1/\alpha, \ldots, x_n/\alpha), \tag{8}$$

where $\alpha > 0$. *Then*

$$\lim_{\alpha\to 0} f_\alpha(x) = \delta(x)\,. \tag{9}$$

Proof. From relations (7) and (8), we obtain the following properties:

(1) $\int_{R_n} f_\alpha(x)dx = 1$,
(2) $\lim_{\alpha\to 0} \int_{|x|>A} f_\alpha(x)\,dx = 0$,
(3) $\lim_{\alpha\to 0} \int_{|x|<A} f_\alpha(x)\,dx = 1$,

for each $A > 0$. These properties can easily be proved for one-dimensional case, and their extension to R_n is obvious. For instance, set $u = x/\alpha$ and get

$$\int_{-\infty}^{\infty} \frac{1}{\alpha} f\left(\frac{x}{\alpha}\right) dx = \int_{-\infty}^{\infty} f(u)du = 1\,,$$

where we have used the hypothesis (7). This proves property (1) in R_1.

The theorem is proved if we show that

$$\lim_{\alpha\to 0} \int_{R_n} f_\alpha(x)\phi(x)dx = \phi(0)\,, \qquad \phi \in D\,,$$

or, equivalently (in view of Property 1).

$$\lim_{\alpha\to 0} \int_{R_n} f_\alpha(x)[\phi(x) - \phi(0)]dx = 0\,, \qquad \phi \in D\,. \tag{10}$$

Division of the region R_n into the two parts, $|x| < A$ and $|x| > A$, gives

$$\left|\int_{R_n} f_\alpha(x)[\phi(x) - \phi(0)]dx\right| \le \left|\int_{|x|<A} f_\alpha(x)[\phi(x) - \phi(0)]dx\right| + \left|\int_{|x|>A} f_\alpha(x)[\phi(x) - \phi(0)]dx\right|\,.$$

Now the boundedness of $\phi(x)$ ensures the boundedness of $[\phi(x) - \phi(0)]$. Let this bound be M. Similarly, let $N(A) = \max_{|x|\le A}[\phi(x) - \phi(0)]$. Then, using the nonnegativity of $f(x)$ and Property 1, we have

$$\left|\int_{R_n} f_\alpha(x)[\phi(x) - \phi(0)]dx\right| \le N(A) + M \int_{|x|>A} f_\alpha(x)dx\,.$$

Since the function $[\phi(x) - \phi(0)]$ is continuous and is zero at $x = 0$, $N(A) \to 0$ as $A \to 0$. Thus for an arbitrary ε, we first choose A sufficiently small so that $N(A) < \varepsilon/2$. Then we appeal to Property 2 and choose α sufficiently small so that $\int_{|\alpha|>A} f_\alpha(x)dx < \varepsilon/2M$. This gives

$$\left|\int_{R_n} f_\alpha(x)[\phi(x) - \phi(0)]\,dx\right| < \varepsilon\,,$$

which proves (10).

Special Case. When $f(x)$ is a function only of $r = (x_1^2 + \cdots + x_n^2)^{1/2}$, that is, $f(x) = g(r)$, (7) reduces to

$$\int_0^\infty r^{n-1} g(r) dr = \frac{1}{S_n(1)}, \tag{11}$$

where $S_n(1)$ is the surface area of the unit n-dimensional sphere and we have used (6). For $n = 3$, (11) and (9) become, respectively,

$$\int_0^\infty r^2 g(r) dr = \frac{1}{4\pi} \tag{12}$$

and

$$\lim_{\alpha \to 0} (1/\alpha^3) g(r/\alpha) = \delta(x_1, x_2, x_3) = \delta(x). \tag{13}$$

An interesting feature of this theorem is that a nonnegative locally integrable function satisfying normalization condition (7), is enough to generate a delta sequence, while in Chapter 1, we constructed the delta sequences with the help of some special functions such as the Gaussian function.

Example 1. Let us prove that

$$\text{(a) for } n = 2, \ \lim_{\alpha \to 0} \frac{\alpha}{2\pi (r^2 + \alpha^2)^{3/2}} = \delta(x_1, x_2); \tag{14}$$

$$\text{(b) for } n = 3, \ \lim_{\alpha \to 0} \frac{\alpha}{\pi (r^2 + \alpha^2)^2} = \delta(x_1, x_2, x_3). \tag{15}$$

In case (a), $g(r) = 1/2\pi(r^2 + 1)^{3/2}$. Because

$$\int_0^\infty \frac{r}{2\pi (r^2 + 1)^{3/2}} = \frac{1}{2\pi},$$

condition (11) is satisfied and the result follows. In case (b),

$$g(r) = 1/\pi^2 (r^2 + 1)^2,$$

and

$$\int_0^\infty \frac{r^2}{\pi^2 (r^2 + 1)^2} dr = \frac{1}{4\pi},$$

which satisfies (12). This proves the result.

Example 2. Recall the test function $\phi_\varepsilon(x)$ as defined in (2.2.3);

$$\phi_\varepsilon(x) = \begin{cases} c_\varepsilon \exp\left(-\frac{\varepsilon^2}{\varepsilon^2 - r^2}\right), & r < \varepsilon, \\ 0, & r \geq \varepsilon, \end{cases}$$

where c_ε is the normalizing factor such that $\int \phi_\varepsilon(x) dx = 1$. By the foregoing analysis it follows that $\phi_\varepsilon(x) \to \delta(\alpha)$ as $\varepsilon \to 0$.

3.4. Fourier Series

It is known [5] that the Fourier series

$$\sum_{m=-\infty}^{\infty} c_m e^{imx}, \tag{1}$$

converges uniformly if, for large m, $|c_m| \le M/m^k$, $k \ge 2$, where M is a constant and k is an integer. It therefore converges distributionally. This series may diverge for other values of k, but by applying Theorem 1 of Section 3.2, we observe that (1) is a distributionally convergent series when k is any integer, because it can be obtained from the uniformly convergent series

$$\sum_{m=-\infty}^{\infty} (im)^{-k-2} c_m e^{imx}, \tag{2}$$

by $k+2$ successive differentiations.

On the other hand, we can expand a periodic and locally integrable function $f(x)$ in a Fourier series. To prove this assertion, let us take the period to be 2π. Since the Fourier coefficients of $f(x)$ are given as

$$c_m = \frac{1}{2\pi}\int_0^{2\pi} f(x) e^{-imx} dx, \qquad m = 0, \pm 1, \pm 2 \ldots$$

we have to prove that the Fourier series

$$\sum_{m=-\infty}^{\infty} c_m e^{imx}, \tag{3}$$

converges distributionally to $f(x)$.

This is proved by considering a partial sum s_{pq} of the series (1):

$$s_{pq}(x) = \sum_{m=-p}^{\infty} c_m e^{imx}.$$

Its scalar product with a test function $\phi(x)$ is

$$\langle s_{pq}, \phi \rangle = \sum_{m=-p}^{q} c_m \langle e^{imx}, \phi \rangle \tag{4}$$

or

$$\langle s_{pq}, \phi \rangle = 2\pi \sum_{m=-p}^{q} c_m \overline{d}_m,$$

where the d_m are the Fourier coefficients of ϕ and the bar labels the complex conjugate. We now appeal to the famous Parseval theorem of classical Fourier analysis. It states that if $\{u_m\}_{m=-\infty}^{\infty}$ is a complete set of orthonormal functions, and if the Fourier coefficients of f with respect to u_m are c_m while those of ϕ are d_m, then $\sum_{m=-\infty}^{\infty} c_m \overline{d}_m = \langle f, \phi \rangle$. In the present case, $u_m = e^{imx}/\sqrt{2\pi}$; thus we find that as $p, q \to \infty$, $\langle s_{pq}, \phi\rangle \to \langle f, \phi\rangle$. This proves the result. Accordingly, for a locally integrable periodic function, the classical convergence and the distributional convergence of the Fourier series are the same thing.

The Fourier series of every periodic distribution t converges to t in the distributional sense. Let us illustrate this by expanding $f(x) = \delta(x-y)$, $f(x) = f(x+l)$, $0 < y < \ell$, in terms of the Fourier sine series on the half range $(0, \ell)$:

$$\delta(x-y) = \frac{2}{l}\sum_{m=1}^{\infty} \sin\left(\frac{m\pi x}{l}\right)\sin\left(\frac{m\pi y}{l}\right). \tag{5}$$

The partial sum s_q is

$$s_q(x,y) = \frac{2}{l}\sum_{m=1}^{q} \sin\left(\frac{m\pi x}{l}\right)\sin\left(\frac{m\pi y}{l}\right).$$

Its scalar product with a test function $\phi(y)$ gives

$$\begin{aligned}\langle s_q(x,y), \phi(y)\rangle &= \int_0^l s_q(x,y)\phi(y)dy \\ &= \sum_{m=1}^{q} \frac{2}{l}\int_0^l \sin\left(\frac{m\pi y}{l}\right)\phi(y)dy \sin\left(\frac{m\pi x}{l}\right) \\ &= \sum_{m=1}^{q} d_m \sin\left(\frac{m\pi x}{l}\right) = \sigma_q(x), \text{ say,}\end{aligned}$$

which is the partial sum of the Fourier series of the smooth function ϕ and accordingly converges to ϕ. Therefore,

$$\lim_{q\to\infty}\langle s_q(x,y), \phi(y)\rangle = \lim_{q\to\infty}\sigma_q(x) = \phi(x) = \langle\delta(x-y), \phi(y)\rangle.$$

This proves that the distributional limit of sequence $s_q(x,y)$ is the delta distribution.

Note that because of the assumption of the periodicity the left side of (5) is really an infinite row of delta functions situated at the periodic points. Moreover, we expanded an even function in terms of the sine series in the half-range $(0, l)$. Consequently, we have the odd extension of this function in the full range.

The periodic row of delta functions in the full range $(-l, l)$ is

$$\sum_{m=-\infty}^{\infty} \delta(x - 2ml) = \frac{1}{2l}\sum_{m=-\infty}^{\infty} c_m e^{i\pi mx/l}.$$

The coefficients c_m are given formally by

$$c_m = \int_{-l}^{l} \delta(x)e^{-i\pi mx/l}dx = 1,$$

so that the relation

$$\sum_{m=-\infty}^{\infty} \delta(x-2ml) = \frac{1}{2l} \sum_{m=-\infty}^{\infty} e^{i\pi mx/l} = \frac{1}{2l}\left[1+2\sum_{m=1}^{\infty} \cos\left(\frac{m\pi x}{l}\right)\right]$$

holds true distributionally even though the series on the right side is divergent. Incidentally, for the special case of $l = \pi$, we obtain the Dirac comb

$$\sum_{m=-\infty}^{\infty} \delta(x-2m\pi) = \frac{1}{2\pi}\left[1+2\sum_{m=1}^{\infty} \cos mx\right] . \tag{6}$$

If we restrict x to the interval $(-\pi, \pi)$, then only one of the delta functions in (6) is nonzero, and we have the representation

$$\delta(x) = \frac{1}{2\pi}\left[1+2\sum_{m=1}^{\infty} \cos mx\right] , \qquad -\pi < x < \pi . \tag{7}$$

Thus

$$\delta(x-\xi) = \frac{1}{2\pi}\left[1+2\sum_{m=1}^{\infty} \cos m(x-\xi)\right] , \qquad -\pi < x, \quad \xi < \pi . \tag{8}$$

For a periodic function $f(x)$ with period 2π, we have

$$\begin{aligned} f(x) &= \int_{-\pi}^{\pi} f(\xi)\delta(x-\xi)d\xi \\ &= \frac{1}{2\pi}\int_{-\pi}^{\pi} f(\xi)d\xi + \sum_{m=1}^{\infty}\left(\frac{1}{\pi}\int_{-\pi}^{\pi} f(\xi)\cos m\xi\, d\xi\right)\cos mx \\ &\quad + \sum_{m=1}^{\infty}\left(\frac{1}{\pi}\int_{-\pi}^{\pi} f(\xi)\sin m\xi\, d\xi\right)\sin mx , \end{aligned} \tag{9}$$

which is the Fourier series of $f(x)$.

This analysis can be readily extended to the distributional convergence of Fourier Bessel series (see Exercise 17 at the end of this chapter). We can also extend the notion of the distributional convergence to integrals, as illustrated in Examples 1 and 2 in Section 3.5. Regularization of divergent integrals is discussed in Chapter 4.

3.5. Examples

Example 1. It is easily verified that the integral

$$f(x) = \int_{1}^{\infty} \frac{\cos yx}{y^2} dy ,$$

exists for every value of x and that the function

$$f_\alpha(x) = \int_1^\alpha \frac{\cos yx}{y^2} dy \tag{1}$$

converges uniformly to $f(x)$ in every bounded region of x as $\alpha \to \infty$. Thus $f_\alpha(x)$ converges in the sense of distributions as well, and we can differentiate it as many times as we desire. For instance, by taking the second derivative of (1) we find that

$$-\int_1^\alpha \cos yx\, dy \tag{2}$$

is the limit of the convergent distributions $\{d^2 f_\alpha(x)/dx^2\}$ as $\alpha \to \infty$, although the integral (2) does not exist in the classical sense as $\alpha \to \infty$.

Example 2. To establish the relation

$$\delta(x-\xi) = (1/2\pi) \int_{-\infty}^{\infty} \exp\{i(x-\xi)s\} ds\,, \tag{3}$$

we set

$$t_\alpha(x-\xi) = (1/2\pi) \int_{-\alpha}^{\alpha} \exp\{i(x-\xi)s\} ds\,.$$

Then

$$\langle t_\alpha, \phi \rangle = (1/2\pi) \int_{-\infty}^{\infty} \phi(\xi) d\xi \int_{-\alpha}^{\alpha} \exp\{i(x-\xi)s\}\, ds\,.$$

Because $\phi(\xi)$ has a compact support, the integral with respect to ξ is finite, and we can change the order of integration. Thus

$$\langle t_\alpha, \phi \rangle = (1/2\pi) \int_{-\alpha}^{\alpha} \exp(ixs) ds \int_{-\infty}^{\infty} \phi(\xi) \exp(-i\xi s) d\xi\,.$$

Appealing to the classical theory of Fourier transform, we obtain from this formula $\lim_{\alpha\to\infty} \langle t_\alpha, \phi \rangle = \phi(x) = \langle \delta(x-\xi), \phi(\xi) \rangle$, which proves (3). In particular,

$$\delta(x) = (1/2\pi) \int_{-\infty}^{\infty} e^{ixs} ds\,. \tag{4}$$

An alternative derivation of (4) is simple and interesting. Recall from advanced calculus that

$$\int_{-\infty}^{\infty} \frac{\sin xs}{s} ds = \begin{cases} \pi\,, & x > 0\,, \\ -\pi\,, & x < 0\,, \end{cases}$$

and

$$\int_{-\infty}^{\infty} \frac{\cos xs}{s} ds = 0\,.$$

With the help of these formulas we have

$$\int_{-\infty}^{\infty} \frac{e^{ixs}}{is} ds = \int_{-\infty}^{\infty} \left(\frac{\cos xs}{is} + i \frac{\sin xs}{is} \right) ds$$
$$= \int_{-\infty}^{\infty} \frac{\sin xs}{s} ds = \begin{cases} \pi\,, & x > 0\,, \\ -\pi\,, & x < 0\,. \end{cases}$$

Thus

$$\frac{1}{2\pi} \int_{-\infty}^{\infty} \frac{e^{ixs}}{is} ds = \left\{ \begin{array}{ll} \frac{1}{2}, & x > 0 \\ -\frac{1}{2}, & x < 0 \end{array} \right\} = \frac{1}{2} \operatorname{sgn} x = H(x) - \frac{1}{2}\,.$$

Differentiating both sides of this relation with respect to x, we recover (4).

Similarly, it can be shown that

$$\delta(x) = \frac{1}{2\pi} \int_{-\infty}^{\infty} e^{-ixs} ds\,. \tag{5}$$

The addition of (4) and (5) yields

$$\delta(x) = \frac{1}{\pi} \int_{-\infty}^{\infty} \cos xs\, ds = \frac{2}{\pi} \int_{0}^{\infty} \cos xs\, ds\,. \tag{6}$$

Many more relations can be derived from these formulas. For instance, differention of (4) and (5) with respect to x yields a representation for $\delta'(x)$:

$$\delta'(x) = -\frac{1}{2\pi i} \int_{-\infty}^{\infty} s e^{ixs} ds = \frac{1}{2\pi i} \int_{-\infty}^{\infty} s e^{-ixs} ds\,. \tag{7}$$

We shall have more to say about these ideas in Chapter 6.

Example 3. Show that

$$\lim_{\alpha \to 0} \left(\frac{1}{\sqrt{\pi x}} \right)^{n} e^{-r^2/\alpha^2} = \delta(x_1, \ldots, x_n)\,. \tag{8}$$

To prove this result we appeal to the relations

$$\int_{0}^{\infty} r^{2n} e^{-r^2} dr = \pi^{1/2} 2^{-2n} \frac{(2n-1)!}{(n-1)!}\,, \tag{9}$$

and

$$\int_{0}^{\infty} r^{2n+1} e^{-r^2} dr = \frac{n!}{2} \tag{10}$$

from calculus. Next we use (3.3.5) and obtain

$$(2n-1)! = 2^{2n} (n-1)! \pi^{n} / S_{2n+1}(1)\,.$$

Thus (9) becomes

$$\int_{0}^{\infty} r^{(2n+1)-1} e^{-r^2} dr = \frac{\pi^{(2n+1)/2}}{S_{2n+2}(1)}\,. \tag{11}$$

Similarly, $n! = 2\pi^{n+1}/S_{2n+2}(1)$, and (10) takes the form

$$\int_0^\infty r^{(2n+2)-1} e^{-r^2} dr = \frac{\pi^{n/2}}{S_{n+2}(1)} \,. \tag{12}$$

Combining (11) and (12), we find that

$$\int_0^\infty r^{n-1} e^{-r^2} dr = \frac{\pi^{n/2}}{S_n(1)} \,, \tag{13}$$

is valid for all positive integers n.

Finally, we set $g_\alpha(r) = (1/\pi^{n/2})(1/\alpha^n)e^{-r^2/\alpha^2}$, appeal to Theorem 1 of Section 3.3, and obtain

$$\lim_{\alpha\to 0} g_\alpha(r) = \delta(x_1, \ldots, x_n) \,.$$

Example 4. The sequence $\{f_m(x)\} = \{\sin mx\}$ is not a convergent sequence in the classical sense unless x is a multiple of π. However, if we take the sequence $g_m(x) = \{-\cos(mx)/m\}$, which converges to zero uniformly in $-\infty < x < \infty$, then $\{g'_m(x)\} = \{f_m(x)\}$ converges to zero in the sense of distributions.

Example 5. In Section 1.4 we studied the sequence

$$t_m(x) = 2m^3 x/\pi(1 + m^2x^2)^2 \,,$$

which describes the dipole distribution. This sequence converges pointwise but not uniformly to zero in $-\infty < x < \infty$. However, it is the derivative of the uniformly convergent sequence $-m/\pi(1 + m^2x^2)$, which converges to the delta sequence, as proved in Example 2 of Section 1.3. Accordingly, we can appeal to Theorem 1 of Section 3.2 and get

$$t_m(x) \to -\delta'(x) \qquad \text{or} \qquad \langle t_m, \phi\rangle = \phi'(0) \,. \tag{14}$$

3.6. The Delta Function as a Stieltjes Integral

Let us examine the Stieltjes integral

$$\int_a^b f(x) dH_c(x - \xi) \,, \tag{1}$$

where $f(x)$ is a function continuous on the interval $[a, b]$, ξ is some fixed number in the interval (a, b), and H_c is the Heaviside function that takes the value c, $0 \le c \le 1$, at $x = \xi$. Let us partition $[a, b]$ into n parts such that

$$a = x_0 < x_1 < \cdots < x_{n-1} < x_n = b \,,$$

and let the quantity $\Delta_k g = g(x_k) - g(x_{k-1})$, where $g(x)$ is some fixed, monotone increasing function. Consequently, we have two possibilities:

(1) ξ is an interior point of some subinterval, say, $x_{k-1} < \xi < x_k$, so that

$$\Delta_k g = H_c(x_k - \xi) - H_c(x_{k-1} - \xi) = 1 - 0 = 1\,,$$
$$\Delta_p g = 0\,, \qquad \text{for every} \qquad p \neq k\,.$$

Thus

$$\sum_{p=1}^{n} f(x_p)\Delta_p g = f(x_k)\,, \qquad \text{where} \qquad x_{k-1} \leq x_k \leq x_k\,. \tag{2}$$

(2) ξ is a boundary point of two adjacent subintervals, say, $\xi = x_k$, so that

$$\Delta_k g = H_c(x_k - \xi) - H_v(x_{k-1} - \xi) = c - 0 = c\,,$$
$$\Delta_{k+1} g = H_c(x_{k+1} - \xi) - H_c(x_k - \xi) = 1 - c\,,$$

while all other terms $\Delta_p g$ vanish. Thus

$$\sum_{p=1}^{n} f(x_p)\Delta_p g = cf(x_k) + (1-c)f(x_{k+1}) = f(x_{k+1}) + c[f(x_k) - f(x_{k+1})]\,, \tag{3}$$

where $x_{k+1} \leq x_k \leq \xi \leq \xi_{k+1} \leq \xi_{k+1}$.

From (1)–(3) and the hypothesis, it follows that, in the limit.

$$\int_a^b f(x) dH_c(x - \xi) = f(\xi)\,, \tag{4}$$

which is precisely the required sifting property of the delta function.

Exercises

1. Show that if $ad - bc \neq 0$, $\delta(ax + by)\delta(cx + dy) = \delta(x)\delta(y)/|ad - bc|$.

2. Evaluate
   (a) $\int_0^1 e^{2x}\delta(2x - 1)dx$,
   (b) $\int_{-1}^0 (\sinh 2x)\delta(5x + 2)dx$,
   (c) $\int_{-\pi}^{\pi} e^{-|x|}\delta(\sin \pi x)dx$,
   (d) $\int_{-1}^1 (\sin^{-1} \delta(4x^2 - 3)dx$.

3. Interpret the expressions $\delta(e^x)$, $\delta(\ln x)$, and $\delta(\sin |x|)$.

4. Show that
   (a) $\delta'(x^3 + 3x) = \frac{1}{9}\delta'(x)$,
   (b) $\delta''(\frac{1}{3}x^3 + x) = \delta''(x) - 2\delta(x)$,
   (c) $\delta(3x - \frac{1}{2}\sin 2x) = \frac{1}{2}\delta(x)$.

5. Prove that for $x > 0$

$$\delta\left(a - \frac{1}{x}\right) = \begin{cases} (1/a^2)\delta(x - 1/a)\,, & a > 0\,, \\ 0\,, & a \leq 0\,. \end{cases}$$

6. Prove that $(x^2 - 1)\delta(x^2 - 1) = 0$.

7. Evaluate

(a) $\delta''[f(x)]$ and (b) $\operatorname{sgn}'[f(x)]$.

8. Show that

(a) $\lim_{\alpha\to\infty} \sin \alpha x/\pi x = \delta(x)$,
(b) $\lim_{\alpha\to\infty}(1/\pi)\sin 62\alpha x/\alpha x^2) = \delta(x)$,
(c) $\lim_{\alpha\to 0}(1/2\left(\sqrt{\pi\alpha}\right)\exp(-x^2/4\alpha) = \delta(x)$.

9. Recall the definitions of $Pf(1/x)$, $\delta(x)$, $\delta^+(x)$, and $\delta^-(x)$ from Chapter 2 and show that

(a) $\delta^{\pm}(x) = \mp \lim_{\varepsilon\to 0}(1/2\pi i)[1/(x \pm i\varepsilon)]$,
(b) $\delta(x) = \lim_{\varepsilon\to 0}(1/\pi)[\varepsilon/(x^2 + \varepsilon^2)]$,
(c) $Pf(1/x) = \lim_{\varepsilon\to 0}[\pi i x/(x^2 + \varepsilon^2)]$,
(d) $\ln|x| + i\pi H(-x) = \lim_{\varepsilon\to 0}\ln(x + i\varepsilon)$.

10. Consider the following sequence of locally integrable functions of the single variable x:

$$f_\alpha(x) = \begin{cases} 0\,, & x < 0\,, \\ \dfrac{\alpha e^{-4x\alpha^2}}{\sqrt{4\pi}\,x^{3/2}}\,, & x > 0\,. \end{cases}$$

Show that $\lim_{\alpha\to 0} f_\alpha(x) = \delta(x)$.

11. Consider the function

$$f_\alpha(x) = \int_0^\alpha 2\cos(2\pi y x)dy = \sin 2\pi\alpha x/\pi x\,,$$

and show that $\int_0^\infty 2\cos(2\pi y x)dy = \delta(x)$.

12. Compute the Fourier series of the distributions $\delta(x - a)$, $\delta'(x - a)$ and $H(x - a)$, $-l < a, x < l$.

13. Show that the series $\sum_{m=1}^\infty m \sin mx$, converges to $-\delta'(x)$. This result can be extended to show that any Fourier series with coefficients that are polynomials in m converges in the distributional sense to a polynomial expression in the derivatives of the delta function.

14. Compute the Fourier series of the following periodic functions:

$$\text{(a)}\quad f(t) = \begin{cases} 1\,, & 0 \leq t < 1\,, \\ 0\,, & 1 \leq t < 2\,, \end{cases}$$
$$f(t) = f(t-2)\,, \qquad -\infty < t < \infty\,.$$

(b) $f(t) = e^t, 0 \leq t < 2\pi$; $f(t) = f(t - 2\pi)$, $-\infty < t < \infty$.

15. The Fourier series

$$\sum_{m=1}^{\infty} \frac{(-1)^{m+1}}{m} \sin mx$$

converges to $x/2$ when $-\pi < x < \pi$ and to 0 when $x = \pm\pi$.

(a) Explain why the convergence cannot be uniform on $[-\pi, \pi]$.

(b) We might hope that term-by-term differentiation would yield $\sum_{m=1}^{\infty}(-1)^{m+1} \cos mx = \frac{1}{2}$ for $-\pi < x < \pi$, but this is not valid. Show that this series diverges for all rational numbers x.

(c) Explain why $\sum_{m=1}^{\infty}(-1)^{m+1} \cos mx$ converges in the distributional sense. To what distribution does it converge?

16. Find a sequence of locally integrable functions $g_m(x)$ converging (in D') to the generalized function $\delta^{(k)}(x)$. *Hint.* Set $g_m(x) = f_m^{(k)}(x)$, where $f_m(x)$ is a delta sequence of infinitely differentiable functions.

17. Given an arbitrary complete system of orthonormal functions $\{\Phi_m(x)\}$ in an interval (a, b), form the Fourier Bessel series with respect to $\{\Phi_m\}$ of a function $f(x)$ that is integrable along with its square and whose support lies in (a, b), namely,

$$\sum_{m=1}^{\infty} c_m \Phi_m(x), \qquad c_m = \int_a^b f(x)\Phi_m(x)dx, \qquad m = 1, 2, \ldots.$$

Show that this series converges to $f(x)$ in the sense of distributions.

18. Show that

$$\sum_{m=1}^{\infty} \frac{\cos mx}{m} = -\ln\left|2\sin\frac{x}{2}\right|, \qquad 0 < x < 2\pi.$$

19. Prove that $\lim_{\alpha\to\infty} \alpha J_\alpha(\alpha a x) = \delta(ax - 1)$, where $J_\alpha(\alpha a x)$ is the Bessel function.

20. Prove that in the distributional sense

$$\sum_{m=1}^{\infty} \sin mx = \frac{1}{2}\cot(x/2), \qquad \sum_{m=1}^{\infty} m\cos mx = -\frac{1}{4}\csc^2(x/2).$$

21. Show that the series

$$u(x) = \frac{1}{2a^2} + \sum_{m=1}^{\infty} \frac{\cos mx}{a^2 + m^2}$$

satisfies the boundary value problem $d^2u/dx^2 - a^2u = -\pi\delta(x)$, $u'(\pi) = u'(-\pi) = 0$. Use the fact that its solution is

$$u(x) = \begin{cases} \dfrac{\pi\cosh a(\pi + x)}{2a\sinh a\pi}, & -\pi < x < 0, \\[2ex] \dfrac{\pi\cosh a(\pi - x)}{2a\sinh a\pi}, & 0 < x < \pi. \end{cases}$$

and show that

$$\frac{\pi}{2a}\coth a\pi = \frac{1}{2a^2} + \sum_{m=1}^{\infty} \frac{1}{m^2 + a^2}.$$

As a particular case, prove that $\sum_{m=1}^{\infty}(1/m^2) = \pi^2/6$.

22. Show that a line source of uniform density σ distributed along the positive z axis can be represented in the notation of Section 3.1 by the density function $(2\pi r^2 \sin\theta)^{-1}\sigma\delta(\theta)$. State the corresponding result for a source distributed along the negaive z axis.

CHAPTER 4

Distributions Defined by Divergent Integrals

4.1. Introduction

In the previous chapters we have defined various singular distributions. One of them is $\mathrm{Pf}(1/x)$, defined in Example 4 of Section 2.4. The function $1/x$ is not integrable on any neighborhood of the origin. We succeeded in regularizing this function by defining the functional $\mathrm{Pf}(1/x)$ by the principal value of the singular integral defined by the quantity $\langle \phi, 1/x \rangle$. The aim of this chapter is to extend this idea and to regularize various singular integrals and thereby define the coresponding distributions. Let us start with a simple example.

The locally integrable function

$$f(x) = \begin{cases} 1/\sqrt{x}, & x > 0, \\ 0, & x < 0, \end{cases}$$
$$= H(x)/\sqrt{x} \tag{1}$$

generates the distribution

$$\langle f, \phi \rangle = \int_0^\infty \frac{1}{\sqrt{x}} \phi(x) dx, \qquad \phi \in D.$$

The distributional derivative of $f(x)$ is

$$\langle f', \phi \rangle = -\langle f, \phi' \rangle = -\int_0^\infty \phi'(x)/\sqrt{x} dx. \tag{2}$$

The function $f(x)$ has the classical derivative (except at $x = 0$)

$$f'(x) = \begin{cases} -\frac{1}{2} x^{3/2}, & x > 0, \\ 0, & x < 0, \end{cases}$$
$$= -\frac{1}{2} H(x)/x^{3/2}. \tag{3}$$

We should very much like to see that the ordinary derivative (3) produces a distribution identical to the distributional derivative (2). However, $f'(x)$ as given by (3) is not a locally integrable function and therefore does not generate a regular distribution. We thus have the following embarrassing situation: Every locally integrable function is identifiable with a distribution and thus has a distributional derivative, but the classical derivative of a locally integrable function may not be identical with the distributional derivative. To remedy this situation, let us look at the distributional derivative (2) a little more carefully. Let us write it as

$$\langle f', \phi \rangle = -\lim_{\varepsilon\to 0} \int_\varepsilon^\infty \frac{\phi'(x)}{\sqrt{x}} dx$$

and integrate by parts. The result is

$$\langle f', \phi \rangle = \lim_{\varepsilon\to 0} \left[\frac{\phi(\varepsilon)}{\sqrt{\varepsilon}} - \frac{1}{2} \int_\varepsilon^\infty \frac{\phi(x)}{x^{3/2}} dx \right]. \tag{4}$$

In view of the mean value theorem, we have

$$\phi(x) = \phi(0) + x\phi'(tx) = \phi(0) + x\psi(x), \qquad 0 < t < 1, \tag{5}$$

where $\psi(x) = \phi'(tx)$. Thus $\phi(\varepsilon)/\sqrt{\varepsilon} = \phi(0)/\sqrt{\varepsilon} + \sqrt{\varepsilon}\phi'(t\varepsilon)$ and $\phi(\varepsilon)/\sqrt{\varepsilon} \to \phi(0)/\sqrt{\varepsilon}$ as $\varepsilon \to 0$. Then (4) becomes

$$\langle f', \phi \rangle = \lim_{\varepsilon\to 0} \left[\frac{\phi(0)}{\sqrt{\varepsilon}} - \frac{1}{2} \int_\varepsilon^\infty \frac{\phi(x)}{x^{3/2}} dx \right]. \tag{6}$$

The function $\psi(x)$ defined by (5) is the remainder in Taylor's formula for the function $\phi(x)$. It is therefore a continuous function of x for all $x \neq 0$ and can be extended so that it is continuous at $x = 0$. If R is a real number sufficiently large that $\phi(x) = 0$ for $x > R$, then from (5) it follows that

$$\lim_{\varepsilon\to 0} \frac{1}{2} \int_\varepsilon^R \frac{\phi(x)}{x^{3/2}} dx = \lim_{\varepsilon\to 0} \left[\frac{\phi(\varepsilon)}{\sqrt{\varepsilon}} - \frac{\phi(0)}{\sqrt{R}} + \frac{1}{2} \int_\varepsilon^R \frac{\psi(x)}{\sqrt{x}} dx \right]$$

or

$$\lim_{\varepsilon\to 0} \left[\frac{\phi(0)}{\sqrt{\varepsilon}} - \frac{1}{2} \int_\varepsilon^R \frac{\phi(x)}{x^{3/2}} dx \right] = \lim_{\varepsilon\to 0} \left[\frac{\phi(0)}{\sqrt{R}} - \frac{1}{2} \int_\varepsilon^R \frac{\psi(x)}{\sqrt{x}} dx \right]. \tag{7}$$

The first term on the right side of (7) is a real number, while the second term remains finite because $\psi(x)$ is continuous at $x = 0$. Now we let $R \to \infty$ so that the left side of (7) becomes the right side of (6). Hence, the right side of (6) is the finite limit of the difference of two terms each of which tends to infinity. This is the finite part, in the sense of Hadamard, of the divergent integral

$$-\frac{1}{2} \int_0^\infty \frac{\phi(x)}{x^{3/2}} dx, \tag{8}$$

and is written

$$-\frac{1}{2} FP \int_0^\infty \frac{\phi(x)}{x^{3/2}} dx = \lim_{\varepsilon\to 0} \left[\frac{\phi(0)}{\sqrt{\varepsilon}} - \frac{1}{2} \int_\varepsilon^\infty \frac{\phi(x)}{x^{3/2}} dx \right]. \tag{9}$$

It can be easily shown that it is a linear continuous functional (Exercise 1); it is the pseudofunction and is a regularization of the divergent integral.

Finally, we combine (6) and (9) to write

$$-\frac{1}{2} \int_{-\infty}^\infty \phi(x) \mathrm{Pf}\,[H(x)x^{-3/2}] = -\int_{-\infty}^\infty \phi'(x) H(x) x^{-1/2} dx$$

or

$$\langle -\frac{1}{2}\text{Pf}\,(H(x)x^{-3/2}), \phi(x)\rangle = -\langle H(x)x^{-1/2}, \phi'(x)\rangle\,. \tag{10}$$

Thus we have the formula for the distributional derivative

$$\frac{d[H(x)x^{-1/2}]}{dx} = -\frac{1}{2}\text{Pf}\,(H(x)x^{-3/2})\,, \tag{11}$$

which reconciles relations (2) and (3).

This analysis can be generalized to include functions

$$f(x) = H(x)x^{-(\alpha+1)}\,, \qquad 0 < \alpha < 1\,. \tag{12}$$

Indeed, in this case, due to (5), we have

$$\begin{aligned}\int_\varepsilon^\infty \phi(x)x^{-(\alpha+1)}dx &= \int_\varepsilon^\infty \left[\frac{\phi(0)}{x^{\alpha+1}} + \frac{\phi'(tx)}{x^\alpha}\right]dx \\ &= \int_\varepsilon^\infty \frac{\phi'(tx)}{x^\alpha}dx + \phi(0)\left[\frac{x^{-\alpha}}{-\alpha}\right]_\varepsilon^\infty = \int_\varepsilon^\infty \frac{\phi'(tx)}{x^\alpha}dx + \frac{\phi(0)}{\alpha\varepsilon^\alpha}\,,\end{aligned}$$

for any test function $\phi(x)$. The integral term on the right of the proceding relation converges absolutely as $\varepsilon \to 0$, while the second term tends to $\pm\infty$ unless $\phi(0)$ vanishes. Accordingly, the integral term is the required Hadamard finite part, and we have

$$\begin{aligned}FP\int_{-\infty}^\infty \phi(x)H(x)x^{-(\alpha+1)}dx &= \lim_{\varepsilon\to 0}\int_\varepsilon^\infty \frac{\phi'(tx)}{x^\alpha}dx \\ &= \lim_{\varepsilon\to 0}\int_\varepsilon^\infty \frac{\phi(x)-\phi(0)}{x^{\alpha+1}}dx\,.\end{aligned} \tag{13}$$

Now

$$\begin{aligned}\int_\varepsilon^\infty \frac{\phi(x)-\phi(0)}{x^{\alpha+1}}dx &= \left[\frac{x^{-\alpha}}{-\alpha}\{\phi(x)-\phi(0)\}\right]_\varepsilon^\infty + \frac{1}{\alpha}\int_\varepsilon^\infty \phi'(x)x^{-\alpha}dx \\ &= \int_\varepsilon^\infty \{-\phi'(x)\}\left(\frac{x^{-\alpha}}{-\alpha}\right)dx + \frac{1}{\alpha}\left(\frac{\phi(\varepsilon)-\phi(0)}{\varepsilon^\alpha}\right)\,,\end{aligned} \tag{14}$$

where the last term vanishes with ε because $(\phi(\varepsilon)-\phi(0))/\varepsilon^\alpha = \varepsilon^{1-\alpha}\phi'(t\varepsilon)$, and $1-\alpha > 0$. Thus from (13) and (14) we have

$$-\alpha\int_{-\infty}^\infty \phi(x)\text{Pf}\,(H(x)x^{-(\alpha+1)})dx = \int_{-\infty}^\infty (-\phi'(x))(H(x)x^{-\alpha})dx\,,$$

or

$$\langle\phi(x)\,, -\alpha\text{Pf}\,(H(x)x^{-(\alpha+1)})\rangle = -\langle\phi'(x)\,,\; H(x)x^{-\alpha}\rangle\,,$$

which means that

$$\frac{d}{dx}(H(x)x^{-\alpha}) = -\alpha\text{Pf}\,(H(x)x^{-(\alpha+1)})\,. \tag{15}$$

Let us now examine the function

$$f(x) = H(x)x^{-(\alpha+2)}, \qquad 0 < \alpha < 1. \tag{16}$$

To take its inner product with a test function $\phi(x)$, we need the Taylor expansion for $\phi(x)$,

$$\phi(x) = \phi(0) + x\phi'(0) + \frac{1}{2}x^2\phi''(tx), \qquad 0 < t < 1, \tag{17}$$

so that

$$\begin{aligned}\int_\varepsilon^\infty \frac{\phi(x)}{x^{\alpha+2}}dx &= \int_\varepsilon^\infty \left[\frac{\phi(0)}{x^{\alpha+2}} + \frac{\phi'(0)}{x^{\alpha+1}} + \frac{\phi''(tx)}{2x^\alpha}\right]dx \\ &= \int_\varepsilon^\infty \frac{\phi''(tx)}{2x^\alpha}dx + \frac{\phi'(0)}{\alpha\varepsilon^\alpha} + \frac{\phi(0)}{(\alpha+1)\varepsilon^{\alpha+1}}.\end{aligned}$$

Here again, the integral term converges as $\varepsilon \to 0$, so that the Hadamard finite part is

$$\begin{aligned}FP\int_{-\infty}^\infty \phi(x)H(x)x^{-(\alpha+2)})dx &= \lim_{\varepsilon\to 0}\int_\varepsilon^\infty \frac{\phi''(tx)}{2x^\alpha}dx \\ &= \int_0^\infty \left[\frac{\phi(x) - \phi(0) - x\phi'(0)}{x^{\alpha+2}}\right]dx. \end{aligned}\tag{18}$$

The action of $f(x)$ on $\phi(x)$ as given by (18) defines the required pseudofunction $Pf(H(x)x^{-(\alpha+2)})$. To prove that

$$\frac{d}{dx}[H(x)x^{-(\alpha+1)}] = -(\alpha+1)Pf(H(x)x^{-(\alpha+2)}), \tag{19}$$

we integrate the middle integral of (18) by parts:

$$\begin{aligned}\langle\phi(x), Pf(H(x)x^{-(\alpha+2)})\rangle &= FP\int_{-\infty}^\infty \phi(x)H(x)x^{-(\alpha+2)}dx \\ &= \int_\varepsilon^\infty \frac{\phi(x) - \phi(0) - x\phi'(0)}{x^{\alpha+2}}dx \\ &= \left[\frac{x^{-(\alpha+1)}}{-(\alpha+1)}\{\phi(x) - \phi(0) - x\phi'(0)\}\right]_\varepsilon^\infty \\ &\quad + \frac{1}{\alpha+1}\int_\varepsilon^\infty \left(\frac{\phi'(x) - \phi'(0)}{x^{\alpha+1}}\right)dx \\ &= \int_\varepsilon^\infty \left(\frac{\phi'(x) - \phi'(0)}{x^{\alpha+1}}\right)dx \\ &\quad + \frac{1}{\alpha+1}\left[\frac{\phi(x) - \phi(0) - \varepsilon\phi'(0)}{\varepsilon^{\alpha+1}}\right].\end{aligned}$$

As $\varepsilon \to 0$, the second term on the right vanishes because $\phi(\varepsilon) - \phi(0) - \varepsilon\phi'(0) = \frac{1}{2}\varepsilon^2\phi(t\varepsilon)$, while the first becomes

$$\int_0^\infty \left(\frac{\phi'(x) - \phi'(0)}{(\alpha+1)x^{\alpha+1}}\right) dx = \int_{-\infty}^\infty (-\phi'(x) + \phi'(0)) \left(\frac{-H(x)}{(\alpha+1)x^{\alpha+1}}\right) dx$$
$$= \left\langle Pf\left(\frac{H(x)x^{-(\alpha+1)}}{\alpha+1}\right), \phi'(x)\right\rangle,$$

and (19) follows.

4.2. The Pseudofunction $\mathbf{H(x)/x^n, n = 1, 2, 3, \ldots}$

Let us now examine the function

$$x_+^{-n} = \begin{cases} x^{-n}, & x > 0, \\ 0, & x < 0, \end{cases}$$
$$= H(x)/x^n, \tag{1}$$

where n is a positive integer. The interesting feature of this function is that upon regularizing it in the manner of Section 4.1 there appear the delta function and its derivatives for $n \geq 2$. To illustrate this, we first consider the simple case $n = 1$ and therefore study the integral

$$\int_{-\infty}^\infty \frac{H(x)\phi(x)}{x} dx = \int_0^\infty \frac{\phi(x)}{x} dx.$$

Since $x = 0$ is the point of singularity, we examine the integral

$$\int_\varepsilon^\infty \frac{\phi(x)}{x} dx = \int_\varepsilon^1 \frac{\phi(x)}{x} dx + \int_1^\infty \frac{\phi(x)}{x} dx.$$

If we put $\phi(x) = \phi(0) + x\phi'(tx)$, $0 < t < 1$, this relation becomes

$$\int_\varepsilon^\infty \frac{\phi(x)}{x} dx = \int_1^\infty \frac{\phi(x)}{x} dx + \int_\varepsilon^1 \left[\frac{\phi(0)}{x} + \phi'(tx)\right] dx$$
$$= \int_1^\infty \frac{\phi(x)}{x} dx + \int_\varepsilon^1 \phi'(tx)dx - \phi(0)\ln\varepsilon,$$

from which it immediately follows that the finite part is

$$FP \int_{-\infty}^\infty \frac{H(x)\phi(x)}{x} dx = \lim_{\varepsilon\to 0}\left[\int_1^\infty \frac{\phi(x)}{x} dx + \int_\varepsilon^1 \left\{\frac{\phi(x) - \phi(0)}{x}\right\} dx\right]. \tag{2}$$

When we integrate by parts, (2) becomes

$$FP \int_{-\infty}^\infty \frac{H(x)\phi(x)}{x} dx = \lim_{\varepsilon\to 0}\left[\int_\varepsilon^\infty (-\phi'(x)) \ln x \, dx - \{\phi(\varepsilon) - \phi(0)\} \ln \varepsilon\right].$$

Since $\phi(\varepsilon) - \phi(0) = \varepsilon\phi'(t\varepsilon)$, $|\phi'(t\varepsilon)|$ is bounded, and $\lim_{\varepsilon\to 0}(\varepsilon \ln \varepsilon) = 0$, we finally have the required value of the $Pf(H(x)/x)$:

$$\int_{-\infty}^{\infty} \phi(x)\text{Pf}\left(\frac{H(x)}{x}\right) dx = FP\int_{-\infty}^{\infty} \frac{H(x)\phi(x)}{x} dx$$
$$= \int_{-\infty}^{\infty} (-\phi'(x))H(x)\ln|x|dx\,.$$

Hence,

$$\text{Pf}\,(H(x)/x) = (d/dx)(H(x)\ln|x|)\,. \tag{3}$$

We shall rederive this formula in a different manner in Example 1 of Section 4.4.

For $n = 2$ we use the relation

$$\phi(x) = \phi(0) + x\phi'(0) + \frac{1}{2}x^2\phi''(tx)\,, \qquad 0 < t < 1\,,$$

and proceed as before, so

$$\int_{\varepsilon}^{\infty} \frac{\phi(x)}{x^2} dx = \int_1^{\infty} \frac{\phi(x)}{x^2} dx + \int_{\varepsilon}^{1} \left\{\frac{\phi(0)}{x^2} dx + \frac{\phi'(0)}{x} + \frac{\phi''(tx)}{2}\right\} dx$$
$$= \int_1^{\infty} \frac{\phi(x)}{x^2} dx + \int_{\varepsilon}^{1} \frac{\phi''(tx)}{2} dx - \phi(0) + \left[\frac{\phi(0)}{\varepsilon} - \phi'(0)\ln\varepsilon\right]\,.$$

Thus the finite part is

$$FP\int_{-\infty}^{\infty} \frac{H(x)\phi(x)}{x^2} dx = \int_1^{\infty} \frac{\phi(x)}{x^2} - \phi(0) + \lim_{\varepsilon\to 0}\int_{\varepsilon}^{1} \frac{\phi''(tx)}{2} dx\,,$$
$$= \int_1^{\infty} \frac{\phi(x)}{x^2} dx - \phi(0)$$
$$+ \lim_{\varepsilon\to 0}\int_{\varepsilon}^{1} \frac{\phi(x) - \phi(0) - x\phi'(0)}{x^2} dx\,,$$

which, when integrated by parts, yields

$$FP\int_{-\infty}^{\infty} \frac{H(x)\phi(x)}{x^2} dx = \int_1^{\infty} \frac{\phi'(x)}{x} dx$$
$$+ \lim_{\varepsilon\to 0}\left[\int_{\varepsilon}^{1} \frac{\phi'(x) - \phi'(0)}{x} dx + \frac{\phi(\varepsilon) - \phi(0)}{\varepsilon}\right]$$
$$= \int_0^{\infty} \frac{\phi'(x)}{x} dx + \int_0^{1} \frac{\phi'(x) - \phi'(0)}{x} dx + \phi'(0)$$
$$= -\int_{-\infty}^{\infty} \{-\phi'(x)\}\left(\frac{H(x)}{x}\right) dx - \int_{-\infty}^{\infty} \phi(x)\delta'(x)dx\,.$$

This finally yields the required relation:

$$\text{Pf}\,(H(x)/x^2) = -(d/dx)\text{Pf}\,(H(x)/x) - \delta'(x)\,. \tag{4}$$

We can continue this process and prove (see Exercise 2)

$$\mathrm{Pf}\left(\frac{H(x)}{x^{m+1}}\right) = -\frac{d}{dx}\mathrm{Pf}\left(\frac{H(x)}{mx^m}\right) + \frac{(-1)^m}{m!}\frac{\delta^m(x)}{m}, \qquad m \geq 1. \tag{5}$$

Observe that we can obtain formula (4) by differentiating the function $H(x)/x$. Indeed,

$$\frac{d}{dx}\left(\frac{H(x)}{x}\right) = -\frac{H(x)}{x^2} + \frac{1}{x}\delta(x).$$

Next, we appeal to formula (2.6.26) and find that $x\delta'(x) = -\delta(x)$, so that the above relation agrees with (4). Relation (5) follows similarly. However, the previous analysis displays the technique for dealing with pseudofunctions.

Relation (5) can be iterated to yield another interesting formula. For this purpose, we substitute the value of $H(x)/x^m$ with m replaced by $m-1$, so that we have

$$\begin{aligned}\mathrm{Pf}\,\frac{H(x)}{x^{m+1}} &= -\frac{d}{dx}\left[\frac{1}{m}\left\{-\frac{d}{dx}\mathrm{Pf}\,\frac{H(x)}{(m-1)x^{m-1}} + \frac{(-1)^{m-1}}{(m-1)!}\delta^{m-1}(x)\right\}\right] \\ &\quad + (-1)^m\frac{\delta^m(x)}{m} \\ &= (-1)^2\frac{d^2}{dx^2}\left(\mathrm{Pf}\,\frac{H(x)}{m(m-1)}\right) + \left[(-1)^m\left(\frac{1}{m} + \frac{1}{m-1}\right)\right]\delta^m(x).\end{aligned}$$

Next, we substitute formula (5) with $m-2$, in the above relation and then continue this process. The result is

$$\mathrm{Pf}\,\frac{H(x)}{x^{m+1}} = \frac{(-1)^m}{m!}\left\{\left[\frac{d^m}{dx^m}\mathrm{Pf}\left(\frac{H(x)}{x}\right)\right] + (\psi(m+1) + \gamma)\delta^m(x)\right\}, \tag{6}$$

where

$$\psi(m+1) + \gamma = \sum_{k=1}^{m}\frac{1}{k}, \qquad \psi(1) = -\gamma. \tag{7}$$

We end this section by introducing the concept of an associated homogeneous function [6]. In relation (2.6.9) we found that $\delta(x)$ is a homogeneous function because $\delta(cx) = (1/|c|)\delta(x)$ in R_1. There are generalized functions which have the same property as given above for $\delta(x)$ except that there is an additional term on the right hand side. To illustrate this we use formula (3) by replacing $H(x)/x$ with the function $H(cx)/(cx)$ for $c > 0$;

$$\frac{H(cx)}{cx} = \frac{1}{c}\frac{d}{dx}\{H(x)\ln(cx)\},$$

because $H(cx) = H(x)$, for $c > 0$. Thus

$$\begin{aligned}\frac{H(cx)}{cx} &= \frac{1}{c}\frac{d}{dx}\{H(x)(\ln x + \ln c)\} \\ &= \frac{1}{c}\frac{H(x)}{x} + \frac{1}{c}\delta(x)\{\ln x + \ln c\}.\end{aligned}$$

Taking the finite part of the right hand side we get

$$\frac{H(cx)}{cx} = \frac{1}{c}\frac{H(x)}{x} + \frac{\ln c}{c}\delta(x) . \tag{8}$$

The first term of the above equation is the same as that of a homogeneous function. Because of the additional term $(\ln(c)/c)\delta(x)$, the function $H(cx)/cx$ is called an associated homogeneous function.

4.3. Functions with Algebraic Singularity of Order m

Fortunately, the ideas of the previous section can be extended to a rather large class of functions.

A function $f(x)$ has an algebraic singularity at $x = a$ if for a certain integer $m > 0$ the function $f(x)|x-a|^m$, that is, the function

$$f(x_1, x_2, \ldots, x_n)((x_1 - a_1)^2 + (x_2 - a_2)^2 + \cdots + (x_n - a_n)^2)^{m/2} , \tag{1}$$

is locally integrable in the neighborhood of the point $x = a$. The smallest nonnegative integer m for which (1) is locally integrable is called the order of the algebraic singularity of $f(x)$. For instance, the order of the algebraic singularity of the function $f(x)$ defined by relation (4.1.3) is 1 because $(-1/2x^{-3/2})x = -1/2x^{-1/2}$ which is locally integrable.

Let us, for simplicity, confine ourselves to the functions that have only one algebraic singularity, occurring, say, at the origin. Then the foregoing discussion leads us to the following definition for the regularization principle.

Definition. A *regularization* of the divergent improper integral

$$\int_{R_n} f(x)\phi(x)\,dx , \tag{2}$$

is a distribution identifiable with the function $f(x)$ in every region that does not contain the origin. In the neighborhood of the origin it is the finite part, in the sense of Hadamard.

The following theorem generalizes the concepts of Section 4.1 to the case of a function $f(x)$ that has an algebraic singularity of order m at the origin.

Theorem 1. *Let the function $f(x)$ have an algebraic singularity of order m at the origin. Then its regularization is*

$$\langle f, \phi\rangle = \int_{R_n} f(x)\left\{\phi(x) - \left[\phi(0) + \sum_{i=1}^{n}\left(\frac{\partial\phi}{\partial x_i}\right)_0 x_i \right.\right.$$
$$+ \frac{1}{2!}\sum_{i,j=1}^{n}\left(\frac{\partial^2\phi}{\partial x_i\partial x_j}\right)_0 x_i x_j + \cdots + \frac{1}{(m-1)!}$$
$$\left.\left.\times \sum_{i,j\ldots,k=1}^{n}\left(\frac{\partial^{m-1}\phi}{\partial x_i\partial x_j\cdots\partial x_k}\right) x_i x_j\cdots x_k\right] H\left(1-\frac{|x|}{\varepsilon}\right)\right\} dx . \tag{3}$$

Proof. The expression in square brackets is the Taylor polynomial of degree $m-1$. Hence,

$$\begin{aligned}\psi(x) =&\phi(x) - \left[\phi(0) + \sum_{i=1}^{n}\left(\frac{\partial\phi}{\partial x_i}\right)_0 x_i + \frac{1}{2!}\sum_{i,j=1}^{n}\left(\frac{\partial^2}{\partial x_i \partial x_j}\right)_0 x_i x_j\right.\\ &\left.+\cdots+\sum_{i,j,\ldots k=1}^{n}\frac{1}{(m-1)!}\left(\frac{\partial^{m-1}}{\partial x_i \partial x_j \cdots \partial x_k}\right)_0 x_i x_j \cdots x_k\right]\\ =& O(|x|^m),\end{aligned}$$

where O is the usual symbol for the order of magnitude [6]. Accordingly, the product $f(x)\psi(x)$ is integrable in the neighborhood of the origin. Moreover, inside the sphere $|x| < \varepsilon$, $H(1-|x|/\varepsilon) = 1$, while for $|x| > \varepsilon$, it is equal to zero, so the expression in square brackets disappears, and the integral is convergent for all $\phi(x) \in D$.

If, for a bounded region R not containing the origin, we have a test function belonging to D_R, then (3) is transformed into (2). Accordingly, (2) generates a distribution that is regular in R and is thus identifiable with $f(x)$.

We leave it to the reader (see Exercise 3) to prove that (3) defines a linear continuous functional.

Several important questions remain to be answered. First, what kinds of functions $f(x)$ do not admit a regularization? Our contention is that if

$$|f(x)| > A_m/|x|^m \tag{4}$$

in the neighborhood of the origin for $m = 0, 1, 2, \ldots$, then it admits no regularization (see Exercise 4).

The second question relates to the uniqueness of the regularization. Since the positive constant ε can be chosen arbitrarily in (3), we observe that the improper integral (1) has infinitely many regularizations. The following theorem clarifies the situation.

Theorem 2. *A regularization, if it exists, is uniquely determined apart from a linear combination of the delta function and its derivatives concentrated at the origin* [*the point of the singularity of* $f(x)$].

Proof. Let the distribution t be a regularization of (2). Then $t + t_0$, where t_0 is the sum of the delta function and its derivatives concentrated at $x = 0$, is also a distribution. This distribution can be identified with $f(x)$ in every region R that does not contain $x = 0$. For $\phi(x) \in D_R$, $\phi(x)$ and all its derivatives vanish at $x = 0$. This gives

$$\langle (t + t_0), \phi\rangle = \langle t, \phi\rangle + \langle t_0, \phi\rangle = \langle t, \phi\rangle\,,$$

which is a regular distribution generated by $f(x)$ in view of the foregoing discussion.

On the other hand, if the distributions t_1 and t_2 are two different regularizations of (2), then $t_1 - t_2$ is also a distribution. This means that for $\phi(x) \in D_R$ we have

$$\langle t_1 - t_2, \phi\rangle = \langle t_1, \phi\rangle - \langle t_2, \phi\rangle = 0\,,$$

because both t_1 and t_2 are identifiable with $f(x)$ in R. Consequently, $t_1 - t_2 = t_0$ is a distribution concentrated at the origin and is, therefore, a linear combination of the delta function and its derivatives. This means that whenever we want to write the general form of the regularization of the divergent integral (2), we arbitrarily choose a small value of ε and add the distribution t_0 (as defined here) to it.

The third important question is that if $f(x)$ has a regularization and possesses the classical derivative $f'(x)$ except at $x = 0$, then is the distributional derivative of a regularization of $f(x)$ a regularizaiton of $f'(x)$? As we found out for the function (4.1.1), the answer is in the affirmative. Let us prove this for a single variable; the generalization to higher dimensions is immediate.

Let Df denote the distributional derivative of $f(x)$. Then for all $\phi \in D$

$$\begin{aligned}\langle Df, \phi\rangle &= -\langle f, \phi'\rangle \\ &= -\int_{-\infty}^{\infty} f(x)\Big\{\phi'(x) - \Big[\phi'(0) + \phi''(0)x \\ &\quad + \cdots + \frac{\phi^m(0)}{(m-1)!}x^{m-1}\Big] H\left(1 - \frac{|x|}{\varepsilon}\right)\Big\}dx \\ &= -\int_{-\infty}^{-\varepsilon} f(x)\phi'(x)dx - \int_{\varepsilon}^{\infty} f(x)\phi'(x)dx \\ &\quad - \int_{-\varepsilon}^{\varepsilon} f(x)\left\{\phi'(x) - \left[\phi'(0) + \cdots + \frac{\phi^m(0)}{(m-1)!}x^{m-1}\right]\right\}dx\end{aligned}$$

is a regularization of $f(x)$ where we have used (3) with ϕ replaced by ϕ'. Integrating by parts, we have

$$\begin{aligned}\langle Df, \phi\rangle &= [-f(x)\phi(x)]_{-\infty}^{\varepsilon} - [f(x)\phi(x)]_{\varepsilon}^{\infty} + \int_{-\infty}^{-\varepsilon} f'(x)\phi(x)dx \\ &\quad + \int_{\varepsilon}^{\infty} f'(x)\phi(x)dx \\ &\quad - \left\{f(x)\left[\phi(x) - \left(\phi(0) + \phi'(0)x + \cdots + \frac{\phi^{(m)}(0)}{m!}x^m\right)\right]\right\}_{-\varepsilon}^{\varepsilon} \\ &\quad + \int_{-\varepsilon}^{\varepsilon} f'(x)\left[\phi(x) - \left(\phi(0) + \phi'(0)x + \cdots + \frac{\phi^{(m)}}{m!}x^m\right)\right]dx\end{aligned}$$

$$
\begin{aligned}
&= \int_{|x|>\varepsilon} f'(x)\phi(x)dx \\
&+ \int_{-\varepsilon}^{\varepsilon} f'(x)\left[\phi(x) - \left(\phi(0) + \phi'(0)x + \cdots + \frac{\phi^{(m)}(0)}{m!}x^m\right)\right] dx \\
&+ f(-\varepsilon)\left[\phi(0) - \phi'(0)\varepsilon + \cdots + (-1)^m \frac{\phi^m(0)}{m!}\varepsilon^m\right] \\
&- f(\varepsilon)\left[\phi(0) + \phi'(0)\varepsilon + \cdots + \frac{\phi^{(m)}(0)}{m!}\varepsilon^m\right] \\
&= \int_{-\infty}^{\infty} f'(x)\left[\phi(x) - (\phi(0) + \phi'(0)x + \cdots \right. \\
&\left. + \frac{\phi^{(m)}(0)}{m!}x^m\right) H\left(1 - \frac{|x|}{\varepsilon}\right)\Bigg] dx \\
&+ \alpha_0\phi_0 + \alpha_1\phi'(0) + \cdots + \alpha_m\phi^{(m)}(0)\,,
\end{aligned} \tag{5}
$$

where $\alpha_0, \alpha_1, \ldots, \alpha_m$ are constants. The right side of this relation gives the regularization of $f'(x)$ plus the distribution $\alpha_0\delta - \alpha_1\delta' + \cdots + \alpha_m(-1)^m\delta^{(m)}$ which is concentrated at $x = 0$. This proves our assertion.

The fourth question is concerned with the function $f(x)$, which has several singular parts. Indeed, if $f(x)$ has a finite or countably infinite number of singular parts such that they are not contained in a finite interval, then we can represent it as $f(x) = \sum_l f_l(x)$, where $f_l(x)$ possesses only a singular point. The regularization of $f(x)$ is obtained by summing the regularizations of $f_l(x)$.

Following the previous notation, we call the regularization of (2) a pseudofunction and denote it by

$$
\text{Pf} \int f(x)\phi(x)\,dx\,. \tag{6}
$$

Thus in the case when $f(x)$ is not locally integrable, it is possible to define a distribution Pf $(f(x))$ by

$$
\langle Pf(f(x)), \phi(x)\rangle = Pf \int f(x)\phi(x)dx\,. \tag{7}
$$

Many examples are presented in the next section to illustrate these concepts.

4.4. Examples

Example 1. The locally integrable function

$$
\begin{aligned}
f(x) = (\ln x)_+ &= \begin{cases} \ln x\,, & x > 0\,, \\ 0\,, & x < 0\,, \end{cases} \\
&= H(x)\ln x
\end{aligned} \tag{1}
$$

defines the distribution

$$\langle f, \phi \rangle = \int_0^\infty \phi(x) \ln x \, dx \,, \qquad \phi \in D \,. \tag{2}$$

Its distributional derivative is

$$\langle f', \phi \rangle = \langle -f, \phi' \rangle = -\int_0^\infty \phi'(x) \ln x \, dx \,, \tag{3}$$

while the derivative of the function defined by (1) is $H(x)/x$ as defined by (4.2.1) with $n = 1$. However, we know that this function is not locally integrable. We can use the analysis of Section 4.3 and prove that the distributional derivative (3) coincides with Pf $(H(x)/x)$, as we proved in Section 4.2, by using a slightly different approach. Now, from (3) we have

$$\begin{aligned}
\langle [(\ln x)_+]', \phi \rangle &= -\langle (\ln x)_+, \phi' \rangle = -\int_0^\infty \phi'(x) \ln x \, dx \\
&= -\lim_{\varepsilon \to 0} \int_\varepsilon^\infty \phi'(x) \ln x \, dx \\
&= -\lim_{\varepsilon \to 0} [\phi(x) \ln x]_\varepsilon^\infty + \lim_{\varepsilon \to 0} \int_\varepsilon^\infty \frac{\phi(x)}{x} dx \\
&= \lim_{\varepsilon \to 0} \left[\phi(\varepsilon) \ln \varepsilon + \int_\varepsilon^\infty \frac{\phi(x)}{x} dx \right] \\
&= \lim_{\varepsilon \to 0} \left[\phi(0) \ln \varepsilon + \int_\varepsilon^\infty \frac{\phi(x)}{x} dx \right] .
\end{aligned} \tag{4}$$

Because

$$\phi(0) \ln \varepsilon = -\int_\varepsilon^1 \frac{\phi(0)}{x} dx = -\int_\varepsilon^\infty \frac{\phi(0) H(1-x)}{x} dx \,,$$

(4) becomes

$$\begin{aligned}
\langle [(\ln x)_+]', \phi \rangle &= \lim_{\varepsilon \to 0} \left[-\int_\varepsilon^\infty \frac{\phi(0) H(1-x)}{x} \, dx + \int_\varepsilon^\infty \frac{\phi(x)}{x} \, dx \right] \\
&= \int_0^\infty \frac{1}{x} [\phi(x) - \phi(0) H(1-x)] dx \\
&= \left\langle \text{Pf} \left(\frac{H(x)}{x} \right) , \phi \right\rangle .
\end{aligned}$$

Thus

$$[(\ln x)_+]' = \text{Pf } (1/x)_+ \,, \tag{5}$$

or

$$(d/dx)[H(x) \ln x] = \text{Pf } [H(x)/x] \,,$$

in the sense of distributions.

Similarly (see Exercise 5)

$$[(\ln x)_-]' = \text{Pf } (1/x)_- \,, \tag{6}$$

where

$$(\ln x)_- = \begin{cases} 0\,, & x > 0\,, \\ \ln(-x)\,, & x < 0\,, \end{cases} \tag{7}$$

and

$$\mathrm{Pf}\left(\frac{1}{x}\right)_- = \begin{cases} 0\,, & x > 0\,, \\ -1/x\,, & x < 0\,. \end{cases} \tag{8}$$

Example 2. In Example 4 of Section 2.4 we defined the singular distribution Pf $(1/x)$ as the principal value of $1/x$. Let us examine this distribution in the light of the present discussion. The function $1/x$ has an algebraic singularity of order $m = 1$ at $x = 0$. Therefore, (4.3.3) gives a regularization of $1/x$;

$$\begin{aligned} \left\langle \frac{1}{x}, \phi \right\rangle = \left\langle \mathrm{Pf}\left(\frac{1}{x}\right), \phi \right\rangle &= \int_{-\infty}^{\infty} \frac{1}{x}\left\{\phi(x) - \phi(0)H\left(1 - \frac{|x|}{\varepsilon}\right)\right\} dx \\ &= \int_{-\varepsilon}^{\varepsilon} \frac{\phi(x) - \phi(0)}{x}\, dx + \int_{|x|>\varepsilon} \frac{\phi(x)}{x}\, dx\,, \end{aligned} \tag{9}$$

and both terms on the right side of (9) are convergent and agree with the known result. Thus the regularization of the function coincides with the Cauchy principal value.

For the locally integrable function $f(x) = \ln|x|$, we have $f'(x) = 1/x$, for all $x \neq 0$, in the ordinary sense of the differentiation. Hence $[\ln|x|]'$ is equal to a regularization of $1/x$ in the distributional sense (the proof is similar to that in Example 1); that is,

$$\langle [\ln|x|]', \phi \rangle = \langle \mathrm{Pf}\,(1/x), \phi \rangle\,. \tag{10}$$

Also

$$\ln|x| = (\ln x)_+ + (\ln x)_-\,,$$

and the previous example yields

$$\mathrm{Pf}\,(1/x) = \mathrm{Pf}\,(1/x)_+ + \mathrm{Pf}\,(1/x)_-\,. \tag{11}$$

Example 3. The derivative of the function Pf $(1/x)$ can also be evaluated by this method. Indeed, we proved that if a function $f(x)$ has a regularization and possesses a classical derivative $f'(x)$ at $x = 0$, then the distributional derivative of a regularization is a regularization of $f'(x)$. Accordingly,

$$\left\langle \mathrm{Pf}\left(\frac{1}{x}\right)', \phi \right\rangle = -\left\langle \mathrm{Pf}\left(\frac{1}{x^2}\right)', \phi \right\rangle \quad \text{or} \quad \left[\mathrm{Pf}\left(\frac{1}{x}\right)\right]' = -\mathrm{Pf}\left(\frac{1}{x^2}\right). \tag{12}$$

We can continue this process and obtain the following expression for the mth derivative of the principal value distribution:

$$(\mathrm{Pf}\,(1/x))^{(m)} = (-1)^m m!\mathrm{Pf}\,(1/x^{m+1})\,, \tag{13}$$

where Pf $(1/x^{m+1})$ is a regularization of $1/x^{m+1}$, namely,

$$\left\langle \text{Pf}\left(\frac{1}{x^{m+1}}\right), \phi\right\rangle = \int_{-\infty}^{\infty} \frac{1}{x^{m+1}} \left\{\phi(x) = \left[\phi(0) + \cdots + \frac{\phi^{(m)}(0)}{m!} x^m\right] H\left(1 - \frac{|x|}{\varepsilon}\right)\right\} dx \,. \tag{14}$$

Example 4. Recall the Heisenberg delta distributions (2.4.10)

$$\delta^{\pm}(x) = \frac{1}{2}\delta(x) \mp (1/2\pi i)\text{Pf}\,(1/x)\,.$$

We can combine the results obtained for the derivatives of the delta function and the principal value distributions. This leads us to the useful relations

$$\begin{aligned} \delta^{\pm(m)}(x) &= \frac{1}{2}\delta^{(m)}(x) \mp (1/2\pi i)(\text{Pf}\,(1/x))^{(m)} \\ &= \frac{1}{2}\delta^{(m)}(x) \mp [(-1)^m m!/2\pi i]\text{Pf}\,(1/x^{m+1})\,. \end{aligned} \tag{15}$$

Next, we discuss the distributions x_+^λ, x_-^λ, $|x|^\lambda$, $|x|^\lambda \text{sgn}\, x$.

Example 5(a). The function

$$x_+^\lambda = \begin{cases} x^\lambda, & x > 0, \\ 0, & x < 0, \end{cases} \tag{16}$$

is locally integrable for Re $\lambda > -1$ and defines the regular distribution $\langle x_+^\lambda, \phi\rangle = \int_0^\infty x^\lambda \phi(x)dx$, whose distributional derivative is

$$\langle (x_+^\lambda)', \phi\rangle = -\langle x_+^\lambda, \phi'\rangle\,. \tag{17}$$

On the other hand, the classical derivative of (16) is

$$\frac{dx_+^\lambda}{dx} = \begin{cases} \lambda x^{\lambda-1}, & x > 0, \\ 0, & x < 0, \end{cases} \tag{18}$$

which is not locally integrable. However, by using the method of Section 4.3, we can show that the distributional derivative (17) coincides with a regularization of

$$\int_0^\infty \lambda x^{\lambda-1}\phi(x)dx\,. \tag{19}$$

From relation (17) we have

$$\langle (x_+^\lambda)', \phi\rangle = -\langle x_+^\lambda, \phi'\rangle = -\int_0^\infty x^\lambda \phi'(x)dx = -\lim_{\varepsilon\to 0}\int_\varepsilon^\infty x^\lambda \phi'(x)dx\,,$$

which can be integrated by parts to yield

$$\begin{aligned}\langle (x_+^\lambda)', \phi\rangle &= -\lim_{\varepsilon\to 0}[x^\lambda(\phi(x)-\phi(0))]_\varepsilon^\infty \\ &\quad + \lim_{\varepsilon\to 0}\int_\varepsilon^\infty \lambda x^{\lambda-1}[\phi(x)-\phi(0)]dx \\ &= \int_0^\infty \lambda x^{\lambda-1}[\phi(x)-\phi(0)]dx\,. \end{aligned} \tag{20}$$

The integral on the right side of (20) converges both for $x = 0$ and $x = \infty$ ($-1 < \operatorname{Re}\lambda < 0$) and is a regularization of (19). It even coincides with (19) if we choose a test function $\phi(x)$ such that $\phi(0) = 0$.

Now let $-2 < \operatorname{Re}\lambda < -1$. For this case, x_+^λ has an algebraic singularity of order $m = 1$ at the origin. Accordingly, from (4.3.3) we derive

$$\begin{aligned}\langle x_+^\lambda, \phi\rangle &= \int_0^\infty x^\lambda\left\{\phi(x)-\phi(0)H\left(1-\frac{x}{\varepsilon}\right)\right\}dx \\ &= \int_0^\varepsilon x^\lambda[\phi(x)-\phi(0)]dx + \int_\varepsilon^\infty x^\lambda\phi(x)dx\,.\end{aligned}$$

Since the two regularizations differ from one another by a distribution t_0 concentrated at the origin, we put $\varepsilon = 1$ and obtain the general form

$$\langle x_+^\lambda, \phi\rangle = \int_0^1 x^\lambda[\phi(x)-\phi(0)]dx + \int_0^\infty \phi(x)dx + \langle t_0, \phi\rangle\,. \tag{21}$$

Next, we select a particular value of t_0 by writing $\langle x_+^\lambda, \phi\rangle$ in the following way:

$$\begin{aligned}\langle x_+^\lambda, \phi\rangle &= \int_0^\infty x^\lambda\phi(x)dx = \int_0^1 x^\lambda\phi(x)dx + \int_1^\infty \phi(x)x^\lambda dx \\ &= \int_0^1 x^\lambda[\phi(x)-\phi(0)]dx + \int_0^\infty x^\lambda\phi(x)dx + \int_0^1 \phi(0)x^\lambda dx \\ &= \int_0^1 x^\lambda[\phi(x)-\phi(0)]dx + \int_1^\infty x^\lambda\phi(x)dx + \frac{\phi(0)}{\lambda+1}\,,\end{aligned} \tag{22a}$$

or

$$\langle x_+^\lambda, \phi\rangle = \int_0^\infty x^\lambda\{\phi(x)-\phi(0)H(1-x)\}dx + \frac{1}{\lambda+1}\langle\delta, \phi\rangle\,. \tag{22b}$$

Comparing (21) and (22), we note that $\langle t_0, \phi\rangle = \langle \delta/(\lambda+1), \phi\rangle$.

Our contention is that (22) yields a particular regularization of x_+^λ in the strip $-2 < \operatorname{Re}\lambda < -1$. Indeed, the first term on the right of (22a) is defined for $\operatorname{Re}\lambda < -2$, the second term for any λ, and the third for $\lambda \neq -1$. Thus, (22) has yielded a regularization of x_+^λ in the entire half-plane $\operatorname{Re}\lambda > -2$ except at $\lambda = -1$. Before we continue this process, let us give the following new interpretation to this discussion.

The function

$$F(\lambda) = \int_0^\infty x^\lambda\phi(x)dx\,, \tag{23}$$

is analytic in λ in the half-plane $\operatorname{Re}\lambda > -1$ since it has the derivative with respect to λ given by

$$F(\lambda) = \int_0^\infty x^\lambda \ln x\phi(x)dx\,.$$

The foregoing analysis shows that if we seek the analytic continuation of the function $F(\lambda)$ in the strip $-2 < \operatorname{Re}\lambda < -1$, we obtain a regularization of the improper integral

$$\int_0^\infty x^\lambda\phi(x)dx \qquad (-2 < \operatorname{Re}\lambda < -1)\,. \tag{24}$$

Furthermore, (22) shows that $F(\lambda)$ has a simple pole at $\lambda = -1$, where its residue is $\phi(0) = \langle\delta, \phi\rangle$.

Proceeding in this manner, we can continue the functional x_+^λ analytically to the domain $\operatorname{Re}\lambda > -n-1$, $\lambda \neq -1, -2, \ldots, -n$. The result is

$$\langle x_+^\lambda, \phi\rangle = \int_0^1 x^\lambda\left[\phi(x) - \phi(0) - x\phi'(0) - \cdots - \frac{\lambda^{n-1}}{\lambda}\frac{\phi^{(n-1)}(0)}{(n-1)!}\right]dx$$
$$+ \int_1^\infty x^\lambda\phi(x)dx + \sum_{m=1}^n \frac{\phi^{(m-1)}(0)}{(m-1)!}\,\frac{1}{\lambda+m}\,. \tag{25}$$

The right side of this formula provides a regularization of the improper integral (24) for $\operatorname{Re}\lambda > -n-1$, $\lambda \neq -1, -2, \ldots, -n$, because x_+^λ has the algebraic singularity of nth order at $x = 0$. Equivalently, it provides the analytic continuation of the function $F(\lambda)$ in this domain, and we observe that it has simple poles at the points $\lambda = -1, -2, \ldots, -n$. The residue at $\lambda = -l$ $(l > 0)$ follows from

$$\frac{\phi^{(l-1)}(0)}{(l-1)!} = \frac{(-1)^{l-1}}{(l-1)!}\langle\delta^{(l-1)}(x), \phi(x)\rangle\,,$$

to be

$$\frac{(-1)^{l-1}}{(l-1)!}\,\delta^{(l-1)}(x)\,. \tag{26}$$

Because for $1 \le l \le n$, $\int_1^\infty x^{\lambda+l-1}dx = -1/(\lambda+l)$, (25) can be converted to a simple form in the strip $-n-1 < \operatorname{Re}\lambda < -n$;

$$\langle x_+^\lambda, \phi\rangle = \int_0^\infty x^\lambda\left[\phi(x) - \phi(0) - x\phi'(0) - \cdots - \frac{x^{n-1}}{(n-1)!}\phi^{(n-1)}(0)\right]dx\,. \tag{27}$$

In addition, (18), which is valid for $\operatorname{Re}\lambda > 0$, can be continued analytically throughout the λ plane (except at the points $0, -1, \ldots$).

Incidentally, this discussion explains the special behavior of the pseudofunction Pf $(H(x)x^\lambda)$ when λ is a negative integer, as we found in Section 4.2.

Example 5(b). In a similar fashion we can define the distribution x_-^λ, which corresponds to the function

$$x_-^\lambda = \begin{cases} 0\,, & x > 0\,, \\ (-x)^\lambda\,, & x < 0\,; \end{cases} \tag{28}$$

that is,

$$\langle x_-^\lambda, \phi \rangle = \int_{-\infty}^{0} |x|^\lambda \phi(x) dx \,. \tag{29}$$

Since we can write

$$\langle x_-^\lambda, \phi(x) \rangle = \langle x_-^\lambda, \phi(-x) \rangle \,, \tag{30}$$

it follows from the foregoing analysis that x_-^λ can be defined, by analytic continuation, for all complex values of λ with the exception of the points $\lambda = -l$ $(l = 1, 2, \ldots)$, where x_0^λ has simple poles. This analytic continuation is equal to a regularization of this function. For instance, the value of this functional in the strip $-n-1 < \operatorname{Re}\lambda < -n$ is given by

$$\langle x_-^\lambda, \phi(x) \rangle$$
$$= \int_0^\infty x^\lambda \left[\phi(-x) - \phi(0) + x\phi'(0) - \cdots - \frac{(-1)^{n-1}x^{n-1}}{(n-1)!} \phi^{(n-1)}(0) \right] dx \,. \tag{31}$$

To find the residue at the poles we note that because we replace $\phi(x)$ by $\phi(-x)$ in the analysis of x_+^λ we have to replace the quantities $\phi^{(j)}(0)$ by $(-1)^j \phi^j(0)$. Accordingly, we find as in (26) that the residue of x_-^λ at the pole $\lambda = -l$ is

$$\frac{\delta^{(l-1)}(x)}{(l-1)!} \,. \tag{32}$$

Example 5(c). From the distributions x_+^λ and x_-^λ we can form a new distribution

$$|x|^\lambda = x_+^\lambda + x_-^\lambda \,, \tag{33}$$

which is even because $\langle |x|^\lambda, \phi(x) \rangle = \langle |x|^\lambda, \phi(-x) \rangle$. It follows from the discussion in Examples 5(a) and 5(b) that $|x|^\lambda$ can be continued analytically in the entire λ plane except at certain poles; its analytic continuation is a regularization of the improper integral $\int_{-\infty}^{\infty} |x|^\lambda \phi(x) dx$, $\operatorname{Re}\lambda < -1$.

Furthermore, the functions x_+^λ and x_-^λ have poles at $\lambda = -l$ with residues (26) and (32), respectively. It follows that the poles at $\lambda = -2l$ $(l = 1, 2, \ldots)$ cancel each other, and $|x|^\lambda$ has poles just at $\lambda = -1, -3, \ldots, -2l-1$. The residue at $\lambda = -2l-1$ is

$$2\delta^{2l}(x)/2l!. \tag{34}$$

At the points $\lambda = -2l$, $|x|^\lambda$ is well defined and is written x^{-2l}.

We can readily obtain the explicit expression for $|x|^\lambda$ from those for x_+^λ and x_-^λ. For example, we can derive the explicit expression for $|x|^\lambda$ in the strip $-2m-1 < \operatorname{Re}\lambda < -2m+1$ by substituting $2m$ for n in (27) and (31) and adding the two relations. The result is

$$\langle |x|^\lambda, \phi \rangle = \int_0^\infty x^\lambda \Bigg\{ \phi(x) + \phi(-x)$$
$$- 2\left[\phi(0) + \frac{x^2}{2}\phi''(0) + \cdots + \frac{x^{2m-2}}{(2m-2)!}\phi^{(2m-2)}(0) \right] \Bigg\} dx \,. \tag{35}$$

Furthermore,

$$\begin{aligned}\frac{d}{dx}|x|^\lambda &= \frac{d}{dx}(x_+^\lambda + x_-^\lambda) = \lambda x_+^{\lambda-1} - \lambda x_-^{\lambda-1}\\ &= \lambda |x|^\lambda \operatorname{sgn} x\,, \qquad \lambda \neq 0, -1, \ldots,\end{aligned} \tag{36}$$

both sides of which admit analytic continuation to negative even values of λ.

Example 5(d). Similarly, we can form the distribution

$$|x|^\lambda \operatorname{sgn} x = x_+^\lambda - x_-^\lambda\,, \qquad \operatorname{Re}\lambda < -1\,. \tag{37}$$

It is an odd distribution because

$$\langle |x|^\lambda \operatorname{sgn} x\,,\ \phi(-x)\rangle = -\langle |x|^\lambda \operatorname{sgn} x\ \phi(x)\rangle\,.$$

Proceeding as in Example 5(c), we find that, when we analytically continue $|x|^\lambda \operatorname{sgn} x$, the poles of x_+^λ and x_-^λ at $\lambda = -(2l+1), l = 0, 1, 2, \ldots$, cancel, and therefore this distribution is meaningful for these values of λ. It has simple poles at $\lambda = -2l, l = 1, 2, \ldots$, with residues

$$-2\delta^{2l-1}(x)/(2l-1)!\,. \tag{38}$$

The explicit expression for $|x|\operatorname{sgn} x$ in the strip $-2m-2 < \operatorname{Re}\lambda < -2m$ follows on substituting $n = 2m$ in (27) and (30) and subtracting the two. That is,

$$\begin{aligned}\langle |x|^\lambda \operatorname{sgn} x, \phi(x)\rangle = \int_0^\infty x^\lambda \Big\{\phi(x) - \phi(-x) - 2\Big[x\phi'(0) + \frac{x^3}{3!}\phi'''(0)\\ + \cdots + \frac{x^{2m-1}}{(2m-1)!}\phi^{(2m-1)}(0)\Big]\Big\}\, dx\,.\end{aligned} \tag{39}$$

Also,

$$\frac{d}{dx}[|x|^\lambda \operatorname{sgn} x] = \frac{d}{dx}(x_+^\lambda - x_-^\lambda) = \lambda x_+^{\lambda-1} + \lambda x_-^{\lambda-1} = \lambda |x|^{\lambda-1}\,, \tag{40}$$

except at $\lambda = 0, -1, \ldots$. Both sides of (37) admit analytic continuation to the negative values of λ.

Example 5(e). The distribution x^{-n}, $n = 1, 2, \ldots$. Combining the results of Examples 5(c) and 5(d), we find that the distribution x^{-n} is meaningful for all integral values of n. Indeed, putting $\lambda = -2m$ in (35) and $\lambda = -2m-1$ in (39), we obtain

$$\begin{aligned}\langle x^{-2m}, \phi\rangle &\equiv \langle |x|^{-2m}, \phi\rangle\\ &= \int_0^\infty x^{-2m} \Big\{\phi(x) + \phi(-x) - 2\Big[\phi(0) + \frac{x^2}{2}\phi''(0)\\ &\quad + \cdots + \frac{x^{2m-1}}{(2m-2)!}\phi^{(2m-2)}(0)\Big]\Big\}\, dx\,,\end{aligned} \tag{41}$$

and

$$
\begin{aligned}
\langle x^{-2m-1}, \phi \rangle &= \langle |x|^{-2m-1}, \operatorname{sgn} x\, \phi \rangle \\
&= \int_0^\infty x^{-2m-1} \Bigg\{ \phi(x) + \phi(-x) - 2 \Bigg[x\phi'(0) + \frac{x^3}{3!}\phi'''(0) \\
&\quad + \cdots + \frac{x^{2m-1}}{(2m-1)!}\, \phi^{(2m-1)}(0) \Bigg] \Bigg\}\, dx\,, \qquad (42)
\end{aligned}
$$

respectively. The distributions x^{-n}, $n = 1, 2, \ldots$, are defined by (41) and (42).

Relations given in Examples 2 and 3 agree with these results. For instance, from (36) and (40) it follows that $(d/dx)(x^{-n}) = -nx^{-n-1}$.

Example 5(f). Multiplication by a Function. In the formula

$$x^m x_+^\lambda = x_-^{m+\lambda}, \qquad (43)$$

the left and right sides have independent meanings. To prove that they define the same distribution, we observe that both sides of (43) are analytic in λ for $\operatorname{Re}\lambda > -1$ and coincide for these values of λ. Therefore, they coincide over their full region of analyticity, that is, for all λ except $\lambda = -1, -2, \ldots$. However, the right side is also analytic for $\lambda = -1, -2, \ldots, -m$. This implies that the factor on the left side is also analytic at these values of λ. In particular, it follows that

$$\lim_{\lambda \to -l} x^m x_+^\lambda = x_+^{m-l}\,, \qquad l = 1, 2, \ldots, m\,,$$

so that (43) is valid for these values of λ if interpreted as an appropriate limit.

By similar arguments we have

$$x^m x_-^\lambda = (-1)^m x_-^{m+\lambda}\,, \qquad (44)$$

$$x^m |x|^\lambda = |x|^{m+\lambda} (\operatorname{sgn} x)^m\,, \qquad (45)$$

$$x^m |x|^\lambda \operatorname{sgn} x = |x|^{m+\lambda} (\operatorname{sgn} x)^{m+1}\,, \qquad (46)$$

for all λ except $\lambda = -m-1, -m-2, \ldots$.

This concept can be used to multiply these functions by a function $f(x)$ that is infinitely differentiable and has an m-fold zero at $x = 0$; i.e., $f(x) = x^m g(x)$, $g(0) \neq 0$. For example,

$$f(x) x_+^\lambda = g(x) x^m x_+^\lambda = g(x) x_+^{m+\lambda}\,. \qquad (47)$$

Example 5(g). The distributions x_+^λ, x_+^λ, $|x|^\lambda$, and $|x|^\lambda \operatorname{sgn} x$ may be normalized by dividing them by an appropriate gamma function, since the latter has the same kind of singularities. Indeed, the distributions

$$x_+^\lambda / \Gamma(\lambda+1)\,, \quad x_-^\lambda / \Gamma(\lambda+1)\,, \; |x|^\lambda \Big/ \Gamma\left(\frac{\lambda+1}{2}\right), \; |x|^\lambda \operatorname{sgn} x \Big/ \Gamma\left(\frac{\lambda+2}{2}\right) \qquad (48)$$

have the property that functionals defined by them, such as

$$\langle x_+^\lambda/\Gamma(\lambda+1)\,,\phi(x)\rangle\,, \tag{49}$$

are entire functions of the complex variable λ. The proof is as follows: Since the residue of the gamma function $\Gamma(\lambda+1)=\int_0^\infty x^\lambda e^{-x}dx$, at $\lambda=-l$ is $(-1)^{l-1}/(l-1)$, we find from (26) and (31), that the first two generalized functions in (48) are well defined at those points where x_+^λ, x_-^λ have poles. Indeed,

$$\left.\frac{x_+^\lambda}{\Gamma(\lambda+1)}\right|_{\lambda=-1}=\delta^{(l-1)}(x)\,,\qquad \frac{x_-^\lambda}{\Gamma(\lambda+1)}=(-1)^{l-1}\delta^{(l-1)}(x)\,, \tag{50}$$

for $l=1,2,\ldots$.

The residue of $\Gamma\left((\lambda+1)/2\right)$ at a pole $-2l-1$ can be determined from the functional

$$\begin{aligned}\Gamma\left(\frac{\lambda+1}{2}\right)&=\frac{2}{\lambda+1}\Gamma\left(\frac{\lambda+3}{2}\right)\\&=\frac{2}{\lambda+1}\,\frac{2}{\lambda+3}\Gamma\left(\frac{\lambda+5}{2}\right)\\&\;\vdots\\&=2^{l+1}\Gamma\left(\frac{\lambda+2l+3}{2}\right)\Big/(\lambda+1)(\lambda+3)\cdots(\lambda+2l+1)\,.\end{aligned}$$

Thus

$$\operatorname{Res}_{\lambda=-2l-1}\Gamma\left(\frac{\lambda+1}{2}\right)=\frac{2(-1)^l}{l!}\,, \tag{51}$$

where Res stands for the residue.

Combining (51) and (34), we find that

$$\left.|x|^\lambda\Big/\Gamma\left(\frac{\lambda+1}{2}\right)\right|_{\lambda=-2l-1}=(-1)^l\frac{l!}{2l!}\delta^{(2l)}(x)\,. \tag{52}$$

Similarly (see Exercise 6),

$$\left.|x|^\lambda\operatorname{sgn}x\Big/\Gamma\left(\frac{\lambda+2}{2}\right)\right|_{\lambda=-2l}=(-1)^l\frac{(l-1)!}{(2l-1)!}\delta^{(2l-1)}(x)\,. \tag{53}$$

Example 5(h). The other interesting combinations of the distributions x_+^λ and x_-^λ are (see Exercise 9)

$$(x\pm i0)^\lambda=\lim_{\varepsilon\to\pm0}(x^2+\varepsilon^2)^{\lambda/2}e^{i\lambda\arg(x+i\varepsilon)}=x_+^\lambda+e^{\pm i\pi\lambda}x_-^\lambda\,, \tag{54}$$

which are valid for Re $\lambda>0$. These distributions are again defined by analytic continuation for other values of λ. By expanding $\langle x_+^\lambda,\phi\rangle$ and $\langle e^{\pm i\pi\lambda}x_-^\lambda,\phi\rangle$ (for algebraic details see [7]) we can prove that the poles originating from both terms on the right side of (52) cancel. This leads to the important formula

$$\lim_{\varepsilon\to 0}(x \pm i\varepsilon)^{-m-1} = (x \pm i0)^{-m-1} = x^{-(m+1)} \mp \frac{i\pi(-1)^{m-1}}{m!}\delta^{(m)}(x)\,, \tag{55}$$

which agrees with (2.4.18,19) for $m = 0$. Combining (15) and (55), we have

$$\delta^{\pm m}(x) = (-1)^m \frac{2\pi i}{m!}(x \pm i0)^{-m-1}\,. \tag{56}$$

Example 6. The Distribution $r^\lambda = (x_1^2 + x_2^2 + \cdots + x_n^2)^{\lambda/2}$. For Re $\lambda > -n$, r^λ is locally integrable and thus defines the distribution

$$\langle r^\lambda, \phi\rangle = \int r^\lambda \phi(x)dx = \int_{-\infty}^{\infty}\cdots\int_{-\infty}^{\infty}(x_1^2 + x_2^2 + \cdots + x_n^2)^{\lambda/2} \times \phi(x_1, x_2, \ldots, x_n)dx_1 dx_2 \cdots dx_n\,. \tag{57}$$

Since formal differentiation yields

$$\frac{d}{d\lambda}\langle r^\lambda, \phi\rangle = \int r^\lambda \ln r\phi(x)dx\,,$$

and $r^\lambda \ln r$ is locally integrable, r^λ represents an analytic function of λ for Re $\lambda > -n$. For Re $\lambda \le -n$, r^λ is not locally integrable, and we use analytic continuation to define it. This can be done because r^λ has an algebraic singularity at the origin. We can reduce the distribution r^λ to x_+^λ and then use the results of the previous example.

It is convenient to use polar coordinates $r, \vartheta_1, \vartheta_2, \ldots, \vartheta_{n-1}$. Then (57) becomes

$$\langle r^\lambda, \phi\rangle = \int_0^\infty r^\lambda \left[\int_S \phi(x)dS\right] r^{n-1}dr\,, \tag{58}$$

where dS is the surface element on the surface S of the unit sphere in R_n. By virtue of the mean value theorem of integral calculus, we can express the inner integral

$$\int_S \phi(x)dS = S_n(1)\phi(r, \vartheta_1^{(0)}, \vartheta_2^{(0)}, \ldots, \vartheta_{n-1}^{(0)})\,, \tag{59}$$

where $S_n(1)$ denotes the surface area of the unit sphere (see Section 3.3), and $\vartheta_1^{(0)}, \vartheta_2^{(0)}, \ldots, \vartheta_{n-1}^{(0)}$ are certain fixed values of the polar angle (depending only on the test function ϕ). If we write

$$\phi(r, \vartheta_1^{(0)}, \vartheta_2^{(0)}, \ldots, \vartheta_{n-1}^{(0)}) = \Omega_\phi(r)\,, \tag{60}$$

we find that (58) and (59) become

$$\langle r^\lambda, \phi\rangle = S_n(1)\int_0^\infty r^{\lambda+n-1}\Omega_\phi(r)dr\,, \tag{61}$$

$$\int_S \phi(x)dS = S_n(1)\Omega_\phi(r)\,. \tag{62}$$

Our contention is that $\Omega_\phi(r)$ has compact support and derivatives of all orders. Moreover, all its derivatives of odd order vanish at $r = 0$. Then (61) is well defined because $\Omega_\phi(r)$ is a test function, and we can transfer the results of Example 5 to this case. This assertion is proved as follows: Since $\phi(x)$ vanishes for sufficiently large r, so does its mean value $\Omega_\phi(r)$. Thus $\Omega_\phi(r)$ has compact support. For $r > 0$, the differentiability of $\Omega_\phi(r)$ follows from definition (60) and the fact that ϕ has derivatives of all orders. In order to prove the differentiability part of the assertion for $r = 0$, we use Taylor's theorem with the remainder term and expand $\phi(x)$ through terms of order r^{2m}. Then (62) yields

$$\Omega_\phi(r) = \frac{1}{S_n(1)} \int_S \Bigg[\phi(0) + \sum_{i=1}^{m} \frac{\partial \phi(0)}{\partial x_i} x_i + \frac{1}{2!} \sum_{i=1}^{m} \sum_{j=1}^{m} \frac{\partial^2 \phi(0)}{\partial x_i \partial x_j} x_i x_j$$

$$+ \frac{1}{3!} \sum_{i=1}^{m} \sum_{j=1}^{m} \sum_{l=1}^{m} \frac{\partial^3 \phi(0)}{\partial x_i \partial x_j \partial x_l} x_i x_j x_l + \cdots + R_{2m} \Bigg] dS \,,$$

where R_{2m} is the remainder term. Because each term in the integrand containing an odd number of factors x_i is an odd function (except the remainder term), its integral vanishes in the course of integration. On the other hand, the term containing an even number, $2m$, say, of factors x_i yields a term of the form $a_m r^{2m}$. Accordingly, we have

$$\Omega_\phi(r) = \phi(0) + a_1 r^2 + a_2 r^4 + \cdots + a_{2m} r^{2m} + o(r^{2m}) \,,$$

where the little o has its usual meaning in measuring the magnitude of a term [6]. This shows that we can differentiate $\Omega_\phi(r) 2m$ times at $r = 0$ and that the odd derivatives vanish. This completes the proof. Consequently, $\Omega_\phi(r)$ is an even function of r in D, and the integral (61) is a well-defined function.

To transfer the results of Example 5, let us write (61) as (note that $r > 0$)

$$\langle r^\lambda, \phi \rangle = S_n(1) \langle x_+^\mu, \Omega_\phi(x) \rangle, \qquad \mu = \lambda + n - 1 \,, \tag{63}$$

which is an analytic function of μ for Re $\mu < -1$ or Re $\lambda < -n$. Its analytic continuation to the rest of the μ plane follows from the discussion of Example 5. The simple poles of x_+^μ occur at the points $\mu = \lambda + n - 1 = -1, -2, -3, \ldots$, or $\lambda = -n, -n-1, \ldots$. The value of the residue at the pole $\mu = m$, readily derived from (26), is

$$S_n(1)(-1)^{m-1} \langle \delta^{(m-1)}(x), \Omega_\phi(x) \rangle / (m-1)! \,. \tag{64}$$

Since the derivatives of odd orders of $\Omega_\phi(x)$ vanish at $x = 0$, no poles exist for even numbers m.

To sum up, we have the following result: The distribution $\langle r^\lambda, \phi \rangle$ can be defined in the whole complex λ plane with the exception of the points $\lambda = -n - 2l (l = 0, 1, 2, \ldots)$, where this functional has simple poles with residues

$$S_n(1) \Omega_\phi^{(2l)}(0) / 2l! \,. \tag{65}$$

At the point $\lambda = -n$, i.e., for $l = 0$, this result can be simplified. From (62) we have

$$S_n(1)\Omega_\phi(0) = \phi(0)\int_S dS = S_n(1)\phi(0)\,. \tag{66}$$

Then (65) reduces to $S_n(1)\phi(0)$; that is,

$$\text{Res}_{\lambda=-n} r^\lambda = S_n(1)\delta(x)\,. \tag{67}$$

The distribution r^λ can be normalized by introducing

$$R^\lambda = 2r^\lambda / S_n(1)\Gamma\left(\frac{\lambda+n}{2}\right), \tag{68}$$

because both the numerator and the denominator have poles at $\lambda = -n, -n-2, -n-4$. The residue of $2r^\lambda$ at $\lambda = -n$ is $2S_n(1)\delta(x)$, and the residue of $\Gamma[(\lambda+n)/2]$ at $\lambda = -n$ can be found as follows.

From the functional equation

$$\Gamma\left(\frac{\lambda+n}{2}\right) = \frac{2}{\lambda+n}\Gamma\left(\frac{\lambda+n+2}{2}\right),$$

we have the relation (as $\lambda = -n$),

$$\Gamma\left(\frac{\lambda+n}{2}\right) = \frac{2}{\lambda+n}\Gamma(1) + O(1)\,.$$

Therefore, the residue of $\Gamma[(\lambda+n)/2]$ at $\lambda = -n$ is $2\Gamma(1) = 2$. Accordingly,

$$R^{-n} = \delta(x_1, x_2, \ldots, x_n) = \delta(x)\,. \tag{69}$$

By repeated applications of the Laplace operator $\nabla^2 = r^{1-n}(r^{n-1}d/dr)$ to $R^{\lambda+2l}$, we obtain

$$(\nabla^2)^l R^{\lambda+2l} = 2^l(\lambda+2)(\lambda+4)\cdots(\lambda+2l)R^\lambda\,. \tag{70}$$

This relation is valid for $\lambda > 0$, and hence also for Re $\lambda \le 0$, by the principle of analytic continuation. From (69) and (70) it follows that

$$R^{-n-2l} = \frac{(-1)^l(\nabla^2)^l\delta(x)}{2^l n(n+2)\cdots(n+2l-2)}, \qquad l = 1, 2, \ldots\,. \tag{71}$$

Moreover, by combining (68) and (70) we have

$$(\nabla^2)^l r^{-n+2l} = S_n(1)2^{l-1}(l-1)!(-n+2)(-n+4)\cdots(-n+2l)\delta(x)\,, \qquad l = 1, 2, \ldots, \tag{72}$$

which, for the special case $l = 1$, yields

$$\nabla^2 r^{-n+2} = -S_n(1)(n-2)\delta(x)\,. \tag{73}$$

We shall discuss this result in more detail in Chapter 10.

Example 7. Decomposition of r^λ *and* $\delta(x)$ *into Plane Waves.*

Let $\omega = (\omega_1, \omega_2, \ldots, \omega_n)$ denote a point on the surface S of the unit sphere in R_n and let the scalar product of ω with a point x in R_n be denoted $(\omega \cdot x) = \omega_1 x_2 + \cdots + \omega_n x_n$. We attempt to evaluate the integral

$$\int_S |(\omega \cdot x)|^\lambda dS = \int_S |\omega \cdot x|^\lambda d\omega = \Psi(x, \lambda), \tag{74}$$

which exists as a proper integral for Re $\lambda > 0$ and as an improper integral for Re $\lambda > -1$. The function Ψ is spherically symmetric in x, for if we substitute Ax for x in (74), where A is the matrix describing simple rotation ($A' = A^{-1}$, where A' is the transpose of A), we obtain

$$\Psi(Ax, \lambda) = \int_S |(\omega \cdot Ax)|^\lambda d\omega = \int_S |(A'\omega \cdot x)|^\lambda d\omega = \int_S |\omega \cdot x|^\lambda d\omega .$$

Accordingly, $\Psi(x, \lambda)$ is a function of r and λ only denoted $\Psi(r, \lambda)$. Moreover, $\Psi(r, \lambda)$ is a homogeneous function of degree λ. Indeed, substituting cx for x, $c > 0$, in (74), we have

$$\Psi(r, \lambda) = \int_S |\omega \cdot cx|^\lambda d\omega = c^\lambda \int_S |\omega \cdot x|^\lambda d\omega = c^\lambda \Psi(r, \lambda) .$$

This means that Ψ is proportional to r^λ;

$$\Psi(r, \lambda) = C(\lambda) r^\lambda . \tag{75}$$

To determine $C(\lambda)$ we take x to be the unit vector $x = \widehat{e} = (0, 0, \ldots, 1)$ in (74) and (75). This gives

$$C(\lambda) = \Psi(1, \lambda) = \int_S |\omega_n|^\lambda dS . \tag{76}$$

Now, in spherical coordinates $\vartheta_1, \vartheta_2, \ldots, \vartheta_{n-1}$,

$$\omega_n = \cos \vartheta_{n-1} , \qquad dS = \sin^{n-2} \vartheta_{n-2} dS_{n-1} d\vartheta_{n-1} ,$$

where ϑ_{n-1} is the angle between $\widehat{e}$ and ω, and dS_{n-1} is the area element of the surface of the $(n-1)$-dimensional unit sphere. Then (76) becomes

$$\begin{aligned} C(\lambda) &= \int_S |\omega_n|^\lambda dS = \int_0^\pi \int_{S_{n-1}} |\cos \vartheta_{n-1}|^\lambda \sin^{n-2}(\vartheta_{n-1}) d\vartheta_{n-1} dS_{n-1} \\ &= S_{n-1}(1) \int_0^\pi |\cos \vartheta_{n-1}|^\lambda \sin^{n-2} \vartheta_{n-1} d\vartheta_{n-1} \\ &= 2 S_{n-1}(1) \int_0^{\pi/2} \cos^\lambda \vartheta \sin^{n-2} \vartheta \, d\vartheta . \end{aligned} \tag{77}$$

Since

$$S_{n-1}(1) = 2\pi^{(n-1)/2} \Big/ \Gamma\left(\frac{n-1}{2}\right) ,$$

and

$$\int_0^{\pi/2} \cos \lambda\vartheta \sin^{n-2}\vartheta \, d\vartheta = \frac{1}{2}\beta\left(\frac{n-1}{2}, \frac{\lambda+1}{2}\right) = \frac{1}{2}\frac{\Gamma\left(\frac{\lambda+1}{2}\right)\Gamma\left(\frac{n-1}{2}\right)}{\Gamma\left(\frac{\lambda+n}{2}\right)},$$

(77) yields the following value for $C(\lambda)$:

$$C(\lambda) = \frac{2\pi^{(n-1)/2}\Gamma(\lambda+1)/2)}{\Gamma((\lambda+n)/2)}. \tag{78}$$

Combining (68), (74), (75), and (78), we have

$$\left[\pi^{(n-1)/2}\Gamma\left(\frac{\lambda+1}{2}\right)\right]^{-1} \int_S |x\cdot\omega|^\lambda dS = R^\lambda, \tag{79}$$

which, as already proved, is an analytic function in the entire λ plane. This equation represents the decomposition of R^λ into plane waves, a concept similar to the Fourier decomposition.

In the next stage, let us examine the integral on the left side of (79). From the analysis of Example 5 we know that, for an even integer $\lambda = -2l$, the functional $|x\cdot\omega|^\lambda/[(\lambda+1)/2]$ has no singularities, whereas for odd λ ($\lambda = -(2l+1)$), its value is $(-1)^l l!\delta^{2l}(\omega\cdot x)/2l!$. On the other hand the value of R^λ is $\delta(x)$ for all integral values $\lambda = -n$. Consequently, (79) gives us a plane wave decomposition of the delta function;

$$\delta(x) = \begin{cases} (-1)^{(n-1)/2}\int_S \delta^{(n-1)}((\omega\cdot x))d\omega, & n \text{ odd}, \\ \dfrac{(-1)^{n/2}(n-1)!}{(2\pi)^n}\int_S(\omega\cdot x)^{-n}d\omega, & n \text{ even}. \end{cases} \tag{80}$$

These plane wave expansions solve *Radon's problem*, i.e., the problem of representing a test function ϕ at any point x in terms of averages of ϕ and its derivatives on hyperplanes $\omega\cdot x = a$ constant.

Example 8. In this example we consider a function $f(x)$ that is homogeneous and continuously differentiable outside the origin. Recall that a homogeneous function $f(x)$, $x = (x_1, \ldots, x_n)$, of degree l satisfies the functional relation

$$f(tx_1, tx_2, \ldots, tx_n) = t^l f(x_1, x_2, \ldots, x_n).$$

In our discussion we take $l = -m+1$.

The function $\partial f/\partial x_i$ defines the distribution

$$\begin{aligned} \left\langle \mathrm{Pf}\left(\frac{\partial f}{\partial x_i}\right), \phi\right\rangle &= \lim_{\varepsilon\to 0}\int_{|x|\geq\varepsilon} \frac{\partial f(x)}{\partial x_i}\phi(x)dx = -\lim_{\varepsilon\to 0}\int_{|x|\geq\varepsilon} f\frac{\partial\phi}{\partial x_i}dx \\ &= \lim_{\varepsilon\to 0}\left\{\int_{|x|\geq\varepsilon}\frac{\partial f}{\partial x_i}\phi(x)dx + \int_{|x|=\varepsilon} f\phi n_x dS\right\} \\ &= \left\langle\frac{\partial f}{\partial x_i}, \phi\right\rangle + \lim_{\varepsilon\to 0}\int_{|x|=\varepsilon} f\phi n_x dS, \end{aligned} \tag{81}$$

where $\phi \in D$ and dS is the element of surface on the sphere $|x| = \varepsilon$. Since $n_x = x_i/|x|$,

$$\int_{|x|=\varepsilon} f\phi n_x dS = \int_{|x|=\varepsilon} f(x)\frac{x_i}{|x|}\{\phi(x) - \phi(0)\}dS + \phi(0)\int_{|x|=\varepsilon} f(x)\frac{x_i}{|x|}dS\,. \tag{82}$$

Now, for $|x| = \varepsilon$, $|f(x)| \le \alpha\varepsilon^{-(m-1)}$, $|x_i/|x|| \le 1$, and $|\phi(x) - \phi(0)| < \beta\varepsilon$, where α and β are constants, so that

$$\left|\int_{|x|=\varepsilon} f(x)\frac{x_i}{|x|}\{\phi(x) - \phi(0)\}dS\right| \le \alpha\beta\varepsilon^{-(m-2)}\int_{|x|=\varepsilon} dS$$
$$= S_n(1)\alpha\beta\varepsilon^{-(m-2)}\varepsilon^{m-1} = S_m(1)\alpha\beta\varepsilon = O(\varepsilon)\,.$$

As for the second integral on the right side of (82), we observe that it is independent of ε because the expression $f(x)(x_i/|x|)$ is homogeneous of degree $-(m-1)$ while dS is homogeneous of degree $m-1$. Accordingly, if we let $\varepsilon \to 0$ in the first integral and set $\varepsilon = 1$ in the second integral on the right-hand side of (82), we find that relation (81) becomes

$$\left\langle \mathrm{Pf}\left(\frac{\partial f}{\partial x_i}\right), \phi\right\rangle = \left\langle \frac{\partial f}{\partial x_i}, \phi\right\rangle + \left\langle \delta(x)\int_{|x|=1} f(x)x_i dS, \phi\right\rangle .$$

This means that

$$\mathrm{Pf}\left(\frac{\partial f}{\partial x_i}\right) = \frac{\partial f}{\partial x_i} + c\delta\,, \qquad \text{where} \quad c = \int_{|x|=1} f(x)x_i\, dS\,. \tag{83}$$

From relation (83) we can recover (73) (see Exercise 17).

Exercises

1. Show that the functional defined by (4.1.9) is linear and continuous.

2. Prove (4.2.5).

3. Show that (4.3.3) defines a linear continuous functional.

4. Let $f(x)$ be a locally integrable function except in a neighborhood of the origin where $|f(x)| > A_m/|x|^m$, $0 < x < x_0$, $m = 0, 1, 2, \ldots$. Show that $f(x)$ cannot be regularized.

5. Show that $[(\ln x)_-]' = \mathrm{Pf}\,(1/x)_-$.

6. Establish relations (4.4.53).

7. Show that

$$\Gamma(\lambda) = \int_0^\infty x^{\lambda-1} e^{-x}\,dx$$

$$= \begin{cases} \displaystyle\int_0^1 x^{\lambda-1}\left[e^{-x} - \sum_{m=0}^{\infty}(-1)^m \frac{x^m}{m!}\right]dx + \int_1^\infty x^{\lambda-1} e^{-x}\,dx \\ \displaystyle\quad + \sum_{m=0}^{n} \frac{(-1)^m}{m!(m+\lambda)}, \qquad \operatorname{Re}\lambda > -1, \quad \lambda \neq -1, \ldots, -n, \\ \displaystyle\int_0^\infty x^{\lambda-1}\left[e^{-x} - \sum_{m=0}^{\infty}(-1)^m \frac{x^m}{m!}\right]dx, \qquad -n-1 < \operatorname{Re}\lambda < -n. \end{cases}$$

8. Show that the general solution of $x^m t(x) = 1$, is

$$t(x) = x^{-m} + \sum_{l=1}^{\infty} c_l \delta^{(l-1)}(x).$$

9. Writing $(x + i\varepsilon)^\lambda = e^{\lambda \ln(x \pm i\varepsilon)}$, prove (4.4.54).

10. Discuss Examples 6 and 7 of Section 4.4 for the special case $n = 3$.

11. Establish (4.4.70).

12. Evaluate the finite part of the integral

$$\int_0^\infty \frac{\ln x}{x}\,\phi(x)\,dx, \qquad \phi \in D,$$

and determine the distribution it defines.

13. Show that

$$\lim_{R\to\infty} \frac{1 - \cos Rx}{x} = \operatorname{Pf}\left(\frac{1}{x}\right).$$

14. For $n = 2$, define the function $\operatorname{Pf}(1/r^2)$ as

$$\nabla^2\left(\frac{1}{2}\ln^2 r\right) = \operatorname{Pf}(1/r^2),$$

and show that

$$\left\langle \operatorname{Pf}\left(\frac{1}{r^2}\right), \phi \right\rangle = \lim_{\varepsilon\to 0}\left[\int_{r\geq\varepsilon}\left\{\frac{\phi}{r^2}dx + 2\pi\phi(0)\ln\varepsilon\right\}\right], \qquad \phi \in D.$$

15. Derive the relations of Section 4.2 with the help of the results in Example 5 of Section 4.4.

16. Show that if $-1 < \lambda < 0$, and $k = 1, 2, 3, \ldots$, then

$$\langle \mathrm{Pf}\,(H(x)x^{\lambda-k}), \phi \rangle = \int_0^\infty x^{\lambda-k} \left(\phi(x) - \sum_{m=0}^{k-1} \frac{\phi^{(m)}(0)}{m!}\, x^m \right) dx\,.$$

17. In (4.4.83) set $f(x) = \partial g/\partial x_i$, where $g(x) = |x|^{-(m-2)}$, to derive (4.6.73).

18. Show that

$$(m^2 + P \pm i0)^{-1} = (m^2 + P)^{-1} \pm i\pi\delta(m^2 + P)\,,$$

where m is a nonzero real number and $P = x_1^2 + x_2^2 + x_3^2 - t^2$.

CHAPTER 5

Distributional Derivatives of Functions with Jump Discontinuities

In boundary and initial-value problems relating to the potential, scattering and wave propagation theories, we encounter functions that are defined inside or outside some surface S if the surface is closed, and on both sides of it if it is open. However, these functions or their first- or higher-order derivatives have jumps across S. Classical theory is based on solving such problems on both sides of the boundaries and then attempting to satisfy the boundary conditions or jump conditions across S, as the case may be. There are many problems, however, that cannot be solved by classical techniques.

Our aim is to develop the vector analysis of functions with jump discontinuities across surfaces and boundaries. With the help of this analysis we can solve many unsolved problems in the potential, scattering, and wave propagation theories. Furthermore, problems whose solutions are already known can be solved by this method in a very simple manner [8]. To distinguish between the classical and distributional derivatives we shall put a bar over the latter whenever there is an ambiguity.

5.1. Distributional Derivatives in $\mathbf{R}_1$

Let $F(x)$ be a function of a single variable x that has a jump discontiuity at $x = \xi_1$ of magnitude a_1 but has a continuous derivative everywhere else. Let the derivative in the interval $x < \xi_1$ and $x > \xi_1$ be denoted $f'(x)$. This derivative is undefined at $x = \xi_1$. With the help of generalized functions, however, the distributional derivative $\overline{F}'(x)$ is obtained by setting

$$f(x) = F(x) - a_1 H(x - \xi_1), \tag{1}$$

where H is the Heaviside function. The function $f(x)$ is continuous at $x = \xi_1$. Its derivative coincides with that of $F(x)$ on both sides of ξ_1. Accordingly, we differentiate both sides of (1) and obtain

$$F'(x) = \overline{F}'(x) - a_1\delta(x - \xi_1) \tag{2}$$

or

$$\overline{F}'(x) = F'(x) + a_1\delta(x - \xi_1). \tag{3}$$

Equation (3) is easily generalized to a function $F(x)$ that has jumps of magnitude $a_1, a_2, \ldots, a_l$ at $\xi_1, \xi_2, \ldots, \xi_l$. The result is

$$\overline{F}'(x) = F'(x) + \sum_{j=1}^{l} a_j\delta(x - \xi_j). \tag{4}$$

Let us now consider a function $F(x)$ that admits derivatives up to the second order on both sides of the point ξ_1, that has a jump discontinuity of strength a_1, and whose derivative has a jump discontinuity of strength b_1 at this point. To obtain $\overline{F}''(x)$, we substitute $F'(x)$ for $F(x)$ in (2) and obtain

$$(F')' = (\overline{F}')' - b_1\delta(x-\xi_1) = \overline{F}'' - a_1\delta'(x-\xi_1) - b_1\delta(x-\xi_1),$$

or

$$\overline{F}'' = F'' + a_1\delta'(x-\xi_1) + b_1\delta(x-\xi_1). \tag{5}$$

This process can be continued for higher derivatives and for singularities at several points. Thus, a function $F(x)$ that admits continuous derivatives up to the mth order in each of the intervals (ξ_{j-1}, ξ_j), $j = 1, 2, \ldots, l$, has mth order distributional derivative $\overline{F}^{(m)}$,

$$\begin{aligned}\overline{F}^{(m)}(x) =& F^m(x) + \sum_{j=1}^{l}[a_j\delta^{(m-1)}(x-\xi_j) \\ &+ b_j\delta^{(m-2)}(x-\xi_j) + \cdots + f_j\delta(x-\xi_j)],\end{aligned} \tag{6}$$

where

$$a_j = [F(x)]_{\xi_j}, \qquad b_j = [F'(x)]_{\xi_j}, \qquad \ldots, \qquad f_j = [F^{(m-1)}(x)]_{\xi_j},$$

and $[\ ]_{\xi_j}$ stands for the jump in the quantity across the point ξ_j.

Example 1. The generalized solution of the initial value problem

$$Ly = \left[A_0(x)\frac{d^n}{dx^n} + A_1(x)\frac{d^{n-1}}{dx^{n-1}} + \cdots + A_n\right]y = f(x), \qquad x \geq 0,$$

$$y(0) = y_0, \qquad y'(0) = y_1, \qquad \ldots, \qquad y^{(n-1)}(0) = y_{n-1},$$

is equal to the solution of the inhomogeneous equation

$$\begin{aligned}&\left(A_0(x)\frac{\overline{d}^n}{dx^n} + A_1(x)\frac{\overline{d}^{n-1}}{dx^{n-1}} + \cdots + A_n\right)y \\ &\quad = f(x) + \sum_{k=1}^{n} A_{n-k}(x)\sum_{p=0}^{k-1} y_p\delta^{(k-1-p)}(x).\end{aligned}$$

Since we are interested in the value of $y(x)$ only for positive values of x, we may assume that $y(x)$ is identically zero for negative values of x. The result then follows by using (6) for $m = n, n-1, \ldots, 1$, so that

$$[y^{(k)}(x)]_{x=0} = y_k, \qquad k = 1, 2, \ldots, n-1,$$

and

$$\overline{y}^{(m)}(x) = y^{(m)}(x) + y_0\delta^{(m-1)}(x) + y_1\delta^{(m-2)}(x) + \cdots + y_{m-1}\delta(x).$$

Example 2. Consider the function

$$f(x) = \frac{1}{2} - \frac{x}{2\pi}, \qquad x \in [0, 2\pi], \tag{7}$$

which is a 2π-periodic function; i.e., $f(x + 2\pi) = f(x)$ (see Figure 5.1). It has jump discontinuities at $x = \pm 2m\pi$. The magnitude of the jump at each discontinuity is 1, so the value of the distributional derivative $\overline{f}'(x)$ follows from (4);

$$\overline{f}'(x) = -\frac{1}{2\pi} + \sum_{m=-\infty}^{\infty} \delta(x - 2m\pi). \tag{8}$$

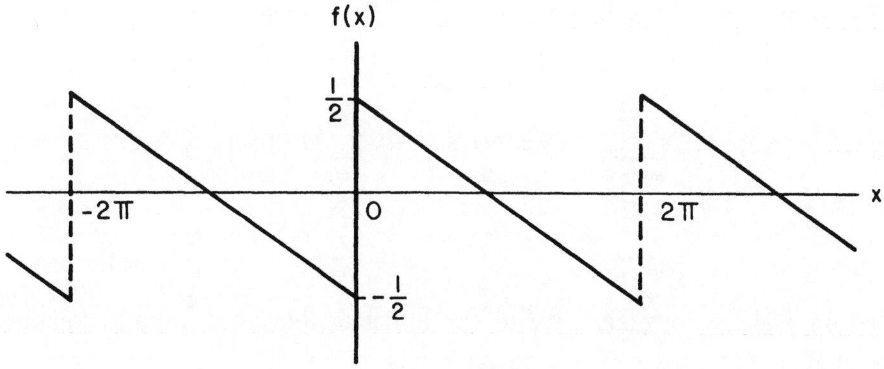

Figure 5.1. $f(x)$ is 2π-periodic.

The Fourier series of the function $f(x)$ is

$$f(x) = -\frac{i}{2\pi} \sum_{m=-\infty}^{\infty} \frac{1}{m} e^{imx}, \qquad m \neq 0, \tag{9}$$

which, in view of the analysis of Section 3.4, can be differentiated term by term in D' any number of times. Thus

$$f'(x) = \frac{1}{2\pi} \sum_{m=-\infty}^{\infty} e^{imx} - \frac{1}{2\pi}. \tag{10}$$

Comparing (8) and (10), we obtain

$$\sum_{m=-\infty}^{\infty} \delta(x - 2m\pi) = \frac{1}{2\pi} \sum_{m=-\infty}^{\infty} e^{imx} = \frac{1}{2\pi} + \frac{1}{\pi} \sum_{m=1}^{\infty} \cos mx,$$

which agrees with (3.4.6). With $x = 2\pi y$ this result becomes

$$\sum_{m=-\infty}^{\infty} \delta\{2\pi(y - m)\} = \frac{1}{2\pi} \sum_{m=-\infty}^{\infty} e^{i2\pi my} = \frac{1}{2\pi} + \frac{1}{\pi} \sum_{m=1}^{\infty} \cos 2\pi my. \tag{11}$$

Using the relation $\delta\{2\pi(y-m)\} = (1/2\pi)\delta(y-m)$, in (11) and relabeling y and x, we obtain

$$\sum_{m=-\infty}^{\infty} \delta(x-m) = \sum_{m=-\infty}^{\infty} e^{i2\pi mx} = 1 + 2\sum_{m=1}^{\infty} \cos 2\pi mx\,. \tag{12}$$

We already encountered this row of deltas in Section 2.4. Relation (12) is a distributional formula in the sense that

$$\sum_{m=-\infty}^{\infty} \langle \delta(x-m), \phi \rangle = \sum_{m=-\infty}^{\infty} \langle e^{i2\pi mx}, \phi \rangle\,, \qquad \phi \in D\,.$$

Since every test function vanishes outside a finite interval, the summation on the left side of this relation is finite.

By successive differentiations of (12), we derive many more interesting Fourier series. For instance,

$$\sum_{m=-\infty}^{\infty} \delta'(x-m) = -4\pi \sum_{m=1}^{\infty} m \sin 2\pi mx\,,$$

$$\sum_{m=-\infty}^{\infty} \delta''(x-m) = -8\pi^2 \sum_{m=1}^{\infty} m^2 \cos 2\pi mx\,,$$

and so on. These formulas give us the effect of the poles, dipoles, quadrupoles, etc. distributed at the points $x = \pm m$ on the x axis.

Note that we could have started with the function $g(x)$ whose kth derivative is the function $f(x)$ as defined in (7). For instance, the function

$$g(x) = -\frac{1}{2\pi} \sum_{m=-\infty}^{\infty} \frac{1}{m^2} e^{imx}\,, \tag{13}$$

is such that $g'(x) = f(x)$. The series given by (13) is the Fourier series for a periodic function consisting of parabolic arcs. Comparing (10) and (13), we find that

$$\sum_{m=-\infty}^{\infty} e^{imx} = 1 + 2\pi \frac{d^2 g}{dx^2}\,. \tag{14}$$

Example 3. Let us prove the *Leibnitz formula* with the help of the analysis of this section, i.e.,

$$\begin{aligned}\frac{d}{dt}\int_{a(t)}^{b(t)} f(x,t)dx = \int_{a(t)}^{b(t)} \frac{\partial f(x,t)}{\partial t}dx + f(b(t),t)b'(t) \\ - f(a(t),t)a'(t)\,,\end{aligned} \tag{15}$$

for the differentiation of the integrals with varying limits of integration.

Let us assume that $f(x,t)$ is continuously differentiable in the closed region $a(t) \leq x \leq b(t)$ and then extend this function to all (x,t) by setting $f(x,t) = 0$, whenever

$x < a(t)$ and $x > b(t)$. This produces jump discontinuities in the function $f(x, t)$ at $x = a(t)$ and $x = b(t)$ and we can write

$$f(x,t) = f(x,t)H(x - a(t))H(b(t) - x). \tag{16}$$

Thus,

$$F(t) = \int_{-\infty}^{\infty} f(x,t)dx = \langle f(x,t), 1\rangle_x, \tag{17}$$

where 1 denotes the function of x identically equal to one and $\langle\ ,\ \rangle_x$ means that we are evaluating the integral with respect to x. Thus,

$$F'(t) = \left\langle \frac{\overline{\partial}}{\partial t} f(x,t), 1 \right\rangle_x. \tag{18}$$

But from (16) we have

$$\begin{aligned}\frac{\overline{\partial}}{\partial t} f(x,t) &= \left(\frac{\partial}{\partial t} f(x,t)\right) H(x - a(t))H(b(t) - x) \\ &\quad - f(a(t),t)a'(t)\delta(x - a(t)) + f(b(t),t)b'(t)\delta(x - b(t)),\end{aligned} \tag{19}$$

where we have used the fact that $g(t)\delta(t - \xi) = g(\xi)\delta(t - \xi)$. Finally, we substitute (19) in (18), use the sifting property of the delta function and get the required formula (15).

Example 4. The analysis of this section can be used to study a continuous piecewise linear function $F(x)$. Then $F'(x)$ is a piecewise constant function with jumps at the end points of its pieces. Let $\xi_1 < \xi_2 < \cdots < \xi_k$ and let $f'(x)$ vanish outside the interval $[\xi_1, \xi_k]$. Suppose that b_j is the jump in the slopes of $f(x)$ at ξ_j. Then from relation (5) we obtain

$$\overline{F}''(x) = \sum_{j=1}^{k} b_j \delta(x - \xi_j). \tag{20}$$

5.2. Moving Surfaces of Discontinuity in $\mathbf{R_n}$, $\mathbf{n \geq 2}$

The previous analysis can be extended to higher-dimensional spaces. Our aim is to develop systematically the distributional partial derivative of a function $F(x)$, $x = (x_1, \ldots, x_n)$, in terms of the jumps $[F]$, $[\text{grad } F]$, and $[\text{grad grad } F]$ as well as the surface distributions. Since these partial derivatives play an important part in the theory of wave propagation, we shall include the time variable t in our derivation. Accordingly, in this section we present the kinematics of a moving surface.

The analysis is presented for the general case of $n > 2$. The results for $n = 2$ are presented in Section 5.8.

Let $\Sigma(t)$ be a moving surface in the p-dimensional space R_p. Such a surface can be represented locally either in an implicit equation of the form

$$u(x_1, \ldots, x_p, t) = 0, \tag{1}$$

or in terms of the curvilinear Gaussian coordinates $v_1, \ldots, v_{p-1}$ on the surface:

$$x_i = x_i(v_1, \ldots, v_{p-1}, t). \tag{2}$$

We shall assume that the surface $\Sigma(t)$ is regular so that all functions in (1) and (2) have derivatives of all orders with respect to each of their arguments and that for all fixed values of t the corresponding Jacobian matrices of transformation have appropriate ranks; that is,

$$\nabla u \neq 0, \tag{3}$$

where ∇ stands for the gradient, the vector of components $\partial/\partial x_i$, $i = 1, \ldots, p$, and

$$\text{rank}\,(\partial x_i/\partial v_j) = p - 1. \tag{4}$$

Furthermore, this wave front divides the space in two parts, which we shall call positive and negative.

It should be remarked that $\Sigma(t)$ could be considered as a submanifold of the $(p+1)$-dimensional spacetime. Although we shall sometimes use this interpretation, it is in general more convenient to distinguish the time t from the space coordinates.

Our regularity assumptions imply that at every point of the surface there exists a well-defined tangent plane and a well-defined normal line. We shall denote by $\widehat{n}(n_i;\ i = 1, \ldots, p)$ the unit normal vector pointing in the positive direction such that

$$\partial u/\partial x_i = |\nabla u| n_i. \tag{5}$$

Let f be a funciton of t defined on the surface $\Sigma(t)$ in some interval, and let $\delta f/\delta t$ denote the derivative with respect to time as it would be computed by an observer moving with the surface. This δ derivative has the following geometrical interpretation. Let P_0 be a point on the surface at time $t = t_0$. Construct the normal line to the surface at P_0. At time $t = t_0 + \Delta t$, Δt an infinitesimal, this normal meets the surface $\Sigma(t + \Delta t)$ at the point $P_1 = P_1\ (t_0 + \Delta t)$. Then the δ derivative is defined as

$$\frac{\delta f(P_0, t)}{\delta t} = \lim_{\Delta t \to 0} \frac{f(P_1) - f(P_0)}{\Delta t}. \tag{6}$$

This means that if G denotes the speed of the wave front $\Sigma(t)$, i.e.,

$$G = \lim_{\Delta t \to 0} \Delta s/\Delta t, \tag{7}$$

where Δs is the distance between P_0 and P_1, then

$$\frac{\delta x_i}{\delta t} = \lim_{\Delta t \to 0} \frac{\Delta x_i}{\Delta t} = \lim_{\Delta t \to 0} \frac{\Delta s}{\Delta t}\frac{\Delta x_i}{\Delta s} = G n_i, \tag{8}$$

and

$$0 = \frac{\delta u}{\delta t} = \frac{\delta u}{\delta x_i}\frac{\delta x_i}{\delta t} + \frac{\partial u}{\partial t} = G|\nabla u| n_i n_i + \frac{\partial u}{\partial t} = G|\nabla u| + \frac{\partial u}{\partial t}, \tag{9}$$

or

$$\frac{\partial u}{\partial t} = -G|\nabla u|, \tag{10}$$

where we have used the summation convention and shall use it in the sequel. Sometimes we shall interpret (10) by saying that $-G$ is the component of the normal vector in the time direction. Let us note that the essential feature of the δ derivative is that it is computed on the surface and relation (9) implies that u remains constant on the surface.

Let $u_1, \ldots, u_p$ be a local system of coordinates with $u_1 = u$ such that

$$\frac{\partial u_1}{\partial x_i}\frac{\partial u_j}{\partial x_i} = 0, \qquad j - 2, \ldots, p. \tag{11}$$

Because

$$\left(\frac{\partial u_1}{\partial x_i} - |\nabla u_1|^2 \frac{\partial u_i}{\partial x_1}\right)\frac{\partial u_j}{\partial x_i} = 0, \tag{12}$$

for all j and the vectors $\nabla u_1, \ldots, \nabla u_p$ form a basis for R_p, it follows that

$$\frac{\partial u_1}{\partial x_i} = |\nabla u_1|^2 \frac{\partial x_i}{\partial u_1}. \tag{13}$$

Accordingly, if $F(x, t)$, $x = (x_1, \ldots, x_p)$, is a function defined in a neighborhood of $\Sigma(t)$, we have

$$\frac{\partial F}{\partial x_i} n_i = \frac{\partial F}{\partial x_i}\frac{1}{|\nabla u|}\frac{\partial u}{\partial x_i} = \frac{\partial F}{\partial x_i}\frac{\partial x_i}{\partial u}|\nabla u_1| \quad \text{or} \quad \frac{dF}{dn} = \frac{\partial F}{\partial u}|\nabla u_1|. \tag{14}$$

Suppose that a function f is defined only on $\Sigma(t)$ and that it is equal to the restriction of F to the surface. Then we have

$$\frac{\partial F}{\partial t} = \frac{\delta f}{\delta t} + \frac{\partial F}{\partial u}\frac{\partial u}{\partial t} = \frac{\delta f}{\delta t} - G\frac{dF}{dn}, \tag{15}$$

or

$$\frac{\delta f}{\delta t} = \frac{\partial F}{\partial t} + G\frac{dF}{dn}. \tag{16}$$

Since (16) holds irrespective of the way F extends f, we use it as our formal definition of δ time derivative. Observe also that this definition does not depend on the local system we take.

It is also convenient to introduce δ derivatives with respect to the space variables. They are defined by

$$\frac{\delta f}{\delta x_i} = \frac{\partial F}{\partial x_i} - n_i\frac{dF}{dn}, \tag{17}$$

which is analogous to (16) and implies that $\delta f/\delta x_i$ is tangent to $\Sigma(t)$. Indeed,

$$\frac{\delta f}{\delta x_i} = g^{\alpha\beta}\frac{\partial x_i}{\partial v_\alpha}\frac{\partial f}{\partial v_\beta}, \tag{18}$$

where

$$g_{\alpha\beta} = \frac{\partial x_i}{\partial v_\alpha}\frac{\partial x_i}{\partial v_\beta} = x^i_\alpha x^i_\beta \,, \tag{19}$$

are the components of the first fundamental form of the surface and where the Greek indices assume the values from 1 to $p-1$. Observe that $g^{\alpha\beta}$ is the inverse of the matrix $g_{\alpha\beta}$ so that $g_{\alpha\beta}g^{\beta\gamma} = \delta^\gamma_\alpha$. Also the element $d\Sigma$ on the surface Σ satisfies the relation $d\Sigma^2 = g_{\alpha\beta}dv^\alpha dv^\beta$.

The quantities

$$\mu_{ij} = \delta n_i/\delta x_j \,, \tag{20}$$

will play an important role and are called components of the second fundamental form of the surface. Observe that μ_{ij} is a symmetric surface tensor; that is,

$$\mu_{ij} = \mu_{ji} \,, \qquad \mu_{ij}n_j = 0 \,.$$

The symmetry is preserved even with respect to time, since

$$\mu_{it} = \delta n_i/\delta t = \delta(-G)/\delta x_i = \mu_{ti} \,. \tag{21}$$

We shall denote by $\mu^{(r)}_{ij}$ the entries of the rth power of the matrix μ, and set

$$\mu^{(0)}_{ij} = \delta_{ij} - n_i n_j \,, \tag{22}$$

so that

$$\mu^{(1)}_{ij} = \mu_{ij} = \frac{\delta n_i}{\delta x_j} \,, \qquad \mu^{(2)}_{ij} = \mu_{ik}\mu_{kj} \,, \qquad \mu^{(3)}_{ij} = \mu^{(2)}_{ik}\mu_{kj} \,.$$

In general

$$\mu^{(n)}_{ij} = \mu^{(n-m)}_{ik}\mu^{(m)}_{kj} \,. \tag{23}$$

The trace of the matrix $\mu^{(r)}_{ij}$ is denoted ω_r:

$$\omega_r = \mu^{(r)}_{ii} \,, \tag{24}$$

and we shall usually write -2Ω instead of ω_1, because in the three-dimensional case $-\omega_1/2$ is equal to the mean curvature Ω of Σ.

The differential operator $\delta/\delta x_i$, defined initially for smooth functions f, can be extended to the distributions defined on Σ as follows: Let $D(\Sigma)$ denote the space of test functions defined on Σ. Let $D'(\Sigma)$ be its dual space. If f is a smooth function defined on Σ, we define the adjoint operator $\delta^*/\delta x_i$ as

$$\frac{\delta^* f}{\delta x_i} = \frac{\delta f}{\delta x_i} - n_i\omega_1 f \,. \tag{25}$$

If both f and g are smooth and at least one of them belongs to $D(\Sigma)$, then

$$\left\langle \frac{\delta f}{\delta x_i}, g \right\rangle = -\left\langle f, \frac{\delta g}{\delta x_i} \right\rangle \,. \tag{26}$$

This helps us in defining the operator $\delta f/\delta x_i$ as

$$\left\langle \frac{\delta f}{\delta x_i}, \phi \right\rangle = -\left\langle f, \frac{\delta^* \phi}{\delta x_i} \right\rangle, \qquad \phi \in D(\Sigma). \tag{27}$$

Observe that we can also define $\delta^* f/\delta x_i$ for $f \in D(\Sigma)$, either by duality or by using (25). It is interesting to observe that

$$\frac{\delta^* f}{\delta x_i} = \frac{\delta}{\delta x_j}(\mu_{ij}^0 f). \tag{28}$$

The differential operators $\delta/\delta x_i$ and $\delta/\delta x_j$ do not commute in general. A direct computation using (16) and (17) gives

$$\frac{\delta^2 f}{\delta x_i \delta t} = \frac{\delta^2 f}{\delta t \delta x_i} + \mu_{ik} G \frac{\delta f}{\delta x_k} + \mu_{tk} n_i \frac{\delta f}{\delta x_n}, \tag{29}$$

$$\frac{\delta^2 f}{\delta x_i \delta x_j} = \frac{\delta^2 f}{\delta_j \delta x_i} - \mu_{ik} n_j \frac{\delta f}{\delta x_k} + \mu_{jk} n_i \frac{\delta f}{\delta x_k}. \tag{30}$$

This necessitates the definition of the symmetric second order differential operator D^2 as

$$D_{ij}^2 = \frac{\delta^2}{\delta x_i \delta x_j} - n_i \mu_{jk} \frac{\delta}{\delta x_k}. \tag{31}$$

In passing we observe that if we multiply equations (29) and (30) by n_i and sum on i we obtain

$$\frac{\delta^2 f}{\delta t \delta x_i} n_i = -\mu_{tk} \frac{\delta f}{\delta x_k}, \tag{32}$$

$$\frac{\delta^2 f}{\delta x_i \delta x_j} n_i = -\mu_{jk} \frac{\delta f}{\delta x_k}. \tag{33}$$

We shall extend these results in Chapter 12 to study various concepts in wave propagation.

5.3. Surface Distributions

In this section we study certain distributions defined on the wave front $\Sigma(t)$ and their extensions to the whole space. The basic distribution concentrated on $\Sigma(t)$ is the Dirac delta function, whose action on a test function $\phi(x, t)$ is given by

$$\langle \delta(\Sigma), \phi \rangle = \int_{-\infty}^{\infty} \int_{\Sigma(t)} \phi(x, t) dS(x) dt, \tag{1}$$

where $dS(x)$ is the surface element. Observe the special treatment of time in (1). The integration with respect to the space variables is surface integration while that with respect to time is ordinary integration. If t is treated as an ordinary variable, we shall obtain a different delta function $\widehat{\delta}(\Sigma)$, which is related to $\delta(\Sigma)$ by

$$\delta(\Sigma) = (1 + G^2)^{-1/2}\widehat{\delta}(\Sigma), \tag{2}$$

because going from δ to $\widehat{\delta}$ amounts to considering the unit normal on R_{p+1}, which is obtained from $n_1, \ldots, n_p, -G$ upon multiplication by $(1 + G^2)^{-1/2}$.

Given any distribution on Σ, there are several ways to extend its action to the whole space. For our purposes this extension is achieved as follows: If A is a distribution on Σ and $\phi(x, t)$ is a test function on R_{p+1}, then A can act on ϕ by the transposition,

$$\langle \widehat{A}, \phi \rangle = \langle A, \phi/\Sigma \rangle . \tag{3}$$

This extension $\widehat{A}$ is usually denoted $A\delta(\Sigma)$, since

$$\langle \widehat{A}, \phi \rangle = \int_{\Sigma(t)} A(x, t)\phi(x, t)dS(x, t), \tag{4}$$

whenever A is a locally integrable function.

In the present study we shall obtain the extension of a distribution A by multiplying it with $\delta(\Sigma)$ so that, using (2), we have

$$A\delta(\Sigma) = (1 + G^2)^{-1/2}\widehat{A}. \tag{5}$$

Observe that $\delta(\Sigma)$ and $\widehat{\delta}(\Sigma)$ have two different meanings, either that of distributions on R_{p+1} or that of extension operators, and we shall distinguish them clearly as we go along.

Another basic distribution concentrated on Σ is the normal derivative of the delta function:

$$\delta'(\Sigma) = \frac{\overline{\partial}}{\partial x_i} [\delta(\Sigma)] n_i , \tag{6}$$

where the bar labels the distributional derivative as mentioned previously. Note that $\delta'(\Sigma)$ is not a normal derivative operator. Indeed, the normal derivative operator will be denoted $d_n\delta(\Sigma)$ and its action given by

$$\langle d_n\delta(\Sigma), \phi \rangle = \int_{-\infty}^{\infty} \int_{\Sigma(t)} -\frac{d\phi}{dn}\, dS\, dt . \tag{7}$$

From their definitions we find that

$$\begin{aligned}
\langle \delta'(\Sigma), \phi \rangle &= \left\langle \frac{\overline{\partial}\delta(\Sigma)}{\partial x_i} n_i, \phi \right\rangle = \left\langle \frac{\overline{\partial}\delta(\Sigma)}{\partial x_i} n_i, \phi \right\rangle = -\left\langle \delta(\Sigma), \frac{\partial}{\partial x_i}(n_i\phi) \right\rangle \\
&\quad - \left\langle \delta(\Sigma), \frac{\partial \phi}{\partial x_i} n_i + \phi \frac{\delta n_i}{\delta x_i} \right\rangle = -\left\langle \delta(\Sigma), \frac{d\phi}{dn} - 2\Omega\phi \right\rangle \\
&= \langle d_n\delta(\Sigma) + 2\Omega\delta(\Sigma), \phi \rangle ,
\end{aligned}$$

where Ω is the mean curvature of the surface $\Sigma(t)$ and is given by

$$-2\Omega = \delta n_i/\delta x_i , \tag{8}$$

as mentioned earlier. Hence

$$\delta'(\Sigma) = 2\Omega\delta(\Sigma) + d_n\delta(\Sigma)\,. \tag{9}$$

More generally, let $Q(x_1, x_2, \ldots, x_p)$ be any distribution concentrated on the surface $\Sigma(t)$. We define its normal derivative as

$$Q' = (\bar{\partial}\, Q/\partial x_i)n_i\,, \tag{10}$$

and by iteration, the rth-order derivative as

$$Q^{(r)} = (Q^{(r-1)})'\,. \tag{11}$$

Then the following result holds:

Lemma 1. *If Q is any distribution concentrated on Σ, then*

$$Q^{(r)} = \frac{\bar{\partial}^r Q}{\partial x_{i_1} \cdots \partial x_{i_r}} n_{i_1} \cdots n_{i_r}\,. \tag{12}$$

Proof. The result is true for $r = 1$ because of (10). Let us suppose that it holds for r, so that

$$\begin{aligned}
Q^{(r+1)} = (Q^{(r)})' &= \left(\frac{\bar{\partial}^r Q}{\partial x_{i_1} \cdots n_{i_r}} n_{i_1} \cdots n_{i_r}\right)' \\
&= \frac{\bar{\partial}}{\partial x_k}\left(\frac{\bar{\partial}^r Q}{\partial x_{i_1} \cdots n_{i_r}} n_{i_1} \cdots n_{i_r}\right) n_k \\
&= \frac{\bar{\partial}^{r+1} Q}{\partial x_k \partial x_{i_1} \cdots n_{i_r}} n_{i_1} \cdots n_{i_r} + \frac{\bar{\partial}^r Q}{\partial x_{i_1} \cdots n_{i_r}} \frac{\delta(n_{i_1} \cdots n_{i_r})}{\delta x_k} n_k \\
&= \frac{\bar{\partial}^{r+1} Q}{\partial x_k \partial x_{i_1} \cdots \partial x_{i_r}} n_i \cdots n_{i_r} n_k\,,
\end{aligned}$$

because $(\delta/\delta x_k)n_k$ is zero for any distribution on $\Sigma(t)$. Letting $k = r + 1$, the lemma follows.

Let us now assume that

$$\bar{\partial} Q/\partial x_i = Q' n_i\,, \tag{13}$$

then

$$\frac{\bar{\partial}^2 Q}{\partial x_i \partial x_j} = \frac{\bar{\partial} Q'}{\partial x_j} n_i + Q' \frac{\delta n_i}{\delta x_j}\,. \tag{14}$$

By symmetry, then, it follows that

$$\frac{\bar{\partial} Q'}{\partial x_i} n_j = \frac{\bar{\partial} Q'}{\partial x_j} n_i\,. \tag{15}$$

Multiplying both sides of (15) by n_j and summing on j yields

$$\overline{\partial} Q'/\partial x_i = Q'' n_i . \tag{16}$$

We thus obtain, by induction, the following lemma:

Lemma 2. *If Q is a distribution concentrated on Σ that satisfies the relation $\overline{\partial} Q/\partial x_i = Q' n_i$, then for any $r \geq 0$*

$$\overline{\partial} Q^{(r)}/\partial x_i = Q^{(r+1)} n_i . \tag{17}$$

5.4. Various Other Representations

A few singular distributions closely related to $\delta(\Sigma)$ have been considered by Gelfand and Shilov [7], Jones [9], and DeJager [10]. Let us compare them and connect them with the results of Section 5.3.

Let us suppose for the moment that time is absent in our analysis. Let $P(x_1, \ldots, x_p)$ be a smooth function such that the surface $P = 0$, coincides with our regular surface Σ and such that $\nabla P \neq 0$ on Σ. Then, according to Gelfand and Shilov [7], the distribution $\delta(P)$ is defined as follows: Let $u_1, \ldots, u_p$ be a coordinate system with $u_1 = P$, and let $\phi(x)$ be a test function. Then, performing a formal change of variables, we have

$$\begin{aligned}\langle \delta(P), \phi \rangle &= \int \delta(P)\phi(x)dx = \int \delta(u_1)\psi(u_1, \ldots, u_p)J\, du_1 \cdots du_p \\ &= \int_{u_1=0} \psi(0, u_2, \ldots, u_p)J(0, u_2, \ldots, u_p)\, du_2 \cdots du_p ,\end{aligned} \tag{1}$$

where $\psi(u_1, \ldots, u_p) = \phi(x_1, \ldots, x_p)$ and J is the Jacobian of the transformation. Gelfand and Shilov prove that the value of the integral in (1) is independent of the coordinate system and accordingly it defines a distribution concentrated on Σ.

The relation between $\delta(P)$ defined this way and $\delta(\Sigma)$ of Section 5.3 is

$$\langle \delta(P), \phi \rangle = \int_\Sigma \phi(y)dS(y)\frac{1}{|\nabla P|} , \tag{2}$$

where dS is surface measure on Σ. Thus

$$\delta(P) = \delta(\Sigma)/|\nabla P| . \tag{3}$$

With the help of (3) and the relation $\partial P/\partial x_i = \pm|\nabla P| n_i$, we can write many of the formulas with $\delta(P)$ instead of $\delta(\Sigma)$. For example,

$$n_i \delta(\Sigma) = (\partial P/\partial x_i)\delta(P) .$$

We shall use this representation occasionally in subsequent chapters.

Note that while the distribution $\delta(\Sigma)$ depends only on the surafce Σ, the distribution $\delta(P)$ depends on P, that is, on the way the surface Σ is represented. Let $Q(x)$ be a nowhere vanishing smooth function, then $Q(x)P(x) = 0$ also represents Σ, while, according to (3), we have

$$\delta(QP) = (1/Q)\delta(P)\,. \tag{4}$$

Another way of introducing the distribution $\delta(P)$ is used by DeJager [10];

$$\langle \delta(P), \phi \rangle = \lim_{c \to 0} \frac{1}{c} \int_{0 \le P \le c} \phi(x)\, dx\,. \tag{5}$$

Similarly, its higher derivatives can be defined as

$$\delta_k(P) = \lim_{c \to 0} \frac{1}{c}[\delta_{k-1}(P+c) - \delta_{k-1}(P)]\,, \qquad k = 1, 2, \ldots, \tag{6}$$

which are denoted in the literature as $\delta^{(k)}(P)$. However, they are not the normal derivatives as defined in Section 5.3. Indeed,

$$\frac{\bar{\partial}}{\partial x_i}\delta_k(P) = \delta_{k+1}(P)\frac{\partial P}{\partial x_i}\,, \tag{7}$$

and it follows from (1) and (5.5.8) (in the next section) that

$$\frac{\bar{\partial}}{\partial x_i}\delta^{(k)}(\Sigma) = \delta^{(k+1)}(\Sigma)n_i\,. \tag{8}$$

A few other related concepts are also discussed in the literature. For instance, the distributions concentrated on manifolds of smaller dimensions, such as those on lines and curves, can be represented in terms of the delta function of several arguments. Similarly, the distributions on intersecting surfaces can be considered [9].

5.5. First-Order Distributional Derivatives

The main objective of this work is to study the behavior of functions that are singular on the surface $\Sigma(t)$. The word "singular," of course, has many meanings, and hence our first aim is to clarify its definition.

Let $u_1, \ldots, u_p$ be a local system of coordinates with $u_1 = u$ as described in Section 5.2. For a small $\varepsilon > 0$, the equation

$$u(x_1, \ldots, x_p, t) = \varepsilon\,, \tag{1}$$

represents a moving surface, which we shall designate Σ_ε and which can be described by the same curvilinear coordinates $u_1, \ldots, u_p$ as on $\Sigma(t)$. Given a function F defined on one side of the surface, say, in $u > 0$, we can form, by the relation

$$T_\varepsilon = F\big|_{\Sigma_\varepsilon}\,, \tag{2}$$

a sequence of regular distributions on $\Sigma(t)$ that act on test functions whose support is contained in the open set where the local system is defined. If this sequence converges to a distribution A, we say that A is the boundary value of F in the open set. The distribution A is the boundary value of F if it is in every open set of the above form.

Definition. A function F defined on R_{p+1} will be called a *regular singular function* with respect to $\Sigma(t)$ if

(1) F has derivatives of all orders outside $\Sigma(t)$ and
(2) F and all its derivatives have boundary values from both sides of $\Sigma(t)$.

The set of regular singular functions is a vector space closed under multiplication by C^∞ functions as well as under ordinary (classical) differentiation. We shall denote this vector space by $E(\Sigma)$ or merely E when the surface is clear from the context. Given $F \in E$, its jump is the distribution

$$[F] = F_+ - F_-, \tag{3}$$

where F_+ and F_- are the boundary values of F from the positive and the negative sides, respectively, of Σ (see Figure 5.2).

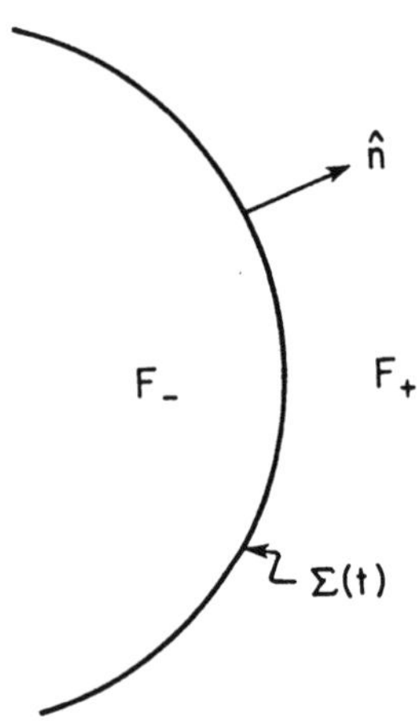

Figure 5.2. Wave front $\Sigma(x, t)$.

Every $F \in E$ is a regular distribution on R_p, and hence we can apply both the ordinary and generalized derivatives to it. As before we shall indicate generalized derivatives with a bar. With this notation we have

Theorem 1. *If* $F \in E$,

$$\frac{\overline{\partial} F}{\partial x_i} = \frac{\partial F}{\partial x_i} + [F] n_i \delta(\Sigma), \tag{4}$$

$$\frac{\overline{\partial} F}{\partial t} = \frac{\partial F}{\partial t} - [F] G \delta(\Sigma). \tag{5}$$

Proof. It is sufficient to prove these results locally. If U is an open set that does not meet $\Sigma(t)$, the results are true because $\delta(\Sigma) = 0$ there. Near the surface we take a local coordinate system $u_1, \ldots, u_p$ defined on an open set U and observe that for a test function $\phi(x, t)$, with support contained in U, we have

$$\begin{aligned}
\left\langle \frac{\overline{\partial} F}{\partial x_i}, \phi \right\rangle &= -\left\langle F, \frac{\partial \phi}{\partial x_i} \right\rangle = -\int_U F \frac{\partial \phi}{\partial x_i} dV = -\lim_{\varepsilon \to 0} \int_{|u_1| > \varepsilon} F \frac{\partial \phi}{\partial x_i} dV \\
&= \lim_{\varepsilon \to 0} \left[\int_{|u_1| > \varepsilon} \frac{\partial F}{\partial x_i} \phi \, dV + \int_{-\infty}^{\infty} \int_{\Sigma_{+\varepsilon(t)}} F \phi n_i dS \, dt \right. \\
&\qquad \left. - \int_{-\infty}^{\infty} \int_{\Sigma_{-\varepsilon(t)}} F \phi n_i dS \, dt \right] \\
&= \int_U \frac{\partial F}{\partial x_i} \phi \, dV + \int_{-\infty}^{\infty} \int_{\Sigma(t)} [F] \phi n_i dS \, dt \\
&= \left\langle \frac{\partial F}{\partial x_i} + [F] n_i \delta(\Sigma), \phi \right\rangle ,
\end{aligned}$$

which proves (4).

Relation (5) can be proved similarly. However, it is more interesting to realize that it is a particular case of the first result. Indeed, if we consider $\Sigma(t)$ as a submanifold of R_{p+1}, it follows from the foregoing result that

$$\frac{\overline{\partial} F}{\partial x_i} = \frac{\partial F}{\partial x_i} + [F] m_i \widehat{\delta}(\Sigma) , \tag{6}$$

where $m_1, \ldots, m_p, m_t$ are the components of the unit normal vector to $\Sigma(t)$ in R_{p+1}. Clearly,

$$m_i = (1 + G^2)^{-1/2} n_i , \quad \text{and} \quad m_t = -(1 + G^2)^{-1/2} G . \tag{7}$$

Thus, for $i = t$, (6) reduces to (5). It is also true for the most of the results that follow so that no separate consideration of t is required.

From Theorem 1 we can draw some very basic and interesting conclusions.

Theorem 2.

$$\frac{\overline{\partial}}{\partial x_i} \delta(\Sigma) = \delta'(\Sigma) n_i . \tag{8}$$

$$\frac{\overline{\partial}}{\partial t} \delta(\Sigma) = -G \delta'(\Sigma) . \tag{9}$$

Proof. Let F, the function defined in Theorem 1, be such that it is 1 on the positive side of $\Sigma(t)$f and 0 on the negative side. Then Theorem 1 yields

$$\frac{\overline{\partial} F}{\partial x_i} = n_i \delta(\Sigma) . \tag{10}$$

When we take derivatives of both sides of this relation with respect to x_j and recall that generalized derivatives commute, we obtain

$$\frac{\bar{\partial}^2 F}{\partial x_i \partial x_j} = \frac{\bar{\partial}}{\partial x_j}[n_i \delta(\Sigma)] = \frac{\delta n_i}{\delta x_j}\delta(\Sigma) + n_i \frac{\bar{\partial}}{\partial x_j}\delta(\Sigma)$$

$$= \frac{\delta n_j}{\delta x_i}\delta(\Sigma) + n_j \frac{\bar{\partial}}{\delta x_i}\delta(\Sigma) = \frac{\bar{\partial}^2 F}{\partial x_j \partial x_i}\,.$$

However, $\delta n_i/\delta x_j$ is symmetric, and it follows that

$$n_i \frac{\bar{\partial}}{\partial x_j}\delta(\Sigma) = n_j \frac{\bar{\partial}}{\partial x_i}\delta(\Sigma)\,. \tag{11}$$

Multiplying (11) by n_i and summing over i, we obtain (8). Relation (9) is proved in a similar fashion.

Another basic result that follows from Theorem 1 is that if $[F] = 0$, we should have

$$[\partial F/\partial x_i] = Bn_i\,, \quad \text{and} \quad [\partial F/\partial t] = -BG\,, \tag{12}$$

where

$$B = [dF/dn] = [\partial F/\partial x_j]n_j\,. \tag{13}$$

Indeed,

$$\frac{\bar{\partial}^2 F}{\partial x_i \partial x_j} = \frac{\partial^2 F}{\partial x_i \partial x_j} + \left[\frac{\partial F}{\partial x_j}\right] n_i \delta(\Sigma) = \frac{\partial^2 F}{\partial x_j \partial x_i} + \left[\frac{\partial F}{\partial x_i}\right] n_j \delta(\Sigma) = \frac{\bar{\partial}^2 F}{\partial x_j \partial x_i}\,,$$

because the second-order generalized derivatives commute. Hence,

$$[\partial F/\partial x_j]n_i = [\partial F/\partial x_i]n_j\,,$$

and (13) follows by multiplying both sides of the foregoing relation by n_i and summing over i.

When $[F]$ does not vanish, the jump in the gradient of F contains not only a normal component but a tangential component as well. Consequently, we can write it in two parts,

$$F(u_1, \ldots, u_p, t) = \mathcal{F}(u_1, \ldots, u_p, t) + H(u_1)A(u_2, \ldots, u_p, t)\,, \tag{14}$$

where $A = [F]$ and H is the Heaviside function. Observe that the partial derivative with respect to u_1 is a derivative in the normal direction and therefore

$$B = [dF/dn] = [d\mathcal{F}/dn]\,. \tag{15}$$

Also

$$\frac{\partial F}{\partial u_i} = \frac{\partial \mathcal{F}}{\partial u_i} + H(u_1)\frac{\partial A}{\partial u_i}\,, \tag{16}$$

and hence by the chain rule we have, in Cartesian coordinates,

$$[\partial F/\partial x_i] = [\partial \mathcal{F}/\partial x_i] + \partial A/\partial x_i\,. \tag{17}$$

Since $[\mathcal{F}] = 0$, we obtain

$$[\partial F/\partial x_i] = [d\mathcal{F}/dn]n_i = Bn_i \,, \tag{18}$$

and therefore

$$[\partial F/\partial x_i] = Bn_i + \delta A/\delta x_i \,. \tag{19}$$

There is an alternative way in which (19) can be deduced. Indeed, by differentiating $A = [F]$ along the surface, we have

$$A_{.\alpha} = [F_{.\alpha}] = \left[\frac{\partial F}{\partial x_i}\right] x^i_{\alpha} \,, \tag{20}$$

where $A_{.\alpha} = \partial A/\partial v^{\alpha}$. When we multiply both sides of (20) by $g^{\alpha\beta} x^j_{\beta}$, sum on the repeated indices, and use the relation

$$g^{\alpha\beta} x^i_{\alpha} x^j_{\beta} = \delta_{ij} - n^i n^j \,, \tag{21}$$

where the δ^{ij} are the components of the Kronecker delta function, we obtain the interfacial relation

$$A_{.\alpha} g^{\alpha\beta} x^j_{\beta} = [\partial F/\partial x_i](\delta^{ij} - n^i n^j) \,, \qquad \text{or} \qquad [\partial F/\partial x^i] = B_{n_i} + g^{\alpha\beta} A_{.\alpha} \beta^i_{\beta} \,, \tag{22}$$

which is the same as (19).

Thereby we have proved the following theorem:

Theorem 3. *If $F \in E$, then*

$$[\partial F/\partial x_i] = Bn_i + \delta A/\delta x_i \,, \tag{23}$$

$$[\partial F/\partial t] = BG + \delta A/\delta t \,, \tag{24}$$

where $A = [F]$ and $B = [dF/dn]$.

Many other interesting results can be deduced from the relations of this section. For instance, from (4) we find that for a vector F we have

$$\overline{\operatorname{div}} F = \operatorname{div} F + \widehat{n} \cdot [F]\delta(\Sigma) \,, \tag{25}$$

and

$$\overline{\operatorname{curl}} F = \operatorname{curl} F + \widehat{n} \times [F]\delta(\Sigma) \,. \tag{26}$$

As mentioned earlier, the formulas of this section can also be obtained by considering $\Sigma(x, t)$ as a p-dimensional submanifold swept out in R_{p+1} by the moving wave front. We briefly present the corresponding definitions.

Let $f(x, t)$ and $F(x, t)$ denote scalar and vector distributions, respectively. Similarly, let $\psi(x, t)$ and $\phi(x, t)$ denote scalar and vector test functions, respectively. Then we define the following relations:

$$\langle f, \psi \rangle = \int_{R_{p+1}} f\psi \, dV \, dt \,, \tag{27}$$

$$\langle F, \phi \rangle = \int_{R_{p+1}} F \cdot \phi \, dV \, dt \,, \tag{28}$$

$$\langle f\delta(\Sigma), \psi \rangle = \int_{\Sigma} f\psi \, dS \, dt \,, \tag{29}$$

$$\langle F\delta(\Sigma), \phi \rangle = \int_{\Sigma} F \cdot \phi \, dS \, dt \,, \tag{30}$$

$$\langle \overline{\text{grad}} f, \phi \rangle = -\langle f, \text{div}\phi \rangle \,, \tag{31}$$

$$\langle \overline{\text{div}} F, \psi \rangle = -\langle F, \text{grad}\, \psi \rangle \,, \tag{32}$$

$$\langle \overline{\text{curl}} F, \phi \rangle = \langle F, \text{curl}\, \phi \rangle \,, \tag{33}$$

$$\langle \bar{\partial} f/\partial t, \psi \rangle = -\langle f, \partial \psi/\partial t \rangle \,, \tag{34}$$

and

$$\langle \bar{\partial} F/\partial t, \phi \rangle = -\langle F, \partial \phi/\partial t \rangle \,, \tag{35}$$

where the dot denotes the scalar product. It is left as an exercise for the reader to derive formulas (4), (5), (25), and (26) with the help of these definitions.

5.6. Second-Order Distributional Derivatives

In order to generalize the results of Section 5.5 to second-order derivatives it is convenient to introduce some notation that will simplify our formulas. Let us first observe that the tensor formed by the second-order generalized derivatives is symmetric. The same is true for the tensor formed by second-order ordinary derivatives of functions in E. The symmetric behavior of these quantities suggests that the appropriate framework for our study is the algebra of the symmetric tensors. Accordingly, we shall understand that the product of the two tensors will be their product as elements of the symmetric algebra and not the ordinary tensor product. In particular,

$$A \otimes B = (A_i)(B_j) = \frac{1}{2}(A_i B_j + A_j B_i) \,.$$

For $F \in E$, DF is the tensor formed by the first-order derivatives (i.e., the gradient), $D^2 F$ is the tensor formed by the second-order derivatives, and so on. For quantities defined only on the surface, DA is the vector of first-order δ derivatives, while the second order differential $D^2 A$ follows from (5.2.31) to be

$$(D^2 A)_{ij} = \frac{\delta^2 A}{\delta x_i \delta x_j} - \mu_{jk} n_i \frac{\delta A}{\delta x_k} \,. \tag{1}$$

Furthermore, we can consider DF as a p-dimensional or $(p+1)$-dimensional object, according to whether time is one of the variables. Previous remarks show that it really makes no difference so long as we consider $-G$ to be the component of the normal vector in the time direction.

In view of this hypothesis, the theorems of the previous section can be written

$$\overline{D}F = DF + A\widehat{n}\delta(\Sigma)\,, \tag{2}$$

$$\overline{D}\delta(\Sigma) = \delta'(\Sigma)\widehat{n}\,, \tag{3}$$

$$[DF] = B\widehat{n} + DA\,. \tag{4}$$

Theorem 1. *For $F \in E$,*

$$\overline{D}^2 F = D^2 F + (B\widehat{n}^2 + 2D\,A\widehat{n} + AD\widehat{n})\delta(\Sigma) + A\widehat{n}^2\delta'(\Sigma)\,. \tag{5}$$

Proof. Using (2)–(4) we have

$$\begin{aligned}\overline{D}^2 F &= \overline{D}(\overline{D}F) = \overline{D}(DF + A\widehat{n}\delta(\Sigma)) = \overline{D}(DF) + \overline{D}(A\widehat{n}\delta(\Sigma))\\ &= D^2F + \widehat{n}\otimes[DF]\delta(\Sigma) + (DA\otimes\widehat{n} + AD\widehat{n})\delta(\Sigma) + A\widehat{n}D\delta(\Sigma)\\ &= D^2F + (B\widehat{n}\widehat{n} + \widehat{n}\otimes DA + DA\otimes\widehat{n} + AD\widehat{n})\delta(\Sigma) + A\widehat{n}^2\delta'(\Sigma)\\ &= D^2F + (B\widehat{n}^2 + 2DA\widehat{n} + AD\widehat{n})\delta(\Sigma) + A\widehat{n}^2\delta'(\Sigma)\,,\end{aligned}$$

where

$$\widehat{n}^2 = \widehat{n}\widehat{n} = n_in_j\,,\quad (A\otimes B) = \frac{1}{2}(A_iB_j + A_jB_i)\,,\quad\text{and } \widehat{n}\otimes DA + DA\otimes\widehat{n} = 2DA\widehat{n}\,. \tag{6}$$

When t is one of the variables, (5) has the following components:

$$\frac{\overline{\partial}^2 F}{\partial x_i\partial x_j} = \frac{\partial^2 F}{\partial x_i\partial x_j} + \left(Bn_in_j + \frac{\delta A}{\partial x_i}n_j + \frac{\delta A}{\delta x_j}n_i + A\frac{\delta n_i}{\delta x_j}\right)\delta(\Sigma) + An_in_j\delta'(\Sigma)\,, \tag{7a}$$

$$\frac{\overline{\partial}^2 F}{\partial x_i\partial t} = \frac{\partial^2 F}{\partial x_i\partial t} + \left(-Bn_iG + \frac{\delta A}{\partial t}n_i - \frac{\delta A}{\delta x_i}G + A\frac{\delta n_i}{\delta t}\right)\delta(\Sigma) - An_iG\delta'(\Sigma)\,, \tag{7b}$$

$$\frac{\overline{\partial}^2 F}{\partial t^2} = \frac{\partial^2 F}{\partial t^2} + \left(BG^2 - 2G\frac{\delta A}{\delta t} - A\frac{\delta G}{\delta t}\right)\delta(\Sigma) + AG^2\delta'(\Sigma)\,. \tag{7c}$$

In particular,

$$\begin{aligned}\overline{\nabla}^2 F &= \nabla^2 F + (B - 2\Omega A)\delta(\Sigma) + A\delta'(\Sigma)\\ &= \nabla^2 F + B\delta(\Sigma) + Ad_n\delta(\Sigma)\,.\end{aligned} \tag{8}$$

Theorem 2. *If $F \in E$, then*

$$[D^2F] = C\widehat{n}^2 + 2D\,B\widehat{n} + B\,D\widehat{n} + D^2A\,, \tag{9}$$

where

$$C = [d^2F/dn^2] = [\partial^2 F/\partial x_i\partial x_j]n_in_j\,. \tag{10}$$

Proof. It is again convenient to use the notation $F_{,i}$ for $\partial F/\partial x_i$. Then from (4) we have

$$[F_{,ij}] = \left[\frac{dF_{,i}}{dn}\right] n_j + \frac{\delta[F_{,i}]}{\delta x_j} = \left[\frac{dF_{,i}}{dn}\right] n_j + \frac{\delta}{\delta x_j}\left[Bn_i + \frac{\delta A}{\delta x_i}\right]$$
$$= \left[\frac{dF_{,i}}{dn}\right] n_j + \frac{\delta B}{\delta x_{j,}} n_i + B\frac{\delta n_i}{\delta x_j} + \frac{\delta^2 A}{\delta x_j \delta x_i}. \tag{11}$$

Multiplying both sides by n_i and summing over i gives

$$\left[\frac{dF_{,j}}{dn}\right] = Cn_j + \frac{\delta B}{\delta x_j} + \frac{\delta^2 A}{\delta x_j \delta x_i} n_i. \tag{12}$$

Thus if we use this relation to find $[dF_{,i}/dn]$, plug it back in (11), and use (5.2.27), we obtain

$$[F_{,ij}] = Cn_i n_j + \frac{\delta B}{\delta x_i} n_j + \frac{\delta B}{\delta x_j} + \frac{\delta^2 A}{\delta x_i \delta x_j} - \mu_{ik} n_j \frac{\delta A}{\delta x_k}, \tag{13}$$

which is nothing but the component form of the required relation (9).

From (13) we read off the following results:

$$\left[\frac{\partial^2 F}{\partial x_i \partial t}\right] = -CGn_i + \frac{\delta B}{\delta t} n_i - \frac{\delta B}{\delta x_i} G + B\frac{\delta n_i}{\delta t} + \frac{\delta^2 A}{\delta x_i \delta t} + \mu_{ik} G \frac{\delta A}{\delta x_k}, \tag{14}$$
$$\left[\frac{\partial^2 F}{\partial t^2}\right] = CG^2 - 2G\frac{\delta B}{\delta t} - B\frac{\delta G}{\delta t} - B\frac{\delta G}{\delta t} + \frac{\delta^2 A}{\delta t^2} + \mu_{tk} G \frac{\delta A}{\delta x_k}. \tag{15}$$

As a special case we have

$$[\nabla^2 F] = C - 2\Omega B + \delta^2 A/\delta x_i \delta x_j.$$

From (5.3.6) it follows that $Q = \delta(\Sigma)$ satisfies (5.3.13), and hence Lemma 2 yields the following results:

Theorem 3. *Let r be any nonnegative integer. Then if $0 \le s \le r$, we have*

$$\delta^{(r)}(\Sigma) = \frac{\overline{\partial}^r \delta(\Sigma)}{\partial x_{i_1} \cdots \partial x_{i_r}} n_{i_1} \cdots n_{i_r}, \tag{16}$$
$$\delta^{(r)}(\Sigma) = (\delta^{(s)}(\Sigma))^{(r-s)}, \tag{17}$$
$$\overline{D}^{(r)} \delta(\Sigma) = \delta^{r+1}(\Sigma)\widehat{n}, \tag{18}$$
$$\overline{D}^2 \delta^{(r)}(\Sigma) = \widehat{n}\delta^{(r+1)}(\Sigma) + \widehat{n}^2 \delta^{(r+2)}(\Sigma). \tag{19}$$

In particular, we have

$$\frac{\overline{\partial}^2 \delta(\Sigma)}{\partial x_i \partial x_j} = \mu_{ij}\delta'(\Sigma) + \delta''(\Sigma) n_i n_j \,, \tag{20}$$

$$\overline{\nabla}^2 \delta(\Sigma) = -2\Omega\delta'(\Sigma) + \delta''(\Sigma)\,, \tag{21}$$

$$\frac{\overline{\partial}^2 \delta(\Sigma)}{\partial x_i \partial t} = \mu_{it}\delta'(\Sigma) - G n_i \delta''(\Sigma)\,, \tag{22}$$

$$\frac{\overline{\partial}^2 \delta(\Sigma)}{\partial t^2} = -\frac{\delta G}{\delta t}\,\delta'(\sigma) + G^2\delta''(\Sigma)\,. \tag{23}$$

5.7. Higher-Order Distributional Derivatives

The formulas for generalized derivatives of order higher than 2 become very complicated since they contain a rather excessive number of terms. In this section we present the formula for the third-order derivative of a regular singular function [11]. We also obtain formulas for the biharmonic operator

$$\overline{\nabla}^4 F = \overline{\nabla}^2(\overline{\nabla}^2 F) = \overline{\partial}^4 F/\partial x_i \partial x_i \partial x_j \partial x_j\,.$$

Third-Order Derivatives.
Let F be a regular singular function with respect to Σ. Then by using the formulas for the second order derivatives and computing a new derivative we obtain

$$\begin{aligned} \overline{D}^3 F = D^3 F &+ (C\widehat{n}^3 + 3DB\widehat{n}^2 + 3D^2A\widehat{n} + 3B\mu\widehat{n} + 3DA\mu + 3AD\mu)\delta(\Sigma) \\ &+ (B\widehat{n}^3 + 3DA\widehat{n}^2 + 3A\mu\widehat{n})\delta[(\Sigma) + A\widehat{n}^3\delta''(\Sigma)\,, \end{aligned} \tag{1a}$$

the symmetric product is understood. For instance,

$$(DB)\widehat{n}^2 = \frac{1}{3}\left(\frac{\delta B}{\delta x_i} n_j n_k + \frac{\delta B}{\delta x_j} n_i n_k + \frac{\delta B}{\delta x_k} n_i n_j\right)\,,$$

$$D\mu = \frac{1}{3}\left(\frac{\delta \mu_{ij}}{\delta x_k} + \frac{\delta \mu_{ik}}{\delta x_j} + \frac{\delta \mu_{kj}}{\delta x_i}\right)\,.$$

Similarly, we can write this formula in terms of the operators d_n and $(d_n)^2$ as

$$\begin{aligned} \overline{D}^3 F = & D^3 F + ((C + 2\Omega B + (4\Omega^2 - \omega_2)A)\widehat{n}^3 + 3(DB + 2\Omega DA)\widehat{n}^2 \\ &+ 3(B + 2\Omega A)\widehat{n}\mu + 3\mu DA + 3D^2A\widehat{n} + \frac{1}{3}(3D\mu + 2\mu^{(2)}\widehat{n})A)\delta(\Sigma) \\ &+ d_n[((B + 4\Omega A)\widehat{n}^3 + 3DA\widehat{n}^2 + 3A\widehat{n}\mu)\delta(\Sigma)] + (d_n)^2[A\widehat{n}^3\delta(\Sigma)]\,. \end{aligned} \tag{1b}$$

If we take Σ to be a closed surface that encloses a volume V, and F is a function that vanishes outside of $V \cup \Sigma$, then (1b) is actually the abstract form of an integral relation. To illustrate this idea, let us compute $(\overline{\partial}/\partial x_k)(\overline{\nabla}^2 F)$. We multiply it by a test function ϕ and integrate, obtaining

$$\int_V \left(F\nabla^2 \left(\frac{\partial \phi}{\partial x_k} \right) + \frac{\partial}{\partial x_k} (\nabla^2 F)\phi \right) dV = \int_\Sigma \left(R_0\phi + R_1 \frac{d\phi}{dn} + R_2 \frac{d^2\phi}{dn^2} \right) dS, \quad (2)$$

where the surface quantities are given by

$$R_2 = An_k = Fn_k, \quad (3)$$

$$R_1 = (B + 2\Omega A)n_k = \left(\frac{dF}{dn} + 2\Omega F \right) n_k, \quad (4)$$

$$R_0 = (C - \omega_2 A)n_k + \frac{\delta B}{\delta x_k} + \mu_{ik} \frac{\delta^2 A}{\delta x_i^2} n_k. \quad (5)$$

Biharmonic Operator

In the study of the boundary value problems in elasticity one encounters the biharmonic equation $\nabla^4 F = 0$. Indeed, many plane problems of elasticity, when studied with the help of analytic functions, reduce to the solution of the two-dimensional biharmonic equation. Similarly, the discussion of the theory of elastic plates and shells leads to the three-dimensional biharmonic equation [11].

For deriving the higher order distributional derivatives of discontinuous functions we need an extension of the Gaussian different geometry which we shall develop in Chapter 12. However, it is possible to evaluate the distributional biharmonic equation. For this purpose we introduce the following notation. Let $F(x, t)$ be a function defined on both sides of Σ and which has derivatives of all orders there. These derivatives have boundary values from both sides of Σ. Let

$$A^{(Q)} = \left[\frac{d^Q F}{dn^Q} \right] = \left(\frac{d^Q F}{dn^Q} \right)_+ - \left(\frac{d^Q F}{dn^q} \right)_-, \quad Q = 0, 1, 2, \cdots, \quad (6)$$

be the jump of the normal derivative of order Q across Σ. Comparing with the previous notation we have $A^{(0)} = A$, $A^{(1)} = B$ and $A^{(2)} = C$. We shall find this notation very helpful in Chapter 12.

For our present purpose let us set

$$\alpha = [\nabla^2 F], \qquad \beta = \left[\frac{d}{dn} \nabla^2 F \right], \quad \text{and} \quad \gamma = \left[\frac{d^2}{dn^2} \nabla^2 F \right]. \quad (7)$$

Lemma. *If $F \in E$, then*

$$\alpha = A^{(2)} - 2\Omega A^{(1)} + \nabla^2 A^{(0)}, \tag{8}$$

$$\beta = A^{(3)} - 2\Omega A^{(2)} + \nabla^2 A^{(1)} - \omega_2 A^{(1)} + n_j \nabla^2 \left(\frac{\delta A^{(0)}}{\delta x_j} \right), \tag{9}$$

$$\gamma = A^{(4)} - 2\Omega A^{(3)} + \nabla^2 A^{(2)} - 2\omega_2 A^{(2)} + 2n_j \nabla^2 \left(\frac{\delta A^{(1)}}{\delta x_j} \right)$$
$$+ 2\omega_3 A^{(1)} + 2n_j n_k \nabla^2 \left(\frac{\delta^2 A^{(0)}}{\delta x_j \delta x_k} \right) - 2\mu_{jk}^{(2)} \frac{\delta^2 A^{(0)}}{\delta x_j \delta x_k}. \tag{10}$$

Proof. The first result follows by putting $i = j$ in the jump relation (5.6.13). To prove (9), we evaluate $[F_{,iij}]$ in the following two ways. First,

$$[F_{,iij}] = \beta n_j + \delta\alpha/\delta x_j, \tag{11}$$

which is obtained from (5.5.19) by substituting $\nabla^2 F$ for F. Second,

$$[F_{,iij}] = [F_{,iji}] = [dF_{,ij}/dn] n_i + \delta[F_{,ij}]/\delta x_i. \tag{12}$$

Now we multiply both sides of (11) and (12) by n_j and equate the results, producing

$$\beta = \left[\frac{dF_{,ij}}{dn} \right] n_i n_j + \frac{\delta [F_{,ij}]}{\delta x_i} n_j$$
$$= [F_{,ijk}] n_i n_j n_k + \frac{\delta}{\delta x_i} \left[A^{(2)} n_i n_j + \frac{\delta A^{(1)}}{\delta x_i} n_j + \frac{\delta A^{(1)}}{\delta x_j} n_i \right.$$
$$\left. + A^{(1)} \mu_{ij} + \frac{\delta^2 A^{(0)}}{\delta x_i \delta x_j} - \mu_{jk} n_j \frac{\delta A^{(0)}}{\delta x_k} \right] n_j$$
$$= A^{(3)} - 2\Omega A^{(2)} + \nabla^2 A^{(1)} + \frac{\delta \mu_{ij}}{\delta x_i} n_j A^{(1)} + n_j \frac{\delta^3 A^{(0)}}{\delta x_i^2 \delta x_j},$$

which proves (9) on our observing that

$$(\delta \mu_{ij}/\delta x_i) n_j = -\mu_{ij}\mu_{ij} = -\omega_2. \tag{13}$$

Similarly,

$$\gamma = [F_{,iijk}] n_j n_k = [F_{,jkii}] n_j n_k = [\nabla^2 F_{,jk}] n_j n_k$$
$$= \left\{ [F_{,jkil}] n_i n_l - 2\Omega [F_{,jki}] n_i + \frac{\delta^2}{\delta x_i^2} [F_{,jk}] \right\} n_j n_k$$
$$= A^{(4)} - 2\Omega A^{(3)} + \nabla^2 A^{(2)} - 2\omega_2 A^{(2)} + 2n_j \nabla^2 \left(\frac{\delta A^{(1)}}{\delta x_j} \right) + 2\omega_3 A^{(1)}$$
$$+ 2n_j n_k \nabla^2 \left(\frac{\delta^2 a^{(0)}}{\delta x_j \delta x_k} \right) - 2\mu_{jk}^{(2)} \frac{\delta^2 A^{(0)}}{\delta x_j \delta x_k},$$

which proves (10).

We are now ready for the evaluation of $\overline{\nabla}^4 F$;

$$\overline{\nabla}^4 F = \overline{\nabla}^2(\overline{\nabla}^2 F),$$

which with the help of (8) becomes

$$\begin{aligned}\overline{\nabla}^4 F &= \overline{\nabla}^2(\nabla^2 F + (A^{(1)} - 2\Omega A^{(0)})\delta(\Sigma) + A\delta'(\Sigma)) \\ &= \nabla^2(\nabla^2 F) + (\beta - 2\Omega\alpha)\delta(\Sigma) + \alpha\delta'(\Sigma) + \nabla^2(A^{(1)} - 2\Omega A^{(0)})\delta(\Sigma) \\ &\quad + 2\nabla(A^{(1)} - 2\Omega A^{(0)}) \cdot \nabla\delta(\Sigma) + (A^{(0)} - 2\Omega A^{(0)})\overline{\nabla}^2\delta(\Sigma) + \nabla^2 A^{(0)}\delta'(\Sigma) \\ &\quad + 2\nabla A^{(0)} \cdot \nabla\delta'(\Sigma) + A^{(0)}\overline{\nabla}^2\delta'(\Sigma).\end{aligned}$$

Thus we obtain

Theorem. *If $F \in E$, then*

$$\begin{aligned}\overline{\nabla}^4 F &= \nabla^4 F + \Big\{A^{(3)} - 4\Omega A^{(2)} + 2\nabla^2 A^{(1)} + (4\Omega^2 - \omega_2)A^{(1)} \\ &\quad + n_j\nabla^2(\delta A_1/\delta x_j) - 4\Omega\nabla^2 A^{(0)} - 4\frac{\delta\Omega}{\delta x_j}\frac{\delta A^{(0)}}{\delta x_j} - 2(\nabla^2\Omega)A^{(0)}\Big\}\delta(\Sigma) \\ &\quad + \big\{A^{(2)} - 4\Omega A^{(1)} + 2\nabla^2 A^{(0)} - 4\Omega^2 A^{(0)}\big\}\delta'(\Sigma) \\ &\quad + (A^{(1)} - 4\Omega A^{(0)})\delta''(\Sigma) + A^{(0)}\delta'''(\Sigma).\end{aligned} \tag{14}$$

Let us now find the jump $[\nabla^4 F]$. To this end, we replace F by $\nabla^2 F$ in (8) and obtain

$$[\nabla^4 F] = [\nabla^2\nabla^2 F] = \gamma - 2\Omega\beta + \nabla^2\alpha.$$

The use of (8)–(10) in this expression gives us the desired formula,

$$\begin{aligned}[\nabla^4 F] &= A^{(4)} - 4\Omega A^{(3)} + 2\nabla^2 A^{(2)} + (4\Omega^2 - 2\omega_2)A^{(2)} + 2n_j\nabla^2(\delta A^{(1)}/\delta x_j) \\ &\quad - 4\Omega\nabla^2 A^{(1)} - 4\frac{\delta\Omega}{\delta x_j}\frac{\delta A^{(1)}}{\delta x_j} + (2\omega_3 - 2\Omega\omega_2 - 2\nabla^2\Omega)A^{(1)} \\ &\quad + 2n_j n_k\nabla^2\left(\frac{\delta^2 A^{(0)}}{\delta x_j\delta x_k}\right) - 2\Omega n_j\nabla^2\left(\frac{\delta A^{(0)}}{\delta x_j}\right) + \nabla^4 A^{(0)} - 2\mu_{jk}^{(2)}\frac{\delta^2 A^{(0)}}{\delta x_j\delta x_k}.\end{aligned} \tag{15}$$

5.8. The Two-Dimensional Case

Let us consider the Cartesian coordinates $x_1 = x$, $x_2 = y$ in a plane. A curve $\Gamma(x_1, x_2, t)$, where t is time, can be described as [12,13]

$$u(x, y, t) = 0, \tag{1}$$

or, in terms of its arc length s, as

$$x = x_1(s,t)\,, \qquad y = x_2(s,t)\,. \tag{2}$$

The unit tangent vector t_i is defined as

$$t_i = \frac{dx^i}{ds} = \dot{x}^i\,, \tag{3}$$

where the dot stand for d/ds. The outward unit normal $n_i = (n_1, n_2)$ is

$$n_1 = \frac{dx_2}{ds} = \dot{x}_2\,, \qquad n_2 = -\frac{dx_1}{ds} = -\dot{x}\,. \tag{4}$$

These relations can be written as

$$t_i = e_{ik} n_k \tag{5}$$

where e_{ik} is defined as

$$e_{ik} = \begin{cases} 0\,, & i = k\,, \\ 1\,, & i = 1\,, \quad k = 2\,, \\ -1\,, & i = 2\,, \quad k = 1. \end{cases} \tag{6}$$

From relation (4) to (6) we immediately deduce the formula

$$\dot{x}_i \dot{x}_j = t_i t_j = (\delta_{ij} - n_i n_j)\,. \tag{7}$$

If $f(x_1, x_2, t)$ is a function defined on the curve then we denote $\delta/\delta x_i = \dot{x}_i\, d/ds$ so that we have

$$\frac{\delta f}{\delta x_i} = \dot{f}(s)\dot{x}_i\,. \tag{8}$$

Similarly $\frac{\delta f}{\delta t}$ is defined as

$$\frac{\delta f}{\delta t} = \frac{1}{G}\frac{df}{ds}\,, \tag{9}$$

where G is the speed of propagation of the curve.

Next, we recall the Serret-Frenet formulas

$$\frac{dt_i}{ds} = \kappa\, n_i\,, \qquad \frac{dn_i}{ds} = -\kappa\, t_i\,, \tag{10}$$

where κ is curvature of the curve Γ.

With the help of relation (8) to (10) we define the symmetric quantities μ_{ij} as

$$\begin{aligned} \mu_{ij} &= \frac{\delta n_i}{\delta x_j} = \dot{n}_i \dot{x}_j = -\kappa t_i t_j \\ &= -\kappa\,(\delta_{ij} - n_i n_j)\,. \end{aligned} \tag{11}$$

Also

$$\mu_{it} = \frac{\delta n_i}{\delta t} = -G\,\dot{x}_i\,. \tag{12}$$

Next, we write (7) as

$$\mu_{ij}^{(0)} = \dot{x}_i \dot{x}_j = (\delta_{ij} - n_i n_j) . \tag{13}$$

These concepts are extended to the relation

$$\mu_{ij}^{(P)} = (-\kappa)^P \, \dot{x}_i \dot{x}_j , \tag{14}$$

which reduces to (13) and (11) for $P = 0$ and 1.

The trace ω_P of $\mu_{ij}^{(P)}$ is

$$\omega_P = (-\kappa)^P , \tag{15}$$

which for $P = 1$ becomes

$$\frac{\delta n_i}{\delta x_i} = -\kappa . \tag{16}$$

If the curve Γ is given by $y = g(x)$, then $\delta(\Gamma)$ is

$$\delta(\Gamma) = \delta\big(y - g(x)\big) \big/ \big(1 + (g'(x))^2\big)^{1/2} . \tag{17}$$

It is more convenient to follow the notation of the previous sections and start with the definition

$$\langle \delta(x,t), \phi(x,t) \rangle = \int_{-\infty}^{\infty} \int_{-\infty}^{\infty} \int_{-\infty}^{\infty} \phi(x,t) dx_1 dx_2 dt , \tag{18}$$

where $x = (x_1, x_2) = (x, y)$ and $\phi \in D(x, t)$, the class of functions which are infinitely differentiable and have a compact support. Then the delta function concentrated on a curve Γ in the plane (x_1, x_2) is

$$\langle \delta(\Gamma), \phi(x,t) \rangle = \int_{-\infty}^{\infty} \int_{\Gamma(t)} \phi(x,t) ds \, dt . \tag{19}$$

To derive the formula $(\partial/\partial x_i)\delta(\Gamma)$ we evaluate

$$\begin{aligned} \left\langle \frac{\partial}{\partial x_i} \delta(\Gamma), \phi \right\rangle &= -\left\langle \delta(\Gamma), \frac{\partial \phi}{\partial x_i} \right\rangle = -\left\langle \delta(\Gamma), \frac{d\phi}{ds} \, t_i + \frac{d\phi}{ds} \, n_i \right\rangle \\ &= -\left\langle \delta(\Gamma), \frac{d\phi}{ds} \, t_i \right\rangle - \left\langle \delta(\Gamma), \frac{d\phi}{dn} \, n_i \right\rangle \\ &= -\left\langle \delta(\Gamma) t_i, \frac{d\phi}{ds} \right\rangle - \left\langle \delta(\Gamma) n_i, \frac{d\phi}{dn} \right\rangle . \end{aligned} \tag{20}$$

At this stage we need the definitions of two more basic distributions concentrated on Γ. The first is the normal derivative distribution;

$$\delta'(\Gamma) = \frac{\overline{\partial}}{\partial x_i} [\delta(\Gamma)] n_i , \tag{21}$$

and the second is the normal derivative operator $d_n \delta(\Gamma)$;

$$\langle d_n \delta(\Gamma), \phi \rangle = \int_{-\infty}^{\infty} \int_{\Gamma} -\frac{d\phi}{dn} \, ds \, dt . \tag{22}$$

Now we return to (20), using (21) and (22) so that we have,

$$\left\langle \frac{\partial}{\partial x_i}\delta(\Gamma), \phi \right\rangle = \left\langle \frac{d}{ds}(\delta(\Gamma)t_i), \phi \right\rangle + \langle d_n\delta(\Gamma)n_i, \phi\rangle$$
$$= \langle \delta'(\Gamma)n_it_i + \delta(\Gamma)\kappa\, n_i, \phi\rangle + \langle d_n\delta(\Gamma)n_i, \phi\rangle$$
$$= \langle \kappa\, n_i\delta(\Gamma) + n_i\, d_n\delta(\Gamma), \phi\rangle \ ,$$

or

$$\frac{\overline{\partial}}{\partial x_i}\delta(\Gamma) = (\kappa\,\delta(\Gamma) + d_n\delta(\Gamma))n_i\,. \tag{23}$$

When we dot both sides by n_i we obtain the relation between the two distributions defined by (21) and (22), namely,

$$\delta_n'(\Gamma) = d_n\delta(\Gamma) + \kappa\,\delta(\Gamma)\,. \tag{24}$$

We are now able to write down the formulas for the distributional derivatives in the two-dimensional cases which correspond to those in Section 5.5. For this purpose, let us consider a vectorial function $f(x, y) = f_1(x, y)e_x + f_2(x, y)e_y$ so that the gradient, divergence, and curl of f are

$$\text{grad } f = \frac{\partial f_1}{\partial x}e_x + \frac{\partial f_2}{\partial y}e_y, \quad \text{div} f = \frac{\partial f_1}{\partial x} + \frac{\partial f_2}{\partial y} \tag{25}$$

and

$$\text{curl } f = \frac{\partial f_2}{\partial x} - \frac{\partial f_1}{\partial y}\,. \tag{26}$$

Then the distributional derivatives formulas for $f(x, y)$ are

$$\frac{\overline{\partial} f}{\partial x_i} = \frac{\partial f}{\partial x_i} + [f]n_i\delta(\Gamma)\,, \tag{27a}$$

or in vector notation

$$\overline{\text{grad}}\, f = \text{grad } f + [f]\, n\, \delta(\Gamma)\,. \tag{27b}$$

Thus,

$$\overline{\text{div}} f = \text{div} f + [f]\ \cdot n\,\delta(\Gamma)\,, \tag{28}$$

$$\overline{\text{curl}}\, f = \text{curl } f + [f] \times n\,\delta(\Gamma) \text{ and}, \tag{29}$$

$$\frac{\overline{\partial} f}{\partial t} = \frac{\partial f}{\partial t} - [f]\, G\,\delta(\Sigma)\,. \tag{30}$$

From these formulas we immediately get

$$\frac{\overline{\partial}}{\partial x_i}\delta(\Gamma) = \delta'(\Gamma)n_i \text{ and}, \tag{31}$$

$$\frac{\overline{\partial}}{\partial t}\delta(\Gamma) = -G\,\delta'(\Gamma)\,, \tag{32}$$

and the proof is similar to the one given in Theorem 2 of Section 5.5.

To derive the distributional derivatives of the second order for a distribution $f(x, t)$ we need the value of the gradients of the jump $[f] = A$. This is achieved by differentiating this relation along the curve Γ:

$$[f_{,j}]t_j = \frac{dA}{ds} \,. \tag{33}$$

When we multiply both sides of this relation by t_k we have

$$[f_{,j}]t_j t_k = \frac{dA}{ds}\, t_k \,. \tag{34}$$

Next, we use the relation (7) and get

$$[f_{,k}] = \dot{A}\, \dot{x}_k + B\, n_k \,, \tag{35}$$

where $B = df/dn$. The next step is to substitute $f_{,j}$ for f in the relation (27a) and obtain

$$\frac{\overline{\partial}^2 f}{\partial x_i \partial x_j} = \frac{\partial^2 f}{\partial x_i \partial x_j} + (B\, n_i n_j + \dot{f}\, \dot{x}_i n_j + \dot{A} \dot{x}_j n_i - A\kappa \dot{x}_i \dot{x}_j)\delta(\Gamma) + A\, n_i n_j \delta'(\Gamma) \,. \tag{36}$$

Similarly,

$$\frac{\overline{\partial}^2 f}{\partial x_i \partial t} = \frac{\partial^2 f}{\partial x_i \partial t} + \left(-B_i G + \frac{\delta A}{\delta t} - \dot{A}\, \dot{x}_i\, G + A \frac{\delta n_i}{\delta t}\right) \delta(\Gamma) - A\, n_i\, G\, \delta'(\Gamma) \,, \tag{37}$$

$$\frac{\overline{\partial}^2 f}{\partial t^2} = \frac{\partial^2 f}{\partial t^2} + \left(BG^2 = 2G \frac{\delta A}{\delta t} - A\, \frac{\delta G}{\delta t}\right) \delta(\Gamma) + AG^2 \delta'(\Gamma) \,. \tag{38}$$

When we set $i = j$ in (36) and sum on j, we obtain

$$\overline{\nabla}^2 f = \nabla^2 f + (B - \kappa\, A)\delta(\Gamma) + A\, \delta'(\Gamma) \,. \tag{39}$$

Second order interfacial relations across $\Gamma(t)$

To derive the values of $[f_{,ij}]$ across $\Gamma(t)$, we substitute $f_{,i}$ for f in (35) and obtain

$$[f_{,ij}] = \left[\frac{\partial f_{,i}}{dn}\right] n_j + \frac{\delta [f_{,i}]}{\delta x_j} = \left[\frac{\partial f_{,i}}{dn}\right] n_j + (-B\kappa \dot{x}_i + \dot{B} n_i + \dot{A} \dot{x}_i + \dot{A}\kappa n_i)\, \dot{x}_j \,. \tag{40}$$

Multiplying both sides by n_i and summing on i yields

$$\left[\frac{df_{,i}}{dn}\right] = \left[\frac{d^2 f}{dn^2}\right] n_j + (\dot{B} + \dot{A}\kappa)\dot{x}_j \,. \tag{41}$$

Next, we substitute (41), with subscript j changed to i, in (40). Thereby we obtain the required formula

$$[f_{,ij}] = C\, n_i n_j + \dot{B}\, \dot{x}_i n_j + \dot{B}\, n_i \dot{x}_j - B\, \kappa\, n_i\, \dot{x}_j + \dot{A}\, \kappa\, n_j \dot{x}_j + \ddot{A} \dot{x}_i \dot{x}_j\,, \tag{42}$$

where $C = [d^2 f/dn^2]$. Set $i = j$ and sum on j and obtain.

$$[\nabla^2 f] = C - \kappa\, B + \ddot{A}\,. \tag{43}$$

Similarly,

$$[f_{,it}] = -C\, G\, n_i + \frac{\delta B}{\delta t} n_i - \frac{\delta B}{\delta t} G \dot{x}_j - B\, \kappa\, G\, \dot{x}_j + \dot{A}\, \kappa\, G \dot{x}_j + \frac{\delta \dot{A}}{\delta t} \dot{x}_i\,, \tag{44}$$

$$[f_{,tt}] = C\, G^2 - 2G \frac{\delta B}{\delta t} - B \frac{\delta G}{\delta t} - \frac{\delta A}{\delta t} \kappa\, G + \frac{\delta^2 A}{\delta t^2}\,. \tag{45}$$

Let us now present the corresponding formulas for the biharmonic equation. For this purpose we observe that the repeated use of (8) gives

$$\nabla^2 f = \ddot{f}\,,$$

$$n_j \nabla^2 (\delta f/\delta x_j) = 2\kappa\, \ddot{f} + \dot{\kappa}\, \dot{f} \;\text{ and}\,,$$

$$n_j n_k \nabla^2 (\delta^2 f/\delta x_j \delta x_k) = 4\kappa^2 \ddot{f} + 3\kappa \dot{\kappa}\, f\,.$$

If we substitute these expressions in (5.7.14) and recall that $\kappa = 2\Omega$, we obtain

$$\overline{\nabla}^4 F = \nabla^4 F + \{A^{(3)} - 2\kappa\, A^{(2)} + 2\ddot{A}^{(1)} + \kappa^2 A^{(1)} - \dot{\kappa}\, \dot{A}^{(1)} - \ddot{\kappa}\, A^{(0)}\}\delta(\Sigma) + \{A^{(2)} - 2\kappa\, A^{(1)} + 2\ddot{A}^{(0)} + \kappa^2 \dot{A}\}\dot{\delta}(\Sigma) + (A^{(1)} - 2\kappa\, A^{(0)})\delta(\Sigma) + A^{(0)}\, \dddot{\delta}\,(\Sigma)\,. \tag{46}$$

Similarly,

$$[\nabla^4 F] = A^{(4)} - 2\kappa\, A^{(3)} + 2\ddot{A}^{(2)} + (\kappa^2 - 2\kappa) A^{(2)} + 2\kappa\, A^{(1)} - (3\kappa^3 + \kappa) A^{(1)} + A^{(1)} + (6\kappa^2 - 4\kappa) A^{(0)} + 5\kappa\kappa\; A^{(1)}\,. \tag{47}$$

5.9. Examples

Example 1. Integral Theorems of Vector Analysis.

(i) *Divegence Theorem.* Consider a closed three-dimensional region V whose surface is Σ with the unit normal n pointed inwards. Let (e_x, e_y, e_z) be the unit vectors in the Cartesian coordinates so that the gradient of a function $f(x)$ is

$$\nabla f(x) = \frac{\partial f}{\partial x_1} e_x + \frac{\partial f}{\partial x_2} e_y + \frac{\partial f}{\partial x_3} e_z \,.$$

Then

$$\begin{aligned}\int_V (\nabla f) \cdot \phi \, dV &= \int_V \left(\frac{\partial f}{\partial x_1} \phi_1 + \frac{\partial f}{\partial x_2} \phi_2 + \frac{\partial f}{\partial x_3} \phi_3 \right) dV \\ &= - \int_V (\nabla \cdot \phi) f \, dV \,. \end{aligned} \tag{1}$$

For $f(x) = H(x)$, where

$$H(x) = \begin{cases} 1\,, & x \in V\,, \\ 0\,, & \text{otherwise}\,, \end{cases} \tag{2}$$

the formula (5.5.4) yields

$$\nabla H(x) = -n\delta(\Sigma)\,, \tag{3}$$

and the integral relation above yields the divergence theorem,

$$\int_V \nabla \cdot \phi \, dV = \int_V n \cdot \phi \, \delta(\Sigma) dV = \int_\Sigma n \cdot \phi \, d\Sigma \,. \tag{4}$$

(ii) Green's Theorem. Let us now study the corresponding two-dimensional case. In this case we have a planar region Σ bounded by a closed curve Γ with the unit normal n pointed inwards Σ. Then for the two function f and ϕ we have

$$\begin{aligned}\int_\Sigma (\nabla f) \cdot \phi \, d\Sigma &= \int_\Sigma \left(\frac{\partial f}{\partial x} \phi_1 + \frac{\partial f}{\partial y} \phi_2 \right) d\Sigma \\ &= - \int_\Sigma f \left(\frac{\partial \phi_1}{\partial x} + \frac{\partial \phi_2}{\partial y} \right) d\Sigma \,, \end{aligned}$$

or

$$\int_\Sigma (\nabla f) \cdot \phi \, d\Sigma = - \int_\Sigma f (\nabla \cdot \phi) d\Sigma \,. \tag{5}$$

Next, we take

$$f(x) = H(x) = \begin{cases} 1\,, & x \in S\,, \\ 0\,, & \text{otherwise}\,, \end{cases}$$

then from (5.8.27) we have $\nabla H(x) = -n\delta(\Gamma)$, so that (5) becomes

$$\int_\Sigma n \cdot \phi \delta(\Gamma) d\Sigma = \int H \left(\frac{\partial \phi_1}{\partial x} + \frac{\partial \phi_2}{\partial y} \right) d\Sigma \,,$$

or

$$\int_\Gamma \left(-\frac{dy}{ds} \,,\, \frac{dx}{ds} \right) \cdot (\phi_1, \phi_2) d\Gamma = \int_\Sigma \left(\frac{\partial \phi_1}{\partial x} + \frac{\partial \phi_2}{\partial y} \right) d\Sigma \,.$$

Thus

$$\int_\Sigma \left(\frac{\partial \phi_1}{\partial x} + \frac{\partial \phi_2}{\partial y}\right) d\Sigma = \int_\Gamma (\phi_2 dx - \phi_1 dy) d\Gamma\,, \tag{6}$$

which is Green's theorem in a plane.

(iii) Stokes' Theorem. We introduce a new function $\psi = (\psi_1, \psi_2)$ such that $\phi_1 = \psi_2, \phi_2 = -\psi_1$ so that (6) becomes

$$\int_\Sigma \left(\frac{\partial \psi_2}{\partial x} - \frac{\partial \psi_1}{\partial y}\right) d\Sigma = \int_\Gamma \psi_1 dx + \psi_2 dy\,,$$

or

$$\int_\Sigma (\nabla \times \psi) \cdot d\Sigma = \int_\Gamma \psi . d\Gamma\,, \tag{7}$$

which is Stokes's Theorem.

(iv) Green's Second Indentity. For the Laplace operator

$$\nabla^2 = \partial^2/\partial x_1^2 + \partial^2/\partial x_2^2 + \partial^2/\partial x_3^2\,,$$

we found in Section 2.6 that

$$\langle \overline{\nabla}^2 F, \phi\rangle = \langle F, \nabla^2\phi\rangle\,. \tag{8}$$

Substituting (5.6.8), namely, $\overline{\nabla}^2 F = \nabla^2 F + B\delta(\Sigma) + A\, d_n(\Sigma)$, in (8), we obtain

$$\langle \nabla^2 F + B\delta(\Sigma) + A\, d_n(\Sigma)\,, \phi\rangle = \langle F, \nabla^2\phi\rangle\,. \tag{9}$$

For the special case that F vanishes outside Σ, (9) reduces to the classical Green's second identity :

$$\int_V (F\nabla^2\phi - \nabla^2 F) dx = \int_\Sigma \left(F\frac{\partial \phi}{\partial n} - \phi\frac{\partial F}{\partial n}\right) d\Sigma\,. \qquad \phi \in D\,. \tag{10}$$

Example 2. Consider the following distribution defined inR_2:

$$\left\langle \ln\frac{1}{r}, \phi\right\rangle = \int_{-\infty}^{\infty}\int_{-\infty}^{\infty}\left(\ln\frac{1}{r}\right)\phi\, dx_1\, dx_2\,, \tag{11}$$

where $r = |x| = (x_1^2 + x_2^2)^{1/2}$. Then from (8) we have

$$\left\langle \nabla^2 \ln\frac{1}{r}, \phi\right\rangle = \left\langle \ln\frac{1}{r}, \nabla^2\phi\right\rangle\,.$$

Applying Green's theorem to the region lying between a small circle C of radius ε and a circle of sufficiently large radius, we have

$$\int_{r\geq\varepsilon} \ln\frac{1}{r}\nabla^2\phi\, dx_1\, dx_2 = \int_{r\geq\varepsilon} \nabla\left(\ln\frac{1}{r}\right)\phi\, dx_1\, dx_2$$
$$+ \int_{r=\varepsilon} \phi\,\frac{d\ln(1/r)}{dr}\, ds - \int_{r=\varepsilon} \ln\left(\frac{1}{r}\right)\frac{d}{dr}\, ds\,, \tag{12}$$

where ds is the element of length along C. Since $\nabla^2(\ln 1/r) = 0$ for $r \geq \varepsilon$,

$$\lim_{\varepsilon\to 0}\int_{r=\varepsilon}\left(\ln\frac{1}{r}\right)\frac{d\phi}{ds} = 0 \text{ and},$$

$$\lim_{\varepsilon\to 0}\int_{r=\varepsilon}\phi\,\frac{d}{ds}\left(\ln\frac{1}{r}\right)ds = -2\pi\phi(0) = -2\pi\langle\delta,\phi\rangle,$$

(12) reduces to

$$\langle\overline{\nabla}^2\ln(1/r),\phi\rangle = -2\pi\langle\delta,\phi\rangle \qquad \text{or} \qquad \overline{\nabla}^2(\ln 1/r) = -2\pi\delta(r), \tag{13}$$

so that $(1/2\pi)\ln(1/r)$ is the fundamental solution of the two-dimensional Laplace operator $-\overline{\nabla}^2$.

Similarly, it can be shown that

$$\overline{\nabla}^2(1/r^{n-2}) = -(n-2)S_n(1)\delta(r), \tag{14}$$

where $S_n(1)$ is the surface of the sphere of unit radius in n-dimensional space. We have already proved (14) in Example 6 of Section 4.4 in a different context.

Example 3. Let us consider the distribution $\delta(t^2 - r^2)$, where $r^2 = x_1^2 + x_2^2 + x_3^2$ so that $\Sigma(t)$ is given by $t^2 - x_1^2 - x_2^2 - x_3^2 = 0$. Then $\delta(\Sigma)$ is concentrated on a sphere in R_3, and we have

$$\langle\delta(\Sigma),\phi(x_1,x_2,x_3;t)\rangle = \int_{-\infty}^{\infty}dt\int_{\Sigma}\phi(x_1,x_2,x_3;t)dS. \tag{15}$$

Now we introduce the coordinates

$$\begin{aligned}
x_1 &= r\sin\vartheta\cos\varphi = r\omega_1,\\
x_2 &= r\sin\vartheta\sin\varphi = r\omega_2,\\
x_3 &= r\cos\vartheta = r\omega_3,\\
u &= t^2 - r^2.
\end{aligned}$$

Thus

$$dS = r^2\sin\vartheta\,d\vartheta\,d\varphi dr = r^2\,d\Omega\,dr,$$

and $du = 2t\,dt$, or

$$du/2(u+r^2)^{1/2} = dt,$$

where $d\Omega = \sin\vartheta\,d\vartheta\,d\varphi$ is the element of the solid angle. Thus (15) is equivalent to

$$\langle\delta(\Sigma),\phi\rangle = \frac{1}{2}\int_{u\geq 0}\int_{u=0}\phi(r\omega_1,r\omega_2,r\omega_3)r^2\frac{d\Omega\,dr\,du}{(u+r^2)^{1/2}}.$$

Since on $u = 0$, $r^2 = t^2$, i.e., $r = t$, this relation becomes

$$\begin{aligned}
\langle\delta(\Sigma),\phi\rangle &= \frac{1}{2}\int_{r=1}\phi(|t|\omega_1,|t|\omega_2,|t|\omega_3)|t|d\Omega\\
&= \frac{1}{2}|t|\int_{\Omega}\phi(|t|\omega_1,|t|\omega_2,|t|\omega_3)d\Omega.
\end{aligned}$$

Example 4. Consider the four-dimensional distribution $\delta(\Sigma) = \delta(x^2 - m^2)$ with $x^2 = t^2 - x_1^2 - x_1^2 - x_3^2$ and m a real constant. The surface Σ is given by $u = (t^2 - x_1^2 - x_2^2 - x_3^2) - m^2 = 0$.

For a test function $\phi(t, x)$ we have

$$\langle \delta(\Sigma), \phi \rangle = \int_{-\infty}^{\infty} \int_{\Sigma} \phi(ts, x_1, x_2, x_3) dt\, dx_1\, dx_2\, dx_3 . \tag{16}$$

To evaluate this we introduce new variables, as in the previous example,

$$x_1 = r\omega_1, \qquad x_2 = r\omega_2, \qquad x_3 = r\omega_3, \qquad u = t^2 - r^2 - m^2 ,$$

so that

$$dt = du/2(u + r^2 + m^2)^{1/2} \qquad \text{and} \qquad dx_1 dx_2 dx_3 = r^2 d\Omega\, dr .$$

Then (16) can be written as

$$\langle \delta(\Sigma), \phi \rangle = \frac{1}{2} \int_{u \geq 0} \int_{\Sigma} \phi\left(r\omega_1, r\omega_2, r\omega_3, \pm\left(u + r^2 + m^2\right)^{1/2}\right) r^2 \frac{du\, d\Omega\, dr}{(r^2 + m^2)^{1/2}} .$$

But on the surface Σ, $u = 0$ or $x^2 - m^2 = 0$, and the preceding relation becomes

$$\begin{aligned} \langle \delta(x^2 - m^2), \phi \rangle & \\ &= \frac{1}{2} \int_{x^2 - m^2 = 0} (r^2 + m^2)^{-1/2} r^2 \phi(r\omega_1, r\omega_2, r\omega_3, \pm(r^2 + m^2)^{1/2}) d\Omega\, dr \\ &= \frac{1}{2} \int_0^{\infty} \int_{\Omega} (r^2 + m^2)^{-1/2} r^2 \phi(r\omega_1, r\omega_2, r\omega_3, (r^2 + m^2)^{1/2}) d\Omega\, dr \\ &\quad + \frac{1}{2} \int_0^{\infty} \int_{\Omega} (r^2 + m^2)^{-1/2} r^2 \phi(r\omega_1, r\omega_2, r\omega_3, -(r^2 + m^2)^{1/2}) d\Omega\, dr . \end{aligned} \tag{17}$$

Performing the Ω integration and writing

$$\int_{\Omega} \phi(r\omega_1, r\omega_2, , r\omega_3, t) d\Omega = \overline{\phi}(r, t) ,$$

we find that (17) is equivalent to

$$\begin{aligned} \langle \delta(x^2 - m^2), \phi \rangle = \frac{1}{2} \int_0^{\infty} & (r^2 + m^2)^{-1/2} r^2 \overline{\phi}(r, (r^2 + M^2)^{1/2}) dr \\ & + \frac{1}{2} \int_0 (r^2 + m^2)^{-1/2} r^2 \overline{\phi}(r, -(r^2 + m^2)^{1/2}) dr . \end{aligned}$$

The distributions $\delta_+(x^2 - m^2)$ and $\delta_-(x^2 - m^2)$, which are concentrated only on the upper and lower sheets of the hyperboloid $x^2 - m^2 = 0$, are given by

$$\langle \delta_+(x^2 - m^2), \phi(x) \rangle = \frac{1}{2} \int_0^{\infty} (r^2 + m^2)^{-1/2} \overline{\phi}(r, (r^2 + m^2)^{1/2})\, dr$$

and

$$\langle \delta_-(x^2 - m^2), \phi(x)\rangle = \frac{1}{2}\int_0^\infty (r^2 + m^2)^{-1/2}\overline{\phi}(r, -(r^2 + m^2)^{1/2})dr\,,$$

respectively.

Example 5. In aerodynamics it is found that there is a thin layer of vorticity, the vortex sheet, behind the lifting surface. Across this surface there is a jump in the velocity field v. The rest of the flow is irrotational. Let Σ denote the lifting surface together with its vortex sheet. Then from (5.5.26) we have

$$\overline{\text{curl}}\; v = \text{curl}\; v + \widehat{n} \times [v]\delta(\Sigma) = \widehat{n} \times [v]\delta(\Sigma) \tag{18}$$

where we have used the fact the flow is irrotational on both sides of Σ. The solution to (18) follows from vector analysis,

$$v(x) = \frac{1}{4\pi}\int \frac{(\widehat{n} + [v]) \times \widehat{r}}{r^2}\delta(\Sigma)dx' = \frac{1}{4\pi}\int_\Sigma \frac{(\widehat{n} \times [v]) \times \widehat{r}}{r^2}\,dS\,, \tag{19}$$

where $r = |x - x'|, \widehat{r} = (x - x')/|x - x'|$.

This example was given by Farassat [14]. It was in this paper that the notation of putting a bar over the distributional derivatives was first introduced.

Example 6. By the methods outlined in this chapter we can determine the boundary conditions on an obstacle placed in a field. Let us illustrate this approach for the electromagnetic field. For this purpose we write Maxwell equations in terms of the distributional derivatives because of the presence of the singular surface $\Sigma(x, t)$. Thus, we have

$$\overline{\text{curl}}\; E + \frac{\overline{\partial} B}{\partial t} = 0 \tag{20a}$$

$$\overline{\text{div}}\; B = 0 \tag{20b}$$

$$\overline{\text{curl}}\; H - \frac{\overline{\partial} D}{\partial t} = \underset{\sim}{J}\,, \tag{20c}$$

$$\overline{\text{div}}\; D = \rho\,, \tag{20d}$$

where E is the electric field D is the displacement current, B is the density of the magnetic flux, H is the magnetic field, J is the current density and ρ is the charge density. The quantities ρ and J are composed of the volume and surface densities, i.e.,

$$\rho = \rho_V + \rho_S\delta(\Sigma)\,, \quad J = J_V + J_S\delta(\Sigma)\,. \tag{21}$$

Next, we apply formulas (5.5.4), (5.5.5), (5.5.25) and (5.5.26) in system (20) and get

$$\text{curl } E + \frac{\partial B}{\partial t} + (n \times [E] - G[B])\delta(\Sigma) = 0\,, \tag{22}$$

$$\text{div} B + n \cdot [B]\delta(\Sigma) = 0\,, \tag{23}$$

$$\text{curl } H - \frac{\partial D}{\partial t} + (n \times [H] + G[D])\delta(\Sigma) = J_V + J_S\delta(\Sigma)\,, \tag{24}$$

$$\text{Div } D + n \cdot [D]\delta(\Sigma) = \rho_V + \rho_S\delta(\Sigma)\,. \tag{25}$$

Finally, we equate the singular parts on both sides of the equation (22) to (25) and obtain the required matching conditions:

$$n \times [E] = G[B]\,, \tag{26}$$

$$n \cdot [B] = 0\,, \tag{27}$$

$$n \times [H] = J_\Sigma - G[D]\,, \tag{28}$$

$$n \cdot [D] = \rho_\Sigma\,. \tag{29}$$

When the speed G is zero we recover the classic boundary conditions.

The method of this chapter enables us to convert readily many boundary value problems to integral equations [15].

5.10. The Function Pf$\left(\frac{1}{r}\right)$ and its Derivatives

The function Pf $(1/r)$ (we will drop the symbol Pf from now on), where r is the radial distance, plays an important role in many physical and mathematical problems. For instance it defines the gravitational and Coulomb potentials. Various special functions such as Legendre functions are based on this function. The properties of $1/r$ and its derivatives at $r = 0$ play an important role and appear in the analysis which contains this function. These properties become more important in theory of elementary particles because the distance r between two particles is very small necessitating the introduction of distributional derivatives. We present the required analysis in this section [16,17].

Let us first remove the singularity of $1/r$ at $r = 0$ by multiplying it by the Heaviside function $H(r - \varepsilon)$ which is 1 for $r > \varepsilon$ and 0 for $r < \varepsilon$, by defining the function $F(x)$,

$$F(x) = \frac{H(r-\varepsilon)}{r}\,. \tag{1}$$

The corresponding distribution is therefore defined as

$$\left\langle \frac{1}{r}, \phi(x) \right\rangle = \lim_{\varepsilon \to 0} \int_{R^3} \frac{1}{r} H(r-\varepsilon)\phi(x) d^3x\,. \tag{2}$$

To differentiate $1/r$ we observe that the function $F(x)$ defined by (1) has a jump discontinuity of magnitude $1/\varepsilon$ across the sphere $\Sigma(x)$ of radius ε. Accordingly, we can use formula (5.5.4) and get

$$\frac{\partial}{\partial x_j}\left(\frac{H(r-\varepsilon)}{r}\right) = -\frac{x_j}{r^3} + \widehat{r}_j \frac{1}{\varepsilon}\delta(\Sigma), \tag{3}$$

where $\widehat{r_j} = x_j/r_j$ are the components of the unit vector n to the sphere Σ. To examine the second term on the right side of (3), we observe that

$$\lim_{\varepsilon\to 0}\int_\Sigma \phi(x)\frac{1}{\varepsilon}\widehat{r}d\Sigma = \lim_{\varepsilon\to 0}\frac{1}{\varepsilon}\int_{\Sigma(1)}\phi(x)\widehat{r}\varepsilon^2 d\omega = 0, \tag{4}$$

where $\Sigma(1)$ is the unit sphere and ω the solid angle. Thus,

$$\frac{\overline{\partial}}{\partial x_j}\left(\frac{1}{r}\right) = -\frac{x_j}{r^3} = \frac{\partial}{\partial x_j}\left(\frac{1}{r}\right), \tag{5}$$

and we find that the first order distributional and the classical derivatives coincide.

In order to compute the second order distributional derivatives we apply formula (5.5.4) to the function $\partial/\partial x_j(H(r-\varepsilon)/r)$ so that

$$\begin{aligned}\frac{\partial^2}{\partial x_i \partial x_j}\left(\frac{H(r-\varepsilon)}{r}\right) &= \left(\frac{\partial^2}{\partial x_i \partial x_j}\left(\frac{1}{r}\right)\right) H(r-\varepsilon) \\ &\quad + \frac{x_i}{r}\left(-\frac{x_j}{r^2}\right)\delta(\Sigma) \\ &= \left(\frac{3x_i x_j - r^2\delta_{ij}}{r^5}\right) H(r-\varepsilon) - \frac{x_i x_j}{r^4}\delta(\Sigma).\end{aligned} \tag{6}$$

To evaluate the contribution of the second term on the right side of (6) we proceed as in relation (4) and find that

$$\lim_{\varepsilon\to 0}\int_\Sigma \frac{x_i x_j}{r^4}\phi(x)d\Sigma = \frac{4\pi}{3}\delta_{ij}\phi(0) = \frac{4\pi}{3}\delta_{ij}\langle\delta(x),\phi(x)\rangle, \tag{7}$$

where δ_{ij} is the Kronecker delta which is one when $i = j$ and zero otherwise (i.e., the second order identity symetric tensor). Thus, relation (6) takes the form

$$\frac{\overline{\partial}^2}{\partial x_i \partial x_j}\left(\frac{1}{r}\right) = \frac{3x_i x_j - r^2\delta_{ij}}{r^5} - \frac{4\pi}{3}\delta_{ij}\delta(x). \tag{8}$$

This process can be readily continued and extended to the function $(1/r^p)$ in R_n. Indeed, the general formula corresponding to (8) is [17],

$$\overline{D}^N\left(\frac{1}{r^p}\right)=\frac{\overline{\partial}^{N_1+N_2+\cdots+N_n}}{\partial x_1^{N_1}\partial x_2^{N_2}\cdots\partial x_n^{N}}\left(\frac{1}{r^p}\right)$$
$$=\sum_{j=0}^{\left[\frac{N}{2}\right]}\frac{(-1)^{N-j}2^{N-2j}\Gamma\left(\frac{k}{2}+N-j\right)N!}{\Gamma\left(\frac{k}{2}\right)(N-2j)!j!}\Delta^j x^{N-2j}\left(\frac{1}{r^{k+2N-j}}\right)$$
$$-\sum_{j=\frac{|m|-m}{2}}^{\left[\frac{N}{2}\right]}\frac{N!\Gamma\left(\frac{k}{2}+j\right)c_{m+j,n}\beta_{N,j}}{(N-2j)!\Gamma\left(\frac{k}{2}\right)j!(2m+2j)!}\Delta^j D^{N-2j}\nabla^{2m+2}\delta(x) \tag{9}$$

where $\left[\frac{N}{2}\right]$ stands for the greatest integer $\leq N/2$, x^N is the tensor with components $x_{i_1},\ldots,x_{i_N}$, $\Delta=\delta_{ij}$, $\Delta^j=\Delta\cdots\Delta$, j times,

$$c_{m,n}=\frac{2\Gamma\left(m+\frac{1}{2}\right)\pi^{n-\frac{1}{2}}}{\Gamma\left(m+\frac{n}{2}\right)}, \tag{10}$$

and the constants $\beta_{q,p}$ are defined by $\beta_{0,p}=0$ and

$$\beta_{q,p}=\beta_{q,p-1}-\beta_{q-1,p-1}, \qquad q\geq 1. \tag{11}$$

Let us present some simple examples.

Example 1. Let us set $i=j$ in relation (8) and sum over i. The result is

$$\overline{\nabla}^2\frac{1}{r}=\left(\frac{\overline{\partial}^2}{\partial x_1^2}+\frac{\overline{\partial}^2}{\partial x_2^2}+\frac{\overline{\partial}^2}{\partial x_3^2}\right)\frac{1}{r}=-4\pi\delta(x), \tag{12}$$

which states that the fundamental solution of the Laplacian $-\overline{\nabla}^2$ is $1/4\pi r$.

Example 2. For $N=1$, formula (9) reduces to

$$\frac{\overline{\partial}}{\partial x_i}\left(\frac{1}{r^p}\right)=-p\,x_i\left(\frac{1}{r^{p+2}}\right)-\frac{c_{m,n}}{(2m)!\,p}\frac{\overline{\partial}}{\partial x_i}\overline{\nabla}^{2m}\delta(x), \quad p-n=2m. \tag{13}$$

For $p=3$ and $n=3$, $c_{m,n}=4\pi/3$, this formula takes the simple form

$$\frac{\overline{\partial}}{\partial x_i}\left(\frac{1}{r^3}\right)=-3x_i\left(\frac{1}{r^5}\right)+\frac{4\pi}{3}\frac{\partial}{\partial x_i}\delta(x). \tag{14}$$

Example 3. An application to the point-source fields.

The inverse-square field theory in gravitational field, electric charges and in many other physical fields leads to the study of the function [18]

$$\frac{n_{j_1}\cdots n_{j_k}}{r^2}, \qquad n_j=\frac{x_j}{r}. \tag{15}$$

It is required to evaluate the distributional gradient of this function [19]. For this purpose we need formula (13) for $p = k+2$, $n = 3$,

$$\frac{\overline{\partial}}{\partial x_i}\left(\frac{1}{r^{k+2}}\right) = -(k+2)x_i\left(\frac{1}{r^{k+4}}\right) - \frac{4\pi}{(k+2)!!}\frac{\overline{\partial}}{\partial x_i}\overline{\nabla}^{k-1}\delta(x), \tag{16}$$

where the double factorial symbol is defined, for n even and positive, by $n!! = n(n-2)\cdots 1$.

Let us observe an important point here. We have set $p = k+2$, $n = 3$ in formula (13). Thus if k is odd then $k-1$ is even and we get the delta term on the right hand side of (16). When k is even, this term does not appear on the right hand side and we have only the first term which is the classical derivative.

Finally, we derive the distributional derivative of the general expression (15). This is achieved by the following steps. First, we write

$$\frac{\overline{\partial}}{\partial x_i}\left(\frac{n_{j_1}\cdots n_{j_k}}{r^2}\right) = \frac{\overline{\partial}}{\partial x_i}\left(\frac{x_{j_1}\cdots x_{j_k}}{r^{k+2}}\right). \tag{17}$$

As mentioned above, for k even we get only the classical derivative which is

$$\begin{aligned}\frac{\overline{\partial}}{\partial x_i}\left(\frac{n_{j_1}\cdots n_{j_k}}{r^2}\right) &= \sum_{q=1}^{k}\delta_{ij_q}x_{j_1}\cdots\widehat{x}_{j_q}\cdots x_{j_n}r^{-k-2} \\ &\quad -(k+2)x_i x_{j_1}\cdots x_{j_k}r^{-k-4} \\ &= \frac{\sum_{q=1}^{k}\delta_{ij_q}\, n_{j-1}\cdots\widehat{x}_{j_q}\cdots n_{j_k} - (k+2)n_i n_{j_1}\cdots n_{j_k}}{r^3}.\end{aligned} \tag{18}$$

For k odd, we have to add the term

$$-\frac{c_{m,n}}{2m!(k+2)}x_{j_1}\cdots x_{j_k}\frac{\overline{\partial}}{\partial x_i}\overline{\nabla}^{k-1}\delta(x). \tag{19}$$

The next step is to use Formula (2.6.28) and get

$$\begin{aligned}x_{j_1\cdots j_k}\frac{\partial}{\partial x_i}\nabla^{k-1}\delta(x) \\ = -\left(\sum_{\sigma}\delta_{ij_{\sigma(1)}}\delta_{ij_{\sigma(2)}}\cdots\delta_{ij_{\sigma(k+1)}}\delta_{j_{\sigma(n)}}\right)\delta(x)\end{aligned} \tag{20}$$

where the sum is taken over all permutations σ of $\{1,\ldots,k\}$. Accordingly, for k odd we have

$$\begin{aligned}\frac{\overline{\partial}}{\partial x_i}\left(\frac{n_{j_1}\cdots n_{j_k}}{r^2}\right) &= \frac{\left(\sum_{q=1}^{k}\delta_{ij_q}n_{j_1}\cdots\widehat{n}_{j_q}\cdots n_{j_k}\right) - (k+2)n_i\, n_{j_1}\cdots n_{kq}}{r^3} \\ &\quad + \frac{4\pi}{k!(k+2)}\left(\sum_{\sigma}\delta_{ij_{\sigma(1)}}\delta_{j_{\sigma(2)}j_{\sigma(3)}}\cdots\delta_{j_{\sigma(n-1)}j_{\sigma(k)}}\right).\end{aligned} \tag{21}$$

Finally, we observe that these formulas can be easily extended to R_s. In this case, we have

$$\frac{\overline{\partial}}{\partial x_i}\left(\frac{n_j}{r^{s-1}}\right) = \frac{\delta_{ij} - s\, n_i n_j}{r^s} + \frac{\pi^{s/2}}{\Gamma\left(\frac{s+2}{2}\right)}\,\delta(x)\,, \tag{22}$$

and

$$\frac{\overline{\partial}}{\partial x_i}\left(\frac{n_{j_1}\cdots n_{j_k}}{r^{s-1}}\right) = \frac{\left(\sum_{q=1}^{k} \delta_{ij_q} n_{j_1}\cdots n_{j_k}\right) - (k+s-1) n_i n_{j_1}\cdots n_{j_n}}{r^s} + \frac{c_{m,s}}{(2m)!(k+s-1)}\left(\sum_{\sigma} \delta_{ij_{\sigma(1)}} \delta_{j_{\sigma(1)}} \delta_{j_{\sigma(2)}} \cdots \delta_{j_{\sigma(k\cdot)}\, j_{\sigma(n)}}\right)\delta(x)\,. \tag{23}$$

CHAPTER 6

Tempered Distributions and the Fourier Transform

6.1. Preliminary Concepts

In attempting to define the Fourier transform of a distribution $t(x)$, we would like to use the formula (in R_1)

$$\widehat{t}(u) = F(t(x)) = \int_{-\infty}^{\infty} e^{iux} t(x)\, dx\,. \tag{1}$$

However, e^{iux} is not a test function in D, so the action of t on e^{iux} is not defined. We could try Parseval's formula from classical analysis,

$$\int_{-\infty}^{\infty} \widehat{f}(x) g(x) dx = \int_{-\infty}^{\infty} f(x)\widehat{g}(x)\, dx\,, \tag{2}$$

which connects the Fourier transform of two functions $f(x)$ and $g(x)$. That is, we define

$$\langle \widehat{t}, \phi\rangle = \langle t, \widehat{\phi}\rangle\,, \qquad \phi \in D\,. \tag{3}$$

We again run into trouble because $\widehat{\phi}$ may not be a test function even though ϕ is one. These difficulties are circumvented by enlarging the class of test functions and by introducing a new class of distributions.

Test functions of rapid decay

Definition. The space S of *test functions of rapid decay* contains the complex-valued functions $\phi(x) = \phi(x_1, \ldots, x_n)$ having the following properties:

(1) $\phi(x)$ is infinitely differentiable; i.e., $\phi(x) \in C^\infty(R_n)$.

(2) $\phi(x)$, as well as its derivatives of all orders, vanish at infinity faster than the reciprocal of any polynomial.

Property 2 may be expressed by the inequality

$$|x^p D^k \phi(x)| < C_{pk}\,, \tag{4}$$

where $p = (p_1, p_2, \ldots, p_n)$ and $k = (k_1, k_2, \ldots, k_n)$ are n-tuples of nonnegative integers and C_{pk} is a constant depending on p, k, and $\phi(x)$. (Recall that x^p and $D^k\phi(x)$ are short notations for the expressions

$$x^p = x_1^{p_1} x_2^{p_2} \cdots x_n^{p_n}\,, \qquad D^k\phi = \partial^{|k|}\phi(x)/\partial x_1^{k_1}\partial x_2^{k_2} \cdots \partial x_n^{k_n}\,,$$

where $|k| = \sum_{i=1}^{n} k_i$).

It is evident that $S \supset D$, because all test functions in D vanish identically outside a finite interval, whereas those in S merely decrease rapidly at infinity. For instance, the Gaussian function $\exp(-|x|^2/2)$ belongs to S but not to D. The test functions in S form a linear space. Furthermore, if $\phi \in S$, then so is $x^p D^k \phi$ for any n-tuples p and k.

Convergence in S

A sequence of test functions $\{\phi_m(x)\}$ is said to converge to $\phi_0(x)$ if and only if the functions $\phi_m(x)$ and all their derivatives converge to ϕ_0 and the corresponding derivatives of ϕ_0 uniformly with respect to x in every bounded region R of R_n. This means that the numbers C_{pk} occurring in (4) can be chosen independently of x such that

$$|x^p(D^k\phi_m - D^k\phi_0)| < C_{pk} \tag{5}$$

for all values of m. It is not hard to show that $\phi_0 \in S$. Thus the space S is closed with respect to this convergence.

When $\phi_0(x) = 0$, the sequence $\{\phi_m(x)\}$ is called the *null sequence.*

Remark. The space D is dense in S. To prove this assertion, take an arbitrary C^∞ function $a(x)$ that equals 1 for $|x| \leq 1$ and that vanishes for $|x| \geq 2$. When $\phi(x) \in S$, the test functions

$$\phi_m(x) = a(x/m)\phi(x)\,, \qquad m = 1, 2, \ldots\,,$$

are test functions belonging to D such that the sequence $\{\phi_m(x)\}$ converges to $\phi(x)$ in the sense of S.

Functions of slow growth

A function $f(x) = f(x_1, x_2, \ldots, x_n)$ in R_n is of slow growth if $f(x)$, together with all its derivatives, grows at infinity more slowly than some polynomial. This means that there exists constants C, m, and A such that

$$|D^k f(x)| \leq C|x|^m\,, \qquad |x| > A\,. \tag{6}$$

6.2. Distributions of Slow Growth (Tempered Distributions)

A linear continuous functional t over the space S of test functions is called a *distribution of slow growth* or *tempered distribution.* According to each $\phi \in S$, there is assigned a complex number $\langle t, \phi\rangle$ with the properties

(1) $\langle t, c_1\phi_1 + c_2\phi_2\rangle = c_1\langle t, \phi_1\rangle + c_2\langle t, \phi_2\rangle$,

(2) $\lim_{m\to\infty}\langle t, \phi_m\rangle = 0$, for every null sequence $\{\phi_m(x)\} \in S$.

We shall denote by S' the set of all distributions of slow growth.

It follows from the definitions of convergence in D and in S that a sequence $\{\phi_m(x)\}$ converging to the function $\phi(x)$ in the sense of D also converges to $\phi(x)$ in the sense of S. Accordingly, every linear continuous functional on S is also a linear continuous functional on D and, therefore, $S' \subset D'$.

Fortunately, most of the distributions on D that we have discussed in the previous chapters are also distributions on S. Only those distributions on D that grow too rapidly at infinity cannot be extended to S. For instance, the locally integrable function $\exp(x^2) \in D'$ but is not a member of S' (as the reader can easily verify after reading the following analysis). In fact, just as the locally integrable functions formed a special subset of D', the functions of slow growth play that role in S'. The corresponding result is given by the following theorem:

Theorem. *Every function $f(x)$ of slow growth generates a distribution through the formula*

$$\langle f, \phi \rangle = \int f(x)\phi(x)\,dx\,, \qquad \phi \in S\,. \tag{1}$$

Proof. It is clearly a linear functional. To prove continuity, we should show that if $\{\phi_m\}$ is a null sequence in S, then $\langle f, \phi_m \rangle \to 0$ as $m \to \infty$. Now, for each m,

$$\int f(x)\phi_m(x)dx = \int \frac{f(x)}{(1+|x|^2)^l}[(1+|x|^2)^l \phi_m(x)]dx\,,$$

where $l \geq 0$ is an integer. When l is sufficiently large, $f(x)/(1+|x|^2)^l$ is absolutely integrable, and we have

$$\left| \int f(x)\phi_m(x)dx \right| \leq \sup \left| (1+|x|^2)^l \phi_m(x) \right| \int \left| \frac{f(x)}{(1+|x|^2)^l} \right| dx\,.$$

The right side approaches zero as $\phi_m \to 0$. Thus $\langle f, \phi_m \rangle \to 0$ for a null sequence $\{\phi_m\}$, proving continuity.

Note that S' can contain certain locally integrable functions that do not have slow growth. Take, for example, the function $[\cos(e^x)]' = -e^x \sin(e^x)$; it is not a function of slow growth, but it is still a member of S', as can be seen from the formula

$$\langle (\cos e^x)', \phi \rangle = -\int \cos(e^x)\phi'(x)dx\,, \qquad \phi \in S\,.$$

As in the case of D', we define convergence in S' as weak convergence. The exact definition is as follows:

Definition. The sequence $\{t_m\}$ of distributions belonging to S' converges to $t \in S'$ if, for every $\phi \in S$, $\langle t_m, \phi \rangle \to \langle t, \phi \rangle$ as $m \to \infty$.

From this definition, and because $S' \subset D'$, it follows that a sequence of distributions t_m converging in S' to a distribution $t \in S'$, converges also in D' to the distribution t.

Let us sum up the foregoing results in the form of a theorem:

Theorem. *$D \subset S$ and $S' \subset D'$. Furthermore, convergence in D implies convergence in S, and weak convergence in S' implies weak convergence in D'.*

All the singular distributions that we have studied in the previous chapters are in S' as can easily be verified. Moreover, the operations that were defined for distributions in D' remain valid in S' because S' is a subspace of D'. However, the result of some operations on a tempered distribution may not be a tempered distribution. If an operation does produce a tempered distribution, the space S' is said to be closed under that operation. Those operations are as follows:

(1) addition of distributions,
(2) multiplication of a distribution by a constant,
(3) the algebraic operations given in Section 2.5, and
(4) differentiation.

Note that in these operations we now allow the test functions to traverse S.

An example of an operation under which S' is not closed is the multiplication of a distribution by a function that is infinitely smooth. Take, for instance, the distribution $t(x)$,

$$t(x) = \sum_{m=1}^{\infty} \delta(x - m)\,, \tag{2}$$

in S'. But $e^{x^2} t(x)$ is not in S', because, for $\phi(x) = e^{-x^2} \in S$, $\langle \phi(x), e^{x^2} t(x) \rangle = 1 + 1 + \cdots + 1 + \cdots$, which does not converge. On the other hand, if we take $\phi \in D$, then $\langle \phi(x), e^{x^2} t(x) \rangle = \sum_{m=1}^{\infty} e^{m^2} \phi(m)$ possesses only a finite number of nonzero terms and therefore converges.

6.3. The Fourier Transform

Fourier transform of test functions

Let us first consider the Fourier transform of the test functions $\phi \in S$,

$$\widehat{\phi}(u) = F[\phi(x)] = \int e^{iu\cdot x} \phi(x) dx\,, \tag{1}$$

where $u \cdot x = u_1 x_1 + u_2 x_2 + \cdots + u_n x_n$ and $u_1, u_2, \ldots, u_n$ are real numbers. The inverse transform is

$$\phi(x) = F^{-1}[\widehat{\phi}(u)] = \frac{1}{(2\pi)^n} \int e^{-iu\cdot x} \widehat{\phi}(u) du\,. \tag{2}$$

We shall build up the theory for R_1. The extension to n-dimensional space is straightforward. We shall, however, mention the n-dimensional generalization of a result when it is necessary.

Using definition (1) and the definition of the space S, we prove the following theorem:

Theorem 1. *If ϕ is in S, then $\widehat{\phi}(u)$ exists and is also in S.*

Proof. In view of the rapid decay of $\phi(x)$ at $|x| = \infty$, the integral

$$\int_{-\infty}^{\infty} (ix)^k e^{iux}\phi(x)dx\,, \qquad k = 0, 1, 2, \ldots,$$

converges absolutely. This integral is the result of differentiating k times, under the integral sign, the expression for $\widehat{\phi}$. It therefore represents the kth derivative of $\widehat{\phi}$,

$$d^k\widehat{\phi}(u)/du^k = \int_{-\infty}^{\infty} (ix)^k e^{iux}\phi(x)dx = [(ix)^k\phi]^\wedge(u)\,.$$

From this relation it follows that

$$|d^k\widehat{\phi}(u)/du^k| \leq \int_{-\infty}^{\infty} |x^k\phi| dx\,,$$

so the quantity on the left side is bounded for all u. Also

$$\begin{aligned}(iu)^p\frac{d^k\widehat{\phi}(u)}{du^k} &= \int_{-\infty}^{\infty} (ix)^k\phi(x)\frac{d^p}{dx^p}e^{iux}dx \\ &= (-1)^p\int_{-\infty}^{\infty}\left[\frac{d^p}{dx^p}(ix)^k\phi(x)\right]e^{iux}dx\,,\end{aligned}$$

where we have used integration by parts. Since the term in square brackets is in S, the function $[(d^p/dx^p)(ix)^k\phi(x)]e^{iux}$ is absolutely integrable, and therefore $|u^p d^k\widehat{\phi}(u)/du^k|$ is bounded for all u. The numbers p and k being arbitrary, we find that $\widehat{\phi}(u) \in S$.

Thus, the members of S are mapped by Fourier transform into members of S. The inverse transform operation F^{-1} has analogous properties. Moreover, by the inversion formula, we have

$$\int_{-\infty}^{\infty} e^{ixz}\widehat{\phi}(z)dz = 2\pi\phi(-x)\,,$$

or

$$[\widehat{\phi}]^\wedge = 2\pi\phi(-x)\,. \tag{3}$$

This shows that every function in S is a Fourier transform of some function in S. Also, if $\widehat{\phi} = 0$, then $\phi = 0$; i.e., the Fourier transform is unique. Thus, the Fourier transform is a linear mapping of S onto itself. This mapping is also continuous. The proof is as follows.

Let $\phi_m \to \phi$ as $m \to \infty$, in S. Then from the foregoing analysis we find that

$$\begin{aligned}\left|u^p\frac{d^k}{du^k}[\phi_m - \phi]^\wedge(u)\right| &\leq \int_{-\infty}^{\infty}\left|\frac{d^k}{dx^k}[x^p(\phi_m - \phi)]\right|dx \\ &\leq \sup\left|\frac{d^k}{dx^k}[x^p(\phi_m - \phi)]\right|(1 + x^2)^l\int\frac{dx}{(1 + x^2)^l}\,.\end{aligned}$$

From this inequality it follows that

$$u^p \frac{d^k}{dx^k}\widehat{\phi}_m \to u^p \frac{d^k}{dx^k}\widehat{\phi},$$

which was to be proved.

The analogous results hold for the inverse Fourier transform. We can summarize these results in the form of a theorem:

Theorem 2. *The Fourier transform and its inverse are continuous, linear one-to-one mappings of S onto itself.*

In order to obtain the transform of a tempered distribution, we need some specific formulas for the Fourier transform of ϕ. Let us list and prove them.

The transform of $d^k\phi/dx^k$ is

$$\left[\frac{d^k\phi}{dx^k}\right]^\wedge (u) = \int_{-\infty}^{\infty} \frac{d^k\phi(x)}{dx^k} e^{iux}\,dx = (-iu)^k \int_{-\infty}^{\infty} \phi(x)e^{iux}\,dx\,,$$

wherein we have integrated by parts. Thus

$$[d^k\phi/dx^k]^\wedge(u) = (-iu)^k\widehat{\phi}(u)\,, \tag{4a}$$

or

$$[(i\,d/dx)^k\phi]^\wedge(u) = u^k\widehat{\phi}(u)\,. \tag{4b}$$

Let $P(\lambda)$ be an arbitrary polynomial with constant coefficients. Then (4a) and (4b) can be generalized to give

$$[P(d/dx)\phi]^\wedge(u) = P(-iu)\widehat{\phi}(u)\,, \tag{5a}$$

and

$$[P(i\,d/dx)\phi]^\wedge(u) = P(u)\widehat{\phi}(u)\,, \tag{5b}$$

respectively. From the relation

$$\int_{-\infty}^{\infty} (ix)^k\phi(x)e^{iux}\,dx = \frac{d^k}{du^k}\int_{-\infty}^{\infty} \phi(x)e^{iux}\,dx\,,$$

we have the kth derivative of the Fourier transform,

$$[(ix)^k\phi]^\wedge(u) = (d/du)^k\widehat{\phi}(u)\,, \tag{6a}$$

or

$$[x^k\phi]^\wedge(u) = (-i\,d/du)^k\widehat{\phi}(u)\,. \tag{6b}$$

Their generalizations are

$$[P(ix)\phi]^\wedge(u) = P(d/du)\widehat{\phi}(u)\,, \tag{7a}$$

and

$$[P(x)\phi]^\wedge(u) = P(-i\,d/du)\widehat{\phi}(u)\,, \tag{7b}$$

respectively.

A translation of $\phi(x)$ to $\phi(x-a)$ in S leads to the multiplicative factor e^{iau}, as can be readily seen by substituting $x-a$ for x in (1). Indeed,

$$\int_{-\infty}^{\infty} \phi(x-a)e^{iux}dx = e^{iau}\int_{-\infty}^{\infty}\phi(y)e^{iuy}dy\,,$$

or

$$[\phi(x-a)]^\wedge(u) = e^{iau}\widehat{\phi}(u)\,. \tag{8a}$$

On the other hand, the translation of a Fourier transform is

$$[\phi]^\wedge(u+a) = [e^{iax}\phi]^\wedge(u)\,, \tag{8b}$$

as is readily verified by replacing u by $u+a$ in (1).

Similarly, the scale expansion yields

$$[\phi(ax)]^\wedge(u)(1/|a|)\widehat{\phi}(u/a)\,. \tag{9a}$$

In particular, the reflection taking $\phi(x)$ to $\phi(-x)$ has the effect of the reflection $\widehat{\phi}(u)$ of $\widehat{\phi}(-u)$:

$$[\phi(-x)]^\wedge(u) = \widehat{\phi}(-u)\,. \tag{9b}$$

The n-dimensional analogs of the foregoing properties of the Fourier transform are easily written. They are

$$[\widehat{\phi}]^\wedge(u) = (2\pi)^n\phi(-u)\,, \tag{10}$$

$$[D^k\phi]^\wedge(u) = (-iu)^k\widehat{\phi}(u)\,, \tag{11a}$$

$$[(iD)^k\phi]^\wedge(u) = u^k\widehat{\phi}(u)\,, \tag{11b}$$

$$\left[P\left(\frac{\partial}{\partial x_1},\frac{\partial}{\partial x_2},\dots,\frac{\partial}{\partial x_n}\right)\phi(x)\right]^\wedge(u) = P(-iu_1,-iu_2,\dots,-iu_n)\widehat{\phi}(u)\,, \tag{12a}$$

$$\left[P\left(i\frac{\partial}{\partial x_1},i\frac{\partial}{\partial x_2},\dots,i\frac{\partial}{\partial x_n}\right)\phi(x)\right]^\wedge(u) = P(u_1,u_2,\dots,u_n)\widehat{\phi}(u)\,, \tag{12b}$$

$$[(ix)^k\phi]^\wedge(u) = D^k\widehat{\phi}(u)\,, \tag{13a}$$

$$[x^k\phi]^\wedge(u) = (-iD)^k\widehat{\phi}(u)\,, \tag{13b}$$

$$[P(ix_1,ix_2,\dots,ix_n)\phi]^\wedge(u) = P\left(\frac{\partial}{\partial u_1},\frac{\partial}{\partial u_2},\dots,\frac{\partial}{\partial u_n}\right)\widehat{\phi}(u)\,, \tag{14a}$$

$$[P(x_1,x_2,\dots,x_n)\phi]^\wedge(u) = P\left(-i\frac{\partial}{\partial u_1},-i\frac{\partial}{\partial u_2},\dots,-i\frac{\partial}{\partial u_n}\right)\widehat{\phi}(u)\,, \tag{14b}$$

$$[\widehat{\phi}(x-a)]^\wedge(u) = e^{ia\cdot u}\widehat{\phi}(u)\,, \tag{15}$$

$$[\phi]^\wedge(u+a) = [e^{ia\cdot x}\phi]^\wedge(u)\,, \tag{16}$$

and

$$[\phi(Ax)]^\wedge(u) = |\det A|^{-1}\widehat{\phi}((A')^{-1}u)\,, \tag{17a}$$

where A is a nonsingular matrix, A' is its transpose, and $x^k = x_1^{k_1} x_2^{k_2} \cdots x_n^{k_n}$. For a pure rotation of coordinates, $A' = A^{-1}$ and $\det A = 1$. In this case (17a) becomes

$$[\phi(Ax)]^\wedge(u) = \widehat{\phi}(Au) . \tag{17b}$$

Example. We have already observed that the Gaussian function $\phi(x) = \exp(-x^2/2)$ is a member of S. In order to compute its Fourier transform, let us take $n = 1$ and note that $\phi(x)$ satisfies the differential equation $\phi'(x) = -x\phi(x)$. Taking Fourier transform of both sides of this equation, we have, from (4a) and (6b),

$$-iu\widehat{\phi}(u) = i(d/du)\widehat{\phi}(u) , \qquad \text{or} \qquad (d/du)[\widehat{\phi}(u)e^{u^2/2}] = 0 .$$

This gives the solution $\widehat{\phi}(u) = C e^{-u^2/2}$, where C is a constant. Since $\widehat{\phi}(0) = \int_{-\infty}^{\infty} \exp(-x^2/2)dx = \sqrt{2\pi}$, it follows that $C = \sqrt{2\pi}$, and we have

$$\widehat{\phi}(u) = (2\pi)^{1/2}\phi(u) . \tag{18a}$$

That is, the Gaussian function is its own inverse except for a multiplicative factor, (this factor disappears by a slight change in the definition of the Fourier transform). This property also holds for $n > 1$, as the reader can readily verify.

From formula (18a) we can derive many related formulas. For instance, we find that

$$[e^{-ax^2}]^\wedge = \sqrt{\frac{\pi}{a}} e^{-u^2/4a} . \tag{18b}$$

Secondly, by appealing to formulas (11a) and (13b) we can evaluate the Fourier transforms of the functions

$$x^k e^{-x^2/2} \text{ and } \left(\frac{d}{dx}\right)^k e^{-x^2/2} ,$$

for all values of k. Similarly, we can evaluate the Fourier transform of the function exp $(-\alpha x^2 + \beta x + \gamma)$ by completing the square of the quadratic $(-\alpha x^2 + \beta x + \gamma)$.

Fourier transform of tempered distributions

In Section 6.1 we noted the difficulties we ran into by using Parseval's formula to define the Fourier transform of a distribution in D. When we traverse the space of test functions in S and use tempered distributions, these difficulties disappear. We then have the following definition:

Definition. The Fourier transform $\widehat{t}(u)$ of a tempered distribution $t(x)$ is provided by Parseval's formula:

$$\langle \widehat{t}, \phi \rangle = \langle t, \widehat{\phi} \rangle , \qquad \phi \in S . \tag{19}$$

The functional on the right side of (19) is well defined because $\widehat{\phi} \in S$. It is clearly linear. To prove continuity, we observe from Theorem 1 that, whenever $\phi_m \to 0$, then

$\widehat{\phi}_m \to 0$ also. Thus, for all $\phi_m \in S$, whenever $\phi_m \to 0$,

$$\langle t, \widehat{\phi}_m \rangle \to 0 \qquad \text{as} \qquad m \to \infty .$$

From (3) and definition (19) we derive the result $[\widehat{t}]^\wedge = 2\pi t(-x)$. Therefore, every distribution in S' is a Fourier transform of some member of S'. Furthermore, this relation implies

$$\langle \widehat{t}(x), \widehat{\phi}(x) \rangle = \langle t(x), \widehat{\widehat{\phi}}(x) \rangle = 2\pi \langle t(x), \phi(-x) \rangle .$$

This relation is also frequently used as a definition of the Fourier transform instead of (19).

Let a sequence $\{t_m(x)\}$ of distributions in S' converge weakly in S' to a distribution $t \in S'$. Then the sequence $\{\widehat{t_m}(u)\}$ also converges in S', so that its distributional limit is $\widehat{t}(u)$. Indeed,

$$\lim_{m\to\infty} \langle \widehat{t_m}(x), \phi(x) \rangle = \lim_{m\to\infty} \langle t_m(u), \widehat{\phi}(u) \rangle = \langle t(u), \widehat{\phi}(u) \rangle = \langle \widehat{t}(x), \phi(x) \rangle ,$$

as was to be proved. Summarizing, we have a theorem analogous to Theorem 2.

Theorem 3. *The Fourier transform is a continuous linear mapping of S' onto itself. The same is true for the inverse Fourier transform, defined as*

$$\langle F^{-1}(t), \phi \rangle = \langle t, F^{-1}(\phi) \rangle . \tag{20}$$

The importance of this result lies in the fact that, in the classical theory, the Fourier transform of one class of functions is, in general, not the same class of functions.

It is vital that definition (19) be consistent with the classical definition whenever the latter is applicable, hence we give the following theorem.

Theorem 4. *Definition (19) is consistent with the classical definition for the Fourier transform of an ordinary function $f(x)$.*

Proof.

$$\begin{aligned}\langle \widehat{f}, \phi \rangle = \langle f, \widehat{\phi} \rangle &= \int_{-\infty}^{\infty} f(x)dx \int_{-\infty}^{\infty} e^{ixy}\phi(y)dy \\ &= \int_{-\infty}^{\infty} \phi(y)dy \int_{-\infty}^{\infty} f(x)e^{ixy}dx = \int_{-\infty}^{\infty} F(y)\phi(y)dy ,\end{aligned}$$

where $F(x)$ is the classical Fourier transform of $f(x)$. This proves the theorem.

From definition (19) it follows that (4)–(9) carry over to the tempered distributions by tansposition. The corresponding formulas are

$$[d^k t/dx^k]^\wedge(u) = (-iu)^k \widehat{t}(u) , \tag{21a}$$

$$[(i\, d/dx)^k t]^\wedge(u) = u^k \widehat{t}(u) , \tag{21b}$$

$$[P(d/dx)t]^\wedge(u) = P(-iu)\widehat{t}(u) , \tag{22a}$$

$$[P(i\, d/dx)t]^\wedge(u) = P(u)\widehat{t}(u) , \tag{22b}$$

$$[(ix)^k t]^\wedge(u) = (d/du)^k \widehat{t}(u)\,, \tag{23a}$$

$$[x^k t]^\wedge(u) = (-i\,d/du)^k \widehat{t}(u)\,, \tag{23b}$$

$$[P(ix)t]^\wedge(u) = P(d/du)\widehat{t}(u)\,, \tag{24a}$$

$$[P(x)t]^\wedge(u) = P(-i\,d/du)\widehat{t}(u)\,, \tag{24b}$$

$$[t(x-a)]^\wedge(u) = e^{iau}\widehat{t}(u)\,, \tag{25a}$$

$$[t]^\wedge(u+a) = [e^{iax}t]^\wedge(u)\,, \tag{25b}$$

$$[t(ax)]^\wedge(u) = (1/|a|, \widehat{t}(u/a)\,, \tag{26a}$$

$$[t(-x)]^\wedge(u) = \widehat{t}(-u)\,. \tag{26b}$$

Let us prove one of the above results, say (21a):

$$\begin{aligned}
\langle [t^{(k)}]^\wedge(u), \phi(u)\rangle &= \langle t^{(k)}(u), \widehat{\phi}(u)\rangle \\
&= (-1)^k \langle t(u), (d^k/du^k)\widehat{\phi}(u)\rangle \\
&= \langle t(u), [(-iu)^k \phi]^\wedge(u)\rangle \\
&= \langle \widehat{t}(u), (-iu)^k \phi(u)\rangle \\
&= \langle (-iu)^k \widehat{t}(u), \phi(u)\rangle\,,
\end{aligned}$$

which is (21a).

The n-dimensional analogs of the foregoing formulas follow from the corresponding relations (10)–(17). We list them here for completeness and future reference:

$$[\widehat{t}]^\wedge(u) = (2\pi)^n(-x)\,, \tag{27}$$

$$[D^k t]^\wedge(u) = (-iu)^k \widehat{t}(u)\,, \tag{28a}$$

$$[(iD)^k t]^\wedge(u) = u^k \widehat{t}(u)\,, \tag{28b}$$

$$\left[P\left(\frac{\partial}{\partial x_1}, \frac{\partial}{\partial x_2}, \ldots, \frac{\partial}{\partial x_n}\right) t(x)\right]^\wedge (u) = P(-iu_1, -iu_2, \ldots, -iu_n)\widehat{t}(u)\,, \tag{29a}$$

$$\left[P\left(i\frac{\partial}{\partial x_1}, i\frac{\partial}{\partial x_2}, \ldots, i\frac{\partial}{\partial x_n}\right) t(x)\right]^\wedge (u) = P(u_1, u_2, \ldots, u_n)\widehat{t}(u)\,, \tag{29b}$$

$$[(ix)^k t]^\wedge(u) = D^k \widehat{t}(u)\,, \tag{30a}$$

$$[x^k t]^\wedge(u) = (-iD)^k \widehat{t}(u)\,, \tag{30b}$$

$$[P(ix_1, ix_2, \ldots, ix_n)t]^\wedge(u) = P\left(\frac{\partial}{\partial u_1}, \frac{\partial}{\partial u_2}, \ldots, \frac{\partial}{\partial u_n}\right)\widehat{t}(u)\,, \tag{31a}$$

$$[P(x_1, x_2, \ldots, x_n)t]^\wedge(u) = P\left(-i\frac{\partial}{\partial u_1}, -i\frac{\partial}{\partial u_2}, \ldots, -i\frac{\partial}{\partial u_n}\right)\widehat{t}(u)\,, \tag{31b}$$

$$[t(x-a)]^\wedge(u) = e^{ia\cdot u}\widehat{t}(u)\,, \tag{32}$$

$$\widehat{t}(u+a) = [e^{ia\cdot x}t]^\wedge(u)\,, \tag{33}$$

$$[t(Ax)]^\wedge(u) = |\det A|^{-1}\widehat{t}((A')^{-1}u)\,. \tag{34a}$$

For a pure rotation in R_n, (34a) reduces to

$$[t(Ax)]^\wedge(u) = \widehat{t}(Au)\,. \tag{34b}$$

Thus the Fourier transform of the rotation of a distribution is the rotation of its Fourier transform.

A distribution is called radial or spherically symmetric if its value depends only on $r = |x| = (x_1^2 + x_2^2 + \cdots + x_n^2)^{\frac{1}{2}}$. We find as a special case that the Fourier transform of a spherically symemtric distribution is a spherically symmetric distribution.

We can obtain the Fourier transform for the volume and surface distributions as well. For instance, for the single-layer distribution $\sigma(x)$ spread over a closed bounded hypersurface $S \in R_n$, (2.4.25) yields

$$\begin{aligned}\langle \widehat{t}, \phi\rangle = \langle t, \widehat{\phi}\rangle &= \int_S \sigma(x)\widehat{\phi}(x)\, dS \\ &= \int_S \sigma(x) dS \int \phi(u) e^{iu\cdot x}\, du = \int \phi(u) du \int_S \sigma(x) e^{iu\cdot x} dS\,.\end{aligned}$$

Thus

$$\widehat{t}(u) = \int_S \sigma(x) e^{iu\cdot x} dS\,. \tag{35}$$

From the analysis of this and the previous sections, it follows that the theory of distributions does not provide any quick methods for computing Fourier transforms. It does, of course, provide us with the Fourier transform of generalized functions, for which the classical theory is helpless. We present in the next section various interesting examples of that effect.

6.4. Examples

Example 1. The delta function

$$\begin{aligned}(a)\quad \langle [\delta(x_1, \ldots, x_n)]^\wedge, \phi\rangle = \langle \delta(x), \widehat{\phi}\rangle &= \left\langle \delta(x), \int_{-\infty}^{\infty} \phi(y) e^{ixy} dy\right\rangle \\ &= \int_{-\infty}^{\infty} \phi(y) dy = \langle 1, \phi\rangle\,.\end{aligned}$$

Thus

$$\widehat{\delta}(x) = 1\,.$$

(b) According to (6.3.27) we have

$$[1]^\wedge = [\widehat{\delta}]^\wedge = (2\pi)^n \delta(-x) = (2\pi)^n \delta(x)\,, \tag{2}$$

or

$$F^{-1}[\delta(x)] = 1/(2\pi)^n\,. \tag{3}$$

For $n = 1$ this gives the well-known integral representation formula for the delta function, which we derived in Example 2 of Section 3.5 in a different manner,

$$\delta(x) = \frac{1}{2\pi}\int_{-\infty}^{\infty} e^{ixy}\,dy\,. \tag{4}$$

(c) The application of (6.3.22a) and (6.3.29a) yields

$$[P(d/dx)\delta(x)]^\wedge(u) = P(-iu)\widehat{\delta}(u) = P(-iu) \tag{5}$$

and

$$\left[P\left(\frac{\partial}{\partial x_1}, \frac{\partial}{\partial x_2} \cdots, \frac{\partial}{\partial x_n}\right)\delta(x)\right]^\wedge = P(-iu_1, -iu_2, \ldots, -iu_n)\,. \tag{6}$$

In particular,

$$[\delta^{(k)}(x)]^\wedge(u) = (-iu)^k \tag{7}$$

and

$$[D^k\delta(x)]^\wedge(u) = (-iu)^k\,. \tag{8}$$

For $k = 2m$ and $2m + 1$, (7) becomes

$$[\delta^{(2m)}(x)]^\wedge(u) = (-1)^m u^{2m}\,, \tag{9}$$

and

$$[\delta^{(2m+1)}(x)]^\wedge(u) = (-1)^{m+1}(iu)^{2m+1}\,, \tag{10}$$

respectively.

From (8) we find that

$$[\nabla^2\delta(x)]^\wedge(u) = -\sum_{i=1}^{n} u_i^2\,, \tag{11}$$

where ∇^2 is the n-dimensional Laplacian.

(d) When we appeal to formula (6.3.32) we obtain

$$[\delta(x-a)]^\wedge(u) = e^{ia\cdot u}\,. \tag{12}$$

(e) The result of combining (6.3.33) with (2) is

$$[e^{ia\cdot x}]^\wedge(u) = (2\pi)^n\delta(u+a)\,, \tag{13}$$

or

$$F^{-1}[\delta(u+a)] = (1/2\pi)^n e^{ia\cdot x}\,. \tag{14}$$

For $n = 1$, (13) becomes

$$[e^{iax}]^\wedge(u) = 2\pi\,\delta(u+a)\,. \tag{15}$$

Because $\sin \omega x = (e^{i\omega x} - e^{-i\omega x})/2i$ we observe from (15) that

$$[\sin \omega x]^{\wedge}(u) = -i\pi[\delta(u+\omega) - \delta(u-\omega)]\,. \tag{16}$$

Similarly,

$$[\cos \omega x]^{\wedge}(u) = \pi[\delta(u+\omega) + \delta(u-\omega]\,, \tag{17}$$

$$[\sinh \omega x]^{\wedge}(u) = -\pi[\delta(u+i\omega) - \delta(u-i\omega)]\,, \tag{18}$$

$$[\cosh \omega x]^{\wedge}(u) = \pi[\delta(u+i\omega) + \delta(u-i\omega)]\,. \tag{19}$$

The n-dimensional analogs of these results can also be easily derived. For instance,

$$[\sin(\omega \cdot x)]^{\wedge}(u) = -\frac{1}{2}i(2\pi)^n[\delta(u+\omega) - \delta(u-\omega)]\,. \tag{20}$$

Example 2. The Heaviside function, $n = 1$. Since $H'(x) = \delta(x)$, we find from (6.3.21a) that $[H'(x)]^{\wedge}(u) = -iu\widehat{H}(u)$, or

$$1 = -iu[H(x)]^{\wedge}(u)\,. \tag{21}$$

Now recall from Example 5 of Section 2.4 that the solution of the equation $-iut(u) = 1$ is $t(u) = c\delta(u) + i\ Pf(1/u)$, where c is a constant. Hence, it follows from (21) that

$$[H(x)]^{\wedge}(u) = c\delta(u) + i\ Pf(1/u)\,. \tag{22}$$

Changing x to $-x$ in this formula, we find

$$[H(-x)]^{\wedge}(u) = c\delta(u) - i\ Pf(1/u)\,.$$

To find c, we use the relation $H(x) + H(-x) = 1$. Then

$$[H(x)]^{\wedge}(u) + [H(-x)]^{\wedge}(u) = 2\pi\delta(u)\,,$$

or $c = \pi$. Thus

$$[H(x)]^{\wedge}(u) = \pi\delta(u) + i\ Pf(1/u) = \frac{i}{u+i0} \text{ and}\,, \tag{23}$$

$$[H(-x)]^{\wedge}(u) = \pi\delta(u) - i\ Pf(1/u) = \frac{-i}{u-i0}\,, \tag{24}$$

where we have used formulas (2.4.18–19). If we write (23) as

$$\int_{-\infty}^{\infty} H(x)e^{ixu}dx = \pi\delta(u) + i\ Pf(1/u)$$

and separate real and imaginary parts, we obtain

$$\int_0^{\infty} \cos(ux)dx = \pi\delta(u) \text{ and}\,, \tag{25}$$

$$\int_0^\infty \sin(ux)dx = \text{Pf}\,(1/u)\,. \tag{26}$$

We also find that the function $H(a-|x|)$, where a is a constant, has a Fourier transform in the classical sense:

$$[H(a-|x|)]^\wedge(u) = \int_{-a}^{a} e^{iux}dx = 2\sin(au)/u\,. \tag{27}$$

Example 3. The Signum Function $\text{sgn}\,x$ and x^{-m}, $m > 0$. First recall that $\text{sgn}\,x = H(x) - H(-x)$. Therefore, from (23) and (24), we have

$$[\text{sgn}\,(x)]^\wedge(u) = \widehat{H}(-x) - \widehat{H}(-x) = 2i\,\text{Pf}\,(1/u)\,. \tag{28}$$

Next, we use (6.3.27) to derive

$$[(\text{sgn}\,x)^\wedge(u)]^\wedge(x) = 2\pi\text{sgn}\,(-x)\,, \qquad \text{or} \qquad [2i\,Pf(1/u)]^\wedge(x) = 2\pi\,\text{sgn}\,(-x)\,,$$

which, relabeled, yields

$$[Pf(1/x)]^\wedge(u) = i\pi\,\text{sgn}\,u\,. \tag{29a}$$

Next, we use (6.3.32) and obtain

$$[1/(x-a)]^\wedge(u) = i\pi e^{iau}\text{sgn}\,u\,. \tag{29b}$$

Since

$$\frac{1}{x^m} = \frac{(-1)^{m-1}}{(m-1)!}\frac{d^{m-1}}{dx^{m-1}}\left(\frac{1}{x}\right),$$

we find from (6.3.21a) that

$$[x^{-m}]^\wedge(u) = \frac{(-1)^{m-1}}{(m-1)!}[-iu]^{m-1}\left[\frac{1}{x}\right]^\wedge = \frac{i^m\pi}{(m-1)!}\,u^{m-1}\text{sgn}\,u\,. \tag{30a}$$

Then, with the help of (6.3.32), we obtain

$$\left[\frac{1}{(x-a)^m}\right]^\wedge(u) = i^m\pi\frac{u^{m-1}}{(m-1)!}e^{iau}\text{sgn}\,u\,, \tag{30b}$$

which reduces to (29b) for $m = 1$.

We can derive many interesting results from formulas (30a) and (30b). For instance, setting $m = 2$ in (30a) we get

$$\left[\frac{1}{x^2}\right]^\wedge(u) = -\pi\,u\text{sgn}\,u = -\pi|u|. \tag{30c}$$

Example 4. Heisenberg's delta distributions. The Fourier transforms of Heisenberg's delta functions can be obtained by the combination of the Fourier transforms of other distributions. Since

$$\delta^+(x) = \frac{1}{2}\delta(x) - (1/2\pi i)\,Pf(1/x)\,,$$

we have

$$[\delta^+(x)]^\wedge(u) = \frac{1}{2}[\delta(x)]^\wedge(u) - (1/2\pi i)[Pf(1/x)]^\wedge(u)$$
$$= \frac{1}{2} - (1/2\pi i)i\pi \text{ sgn } u$$
$$= \frac{1}{2}(1 - \text{sgn } u) = H(-u)\,. \tag{31}$$

Similarly,

$$[\delta^-(x)]^\wedge(u) = H(u)\,. \tag{32}$$

Example 5. Let us prove that

(i) $[|x|]^\wedge(u) = -\frac{2}{u^2}$ (33a)

(ii) $[x^n|x|]^\wedge(u) = -2(i)^n n! \frac{1}{u^{n+2}}$ (33b)

(iii) $[Pf(1/|x|)]^\wedge = -2\gamma - 2\ln|u|\,,$ (33c)

(iv) $[\ln|x|]^\wedge(u) = -[\pi \text{Pf} \frac{1}{x} + 2\pi\, \gamma\, \delta(x)],$ (33d)

where γ is the Euler's constant,

$$\gamma = \int_0^1 \frac{1-\cos y}{y}\,dy - \int_1^\infty \frac{\cos y}{y}\,dy$$

and the distribution $Pf[1/|x|]$ is defined as

$$\left\langle \text{Pf}\left(\frac{1}{|x|}\right), \phi \right\rangle = \int_{|x|<1} \frac{\phi(x) - \phi(0)}{|x|}\,dx + \int_{|x|>1} \frac{\phi(x)}{|x|}\,dx\,.$$

The proofs are as follows.

$$(i) \qquad [|x|]^\wedge(u) = [x\text{sgn } x]^\wedge(u) = -i\frac{d}{du}\left(2i\text{Pf } \frac{1}{u}\right),$$
$$= -\frac{2}{u^2}\,,$$

where we have used the relation (28). Observe that we can obtain this result also by taking the Fourier transform of both sides of equation (30c).

(ii) To prove (33b) we apply the formulas (23b) and (33a) and get

$$[x^n|x|]^\wedge(u) = \left(-i\frac{d}{du}\right)^n \left(-\frac{2}{u^2}\right)$$
$$= -2(i)^n\, n!\, \frac{1}{u^{n+2}}\,,$$

as desired. Observe that we can recover formula (30a) from the above formula and vice versa by taking the Fourier transform of both sides of these relations.

(iii) To derive formula (33c) we find that

$$\left\langle \left[\mathrm{Pf}\left(\frac{1}{|x|}\right)\right]^\wedge, \phi \right\rangle = \left\langle \mathrm{Pf}\left(\frac{1}{|x|}\right), \widehat{\phi} \right\rangle = \int_{-1}^{1} \frac{\widehat{\phi}(x) - \widehat{\phi}(0)}{|x|} dx + \int_{|x|>1} \frac{\widehat{\phi}}{|x|}\, dx$$

$$= \int_{-1}^{1} \frac{1}{|x|} \int \phi(u)(e^{iux} - x) du\, dx$$

$$+ \int_{|x|>1} \frac{1}{|x|} \int \phi(u) e^{iux} du\, dx$$

$$= 2 \int_0^1 \phi(u) \frac{\cos xu - 1}{x}\, du\, dx$$

$$+ 2 \int_1^\infty \int \phi(u) \frac{\cos xu}{x}\, du\, dx$$

$$= 2 \int \phi(u) \left[\int_0^{|u|} \frac{\cos y - 1}{y}\, dy + \int_{|u|}^\infty \frac{\cos y}{y}\, dy \right] du$$

$$= 2 \int \phi(u) \left[\int_0^1 \frac{\cos y - 1}{y}\, dy + \int_1^{|u|} \frac{\cos y - 1}{y}\, dy \right.$$

$$\left. + \int_1^\infty \frac{\cos y}{y}\, dy - \int_1^{|u|} \frac{\cos y}{y}\, dy \right]$$

$$= -2 \int \phi(u) \left[\int_0^1 \frac{1 - \cos y}{y}\, dy - \int_0^\infty \frac{\cos y}{y}\, dy \right.$$

$$\left. + \int_1^{|u|} \frac{dy}{y} \right] du = -2 \int \phi(u)[\gamma + \ln|u|]\, du,$$

and (33c) follows.

(iv) To prove (33d) we take the Fourier transform of both sides of (33c) so that

$$\left\{ \left[\mathrm{Pf}\left(\frac{1}{|x|}\right)\right]^\wedge (u) \right\}^\wedge (x) = -2\gamma \widehat{1} - 2[\ln|u|]^\wedge(x).$$

Next, we use formula (6.2.27) and get

$$2\pi\, Pf\left(\frac{1}{|x|}\right) = -2\gamma(2\pi\ \delta(x)) - 2[\ln|u|]^\wedge(x),$$

which yields

$$[\ln|u|]^\wedge(x) = -\pi \mathrm{Pf}\left(\frac{1}{x}\right) - 2\pi\ \gamma\ \delta(x),$$

which we relabel and get (33d).

Example 6. In (6.3.24b), namely, $[P(x)t(x)]^\wedge = P(-id/du)\widehat{t}(u)$, we put $t = 1$ and use (2); the result is

$$[P(x)]^\wedge(u) = 2\pi P(-i\, d/du)\delta(u). \tag{34}$$

Let

$$P(x) = a_0 + a_1 x + \cdots + a_n x^n \,. \tag{35}$$

Then (34) becomes

$$[P(x)]^\wedge(u) = 2\pi(a_0\delta - i a_1 \delta' + \cdots + (-i)^n a_n \delta^{(n)})(u)\,. \tag{36}$$

In particular,

$$\widehat{x} = -2\pi i \delta'(u) \quad [x^2]^\wedge(u) = -2\pi\delta''(u) \,\ldots, [x^n]^\wedge(u) = (-i)^n 2\pi \delta^{(n)}(u)\,. \tag{37}$$

Thus, the Fourier transform of the functions x^n, $n = 1, 2, \ldots$, are multipoles. This is remarkable because these functions have no classical Fourier transform. Let us recall that in Exercise 14 of Chapter 2 we found the functions $\{(-1)^m \delta^{(m)}(x)\}_{m=0}^\infty$ and $\{x^n/x!\}_{n=0}^\infty$ form a biorthogonal set. In the subsequent studies we shall find that both these properties play an important role.

Example 7. Poisson Summation Formula. In Section 3.4 we found that every locally integrable periodic function $f(x)$ can be expanded in the Fourier series

$$f(x) = \sum_{m=-\infty}^{\infty} c_m e^{imx}\,,$$

which converges to $f(x)$ in the distributional sense. Consequently, we can take the Fourier transform of right side of this equation term by term. Using (13), we obtain

$$\begin{aligned}[f(x)]^\wedge(u) &= \sum_{m=-\infty}^{\infty} c_m [e^{imx}]^\wedge \\ &= \sum_{m=-\infty}^{\infty} 2\pi c_m \delta(u+m)\,. \end{aligned} \tag{38}$$

The same result holds for a singular distribution. In particular, let us consider the series

$$\sum_{m=-\infty}^{\infty} \delta(x-m) = 1 + 2\sum_{m=1}^{\infty} \cos 2\pi m x = \sum_{m=-\infty}^{\infty} e^{i2\pi m x}\,, \tag{39}$$

which we studied in Example 2 of Section 5.1. If we set $x = X/\lambda$, where λ is a real number in (39), and use the relation $\delta(X/\lambda - m) = |\lambda|\delta(X - m\lambda)$, we obtain

$$|\lambda| \sum_{m=-\infty}^{\infty} \delta(X - m\lambda) = \sum_{m=-\infty}^{\infty} e^{i2\pi X m/\lambda}\,.$$

Multiplying both sides by an arbitrary test function $\phi(X)$ and integrating from $-\infty$ to ∞, we have

$$\begin{aligned}\sum_{m=-\infty}^{\infty} \phi(m\lambda) &= \frac{1}{|\lambda|} \sum_{m=-\infty}^{\infty} \int_{-\infty}^{\infty} e^{i2\pi X m/\lambda} \phi(X)\,dX \\ &= \frac{1}{|\lambda|} \sum_{m=-\infty}^{\infty} \widehat{\phi}\left(\frac{2\pi m}{\lambda}\right)\,. \end{aligned} \tag{40a}$$

Setting $\lambda = 1$ in (40a), we obtain

$$\sum_{m=-\infty}^{\infty} \phi(m) = \sum_{m=-\infty}^{\infty} \widehat{\phi}(2\pi m) . \tag{40b}$$

This formula is usually written in the distributional form.

$$\left[\sum_{m=-\infty}^{\infty} \delta(x - 2\pi m)\right]^{\wedge} (u) = \sum_{m=-\infty}^{\infty} \delta(u - m) . \tag{40c}$$

If the factor 2π occurring in the definitions of the Fourier transform and its inverse is properly adjusted, then relation (40c) states that the Fourier transform operator takes the Dirac comb to itself — an eigenfunction.

Next, we appeal to properties (6.2.4) and (6.2.8) of the Fourier transform and deduce the following formula from (40a):

$$\frac{\lambda}{2\pi} \sum_{m=-\infty}^{\infty} \phi(m\,\lambda) \exp(e^{-im\lambda u}) = \sum_{m=-\infty}^{\infty} \widehat{\phi}\left(u + \frac{2\pi m}{\lambda}\right) . \tag{40d}$$

Equations (40a)–(40d) are called *Poisson's summation formulas*. They generally transform a slowly converging series to a rapidly converging one. As an example, consider the test function $\phi(x) = e^{-x^2}$. Then $\widehat{\phi}(u) = (\pi)^{1/2} e^{-u^2/4}$, and (40a) becomes

$$\sum_{m=-\infty}^{\infty} e^{-m^2\lambda^2} = \frac{\pi^{1/2}}{|\lambda|} \sum_{m=-\infty}^{\infty} e^{-m^2\pi^2/\lambda^2} . \tag{41}$$

The series on the left side converges rapidly for large λ, that on the right side for small λ.

Fortunately, equations (40) remain valid even for functions of a much wider class than the test functions in S.

Let us make an interesting observation from relation (41). For this purpose we set $\lambda^2 = \pi\mu$ so that (41) becomes

$$g(\mu) = \sum_{m=-\infty}^{\infty} e^{-m^2\pi\mu} = \left(\frac{1}{\mu}\right)^{1/2} \sum_{m=-\infty}^{\infty} e^{-m^2\pi/\mu} .$$

Thus, we have

$$g(\mu) = \left(\frac{1}{\mu}\right)^{1/2} g\left(\frac{1}{\mu}\right) .$$

Next we set $f(\mu) = \sum_{m=1}^{\infty} e^{-m^2\pi\mu}$, and the above formulas yield

$$f(\mu) = -\frac{1}{2} + \frac{1}{2}\left(\frac{1}{\mu}\right)^{1/2} + \left(\frac{1}{\mu}\right)^{1/2} f\left(\frac{1}{\mu}\right) .$$

Example 8. consider the quadratic form

$$\sum_{i,j=1}^{n} a_{ij} x_i x_j = (Ax, x) , \qquad A = (a_{ij}) , \tag{42}$$

which is real and positive definite, that is,

$$(Ax, x) \geq b|x|^2, \qquad b > 0,$$

then

$$[\exp(-(Ax, x))]^\wedge(u) = \pi^{n/2}(\det A)^{-1/2}\exp(-\frac{1}{4}(u, A^{-1}u)). \tag{43}$$

To prove this result we define a nonsingular real transformation $x = By$, which reduces the quadratic form (Ax, x) to a diagonal form such that

$$(Ax, x) = (ABy, By) = (B'ABy, y) = |y|^2,$$

where B' is the transpose of B. This means that $B'AB$ is the identity matrix, and we have

$$A^{-1} = BB', \qquad (\det\ A)(\det\ B)^2 = 1.$$

Thus

$$\begin{aligned}
[\exp(-(Ax, x))]^\wedge(u) &= \int \exp[-(Ax, x) + i(u \cdot x)]dx \\
&= |\det B| \int \exp[-(ABy, By) + i(u, By)]dy \\
&= (\det A)^{-1/2} \int \exp[(-|y|^2) + i(B'u, y)]dy, \\
&= (\det A)^{-1/2} \prod_{j=1}^{n} \int \exp[-y_j^2 + i(B'u)_j y_j]dy_j \\
&= \pi^{n/2}(\det A)^{-1/2}\exp(-\frac{1}{4}|B'u|^2) \\
&= \pi^{n/2}(\det A)^{-1/2}\exp(-\frac{1}{4}(u, BB'u)] \\
&= \pi^{n/2}(\det A)^{-1/2}\exp[-\frac{1}{4}(u, A^{-1}u)],
\end{aligned}$$

as required.

Example 9. From (6.3.18b) it follows that

$$(e^{-t|x|^2})^\wedge = (\pi/t)^{n/2}e^{-|u|^2/4t}. \tag{44}$$

If we substitute $t = -is$, we encounter an ambiguity when n is odd, namely, which square root to take for $(\pi/(-is))^{n/2}$. To remedy this, let us think of t as a complex variable z. Since we do not want $e^{-z|x|^2}$ to grow too fast at infinity, we must keep $\operatorname{Re} z \geq 0$. Now, for real z the square root is positive, so if we require that $-\pi/2 \leq \arg z \leq \pi/2$, then the square root is uniquely determined. That is,

$$-is = \begin{cases} -e^{(\pi/2)i}s = e^{-(\pi/2)i}s, & s > 0, \\ -e^{(i\pi/2)i}s = e^{(\pi/2)i}s, & s < 0. \end{cases}$$

Accordingly,

$$(-\pi/is)^{n/2} = \begin{cases} (\pi/|s|)^{n/2} e^{(\pi n/4)i}, & s > 0, \\ (\pi/|s|)^{n/2} e^{-(\pi n/4)i}, & s < 0. \end{cases}$$

With this choice, (44) becomes

$$(e^{is|x|^2})^\wedge = (-\pi/is)^{n/2} e^{-i|u|^2/4s}. \tag{45}$$

It remains to verify that

$$\langle e^{is|x|^2}, \phi\rangle = (-\pi/is)^{n/2} \langle e^{-i|x|^2/4s}, \phi\rangle,$$

or

$$\int e^{is|x|^2} \widehat{\phi}(x) dx = (-\pi/is)^{n/2} \int e^{-i|x|^2/4s} \phi(x) dx,$$

for all $\phi \in S$. To prove this, we again appeal to (44), which ensures that

$$\int e^{-t|x|^2} \widehat{\phi}(x) dx = (\pi/t)^{n/2} \int e^{-|x|^2/4t} \phi(x) dx, \tag{46}$$

and accomplish the substitution by analytic continuation. For this purpose we consider

$$\Phi(z) = \int e^{-z|x|^2} \widehat{\phi}(x) dx \qquad \text{and} \qquad \Psi(z) = \left(\frac{\pi}{2}\right)^{n/2} \int e^{-|x|^2/4z} \phi(x) dx,$$

for fixed $\phi \in S$. For Re $z > 0$, the integrals converge and can be differentiated with respect to z, so they define analytic functions in Re $z > 0$.

Now from (46) we know that Φ and Ψ are equal if z is real and positive. Since an analytic function is determined by its values for real z, we have

$$\Phi(z) = \Psi(z) \qquad \text{for} \qquad \operatorname{Re} z > 0.$$

Furthermore, both Φ and Ψ are continuous up to the boundary $z = is$ for $s \neq 0$, and the result follows by taking the limits of $\Phi(\varepsilon + is)$ and $\Psi(\varepsilon + is)$ as $\varepsilon \to 0$ for positive values of ε.

For $s = 1$ and $n = 1$, (45) reduces to

$$\begin{aligned}[e^{ix^2}]^\wedge(u) &= (\pi/(-i))^{1/2} e^{-iu^2/4} = \sqrt{i}\sqrt{\pi} e^{-iu^2/4} \\ &= \sqrt{\pi} \exp\left[-\frac{1}{4} i(u^2 - \pi)\right].\end{aligned} \tag{47}$$

Example 10(a). Fourier Transform of x_+^λ. Inasmuch as

$$x_+^\lambda = x^\lambda H(x) = \lim_{t\to 0+} (e^{-tx} x^\lambda H(x)),$$

we have, for $-1 < \operatorname{Re} \lambda < 0$,

$$\begin{aligned}[x_+^\lambda]^\wedge(u) &= \int_{-\infty}^{\infty} x_+^\lambda e^{iux} dx = \lim_{t\to 0+} \int_0^\infty x^\lambda e^{i(u+it)x} dx \\ &= \lim_{t\to 0+} \int_0^\infty x^\lambda e^{isx} dx,\end{aligned} \tag{48}$$

where $s = u + it$. Since $t > 0$, $\operatorname{Im} s > 0$, and therefore $0 < \arg s < \pi$. Let us compute the integral on the right side of (48), by setting

$$isx = -\xi\,, \quad \text{or} \quad x = -\xi/is\,, \quad \text{and} \quad dx = -d\xi/is\,.$$

To find the contour for the resulting integral, we observe that $\xi = 0$ when $x = 0$. Also, when $x \to \infty$, $\xi \to \infty$ such that $\arg \xi = -\pi/2 + \arg s$. Since $\arg s$ lies between 0 and π, we have $-\pi/2 < \arg \xi < \pi/2$. Thus, as $x \to \infty$, $\xi \to \infty$ such that $\operatorname{Re} \xi > 0$, and the contour is the ray L shown in Figure 6.1. Accordingly,

$$\int_0^\infty x^\lambda e^{isx} dx = (i/s)^{\lambda+1} \int_L e^{-\xi} \xi^\lambda \, d\xi\,.$$

Now, for $\operatorname{Re} \xi > 0$, $e^{-\xi}$ is exponentially damped. Hence, by Cauchy's theorem

$$\int_P e^{-\xi} \xi^\lambda d\xi = \int_0^\infty e^{-\xi} \xi^\lambda d\xi = \Gamma(\lambda + 1)\,,$$

and (48) gives

$$\begin{aligned}[x_+^\lambda]^\wedge(u) &= \lim_{t\to 0+} (i/s)^{\lambda+1}\Gamma(\lambda+1) = \lim_{t\to 0+} [e^{i\pi(\lambda+1)/2}(u+it)^{-\lambda-1}\Gamma(\lambda+1)] \\ &= e^{i\pi(\lambda+1)/2}\Gamma(\lambda+1)(u+i0)^{-\lambda-1}\,. \end{aligned} \tag{49}$$

But from (4.4.54) we have

$$(u + i0)^{-\lambda-1} = (u_+)^{-\lambda-1} + e^{i\pi(-\lambda-1)}(u_-)^{-\lambda-1}\,.$$

Combining the above two relations, we find that

$$\begin{aligned}[x_+^\lambda]^\wedge(u) &= \Gamma(\lambda+1)[e^{i\pi(\lambda+1)/2}(u_+)^{-\lambda-1} + e^{i\pi(-\lambda-1)/2}(u_-)^{-\lambda-1}] \\ &= \Gamma(\lambda+1)e^{i\pi(\lambda+1)\operatorname{sgn}(u)/2}|u|^{-\lambda-1}\,. \end{aligned} \tag{50}$$

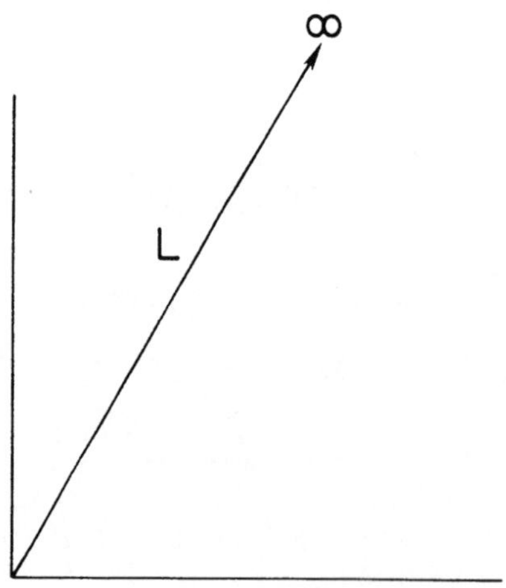

Figure 6.1. The ξ plane.

For the special case $\lambda = -\frac{1}{2}$, we find that

$$\left[\frac{H(x)}{\sqrt{|x|}}\right]^\wedge(u) = \frac{\sqrt{2\pi}}{1-i}\,\frac{1}{\sqrt{u}}\,,$$

for $-1 < \operatorname{Re}\lambda < 0$. Thereafter we use analytic continuation with respect to λ.

Example 10(b). $x_-^\lambda = |x|^\lambda H(-x)$. For $-1 < \operatorname{Re}\lambda < 0$,

$$\begin{aligned}[x_-^\lambda]^\wedge(u) &= \lim_{t\to 0-}\int_{-\infty}^{0}|x|^\lambda e^{iux}e^{-tx}\,dx = \lim_{t\to 0-}\int_0^\infty x^\lambda e^{-iux}e^{tx}\,dx\\ &= \lim_{t\to 0-}\int_0^\infty x^\lambda e^{-isx}\,dx\,, \qquad s = u + it\,.\end{aligned}$$

Proceeding as in Example 10(a), we have

$$\begin{aligned}[x_-^\lambda]^\wedge(u) &= \lim_{t\to 0-}[(-i)^{\lambda+1}s^{-\lambda-1}\Gamma(\lambda+1)] = (-i)^{\lambda+1}(u-i0)^{-\lambda-1}\Gamma(\lambda+1)\\ &= e^{-i\pi(\lambda+1)/2}\Gamma(\lambda+1)(u-i0)^{-\lambda-1}\,. \qquad (51)\end{aligned}$$

Next we use (4.4.54), $(u-i0)^{-\lambda-1} = u_+^{-\lambda-1} + e^{i(\lambda+1)\pi}u_-^{-\lambda-1}$, in (51), obtaining

$$\begin{aligned}[x_-^\lambda]^\wedge(u) &= [e^{-i\pi(\lambda+1)/2}u_+^{-\lambda-1} = u_+^{-\lambda-1} + e^{i\pi(\lambda+1)/2}u_-^{-\lambda-1}]\Gamma(\lambda+1)\\ &= \Gamma(\lambda+1)[e^{-i\pi(\lambda+1)/2}u_+^{-\lambda-1} + e^{i\pi(\lambda+1)/2}u_-^{-\lambda-1}]\\ &= i\Gamma(\lambda+1)[e^{i\pi\lambda/2}u_-^{-\lambda-1} - e^{-i\pi\lambda/2}u_+^{-\lambda-1}]\\ &= i\Gamma(\lambda+1)[e^{-i\pi\lambda\operatorname{sgn}(u)/2}|u|^{-\lambda-1}\operatorname{sgn} u]\\ &= \Gamma(\lambda+1)e^{-i\pi(\lambda+1)\operatorname{sgn}(u)/2}|u|^{-\lambda-1}\operatorname{sgn} u\,. \qquad (52)\end{aligned}$$

Example 10(c). $|x|^\lambda = x_+^\lambda + x_-^\lambda$.

$$\begin{aligned}[|x|^\lambda]^\wedge(u) &= [x_+^\lambda]^\wedge(u) + [x_-]^\lambda(u)\\ &= \Gamma(\lambda+1)[e^{i\pi(\lambda+1)\operatorname{sgn}(u)/2}|u|^{-\lambda-1}\\ &\quad + e^{-i\pi(\lambda+1)\operatorname{sgn}(u)/2}|u|^{-\lambda-1}\operatorname{sgn} u]\\ &= i\Gamma(\lambda+1)[e^{i\lambda\pi/2}u - e^{-i\lambda\pi/2}]|u|^{-\lambda-1}\\ &= -2\Gamma(\lambda+1)\sin(\lambda\pi/2)|u|^{-\lambda-1}\,. \qquad (53)\end{aligned}$$

For the special case $\lambda = -\frac{1}{2}$, we have

$$\left[\frac{1}{\sqrt{|x|}}\right]^\wedge(u) = \frac{\sqrt{2\pi}}{\sqrt{|u|}}$$

Example 10(d). $|x|^\lambda \operatorname{sgn} x = x_+^\lambda - x_-^\lambda$.

$$\begin{aligned}
[|x|^\lambda \operatorname{sgn} x]^\wedge(u) &= [x_+^\lambda]^\wedge(u) - [x_-^\lambda]^\wedge(u) \\
&= \Gamma(\lambda+1)[e^{-i\pi(\lambda+1)\operatorname{sgn}(u)/2}|u|^{-\lambda-1} \\
&\quad - e^{-i\pi(\lambda+1)\operatorname{sgn}(u)/2}|u|^{-\lambda-1}\operatorname{sgn} u] \\
&= i\Gamma(\lambda+1)[e^{-i\lambda\pi\operatorname{sgn}(u)/2}|u|^{-\lambda-1} \\
&\quad + e^{i\lambda\pi\operatorname{sgn}(u)/2}|u|^{-\lambda-1}\operatorname{sgn} u] \\
&= i\Gamma(\lambda+1)[e^{-i\lambda\pi/2} + e^{i\lambda\pi/2}]|u|^{-\lambda-1}\operatorname{sgn} u \\
&= 2i\Gamma(\lambda+1)\cos(\lambda\pi/2)|u|^{-\lambda-1}\operatorname{sgn} u\,. \qquad (54)
\end{aligned}$$

Example 11(a). $x_+^\lambda \ln x_+$. The result of differentiating (49), namely,

$$[x_+^\lambda \ln x_+]^\wedge(u) = ie^{i\lambda\pi/2}\Gamma(\lambda+1)(u+i0)^{-\lambda-1}\,,$$

with respect to λ is

$$\begin{aligned}
[x_+^\lambda \ln x_+]^\wedge(u) &= ie^{i\lambda\pi/2}[\Gamma'(\lambda+1)(u+i0)^{-\lambda-1} \\
&\quad + (i\pi/2)\Gamma(\lambda+1)(u+i0)^{-\lambda-1} \\
&\quad - \Gamma(\lambda+1)(u+i0)^{-\lambda-1}\ln(u+i0)]\,.
\end{aligned}$$

As a special case, we set $\lambda = 0$, obtaining

$$\begin{aligned}
[\ln x_+]^\wedge(u) &= i[\Gamma'(1)(u+i0)^{-1} + (i\pi/2)(u+i0)^{-1} \\
&\quad - (u+i0)^{-1}\ln(u+i0)]\,. \\
&= i\left[\frac{\frac{1}{2}\pi i + \Gamma'(1)}{u+i0} - \frac{\ln(u+i0)}{u+i0}\right]. \qquad (55)
\end{aligned}$$

Example 11(b). $x_-^\lambda \ln x_-$. When we differentiate relation (51), namely,

$$[x_-^\lambda]^\wedge(u) = -ie^{-i\pi\lambda/2}\Gamma(\lambda+1)(u-i0)^{-\lambda-1}\,,$$

we get

$$\begin{aligned}
[x_-^\lambda \ln x_-]^\wedge(u) &= -i[e^{-i\lambda\pi/2}\Gamma'(\lambda+1)(u-i0)^{-\lambda-1} \\
&\quad - (i\pi/2)e^{-i\lambda\pi/2}\Gamma(\lambda+1)(u-i0)^{-\lambda-1} \\
&\quad - e^{-i\lambda\pi/2}\Gamma(\lambda+1)(u-i0)^{-\lambda-1}\ln(u-i0)]\,.
\end{aligned}$$

For the particular case $\lambda = 0$, this becomes

$$\begin{aligned}
[\ln x_-]^\wedge(u) &= -i[\Gamma'(1)(u-i0)^{-1} - (i\pi/2)(u-i0)^{-1} \\
&\quad - (u-i0)^{-1}\ln(u-i0)] \\
&= i\left[\frac{\ln(u-i0)}{u-i0} - \frac{-\frac{1}{2}\pi i + \Gamma'(1)}{u-i0}\right]. \qquad (56)
\end{aligned}$$

Adding and subtracting (55) and (56), we obtain

$$[\ln|x|]^\wedge(u) = i[\Gamma'(1) + i\pi/2](u+i0)^{-1} - i[\Gamma'(1) - i\pi/2](u-i0)^{-1} \\ - i(u+i0)^{-1}\ln(u+i0) - i(u-i0)^{-1}\ln(u-i0) \tag{57}$$

and

$$[\ln|x|\operatorname{sgn} x]^\wedge(u) = i[\Gamma'(1) + i\pi/2](u+i0)^{-1} + i[\Gamma'(1) - i\pi/2](u-i0)^{-1} \\ - i(u+i0)^{-1}\ln(u+i0) - i(u-i0)^{-1}\ln(u-i0)\,, \tag{58}$$

respectively. It is left as an exercise for the reader to reconcile the formulas (33d) and (57).

Example 12. $r^\lambda = (x_1^2 + x_2^2 + \cdots + x_n^2)^{\lambda/2}$. Let

$$g_\lambda(u) = [r^\lambda]^\lambda(u) = \int r^\lambda e^{iu\cdot x}\,dx\,, \tag{59}$$

where $-n < \operatorname{Re}\lambda < 0$. We shall first show that $g_\lambda(u)$ is a homogeneous function of degree $-\lambda - n$, that is,

$$g_\lambda(tu) = t^{-\lambda-n}g_\lambda(u)\,. \tag{60}$$

From (59), for $t > 0$ we have

$$g_\lambda(tu) = \int r^\lambda e^{i(tu\cdot x)}\,dx = \int r^\lambda e^{i(u\cdot tx)}\,dx\,,$$

which upon setting

$$x_j = t^{-1}y_j\,, \qquad j = 1, 2, \ldots, n\,, \qquad dx = t^{-n}y\,,$$
$$r = t^{-1}|y|\,, \qquad |y| = (y_1^2 + \cdots + y_n^2)^{1/2}\,,$$

becomes

$$g_\lambda(tu) = \int |y|^\lambda t^{-\lambda-n}e^{i(u\cdot y)}\,dy = t^{-\lambda-n}g_\lambda(u)\,,$$

which is (60).

Since the Fourier transform of a spherically symmetric (radial) function is also spherically symmetric, we should have

$$g_\lambda(u) = C_\lambda\rho^{-\lambda-n}\,, \qquad \rho = (u_1^2 + u_2^2 + \cdots + u_n^2)^{1/2}\,. \tag{61}$$

To calculate the value of C_λ, we appeal to the relation

$$\langle f(x), \phi(-x)\rangle = [1/(2\pi)^n]\langle\widehat{f}(x), \widehat{\phi}(x)\rangle\,, \qquad \phi \in S\,. \tag{62}$$

For $\phi(x) = e^{-r^2/2}$, we have

$$\widehat{\phi}(u) = \int e^{i(u\cdot x)}e^{-r^2/2}\,dx = \prod_{j=1}^n \int e^{iu_jx_j}e^{-x_j^2/2}\,dx_j$$
$$= \prod_{j=1}^n [(2\pi)^{1/2}e^{-u_j^2/2}] = (2\pi)^{n/2}e^{-\rho^2/2}\,.$$

Then for $f = r^\lambda$ and $\phi = e^{-r^2/2}$, (62) gives

$$(2\pi)^n \int r^\lambda e^{-r^2/2} dx = C_\lambda (2\pi)^{n/2} \int e^{-\rho^2/2} \rho^{-\lambda-n} du \,. \tag{63}$$

Integrals on both sides can be evaluated by transforming to spherical coordinates and writing

$$dx = r^{n-1} dr\, d\omega \,, \qquad du = \rho^{n-1} d\rho\, d\Omega \,,$$

where $d\omega$ and $d\Omega$ are the solid angles in the x and u spaces, respectively. The quantities $\int d\omega$ and $\int d\Omega$ give the areas of the unit sphere in the x and u spaces, respectively. Dividing by the area of the unit sphere, the integrals on both sides of (63) can be replaced by one-dimensional integrals. These integrals and their values are

$$\int_0^\infty e^{-\rho^2/2} \rho^{-\lambda-1} d\rho = 2^{-(\lambda/2)-1} \Gamma(-\lambda/2)$$

and

$$\int_0^\infty e^{-r^2/2} r^{\lambda+n-1} dr = 2^{(\lambda+n-2)/2} \Gamma\left(\frac{\lambda+n}{2}\right)$$

and (63) yields

$$\begin{aligned} C_\lambda &= (2\pi)^{n/2} 2^{\lambda/2+1} 2^{(\lambda+n-2)/2} \Gamma\left(\frac{\lambda+n}{2}\right) \Big/ \Gamma\left(\frac{-\lambda}{2}\right) \\ &= 2^{\lambda+n} \pi^{n/2} \Gamma\left(\frac{\lambda+n}{2}\right) \Big/ \Gamma\left(\frac{-\lambda}{2}\right) . \end{aligned}$$

Consequently, from (59) and (61) we have

$$[r^\lambda]^\wedge(u) = 2^{\lambda+n} \pi^{n/2} \left[\Gamma\left(\frac{\lambda+n}{2}\right) \Big/ \Gamma\left(\frac{-\lambda}{2}\right) \right] \rho^{-\lambda-n} \,. \tag{64}$$

For other values of λ we appeal to the analytic continuation. For instance, for $\lambda = 2-n$, this formula yields

$$\begin{aligned} [1/r^{n-2}]^\wedge &= 4\pi^{n/2} \left[\Gamma\left(\frac{n-2}{2}\right) \right]^{-1} \rho^{-2} \,, \\ F^{-1}[1/\rho^2] &= F^{-1} \left[\frac{1}{u_1^2 + u_2^2 + \cdots + u_n^2} \right] \\ &= \frac{1}{(n-2) S_n(1)} \frac{1}{r^{n-2}} \,, \end{aligned} \tag{65}$$

which for $n = 3$ reduces to

$$F^{-1}[1/\rho^2] = \frac{1}{4\pi r} \,. \tag{66}$$

Example 13. For the case $n = 2$ we introduce the generalized function Pf $(1/r^2)$, $r = (x_1^2 + x_2^2)^{1/2}$, through

$$\left\langle \mathrm{Pf}\left(\frac{1}{r^2}\right), \phi \right\rangle = \int_{r<1} \frac{\phi(x) - \phi(0)}{r^2} + \int_{r>1} \frac{\phi(x)}{r^2}\, dx \,, \qquad \phi \in S \,.$$

Our contention is that

$$[\mathrm{Pf}\,(1/r^2)]^\wedge = -2\pi \ln \rho - 2\pi C\,, \tag{67}$$

where $\rho = |u| = (u_1^2 + u_2^2)^{1/2}$ and

$$C = \int_0^1 \frac{1 - J_0(\xi)}{\xi}\, d\xi - \int_1^0 \frac{J_0(\xi)}{\xi}\, d\xi\,,$$

and J_0 is the Bessel function of order zero. The proof is as follows:

$$\begin{aligned}
\langle [\mathrm{Pf}\,(1/r^2)]^\wedge, \phi\rangle = \langle \mathrm{Pf}\,(1/r^2), \phi\rangle &= \int_{r<1} \frac{\widehat{\phi}(x) - \widehat{\phi}(0)}{r^2}\, dx + \int_{r>1} \frac{\widehat{\phi}(x)}{r^2}\, dx \\
&= \int_{r>1} \frac{1}{r^2}\left(\int \phi(u)(e^{ix\cdot u} - 1)du\right) dx \\
&\quad + \int_{r>1} \frac{1}{r^2}\left(\int \phi(u)e^{ix\cdot u}\right) du\, dx \\
&= \int_0^1 \frac{1}{r} \int \phi(u) \int_0^{2\pi} [\exp(ir\rho\cos\theta)\, d\theta\, du\, dr \\
&\quad + \int_1^\infty \frac{1}{r} \int \phi(u) \int_0^{2\pi} \exp(ir\rho\cos\theta)\, d\theta\, du\, dr \\
&= 2\pi \int_0^1 \frac{1}{r} \int \phi(u)[J_0(r\rho) - 1]\, du\, dr \\
&\quad + 2\pi \int_1^\infty \frac{1}{r} \int \phi(u) J_0(r\rho) du\, dr \\
&= 2\pi \int \phi(u) \left[\int_0^1 \frac{J_0(r\rho) - 1}{r} dr + \int_1^\infty \frac{J_0(r\rho)}{r} dr\right] du \\
&= 2\pi \int \phi(u) \left[\int_0^\rho \frac{J_0(\xi) - 1}{\xi} d\xi + \int_\rho^\infty \frac{J_0(\xi)}{\xi} d\xi\right] du \\
&= -2\pi \int \phi(u)(C + \ln\rho)\, du\,,
\end{aligned}$$

from which we get the required formula (67). Observe the similarity in this example and Example 5(iii).

Example 14. The function $e^{-t|x|}$, $t > 0$, ia rapidly decaying function but is not in S, as it is not differentiable at $x = 0$. In the simple case of $n = 1$, we can find the Fourier transform directly:

$$\begin{aligned}
\widehat{f}(u) &= \int_{-\infty}^{\infty} e^{-t|x|} e^{iux}\, dx = \int_{-\infty}^{0} e^{tx + iux}\, dx + \int_0^\infty e^{-tx + iux}\, dx \\
&= \left[\frac{e^{x(t+iu)}}{t + iu}\right]_{-\infty}^0 + \left[\frac{e^{x(-t+iu)}}{-t + iu}\right]_0^\infty = \frac{1}{t + iu} - \frac{1}{-t + iu} = \frac{2t}{t^2 + u^2}\,.
\end{aligned} \tag{68}$$

The inverse Fourier transform yields

$$e^{-t|x|} = \frac{1}{\pi} \int_{-\infty}^{\infty} \frac{t}{t^2+u^2} e^{-iux}\, du\,. \tag{69}$$

To find the corresponding formula for $n > 1$, we attempt to write $e^{-t|x|}$ as an average of Gaussian functions $e^{-s|x|^2}$, that is, in the form [20]

$$e^{-tr} = \int_0^{\infty} g(t,s) e^{-sr^2} ds\,, \qquad r = |x|\,.$$

Let us begin by computing

$$\int_0^{\infty} e^{-st^2} e^{-su^2} ds = \left[\frac{e^{-s(t^2+u^2)}}{-(t^2+u^2)} \right]_0^{\infty} = \frac{1}{t^2+u^2}\,. \tag{70}$$

From (69) and (70) it follows that

$$\begin{aligned} e^{-tr} &= \frac{1}{\pi} \int_{-\infty}^{\infty} \int_0^{\infty} t e^{-st^2} e^{-su^2} ds\; e^{-iux} du \\ &= \int_0^{\infty} \frac{t}{(\pi s)^{1/2}} e^{-st^2} e^{-r^2/4s} ds\,, \end{aligned}$$

where we have performed the u integration first. Consequently,

$$\begin{aligned} F(e^{-tr})(u) &= \int_0^{\infty} \frac{t}{(\pi s)^{1/2}} e^{-st^2} F(e^{-r^2/4s}) ds \\ &= \int_0^{\infty} \frac{t}{(\pi s)^{1/2}} e^{-st^2} (4\pi s)^{n/2} e^{-s\rho^2} ds \\ &= 2^n \pi^{(n-1)/2} \int_0^{\infty} t s^{(n-1)/2} e^{-s(t^2+\rho^2)} ds\,, \end{aligned}$$

where $\rho = |u|$. To evaluate this integral we set $v = s(t^2+\rho^2)$ so that the preceding relation becomes

$$\begin{aligned} F(e^{-tr})(u) &= \frac{2^n \pi^{(n-1)/2} t}{(t^2+\rho^2)^{(n+1)/2}} \int_0^{\infty} v^{(n-1)/2} e^{-v} dv \\ &= 2^n \pi^{(n-1)/2} \Gamma\left(\frac{n+1}{2}\right) \frac{t}{(t^2+\rho^2)^{(n+1)/2}}\,, \end{aligned} \tag{71}$$

which agrees with (68) for $n = 1$.

Because $F^{-1}(e^{-t\rho}) = [1/(2\pi)^n] F(e^{-t\rho})(-x)$, we find from (71) that

$$F^{-1}(e^{-t|u|}) = \pi^{-(n+1)/2} \Gamma\left(\frac{n+1}{2}\right) \frac{t}{(t^2+|x|^2)^{(n+1)/2}}\,. \tag{72}$$

In the next example we consider a general radial distribution.

Example 15. Recall that a distribution is called radial if its value depends only on $r = |x| = (x_1^2 + x_2^2 + \cdots + x_n^2)^{1/2}$. Since the Fourier integral is invariant with respect to a rotation of orthogonal axes, it follows that the Fourier transform of $f(r)$ is also radial. We have already discussed radial functions in a few examples. In this example our aim is to show that [21]

$$\widehat{f}(\rho) = (2\pi)^{n/2} \int_0^\infty f(r) r^{n/2} \frac{J_{(n-2)/2}(\rho r)}{\rho^{(n-2)/2}}\, dr\,, \tag{73}$$

where $\rho = |u| = (u_1^2 + u_2^2 + \cdots + u_n^2)^{1/2}$.

To prove (73) we evaluate the Fourier transform

$$\widehat{f}(\rho) = \int_{R_n} f(r) e^{iu\cdot x} dx\,, \tag{74}$$

using spherical polar coordinates and taking the polar axis along the u direction, so that $u \cdot x = \rho r \cos\vartheta_1$, where $x_j = r\cos\vartheta_j$, $j = 1, \ldots, n-1$. Then (74) becomes

$$\begin{aligned}\widehat{f}(\rho) = &\int_0^\infty \int_0^\pi \cdots \int_0^\pi \int_0^{2\pi} f(r) r^{n-1} e^{i\rho r\cos\vartheta_1} \\ &\times \sin^{n-2}\vartheta_1 \sin^{n-3}\vartheta_2 \cdots \sin\vartheta_{n-2} d\vartheta_{n-1} d\vartheta_{n-2} \cdots d\vartheta_1 dr\,.\end{aligned}$$

Because

$$\int_0^\pi \sin^k\vartheta\, d\vartheta = \frac{\Gamma((k+1)/2)\pi^{1/2}}{\Gamma((k+2)/2)}\,,$$

the preceding relation reduces to

$$\begin{aligned}\widehat{f}(\rho) = &\int_0^\infty f(r) r^{n-1} 2(\pi)^{(n-1)/2} \frac{1}{\Gamma[(n-1)/2]}\, dr \\ &\times \int_0^\pi e^{i\rho r\cos\vartheta_1} \sin^{n-2}\vartheta_1\, d\vartheta_1\,.\end{aligned} \tag{75}$$

Finally, we use the identity

$$J_v(z) = \frac{(z/2)^v}{\Gamma[(2v+1)/2]\pi^{1/2}} \int_0^\pi e^{iz\cos\vartheta} \sin^{2v}\vartheta\, d\vartheta_1\,,$$

where real part of v is greater than $-\frac{1}{2}$ and find that (75) is equivalent to (73), as desired.

Example 16. Equation (6.3.35), which we may write as

$$\widehat{t}(u) = \int_S \sigma(x) e^{iu\cdot x} dS_n\,, \tag{76}$$

gives the Fourier transform of the single-layer density over a sphere S of radius a. Let $\sigma = 1$ and observe that $u \cdot x = a|u|\cos\vartheta$, and $dS_n = S_{n-1}(1) a^{n-1} \sin^{n-2}\vartheta\, d\vartheta$, where ϑ is the angle between u and x, so that (76) becomes [22]

$$\begin{aligned}\widehat{t}(u) &= a^{n-1} S_{n-1}(1) \int_0^\pi e^{ia|u|\cos\vartheta} \sin^{n-2}\vartheta\, d\vartheta \\ &= 2^{(n-2)/2} |u|^{(2-n)/2} \Gamma\left(\frac{n}{2}\right) J_{(n-2)/2}(|u|)\,,\end{aligned} \tag{77}$$

where we have used relation (3.3.4) for $S_{n-1}(1)$ and the formula

$$\int_0^\pi e^{i|u|\cos\vartheta}\sin^{n-2}\vartheta\,d\vartheta = \frac{2^{(n-2)/2}\Gamma\left(\frac{n}{2}\right)\Gamma\left(\frac{n-1}{2}\right)}{2\,\pi^{(n-1)/2}}\,|u|^{(2-n)/2}\,.$$

For $n = 2$, (77) is

$$\widehat{t}(u) = aS_1(1)\int_0^\pi e^{ia|u|\cos\vartheta}d\vartheta = \frac{1}{2}aS_1(1)\int_{-\pi}^\pi e^{ia|u|\cos\vartheta}d\vartheta = 2aJ_0(a|u|)\,, \qquad (78)$$

where we have used the integral representation formula for $J_0(a|u|)$. Similarly, for $n = 3$, we have

$$\begin{aligned}\widehat{t}(u) &= 2\pi a^2\int_0^\pi e^{ia|u|\cos\vartheta}\sin\vartheta\,d\vartheta = -2\pi a^2\int_0^\pi e^{ia|u|\cos\vartheta}d(\cos\vartheta)\\ &= -2\pi a^2[e^{ia|u|\cos\vartheta}/ia|u|]_0^\pi = 4\pi a\sin a|u|/|u|\,, \qquad (79)\end{aligned}$$

or, in the notation of the previous examples (i.e., $|u| = \rho$),

$$\widehat{t}(u) = 4\pi a\sin a\rho/\rho\,.$$

Example 17. In Chapter 10 we shall use the Fourier transforms to obtain fundamental solutions of partial differential equations. Let us examine here the concepts involved in that process. For this purpose we consider the equation

$$LE(x) = \delta(x)\,, \qquad (80)$$

where L is a differential operator with constant coefficients. Applying the Fourier transformation to both sides of this equation, we get

$$P(u)\widehat{E}(u) = 1\,, \qquad (81)$$

where $P(u)$ is some polynomial. The particular solution $\widehat{E}_p$ of (81) is

$$\widehat{E}_p(u) = 1/P(u)\,, \qquad (82)$$

while the complementary solution $\widehat{E}_c$ is obtained by solving the homogeneous equation

$$P(u)\widehat{E}_c(u) = 0\,. \qquad (83)$$

Accordingly, $\widehat{E}_c$ is the surface distribution $A\delta\{P(u)\}$ where A is an arbitrary constant. Adding this and (82), we derive the general solution:

$$\widehat{E}(u) = 1/P(u) + A\delta\{P(u)\}\,.$$

Finally, we take the inverse Fourier transform and obtain the required fundamental solution:

$$E(x) = F^{-1}[1/P(u) + A\delta\{P(u)\}]\,. \qquad (85)$$

Example 18. Let us use the analysis of Chapter 5 and obtain the Fourier transform of *functions with jump discontinuities.*

In Section 5.1 we studied a function $F(x)$ that has a jump $[F] = a$, at $x = \xi_1$ but has a continuous derivative everywhere else. Relation (5.1.1) then defines a new function $f(x)$,

$$f(x) = F(x) - a_1 H(x - \xi_1)\,, \tag{86}$$

which is continuous at ξ_1 and its derivative coincide with that of $F(x)$ on both sides of ξ_1. Our aim in this example is to find the Fourier transform of the generalized function $F(x)$. For this purpose we take the Fourier transform of both sides of equation (86). When we use formulas (6.3.25) and (6.4.23) we get

$$\widehat{F}(u) = \widehat{f}(u) + a_1 e^{iu\xi_1} \frac{1}{u + i0}\,. \tag{87}$$

In view of the smoothness of the function $f(x)$ we appeal to the Riemann Lebesgue lemma which states that for a function of this smoothness $\widehat{f}(u) \to 0$ as $u \to \infty$ (it is discussed in Section 13.5). Accordingly, relation (87) yields the required formula:

$$\widehat{F}(u) = a_1 \frac{i\, e^{iu\xi_1}}{u + i0} + o\left(\frac{1}{u}\right) \qquad u \to \infty\,. \tag{88}$$

When $F(x)$ has jumps $a_1, \ldots a_\ell$, at points $\xi_1, \cdots \xi_\ell$, the previous formula takes the form

$$\widehat{F}(u) = \sum_{j=1}^{\ell} a_j \frac{i e^{iu\xi_j}}{u + i0} + o\left(\frac{1}{u}\right), \qquad u \to \infty\,. \tag{89}$$

To obtain the corresponding results in R_n, we follow the analysis of Section (5.5). Indeed, let $F(x)$ has the jump $[F]$ across a surface Σ and define a new function

$$f(x) = F(x) - [F]H(\Sigma)\,, \tag{90}$$

where the Heaviside function $H(\Sigma)$ is the characteristic function of the region as displayed in Figure 5.2, namely

$$H(\Sigma) = \begin{cases} 0\,, & \text{in the positive side of } \Sigma\,, \\ 1\,, & \text{in the negative side of } \Sigma\,. \end{cases} \tag{91}$$

Finally, we take the Fourier transform of both sides of (91) and obtain

$$\widehat{F}(u) = [F]\widehat{H}(\Sigma) + o\left(\frac{1}{|u|}\right) \text{ as } |u| \to \infty\,. \tag{92}$$

Exercises

1. Show that if $f(x)$ is a function of slow growth on the real line, then

$$\lim_{\varepsilon\to 0+} \langle f(x)e^{-\varepsilon|x|},\ \phi(x)\rangle = \langle f, \phi\rangle\,.$$

Thus in the distributional sense $\lim_{\varepsilon\to 0+} f(x)e^{-\varepsilon|x|} = f(x)$.

2. Let t_n be a sequence of distributions on S such that $t_n \to t$, where t is a distribution on S. Show that $\widehat{t_n} \to \widehat{t}$.

3. Prove that for $m \geq 0$
   (a) $[x^m H(x)]^\wedge(u) = (-i)^m \pi \delta^m(u) + \mathrm{Pf}\,[m!/(-iu)^{m+1}]$;
   (b) $[x^m \operatorname{sgn} x]^\wedge(u) = 2\ \mathrm{Pf}\,[m!/(iu)^{m+1}]\,.$

4. Find the Fourier transform of the Bessel functions $J_0(2\pi x)$ and $J_1(2\pi x)/2x$.

5. Show that

$$[\{x-(a+ib)\}^{-m}]^\wedge(u) = 2\pi i H(-bu)\operatorname{sgn} b \frac{(-iu)^{m-1}}{(m-1)!} e^{-iu(a+ib)}\,,$$

$$m > 0\,, \qquad b \neq 0\,.$$

Note that the Fourier transform of $(x-a)^{-m}$ is not the limit of that of $\{x-(a+ib)\}^{-m}$ as $b \to 0+$ or $0-$. Rather, it is half the sum of these two limits.

6. Find the Fourier transform of
   (a) $(x^2-4)^{-1}$,
   (b) $[x^2(1+x^2)]^{-1}$,
   (c) $x^3(x^2+4x+3)^{-1}$.
   Hint. Write them as partial fractions.

7. Find the Fourier transform of $(1-x)^{-3/2}H(1-x)$.

8. Find the Fourier transform of $(\sin x - x\cos x)/x^3$.
   Hint. Use relations (5.3.23a) and (5.4.13).

9. Show that
   (a) $\left[\dfrac{\partial^k \delta(x-a)}{\partial x^k}\right]^\wedge (u) = i^{|k|} u^k;\ e^{-ia\cdot u}\,,$
   (b) $\left[x^k e^{-ia\cdot x}\right]^\wedge (u) = (2\pi)^n\ i^{|k|}\ \dfrac{\partial^k \delta(u+\alpha)}{\partial u^k}\,.$
   (c) Find the Fourier transform of $x^m \delta^{(n)}(x)$, m and n being positive integers.
   (d) Use the relation (a) above and the formula (5.1.20), namely

$$\overline{F}''(x) = \sum_{j=1}^{k} b_j\ \delta(x-\xi_j)\,,$$

for the continuous piecewise linear function $F(x)$ and find the Fourier transform of $F(x)$. Illustrate it with the function

$$F(x) = \begin{cases} 1 - |x|, & |x| \le 1 \\ 0, & |x| > 1. \end{cases}$$

10. Show that

$$[\sin|x|]^\wedge(u) = \sqrt{\pi}|u|^{\frac{-3}{2}} \cos\left(\frac{1}{4|u|} + \frac{\pi}{4}\right).$$

11. Show that
(a) $(F^{-1}g)^{(n)} = F^{-1}((-iu)^n g)$,
(b) $f^{-1}(g^{(n)}) = (ix)^n F^{-1}(g)$.

12. Prove that

$$[(\ln x_+)']^\wedge(u) = \frac{1}{2}\pi i + \Gamma'(1) - \ln(u + i0),$$
$$[(\ln x_-)']^\wedge(u) = \frac{1}{2}\pi i - \Gamma'(1) - \ln(u - i0).$$

13. Let k be a positive integer, and let f be a tempered distribution that satisfies the equation $x^k f(x) = 1$, for all x. By using the Fourier transformation, find all possible solutions for f.

14. The generalized function $t(x)$ is said to be even (odd) if

$$\int_{-\infty}^{\infty} t(x)\phi(x)dx = 0,$$

for all odd (even) functions $\phi(x) \in S$. Prove that
(a) Pf $(1/x)$ is odd
(b) $\delta(x)$ is even; $\delta^{(k)}(x)$ is even (odd) if k is even (odd)
(c) if $t(x)$ is even (odd), then $t'(x)$ is odd (even)
(d) if $t(x)$ is even (odd), then $\widehat{t}$ is even (odd).

15. Show that a tempered distribution is the finite-order derivative of a regular function which is $O(|x|^\alpha)$ for some α.

16. (a) Prove that

$$f(x) = (1 + x^2)^{-\alpha} = c_\alpha \int_0^\infty t^{\alpha-1} e^{-t|x|^2} dt,$$

$x \in R_n$, $\alpha > 0$, and c_α is a positive constant. Show that $\widehat{f}(u)$ is a positive function.
(b) Follow the analysis of Example 10(a) of Section 6.4 and show that

$$[H(x)\, x^\lambda\, e^{-sx}]^\wedge(u) = \frac{\Gamma(\lambda + 1)}{(s + iu)^{\lambda+1}}.$$

17. Verify (6.4.69) using the calculus of residues.

18. Let $g(t)$ be the even function of the single variable that for positive t is identifiable with the function $f(x)$ in R_n; that is, $f(x) = g(r), r = |x|$. Since $g(t)$ is even and in R_1, we have

$$\widehat{g}(u) = 2\int_0^\infty g(t)\cos ut\,dt\,, \qquad u \in R_1\,.$$

Now use (6.4.73) and show that for $n = 2k+1$ we have

$$\widehat{f}(u) = \frac{d^k}{ds^k}\widehat{g}\left(\sqrt{s}\right)\,, \qquad u \in R_n\,,$$

where $s = \rho^2$. For $n = 3$ $(k = 1)$, this reduces to $\widehat{f}(u) = -\widehat{g}'(\rho)/2\pi\rho$; derive this formula distributionally.

19. Write (6.4.73) in terms of $J_0(r\sqrt{s})$, $s = \rho^2$, for $n = 2k+2$, by using the recurrence relation

$$\frac{2^k J_k(r\sqrt{s})}{r^k} = \frac{d^k}{ds^k}J_0(r\sqrt{s})\,.$$

Similarly, write (6.4.73) in terms of $J_{-1/2}(r\sqrt{s}) = (1/\pi r\sqrt{s})^{1/2}\cos(r\sqrt{s})$ for $n = 2k+1$.

20. In (6.4.40a) set $\lambda = 1$ and $\phi(x) = e^{-a|x|}$ to show that

$$\frac{1+e^{-a}}{1-e^{-a}} = \sum_{m=-\infty}^{\infty}\frac{2a}{a^2+4\pi^2m^2}\,.$$

21. (a) Prove the following generalization of Poisson's summation formulas (6.4.40a):

$$|b|\sum_{m=-\infty}^{\infty}\phi(a+mb) = \sum_{m=-\infty}^{\infty}e^{-2m\pi ia/b}\widehat{\phi}(2\pi m/b)\,.$$

Use this formula to prove Jacobi's identity

$$\theta(z,\tau) = \sqrt{\frac{i}{\tau}}\exp\left(-i\pi^2\frac{z^2}{t}\right)\theta\left(\frac{z}{\tau}, -\frac{1}{\tau}\right),$$

where $\theta(z,t)$ is the theta function

$$\theta(z,t) = \sum_{m=-\infty}^{\infty}\exp(i\pi m^2\tau + 2i\pi mz)\,,\ \operatorname{Im} z > 0\,.$$

Hint: Use formula (6.3.18b):

$$[\exp(-a\,x^2)]^\wedge = \sqrt{(\pi/a)}\exp(e^{-u^2/4})$$

and relate the quantities a, b with z, τ appropriately.

(b) Consider the formula

$$\sum_{|m|} \delta(A(x) - m) = \sum_{|m|} e^{i2\pi\, m\cdot A(x)},$$

where $A(x)$ is a smooth vector field in R_n. This relation is the n-dimensional analog of equation (6.4.39). Now proceed as in Example 7 of Section 6.4 and derive the Poisson summation formulas which correspond to relations (6.4.40). You shall find formula (3.1.8) helpful in this derivation.

22. Considering the function $e^{-x^2-2\pi iax}$ prove that

$$1+2\sum_{m=1}^{\infty} e^{-m^2\lambda^2}\cos(2m\pi\lambda a) = \frac{\pi^{1/2}}{\lambda} e^{-x^2a^2}\left(1+2\sum_{m=1}^{\infty} e^{-m^2\pi^2/\lambda^2}\cosh\frac{2\pi^2 ma}{\lambda}\right).$$

23. Borrow the formula

$$\frac{\partial}{\partial \bar{z}}\left(\frac{1}{z}\right) = \frac{1}{2\pi}\left(\frac{\partial}{\partial x} + i\frac{\partial}{\partial y}\right)\frac{1}{z} = \delta(x, y)$$

from Section 10.4 and apply formula (6.3.29a) to prove that

$$\left(\frac{1}{z}\right)^{\wedge}(u_1, u_2) = \frac{2\pi}{u_2 - u_1}.$$

24. Show that
(a) $F^{-1}[1/(\rho^2+\sigma^2)] = (1/4\pi)e^{-\sigma r}/r$,
(b) $F^{-1}[1/(\rho^2+4i\sigma u_1)] = e^{2\sigma x_1}e^{-2\sigma r}/4\pi r$,
where $\rho = |u|$, $r = |x|$, and σ is a constant.

25. (a) Use the identity $(\partial/\partial x^j)r = x_j/r$, and formula (6.4.66) to show that

$$\widehat{r}(u) = -\frac{1}{2\pi}\frac{1}{\rho^4}.$$

(b) In Section 5.10 we derived the expansion of $1/r$ in terms of the delta function. Use that analysis and derive the Fourier transform of $D^n(1/r)$ for $n = 2$ and 3. In particular, from formula (5.10.8), deduce the value of

$$\left[\frac{3x_i x_j - r^2\delta_{ij}}{r^5}\right]^{\wedge}(u).$$

26. Show that in R_2,

$$\left[\frac{H(t-|x|)}{\sqrt{t^2-|x|^2}}\right]^{\wedge} = \frac{2\pi\sin(t|u|)}{|u|}.$$

27. Show that

$$[\delta(x_1)\delta(r-a)\widehat{\vartheta}_0]^\wedge(u) = (2\pi i a/U)(\widehat{e}_x \times u)J_1(Ua)\,,$$

where $r = \sqrt{x_2^2 + x_3^2}$, $\widehat{\vartheta}_0$ is the unit azimuthal vector, $\widehat{e}_x$ the unit vector along x_1 axis, a is constant, $U = |\widehat{e}_x \times u|$, and J_1 is a Bessel function.

28. The Hermite functions $h_n(x)$ are defined as

$$h_n(x) = e^{-\pi x^2} H_n(2\sqrt{\pi}x)\,,$$

where

$$H_n(x) = \sum_{n=0}^{[n/2]} \frac{(-1)^j 2^{n-2j} n! x^{n-2}}{(n-2j)!\, j!}$$

are the Hermite polynomials. Show that the Hermite functions are the eigenfunctions of the Fourier transform mapping.

29. **Fourier transform of an integral**. Show that

$$\left[\int_{-\infty}^{x} t(s)ds\right]^\wedge(u) = \frac{i\widehat{t}(u)}{u} + \frac{1}{2}\widehat{t}(0)\delta(u).$$

CHAPTER 7

Direct Products and Convolutions of Distributions

7.1. Definition of the Direct Product

Let R_m and R_n be Euclidean spaces of dimensions m and n respectively, and let $x = (x_1, \ldots, x_m)$ and $y = (y_1, \ldots, y_n)$ denote the generic points in R_m and R_n, respectively. Then a point in the Cartesian product $R_{m+n} = R_m \times R_n$ is $(x, y) = (x_1, \ldots, x_m, y_1, \ldots, y_n)$. Furthermore, let us denote by D_m, D_n, and D_{m+n} the spaces of test functions with compact support in R_m, R_n, and R_{m+n}, respectively. When $f(x)$ and $g(y)$ are locally integrable functions in the spaces R_m and R_n, then the function $f(x)g(y)$ is also locally integrable function in R_{m+n}. It defines the regular distribution:

$$\begin{aligned}\langle f(x)g(y), \phi(x, y)\rangle &= \int f(x) \int g(y)\phi(x, y)dy\, dx \\ &= \langle f(x), \langle g(y), \phi(x, y)\rangle\rangle \end{aligned} \tag{1}$$

or

$$\begin{aligned}\langle g(y)f(x), \phi(x, y)\rangle &= \int g(y) \int f(x)\phi(x, y)dx\, dy \\ &= \langle g(y), \langle f(x), \phi(x, y)\rangle\rangle \end{aligned} \tag{2}$$

for $\phi(x, y) \in D_{m+n}$. Let us denote by $s(x) \otimes t(y)$ the direct product of the distributions $s(x) \in D'_m$ and $t(y) \in D'_n$ according to (1),

$$\langle s(x) \otimes t(y), \phi(x, y)\rangle = \langle s(x), \langle t(y), \phi(x, y)\rangle\rangle\,, \qquad \phi(x, y) \in D_{m+n}\,, \tag{3}$$

and check whether the right side of this equation defines a linear continuous functional over D_{m+n}. For this purpose, we prove the following lemma:

Lemma 1. *The function $\psi(x) = \langle t(y), \phi(x, y)\rangle$, where $t \in D'_n$ and $\phi(x, y) \in D_{m+n}$, is a test function in D_m, and*

$$D^k\psi(x) = \langle t(y), D_x^k\phi(x, y)\rangle \tag{4}$$

for all multiindices k, where D_x^k implies differentiation with respect to $(x_1, x_2, \ldots, x_m)$ only. Also, if the sequence $\{\phi_l(x, y)\} \to \phi(x, y)$ in D_{m+n} as $l \to \infty$, then the sequence $\psi_l(x) = \{\langle t(y), \phi_l(x, y)\rangle\} \to \psi(x)$ in D_m as $l \to \infty$.

Proof. For every point $x \in R_m$, $\phi(x, y)$ is a test function in D_m and as such is well defined in R_m. To prove that it is continuous, we fix x and let a sequence $\{x_l\} \to x$ as $l \to \infty$. Then

$$\phi(x_l, y) \to \phi(x, y)\,, \qquad y \in R_n\,, \tag{5}$$

because the supports of $\phi(x_l, y)$ are bounded in R_n independently of l (see Figure 7.1) and for all q

$$D_y^q\phi(x_l, y) \to D_y^q\phi(x, y) \quad \text{as} \quad l \to \infty, \qquad y \in R_n .$$

Now we appeal to the continuity of the functional $t(y)$ on D_n and find, from (5), that

$$\begin{aligned}\psi(x_l) &= \langle t(y), \phi(x_l, y)\rangle \to \langle t(y), \phi(x, y)\rangle \\ &= \psi(x) \qquad \text{as} \qquad x_l \to x .\end{aligned}$$

This proves that $\psi(x)$ is a continuous function.

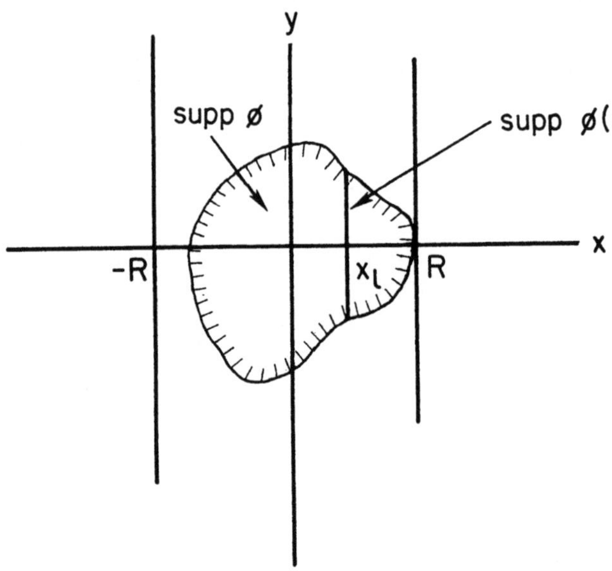

Figure 7.1.

To prove (4) we again fix a point x in R_m and set $h_i = (0, 0, \ldots, h, \ldots, 0)$, where h is located at the ith place in the row. Then

$$\chi^{(i)}(y) = \frac{1}{h}[\phi(x + h_i, y) - \phi(x, y)] \to \frac{\partial\phi(x, y)}{\partial x_i} \qquad \text{as} \quad h \to 0, \tag{6}$$

in D_m. Also, the supports of $\chi^{(i)}$ are bounded in R_n independently of h, and we have, for all q,

$$\begin{aligned}D^q\chi^{(i)}(y) &= (1/h)[D_y^q\phi(x + h_i, y) - D_y^q\phi(x, y)] \\ &\to D_y^q\frac{\partial\phi(x, ky)}{\partial x_i} \qquad \text{as} \quad h \to 0 \quad y \in R_n .\end{aligned}$$

Accordingly, we can use (6) as well as the continuity of $t(y)$ and observe that

$$\begin{aligned}\frac{\psi(x+h_i)-\psi(x)}{h} &= \frac{1}{h}[\langle t(y), \phi(x+h_i, y)\rangle - \langle t(y), \phi(x, y)\rangle] \\ &= \left\langle t(y), \frac{\phi(x+h_i, y)-\phi(x, y)}{h}\right\rangle \\ &= \langle t, \chi^{(i)}\rangle = \langle t(y), \partial\phi(x, y)/\partial x_i\rangle \quad \text{as} \quad h \to 0 .\end{aligned}$$

Thus, (4) is valid for $k = (0, 0, \ldots, 1, \ldots, 0)$, where the 1 is located at ith place. By repeated applications of the preceding steps, we derive (4) in its full generality.

We have thereby proved that $\psi(x) \in C^\infty(R_m)$. It still remains to be proved that $\psi(x)$ has compact support. But this follows from the fact that $\phi(x, y) = 0$ for $|x| > R$ (see Figure 7.1); for these values of x, $\psi(x) = \langle t(y), 0\rangle = 0$. Thus, $\psi(x)$ is a test function.

Finally, to prove the third part it suffices to show that if $\{\phi_l(x, y)\}$ is a null sequence then so is $\{\psi_l(x)\}$. Now, the supports of $\phi_l(x, y)$ are bounded in R_{m+n} independently of l, so the supports of $\{\psi_l(x)\}$ are also bounded independently of l. Accordingly, we have only to prove that

$$D^k \psi_l(x) \to 0 \quad \text{as} \quad l \to \infty, \ x \in R_m . \tag{7}$$

Suppose that (7) does not hold and we can find a number $\varepsilon_0 > 0$, a multiindex k_0, and a sequence of points x_l such that

$$D^{k_0}[\psi_l(x_l)] \geq \varepsilon_0 , \qquad l = 1, 2, \ldots . \tag{8}$$

Since the supports of $\psi_l(x, y)$ are bounded in R_m independently of l, it follows that the sequence $\{x_l\}$ is also bounded in R_m. Appealing to the Bolzano-Weierstrass theorem, we can choose a convergent subsequence $x_{l_j} \to x_0$ as $j \to \infty$. Then,

$$D_x^{k_0}\phi_{l_j}(x_{l_j}, y) \to 0 \quad \text{as} \quad j \to \infty .$$

Consequently, the distribution $t(y)$ satisfies the relation

$$D^{k_0}\psi_{l_j}(x_{l_j}) = \langle t(y), D_x^{k_0}\phi_{l_j}(x_{l_j}, y)\rangle - \langle t(y), 0\rangle = 0 \quad \text{as} \quad j \to \infty . \tag{9}$$

We have the required contradiction, and therefore the lemma is proved.

Returning to definition (3), and using Lemma 1, we find that $\psi(x) = \langle t(y), \phi(x, y)\rangle \in D_m$ for all $\phi \in D_{m+n}$. Thus, the right side of (3), namely, $\langle s, \psi\rangle$, is defined for any distributions s and t and is a functional over D_{m+n}. The linearity of this functional follows from the linearity of the functionals s and t.

To prove the continuity of this functional, let the sequence $\{\phi_1\} \to \phi$ as $l \to \infty$ in D_{m+n}. Then, in view of Lemma 1.

$$\langle t(y), \phi_l(x, y)\rangle \to \langle t(y), \phi(x, y)\rangle \quad \text{in} \quad D_m \quad \text{as} \quad l \to \infty .$$

Since the functional is continuous, we have

$$\langle s(x), \langle t(y), \phi_l(x, y)\rangle\rangle \to \langle s(x), \langle t(y), \phi(x, y)\rangle\rangle , \tag{10}$$

as $l \to \infty$. This proves the continuity of the functional defined by the right side of (3) and $s(x) \otimes t(y)$ is a generalized function in D'_{m+n}.

Some properties of the direct product

Property 1. Commutativity. The direct product is commutative.

Proof. Let the test function $\phi(x, y) \in D_{m+n}$ have the form

$$\sum_{t=1}^{p} \phi_l(x)\psi_l(y), \qquad \phi_l(x) \in D_m, \qquad \psi_l(y) \in D_n. \tag{11}$$

Then according to definition (3) we have the following expression for both $\langle s \otimes t, \phi\rangle$ and $\langle t \otimes s, \phi\rangle$:

$$\sum_{l=1}^{p} \langle s, \phi_l\rangle\langle t, \psi_l\rangle. \tag{12}$$

Now, to prove that this result holds for any test function $\phi(x, y)$, we show that the test functions of form (11) are dense in the space D_{m+n}. For this purpose, let us denote the space of the test function of form (11) $D_m \otimes D_n$ and prove the following lemma:

Lemma 2. *$D_m \otimes D_n$ is dense in D_{m+n}.*

The lemma will be proved if we can show that for any function $\phi(x, y) \in D_{m+n}$ there is a sequence of test functions $\{\phi_l(x, y)\}$ of form (11) converging to $\phi(x, y)$. Suppose that $\phi(x, y)$ has support R ($|x| \le \alpha$, $|y| \le \alpha$); that is, it vanishes outside this block R. Then for a given $\varepsilon = 1/l$, we can construct, by virtue of Weierstrass's theorem, polynomials $P_l(x, y)$ that differ in the region R' ($|x| \le 2\alpha$, $|y| \le 2\alpha$) from $\phi(x, y)$ by less than ε. The same is true for all derivatives of order k. Let $e(x)$ be the function

$$e(x) = \begin{cases} 1, & |x| < \alpha, \\ 0, & |x| > \alpha. \end{cases}$$

Then, $\phi_l(x, y)e(x)e(y) \to \phi(x, y)$ in D_{m+n} as $l \to \infty$.

Property 2. Continuity. When the sequence $\{s_l\} \to s$ in D'_m as $l \to \infty$, then

$$\{s_l(x) \otimes t(y)\} \to s(x)t(y),$$

in D'_{m+n}.

Proof. According to Lemma 1, for $\phi(x, y) \in D_{m+n}$, $\psi(x) = \langle t(y), \phi(x, y)\rangle \in D_m$. Thus,

$$\begin{aligned} \langle s_l(x) \otimes t(y), \phi(x, y)\rangle &= \langle s_l(x), \langle t(y), \phi(x, y)\rangle\rangle = \langle s_l, \psi\rangle \\ &\to \langle s, \psi\rangle = \langle s, \langle t(y), \phi(x, y)\rangle\rangle \\ &= \langle s(x) \otimes t(y), \phi(x, y)\rangle \quad \text{as} \quad l \to \infty. \end{aligned}$$

The theory of the direct product of two distributions can be readily extended to the direct product of any finite number of distributions.

Property 3. Associativity. For $s \in D'_m$, $t \in D'_n$, and $u \in D'_p$ we have

$$s(x) \otimes [t(y) \otimes u(z)] = [s(x) \otimes t(y)] \otimes u(z) . \tag{13}$$

Proof. Let $\phi(x, y, z)$ be a test function in R_{m+n+p}, where p is the dimension of the $z =$ space. Then

$$\begin{aligned}\langle s(x) \otimes [t(y) \otimes u(z)], \phi(x, y, z)\rangle &= \langle s(x), \langle t(y) \otimes u(z), \phi\rangle\rangle \\ &= \langle s(x), \langle t(y), \langle u(z), \phi\rangle\rangle\rangle \\ &= \langle s(x) \otimes t(y), \langle u(z), \phi\rangle\rangle \\ &= \langle [s(x) \otimes t(y)] \otimes u(z), \phi(x, y, z)\rangle .\end{aligned}$$

Property 4. Support.

$$\text{supp}\,(s \otimes t) = (\text{supp}\, s) \times (\text{supp}\, t) . \tag{14}$$

This means that the set of points in supp $(s \otimes t)$ consists of just those points (x, y) in $R_{m,n}$ in which the first coordinate, x, belongs to supp s and the second coordinate, y, belongs to supp t.

Proof. We want to show that when a point x_0 lies outside supp $s(x)$, then every point (x_0, y_0) also lies outside supp $(s(x) \otimes t(y))$, no matter what value y_0 takes. By the definition of the support of the distribution given in Section 2.8, we find that under this premise there is some neighborhood $R(x_0)$ of x_0 such that $\langle s(x), \phi(x)\rangle = 0$ for every $\phi(x) \in D_m$ whose support is contained in this neighborhood. For a test function $\phi(x, y) \in D_{m+n}$ whose support lies in the block $R(x_0) \times R(y_0)$, (i.e., $\phi(x, y)$ is zero whenever x is not in $R(x_0)$), we find that the support of the test function $\psi(t) = \langle t(y), \phi(x, y)\rangle$ is contained in $R(x_0)$. Hence,

$$\langle s(x) \otimes t(y), \phi(x, y)\rangle = \langle s(x), \psi(x)\rangle = 0 .$$

Next we assume that y_0 is outside the support of $t(y)$. By the same arguments we conclude that in this case also (x_0, y_0) is not in the support of $s(x) \otimes t(y)$, no matter what value x_0 takes. Thus, we have proved that

$$\text{supp}\,(s \otimes t) \subset \text{supp}\, s \times \text{supp}\, t . \tag{15}$$

On the other hand, let $x_0 \in \text{supp}\, s$ and $y_0 \in \text{supp}\, t$. Then according to Lemma 2 we can find a test function $\phi(x)\psi(y)$ for any neighborhood $R(x_0, y_0)$ of the point (x_0, y_0), where $\phi(x) \in D_m$ and $\psi(y) \in D_n$, such that the support of $\phi(x)\psi(y)$ is contained in $R(x_0, y_0)$. Thus,

$$\langle s \otimes t, \phi\psi\rangle = \langle s\phi, t\psi\rangle \neq 0 ,$$

which means that

$$\text{supp}\, s \times \text{supp}\, t \subset \text{supp}\,(s \otimes t) . \tag{16}$$

Combining (15) and (16) we have (14).

Property 5. Differentiation.

$$D_x^k[s(x) \otimes t(y)] = [D^k s(x) \otimes t(y)]. \tag{17a}$$

Proof. For $\phi(x, y) \in D_{m+n}$ we have

$$\begin{aligned}\langle D_x^k[s(x) \otimes t(y)], \phi(x, y)\rangle &= (-1)^{|k|}\langle s(x) \otimes t(y), D_x^k \phi(x, y)\rangle \\ &= (-1)^{|k|}\langle t(y), \langle s(x), D_x^k \phi(x, y)\rangle\rangle \\ &= \langle t(y), \langle D^k s(x), \phi(x, y)\rangle\rangle,\end{aligned}$$

which proves (17a). Similarly, it follows (see Exercise 1) that

$$D_x^k D_y^q[s(x) \otimes t(y)] = [D_x^k s(x)] \otimes [D_y^q t(y)]. \tag{17b}$$

Property 6. Multiplication by a C^∞ function. For $a(x) \in C^\infty$ we have

$$a(x)[s(x) \otimes t(y)] = [a(x)s(x)] \otimes t(y). \tag{18}$$

Proof. For $\phi(x, y) \in D_{m+n}$,

$$\begin{aligned}\langle a(x)[s(x) \otimes t(y)], \phi(x, y)\rangle &= \langle s(x) \otimes t(y), a(x)\phi(x, y)\rangle \\ &= \langle s(x), \langle t(y), a(x)\phi(x, y)\rangle\rangle \\ &= \langle s(x), a(x)\langle t(y), \phi(x, y)\rangle \\ &= \langle a(x)s(x), \langle t(y), \phi(x, y)\rangle \\ &= \langle a(x)s(x) \otimes t(y), \phi(x, y)\rangle,\end{aligned}$$

which is equivalent to (18).

Property 7. Translation.

$$(s \otimes t)(x + h, y) = s(x + h) \otimes t(y). \tag{19}$$

Proof. For $\phi(x, y) \in D_{m+n}$,

$$\begin{aligned}\langle (s \otimes t)(x + h, y), \phi(x, y)\rangle &= \langle s(x) \otimes t(y), \phi(x - h, y)\rangle \\ &= \langle t(y), \langle s(x), \phi(x - h, y)\rangle\rangle \\ &= \langle t(y), \langle s(x + h), \phi(x, y)\rangle\rangle \\ &= \langle s(x + h) \otimes t(y), \phi(x, ky)\rangle,\end{aligned}$$

and we have the required result.

Example 1. the direct product of the delta functions over R_m with that over R_n yields the delta function:

$$\delta(x) \otimes \delta(y) = \delta(x, y) = \delta(x)\delta(y),$$

in R_{n+m}, for

$$\begin{aligned}\langle\delta(x)\otimes\delta(y),\phi(x,y)\rangle &= \langle\delta(x),\langle\delta(y),\phi(x,y)\rangle\rangle = \langle\delta(x)\phi(x,0)\rangle \\ &= \phi(0,0) = \langle\delta(x,y),\phi(x,y)\rangle\,. \end{aligned} \tag{20}$$

Example 2. Just as an ordinary function is said to be independent of y if it is of the form $f(x)\otimes 1(y)$, we say that a distribution is independent of y if it is of the form $s(x)\otimes 1(y)$. It acts according to the rule

$$\begin{aligned}\langle s(x)\otimes 1(y),\phi(x,y)\rangle &= \left\langle s(x),\int\phi(x,y)dy\right\rangle \\ &= \int\langle s(x),\phi(x,y)\rangle dy \\ &= \langle 1(y)\otimes s(x),\phi(x,y)\rangle\,, \qquad \phi\in D_{m+n}\,.\end{aligned}$$

In other words, we have the relation

$$\left\langle s(x),\int\phi(x,y)dy\right\rangle = \int\langle s(x),\phi(x,y)\rangle dy\,. \tag{21}$$

Example 3. Let $H(x)$ be the Heaviside function of n variables:

$$H(x) = \begin{cases}1\,, & x_1>0\,,\ x_2>0,\ldots,x_n>0\,,\\ 0 & \text{elsewhere.}\end{cases}$$

This is clearly the direct product $H(x_1)\otimes H(x_2)\otimes\cdots\otimes H(x_n)$. By virtue of (17b) and the relation $dH(x_i)/dx_i=\delta(x_i)$, we have

$$\frac{\partial^n H}{\partial x_1\partial x_2\cdots\partial x_n} = \delta(x_1,x_2,\ldots,x_n)\,. \tag{22}$$

This means that the function $H(x) = H(x_1)H(x_2)\cdots H(x_n)$ defined in Exercise 10 of Chapter 2 coincides with this direct product.

Example 4. The direct product of $\delta(x_1)$ and a locally integrable function f of the variables x_2 and $x_3\in R_3$ is

$$\begin{aligned}&\langle\delta(x_1)f(x_2,x_3),\phi(x_1,x_2,x_3)\rangle \\ &\quad= \langle\delta(x_1),\langle f(x_2,x_3),\phi(x_1,x_2,x_3)\rangle\rangle \\ &\quad= \left\langle\delta(x_1),\int_{R_2}\phi(x_1,x_2,x_3)f(x_2,x_3)dx_2dx_3\right\rangle \\ &\quad= \int_{R_2}\phi(0,x_2,x_3)f(x_2,x_3)dx_2dx_3\,.\end{aligned}$$

Consequently, $\delta(x_1)f(x_2,x_3)$ is the volume source density corresponding to a simple layer of sources spread on the plane $x_1=0$. In particular, $\delta(x_1)1(x_2,x_3)$ corresponds to a simple layer of unit surface density on the plane $x_1=0$.

7.2. The Direct Product of Tempered Distributions

We use the notation and the definitions of the previous section. Let $s(x) \in S'_m$, $x \in R_m$, and $t(y) \in S'_n$, $y \in R_n$. Since $S' \subset D'$, the direct product $s(x) \otimes t(y) \in D_{m+n}$, x, $y \in R_{m+n} = R_m \times R_n$. Our aim is to prove that $s(x) \otimes t(y) \in S'_{m+n}$. In view of the definition of the functional $s(x) \otimes t(y)$, namely,

$$\langle s(x) \otimes t(y), \phi(x, y)\rangle = \langle s(x), \langle t(y), \phi(x, y)\rangle\rangle\,, \tag{1}$$

where $\phi(x, y)$ now traverses the space S, we should show that the right side of (1) is a linear continuous functional on S_{m+n}. We proceed as in the previous section and first state the following lemma, whose proof is analogous to Lemma 1 of Section 7.1.

Lemma. *The function $\psi(x) = \langle t(y), \phi(x, y)\rangle$, where $t \in S'_n$, $\phi \in S_{m+n}$, is a test function in S_m, and*

$$D^k\psi(x) = \langle t(y), D_x^k\phi(x, y)\rangle\,, \tag{2}$$

is valid for all multiindices k. Furthermore, if the sequence in $S_{m+n}\{\phi_l(x, y)\} \to \phi$ as $l \to \infty$, then $\phi_l(x) = \langle t(y), \phi_l(x, y)\rangle \to \langle t(y), \phi(x, y)\rangle$ in S_m.

By virtue of this lemma, (1) defines a linear, continuous functional on S_{m+n}. Thus $s(x) \otimes t(y) \in S'_{m+n}$.

Most of the properties that hold for the direct product in D'_{m+n} hold also in S'_{m+n}. The proof is similar. We state them for the sake of completeness.

Property 1. Commutativity.

$$s(x) \otimes t(y) = t(x) \otimes s(y)\,. \tag{3}$$

In particular,

$$s(x) \otimes 1(y) = 1(y) \otimes s(x)\,. \tag{4}$$

Property 2. Continuity. If $s_l \to s$ in S'_m as $l \to \infty$, then $s_l(x) \otimes t(y) \to s(x) \otimes t(y)$ in S'_{m+n} as $l \to \infty$.

Property 3. Associativity. When $s(x) \in S'_m$, $t(y) \in S'_n$, and $u(z) \in S'_p$,

$$s(x) \otimes [t(y) \otimes u(z)] = [s(x) \otimes t(y)] \otimes u(z)\,. \tag{5}$$

Property 4. Support.

$$\text{supp}\,(s \otimes t) = (\text{supp}\, s) \times (\text{supp}\, t)\,. \tag{6}$$

Property 5. Differentiation.

$$D_x^k[s(x) \otimes t(y)] = D^k[s(x)] \otimes t(y)\,. \tag{7}$$

Property 6. Translation.

$$(s \otimes t)(x+h, y) = s(x+h) \otimes t(y)\,. \tag{8}$$

7.3. The Fourier Transform of the Direct Product of Tempered Distributions

Let $s(x) \in S'_m$ and $t(y) \in S'_n$, then

$$[s \otimes t]^\wedge = \widehat{s} \otimes \widehat{t}\,. \tag{1}$$

For $\phi(v, w) \in S_{m+n}$,

$$\begin{aligned}\langle [s(x) \otimes t(y)]^\wedge(v, w), \phi(v, w)\rangle &= \langle s(x) \otimes t(y), \widehat{\phi}\rangle \\ &= \langle s(x) \otimes t(y), F_w F_v[\phi](x, y)\rangle \\ &= \langle s(x), \langle t(y), F_w F_v[\phi](x, y)\rangle\rangle\,,\end{aligned} \tag{2}$$

where F_v and F_w mean the Fourier transform of the function ϕ for the argument x and y, respectively, so that

$$F_w F_v[\phi](x, y) = \int e^{iwy} \int e^{ivx} \phi(v, w) dv\, dw\,.$$

Thus

$$\begin{aligned}\langle [s(x) \otimes t(y)]^\wedge(v, w), \phi(v, w)\rangle &= \langle s(x), \langle \widehat{t}(w), F_v[\phi](x, w)\rangle\rangle \\ &= \langle s(x) \otimes \widehat{t}(w), F_v[\phi](x, w)\rangle \\ &= \langle F_x[s(x) \otimes \widehat{t}(w)], \phi\rangle \\ &= \langle \widehat{t}(w), \langle s(x), F_v[\phi](x, w)\rangle\rangle \\ &= \langle \widehat{t}(w), \langle \widehat{s}(v), \phi(v, w)\rangle\rangle \\ &= \langle \widehat{s}(v) \otimes \widehat{t}(w), \phi(v, w)\rangle\,,\end{aligned}$$

which proves (1). Note that in this process we have also proved that

$$[s(x) \otimes t(y)]^\wedge = F_v[s(x) \otimes \widehat{t}(w)]\,. \tag{3}$$

Similarly,

$$[s(x) \otimes t(y)]^\wedge = F_w[\widehat{s}(v) \otimes t(y)]\,. \tag{4}$$

Example. For the case $n = 2$, consider the function $H(x, y)$,

$$H(x, y) = \begin{cases} 1\,, & x > 0\,, \quad y > 0\,, \\ 0 & \text{for all other values of } x, y\,, \end{cases}$$

which can be written $H(x, y) = H(x) \otimes H(y)$. When we use (1) for the Fourier transform of the direct product we find that

$$\begin{aligned}[H(x, y)]^\wedge &= [H(x) \otimes H(y)]^\wedge = \widehat{H}(x) \otimes \widehat{H}(y) \\ &= (\pi\delta(v) + i\mathrm{Pf}\,(1/v)) \otimes (\pi\delta(w) + i\mathrm{Pf}\,(1/w))\,.\end{aligned}$$

7.4. The Convolution

The convolution $f * g$ of two functions $f(x)$ and $g(x)$, both in R_n, is defined as

$$(f * g)(x) = \int f(y)g(x-y)dy\,. \tag{1}$$

It is clear that

$$f * g = \int f(y)g(x-y)dy = \int g(y)f(x-y)dy = g * f\,, \tag{2}$$

whenever the convolution exists. Let us assume that functions $f(x)$ and $g(x)$ are locally integrable in R_n. Then $f * g$ is locally integrable in R_n and hence defines a regular distribution $\langle f * g, \phi\rangle$,

$$\begin{aligned}\langle f * g, \phi\rangle &= \int (f * g)(z)\phi(z)dz \\ &= \int \left[\int g(y)f(z-y)dy\right]\phi(z)dz \\ &= \int g(y)\left[\int f(z-y)\phi(z)dz\right]dy \\ &= \int g(y)\left[\int f(x)\phi(x+y)dx\right]dy\end{aligned}$$

or

$$\begin{aligned}\langle f * g, \phi\rangle &= \int f(x)g(y)\phi(x+y)dx\,dy\,, \\ &= \langle f(x)\otimes g(y), \phi(x+y)\rangle\,, \qquad \phi \in D\,.\end{aligned} \tag{3}$$

Equation (3) seems to reveal a property of the convolution that might be used to define the convolution of two distributions. That is, the convolution of two distributions s and t in D is

$$\langle s * t, \phi\rangle = \langle s \otimes t, \phi(x+y)\rangle\,. \tag{4}$$

A small problem arises; The function $\phi(x+y)$ does not have compact support. (Its support is the infinite strip that lies between $x + y = A$ and $x + y = -A$, where the constant A depends on the supports of s and t.) In order to ensure that the formula works we have to make certain assumptions.

We have seen in Section 7.2 that supp $(s \otimes t)$ = supp $s \times$ supp t. Accordingly, (4) will become meaningful if the intersection of supp $(s \otimes t)$ and supp $\phi(x+y)$ is bounded. Indeed, in that case, we replace $\phi(x+y)$ by a finite function $\phi(x, y)$ that is equal to $\phi(x+y)$ in this intersection and vanishes outside it. In the sequel, when we write $\phi(x+y)$, we mean such a function $\phi(x, y)$.

The boundedness of the intersection of the supp $(s \otimes t)$ and supp $\phi(x+y)$ can be achieved in the following two ways:

1. The support of one of the distributions is bounded. Let, for example, the support of t be bounded. In this case the support of $\phi(x+y)$ is contained in a horizontal strip of a finite width $[x, y : |x+y| \leq A, |y| \leq R]$ (see Figure 7.2). Thus, by virtue of the definition of the direct product, we have

$$\langle (s * t), \phi \rangle = \langle (s \otimes t), \phi(x+y) \rangle = \langle s(x), \langle t(y), \phi(x+y) \rangle \rangle . \tag{5}$$

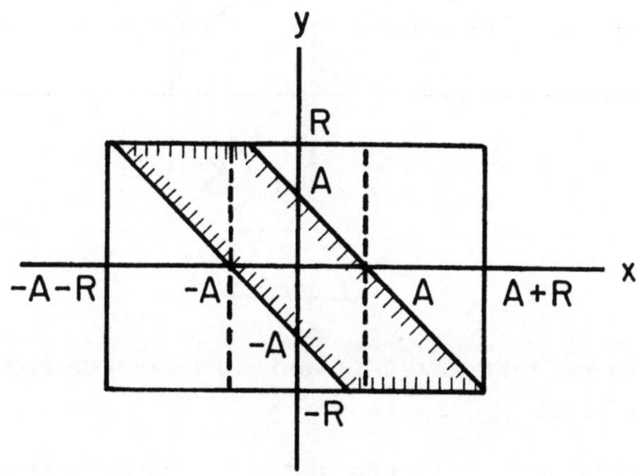

Figure 7.2.

On the other hand, if the support of s is bounded, then the support of $\phi(x+y)$ is contained in a vertical strip of a finite width. Under either of these circumstances, the function $\psi(x) = \langle t(y), \phi(x+y) \rangle$ is a member of D_m, as proved in Section 7.1.

2. Both s and t have supports that are bounded on the same side. For example, let $s = 0$ for $x > R_1$, and let $t = 0$ for $y > R_2$. In this case the support of $\phi(x+y)$ is contained in a quarter-plane lying below some horizontal line and to the left of some vertical line (see Figure 7.3). Therefore, the right side of (5) is again well defined.

Properties of the convolution of distributions

Property 1. Commutativity.

$$s * t = t * s . \tag{6}$$

This is an immediate consequence of the definition (5) and the commutativity of the direct products $s \otimes t$.

Property 2. Associativity.

$$(s * t) * u = s * (t * u) \tag{7}$$

if the supports of the two of these three distributions are bounded or if the supports of all three distributions are bounded on the same side.

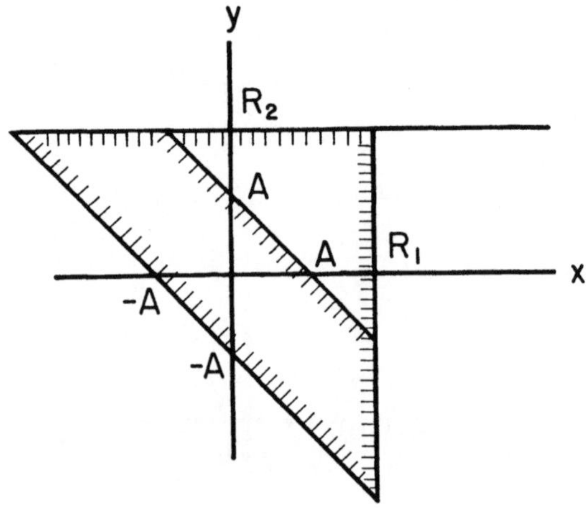

Figure 7.3.

The proof of this result is a straightforward extension of the corresponding proof given for the validity of definition (5).

Property 3. Differentiation. If the convolution $s*t$ exists, then the convolutions $(D^k s)*t$ and $s*(D^k t)$ exist, and

$$(D^k s) * t = D^k(s * t) = s * (D^k t)\,. \tag{8}$$

Proof. It is sufficient to prove that (8) holds for each first derivative $\partial/\partial x_j$, $j = 1, \ldots, n$. For $\phi \in D$,

$$\begin{aligned}\langle \partial/\partial x_j(s * t), \phi\rangle &= (-1)\langle s * t, \partial\phi/\partial x_j\rangle = (-1)\langle s \otimes t, (\partial/\partial x_j)\phi(x+y)\rangle \\ &= \langle s, \langle(-1)t, (\partial/\partial x_j)\phi(x+y)\rangle\rangle \\ &= \langle s, \langle \partial t/\partial x_j, \phi(x+y)\rangle\rangle \\ &= \langle s \otimes \partial t/\partial x_j, \phi(x+y)\rangle = \langle s * \partial t/\partial x_j, \phi\rangle\,,\end{aligned}$$

or

$$(\partial/\partial x_j)(s * t) = s * \partial t/\partial x_j\,. \tag{9}$$

By virtue of the commutativity of the convolution, we interchange s and t in (9) and get

$$\frac{\partial}{\partial x_j}(s * t) = \frac{\partial}{\partial x_j}(t * s) = t * \frac{\partial s}{\partial x_j} = \frac{\partial s}{\partial x_j} * t\,. \tag{10}$$

Combining (9) and (10), we have (8), as required.

If L is a differential operator with constant coefficients, we find from (8) that

$$(Ls) * t = L(s * t) = s * (Lt)\,. \tag{11}$$

These results imply that, in order to differentiate a convolution, it suffices to differentiate any one of the factors.

Property 4. Continuity. In certain cases the convolution is a continuous operator. The following theorem embodies this result.

Theorem. *Let the sequence of distributions* $\{s_l\} \to s$ *as* $l \to \infty$, *then* $\{s_l * t\} \to s * t$ *under each of the following conditions:*

(1) *All distributions* s_l *are concentrated on the same bounded set.*
(2) *The distribution* t *is concentrated on a bounded set.*
(3) *The supports of the distributions* s *and* t *are bounded on the same side by a constant independent of* l.

Proof. In view of (5), we have

$$\langle s_l * t, \phi \rangle = \langle s_l, \langle t, \phi(x+y) \rangle \rangle, \qquad \phi \in D. \tag{12}$$

If condition 1 holds, we can replace $\langle t, \phi(x+y) \rangle$ by a test function $\psi(x)$ that vanishes outside the region on which all the distributions s_l are concentrated. Then

$$\begin{aligned} \langle s_l * t, \phi \rangle &= \langle s_l \otimes t, \phi(x+y) \rangle = \langle s_l, \langle t, \phi \rangle \rangle \\ &= \langle s_l, \psi \rangle \to \langle s, \psi \rangle = \langle s * t, \phi \rangle, \end{aligned} \tag{13}$$

and we have

$$s_l * t \to s * t, \qquad \text{as} \quad l \to \infty,$$

as required.

In the second case, $\psi(x) = \langle t, \phi(x+y) \rangle$ is a test function, and we follow the steps leading to (13) to derive our formula.

Finally, in case 3 we suppose that the support of the distributions s and t are bounded on the left. Then the support of the function $\psi(x) = \langle t, \phi(x+y) \rangle$ is bounded on the right. The rest of the proof proceeds as in the other two cases.

As a corollary to this theorem, we have the continuity of the distribution s_α depending on a parameter α under each of the following condition:

(1) All the s are concentrated on the same bounded set.
(2) The distribution t is concentrated on a bounded set.
(3) The supports of the distributions s and t are bounded on the same side by a constant independent of α.

As a special case of this corollary we find that if $\partial s_\alpha / \partial \alpha$ exists, then

$$(\partial / \partial \alpha)(s_\alpha * t) = \partial s_\alpha / \partial \alpha * t, \tag{14}$$

because the derivative $\partial s_\alpha / \partial \alpha$ is the limit of

$$\frac{1}{\alpha - \alpha_0} (s_\alpha - s_{\alpha_0}) \qquad \text{as} \qquad \alpha \to \alpha_0.$$

Convolution of tempered distributions

The foregoing analysis can be extended to distributions of slow growth. We have the same definition (4) for the convolution of two distributions. The restrictions are also similar. To establish various properties we thus appeal to the direct product of tempered distributions.

In applications, it is the convolution $t * \phi$ of a tempered distribution t and a test function $\phi \in S$ that plays an important part.

7.5. The Role of Convolution in the Regularization of the Distributions

In Chapter 1 we defined the Dirac delta function. In Chapter 2 we defined the class of test functions, which helped us define not only this function but many more generalized functions. Now we study the convolution of a distribution with some test function in D. This operation converts the distribution into a function that is infinitely smooth. Since $\psi \in D$ has compact support, this convolution exists. Moreover, this convolution satisfies an interesting equality. Specifically, we have the following theorem:

Theorem.

$$s * \psi = \langle s(y), \psi(x-y)\rangle \in C^{\infty}(R_n)\,, \qquad \psi \in D\,. \tag{1}$$

Proof. This infinite differentiability of the right side of (1) is established in the same way as in Lemma 1 of Section 7.1.

To prove the equality we appeal to (7.4.5) and find that, for $\phi \in D$,

$$\begin{aligned}
\langle s * \psi, \phi\rangle &= \langle s(y) \otimes \psi(z), \phi(y+z)\rangle \\
&= \langle s(y), \langle \psi(z), \phi(y+z)\rangle\rangle \\
&= \left\langle s(y), \int \psi(z)\phi(y+z)dz\right\rangle \\
&= \left\langle s(y), \int \phi(x)\psi(x-y)dx\right\rangle\,.
\end{aligned} \tag{2}$$

Note that the function $\phi(x)\psi(x-y)$ belongs to $D(R_{2n})$. We can, therefore, use (7.1.21) to obtain from (2) the result

$$\begin{aligned}
\langle s * \psi, \phi\rangle &= \int \phi(x)\langle s(y), \psi(x-y)\rangle dx \\
&= \langle\langle s(y), \psi(x-y)\rangle, \phi\rangle\,,
\end{aligned}$$

as desired.

Let $\phi_\varepsilon \in D$ be the function defined in Section 2.2,

$$\phi_\varepsilon(x) = \begin{cases} c_\varepsilon \exp\left(-\frac{\varepsilon^2}{\varepsilon^2-|x|^2}\right)\,, & |x| \le \varepsilon\,, \\ 0\,, & |x| > \varepsilon\,, \end{cases}$$

where c_ε is such that $\int \phi_\varepsilon(x)dx = 1$. When we use this function in the definition (1), we have the regularization

$$s_\varepsilon(x) = s * \phi_\varepsilon = \langle s(y), \phi_\varepsilon(x-y)\rangle . \tag{3}$$

Exercise 12 of this chapter illustrates this process for the pseudofunction Pf $(H(x)/x)$.

Recall that in Example 2 of Section 3.3 we proved that $\phi_\varepsilon(x) \to \delta(x)$ as $\varepsilon \to 0$. Combining this with the continuity of the convolution $s * \phi_\varepsilon$ with respect to ϕ_ε, we find that

$$s_\varepsilon(x) \to s(x) \qquad \text{as} \quad \varepsilon \to 0 .$$

This means that each distribution is a weak limit of its own regularization.

These considerations lead to the following very important result:

Theorem. *Each distribution s is a weak limit of a test function. That is, the space D is dense in D'.*

This theorem is really proved by the remarks preceding its statement if all the s_ε are of bounded support. Otherwise, we introduce a sequence $\zeta_\varepsilon(x)$ of cutoff factors that are identically 1 for $|x| \le 1/\varepsilon$ and vanish for $|x| \ge 2/\varepsilon$. Then

$$\lim_{\varepsilon\to 0}\langle \zeta_\varepsilon s_\varepsilon(x), \phi(x)\rangle = \lim_{\varepsilon\to 0}\langle s_\varepsilon, \zeta_\varepsilon\phi\rangle = \lim_{\varepsilon\to 0}\langle s_\varepsilon, \phi\rangle = \langle s, \phi\rangle ,$$

for all $\phi \in D$. That is, the sequence $\zeta_\varepsilon(x)s_\varepsilon(x) \to s(x)$ as $\varepsilon \to 0$ in D', and the theorem is proved.

Since $D \subset S \subset S' \subset D'$, it follows that S and S', as well as D, are dense in D'.

7.6. The Dual Spaces E and E′

From the analysis of the previous sections the reader must have observed the important part the distributions of compact support play in this theory. Accordingly, the subspace of D' all of whose members have compact support is given a separate symbol. We shall denote it as E'. It is the linear space of all distributions having bounded support. A sequence of distributions $\{t_m\}$ is said to converge in E' to a limit t if it converges in D' to t and if all the t_m have their supports contained in one fixed bounded region R. Then clearly the limit distribution t is also in E' and has its support contained in R.

The dual space E of E' is the space of test functions $\phi(x)$ that are continuous and have continuous derivatives of all orders. These test functions, unlike those in D, are not required to have compact support on the real axis. However, we need the following definition of convergence for functions $\phi(x) \in E$.

Definition. A sequence $\{\phi_m(x)\}$ of test functions in E is said to *converge to zero* if and only if for every fixed integer p the sequence $\{D^p\phi_m\}$ converges uniformly to zero on every compact subset of the real axis. A sequence of functions $\{\phi_m(x)\}$ in E *converges to another* *function* $\phi(x)$ if the sequence $\{\phi_m(x) - \phi(x)\}$ converges to zero in this sense. It is clear that under this definition of convergence, the space E is complete.

From this definition it is clear that the functions $\phi \in D$ are also in E but that the converse is not true. Thus $E \supset D$. Indeed, it can be shown that D is dense in E.

Let E' be the space of all linear functionals t on E that are continuous with respect to the present definition; that is,

$$\lim_{m\to\infty} \langle t, \phi_m \rangle = \left\langle t, \lim_{m\to\infty} \phi_m \right\rangle .$$

Thus, if t is a distribution on E, then the sequence $\{\langle t, \phi_m \rangle\}$ converges to zero if the sequence $\{\phi_m(x)\}$ does. Thus E' is the dual space of E. As observed before, an interesting feature of the distributions $t \in E'$ is that they have compact support. Otherwise we can always find a function $\phi(x) \in C^\infty$ with sufficient growth at infinity to make the integral $\langle t, \phi \rangle$ divergent. Since $D \subset E$, we have $E' \subset D'$.

From this disucssion we observe that the test functions of compact support lead to the distributions of arbitrary support, whereas the test functions of arbitrary support yield the distributions of compact support.

The Cauchy representation

Let z denote a point located in the upper or lower half-plane but not on the x axis. Then the function

$$[2\pi i(x - z)]^{-1} \tag{1}$$

is continuous and has continuous derivatives of all orders for all values of x and, hence, is a function of the space E. Accordingly, for a distribution $t(x) \in E'$, the operation

$$t_c(z) = \frac{1}{2\pi i}\left\langle t(x), \frac{1}{x - z}\right\rangle , \tag{2}$$

generates a function $t_c(z)$ of the complex variable z for all values of z outside the real axis. This function is defined to be the *Cauchy representation of the distribution* $t(x)$.

Our contention is that $t_c(z)$ is an analytic function of z in the complement of the support of $t(x)$. To prove this, we examine

$$\frac{d}{dz}t_c(z) = \lim_{\varepsilon\to 0} \frac{1}{2\pi i}\left\langle t(x), \frac{1}{\varepsilon}\left(\frac{1}{x - z - \varepsilon} - \frac{1}{x - z}\right)\right\rangle . \tag{3}$$

Because

$$\lim_{\varepsilon\to 0} \frac{1}{\varepsilon}\left[\frac{1}{x - z - \varepsilon} - \frac{1}{x - z}\right] = \frac{1}{(x - z)^2} ,$$

and t is continuous on E, we find from (3) that

$$\frac{d}{dz}t_c(z) = \frac{1}{2\pi i}\left\langle t(x), \frac{1}{(x - z)^2}\right\rangle , \tag{4}$$

so that the complex derivative of $t_c(z)$ exists for all z with Im $z \neq 0$ and hence is analytic in that region. Equation (4) also proves that

$$dt_c(z)/dz = t_c'(z)\,,$$

where t' is the distributional derivative of t.

Example 1. Let $t(x)$ be a locally integrable function that vanishes outside some finite interval on the real axis and is the restriction to this interval of a function $t(z)$ that vanishes for large values of z. Then the regular distribution t, $\langle t, \phi\rangle = \int_{-\infty}^{\infty} t(x)\phi(x)dx$, is a distribution of bounded support. The Cauchy representation of this distribution takes the form of the ordinary Cauchy integral

$$t_c(z) = \frac{1}{2\pi i}\int_{-\infty}^{\infty} \frac{t(x)}{x-z}dx = \begin{cases} t(z)\,, & \text{Im}\, z > 0\,, \\ 0\,, & \text{Im}\, z < 0\,. \end{cases}$$

Example 2. In the case $t(x) = \delta(x)$ we have

$$\delta_c(z) = \frac{1}{2\pi i}\left\langle \delta(x), \frac{1}{x-z}\right\rangle = -\frac{1}{2\pi i z}\,.$$

The Cauchy representation $t_c(z)$ of a distribution t on the space E has far-field behavior $t_c(z) = O(1/z)$ as $z \to \infty$. We observe from this example that $\delta_c(z) = O(1/z)$ as $z \to \infty$.

The Fourier transform of tempered distributions with compact support

In Section 6.1 we attempted to define the Fourier transform of a distribution by the formula

$$t(u) = \langle t, e^{iux}\rangle\,, \tag{5}$$

and got into trouble because the function e^{iux} is not in D. Then we tried the definition $\langle \widehat{t}, \phi\rangle = \langle t, \widehat{\phi}\rangle$, $\phi \in D$, and found that $\widehat{\phi}$ may not be in D. This necessitated the introduction of the space S. However, if we define the Fourier transform of a distribution t of a compact support, then (5) is clearly valid so long as we confine ourselves to the class of test function in S. For example, the delta function is a temperered distribution with compact support, and thus

$$\widehat{\delta}(x) = \langle \delta(x), e^{iux})\rangle = 1\,.$$

The hypothesis of compact support in the representation (5) can be relaxed if we move from the real axis into the complex plane and consider the test functions $e^{i\lambda x}$, where λ is a complex variable. Indeed, if $t(x)$ is a tempered distribution that vanishes for negative values of x,then we define $\widehat{t}(\lambda)$ by

$$\widehat{t}(\lambda) = \langle t, e^{i\lambda x}\rangle \tag{6}$$

for all values of λ in the upper half of the complex plane. It follows by writing $\lambda = u + iv$, $v > 0$, that $e^{i\lambda x} = e^{iux}e^{-vx}$ is in S. Moreover, $\widehat{t}(\lambda)$ as defined by (6), and considered as a function of λ, is analytic in the upper half of the complex plane.

Similarly, if the tempered distribution t vanishes for positive values of x, then $\widehat{t}(\lambda)$ in (6) exists for all values of λ in the lower half of the complex plane, where it defines an analytic function.

The Cauchy representation of the distributions is not the only area of common interest in the theories of distributions and analytic functions. Indeed, we already had a glimpse of an interplay between these theories in the Sokhotski-Plemelj equations presented in Section 2.4, namely

$$\lim_{y\to 0} \frac{1}{x \pm iy} = \mp\pi\delta(x) + \mathrm{Pf}\left(\frac{1}{x}\right) .$$

These formulas yield the boundary values of the functions $1/z$ and $1/\overline{z}$ as they approach the x-axis. For further study in the field the reader may consult references [23-25].

Estrada and Kanwal [12] have presented a theory for distributional boundary values of harmonic and analytic functions by introducing several indicators that measure the growth of a harmonic function near a boundary. They find several relations among different indicators which enable them to characterize the harmonic functions that have distributional boundary values. They prove an extension of the Phragamén-Lindelöf theorem of the maximum principle. They also discuss many more results such as the algebraic properties of the space of real periodic distributions. By introducing a new product, the harmonic product, the boundary conditions involving harmonic functions are transformed into ordinary differential equations.

7.7. Examples

Example 1

(a) We can express the integral

$$f(x) = \int_{-\infty}^{x} \phi(y)dy\,, \qquad \phi \in S\,,$$

as a convolution. Setting $y = x - z$, we find that

$$f(x) = \int_{-\infty}^{x} \phi(y)dy = -\int_{0}^{\infty} \phi(x-z)dz = \int_{-\infty}^{\infty} H(z)\phi(x-z)dz = H * \phi\,.$$

(b) Let t be an arbitrary distribution. The convolution of $\delta(x)$ and $t(x)$ is

$$\langle \delta * t, \phi\rangle = \langle t(x), \langle \delta(z), \phi(x+z)\rangle\rangle = \langle t(x), \phi(x)\rangle\,, \qquad \phi \in D\,,$$

or

$$\delta * t = t\,. \tag{1}$$

Thus the delta function is an identity element in D' for the operation of convolution.

(c)
$$\delta_a * t = \text{sh}_a t\,, \tag{2a}$$

where $\delta_a = \delta(x-a)$ and the symbol sh_a shifts t by a translation of a. The translated distribution is also denoted t_{x-a} or $t(x-a)$; that is,

$$t(x-a) = t(x) * \delta(x-a)\,. \tag{2b}$$

Proof. For all $\phi \in D$ we have

$$\begin{aligned}\langle \delta_a * t, \phi\rangle &= \langle \delta_a(x) \otimes t(y), \phi(x+y)\rangle = \langle t(y), \langle \delta_a(x), \phi(x+y)\rangle\rangle \\ &= \langle t(y), \phi(a+y)\rangle = \langle \text{sh}_a t, \phi\rangle\,,\end{aligned}$$

which is (2a). In particular

$$\delta_a * \delta_b = \delta_{a+b}\,. \tag{3}$$

(d) Next, we combine relations (7.4.8) and (1) to find that

$$\partial\delta/\partial x_j * t = \delta * \partial t/\partial x_j = \partial t/\partial x_j\,. \tag{4}$$

This result can be extended to a differential operator L of order p. Indeed, from (7.4.11) and (2) we have

$$(L\delta) * t = \delta * L(t) = L(t)\,. \tag{5}$$

Thus, every linear differential operator with constant coefficients can be represented as a convolution.

An interesting particular case of (5) is

$$\delta^{(m)} * t = t^{(m)}\,. \tag{6}$$

Example 2. From relation (6) we derive that $\delta' * 1 = 0$. Thus $(1 * \delta') * H = 0$. On the other hand $\delta' * H = \delta$, and therefore $1 * (\delta' * H) = 1$. This means that

$$(1 * \delta') * H \neq 1 * (\delta' * H)\,,$$

so convolution is not necessarily associative.

Example 3. Relation (7.4.8) states that $(D^k s) * t = s * (D^k t)$. The mere existence of the convolutions $D^k s * t$ and $s * D^k t$, is not sufficient for the existence of the convolution $s * t$; specifically, these convolutions may not be equal. For instance, $H' * 1 = \delta * 1 = 1$, while $H * 1' = H * 0 = 0$. The trouble is that $H * 1$ does not exist, because neither the support of $H(x)$ nor that of 1 is bounded.

Example 4. One of the most important uses of convolution theory is to obtain the particular solution of a differential equation

$$Lu = f\,, \tag{7}$$

where $Lu = L(D)u$, is a linear differential operator. This is achieved by appealing to the fundamental solution E, given by

$$LE = \delta . \tag{8}$$

Indeed, our contention is that

$$u = f * E . \tag{9}$$

This follows by applying the operator L to both sides of (9). The result is

$$Lu = L(f * E) = f * LE = f * \delta = f ,$$

as desired.

Let us illustrate this concept by considering Poisson's equation in R_3,

$$\nabla^2 u(x) = -\rho(x) . \tag{10}$$

We find from (4.4.73) that the fundamental solution of the operator $-\nabla^2$ in R_3 is $(1/4\pi|x|)$. Accordingly, the particular solution of (10) is

$$u = \rho * \frac{1}{4\pi|x|} = \frac{1}{4\pi}\int \frac{\rho(y)}{|x-y|}dy . \tag{11}$$

This is called the *Newtonian* or *volume potential* for mass density ρ and is written V_3. Its generalization in R_n is

$$V_n(x) = \rho * \frac{1}{(n-2)S_n(1)|x|^{n-2}} \frac{1}{(n-2)S_n(1)} \int \frac{\rho(y)}{|x-y|^{n-2}}dy , \qquad n \geq 3 , \tag{12a}$$

$$V_2(x) = \rho * \frac{1}{2\pi}\ln\frac{1}{|x|} = \frac{1}{2\pi}\int \rho(y)\ln\frac{1}{|x-y|}dy , \tag{12b}$$

where $S_n(1)$ is the surface area of the unit sphere.

Example 5. Single-layer distribution We have already come across the concept of the single-layer density in previous chapters. With the help of the definition of convolution we can study this concept in more detail.

Let $\sigma(x)$ be a locally integrable function defined over a bounded piecewise smooth two-sided surface S. Then $\sigma(x)\delta(S)$ is the single layer over S with surface density σ. The potential generated by this distribution is

$$V_n^{(0)} = \sigma(x)\delta(S) * \frac{1}{(n-2)S_n(1)|x|^{n-2}} , \qquad n \geq 3 , \tag{13a}$$

$$V_2^{(0)} = \sigma(x)\delta(S) * \frac{1}{2\pi}\ln\left(\frac{1}{|x|}\right) , \tag{13b}$$

and is called the surface potential of the single layer with density σ. It is expressed

$$V_n^{(0)}(x) = \frac{1}{(n-2)S_n(1)}\int_S \frac{\sigma(y)}{|x-y|^{n-2}}\, dS_y , \qquad n \geq 3 , \tag{14a}$$

$$V_2^{(0)}(x) = \frac{1}{2\pi}\int_S \sigma(y)\ln\frac{1}{|x-y|}\, dS_y , \tag{14b}$$

where for $n = 2$, S is a curve.

Let us prove the case $n \geq 3$. For this purpose, we use definition (7.4.5) and find that for all $\phi \in D$

$$\begin{aligned}\left\langle \sigma(x)\delta(S) * \frac{1}{|x|^{n-2}}, \phi \right\rangle &= \left\langle \sigma(y)\delta(S) \otimes \frac{1}{|z|^{n-2}}, \phi(y+z) \right\rangle \\ &= \left\langle \sigma(y)\delta(S), \left\langle \frac{1}{|z|^{n-2}}, \phi(y+z) \right\rangle \right\rangle \\ &= \int_S \sigma(y) \left[\int \frac{1}{|z|^{n-2}} \phi(y+z)dz \right] dS_y \\ &= \int_S \sigma(y) \left[\int \frac{1}{|x-y|^{n-2}} \phi(x)dx \right] dS_y \\ &= \int \phi(x) \int_S \frac{\sigma(y)}{|x-y|^{n-2}}\, dS_y \,, \end{aligned} \tag{15}$$

from which (14a) follows.

We can obtain a more general result by repeating the steps in (15). Indeed, let $f(x)$ be a locally integrable function in R_n and let $\sigma\delta(S)$ be a single layer spread over a bounded smooth surface S with density σ. Then, for $\phi \in D$,

$$\begin{aligned}\langle \sigma(x)\delta(S) * f(x), \phi \rangle &= \langle \sigma(y)\delta(S) \otimes f(z), \phi(y+z) \rangle \\ &= \langle \sigma(y)\delta(S)\,,\ \langle f(z), \phi(y+z)\rangle\rangle \\ &= \int_S \sigma(y) \left[\int f(z)\phi(y+z)dz \right] dS_y \\ &= \int_S \sigma(y) \left[\int f(x-y)\phi(x)dx \right] dS_y \\ &= \int \phi(x) \left[\int_S \sigma(y) f(x-y) dS_y \right] dx \,, \end{aligned}$$

or

$$\sigma(x)\delta(S) * f(x) = \int_S \sigma(y) f(x-y) dS_y \,. \tag{16}$$

Example 6. Double-layer distribution Let $\tau(x)$ be a continuous function over S (as defined in Example 5 for a single layer density) and let $-(\partial/\partial n)[\tau(x)\delta(S)]$ be the double layer over S with density τ. Here n stands for the unit normal to S. The potential generated by this distribution of singularities is

$$V_n^{(1)} = -\frac{\partial}{\partial n}[\tau(x)\delta(S)] * \frac{1}{(n-2)S_n(1)}\,\frac{1}{|x|^{n-2}}\,, \qquad n \geq 3\,, \tag{17a}$$

$$V_2^{(1)} = -\frac{\partial}{\partial n}[\tau(x)\delta(S)] * \frac{1}{2\pi} \ln \frac{1}{|x|}\,, \tag{17b}$$

and is called the surface potential of the double layer. The function $V_n^{(1)}$ is a locally integrable function in R^n and can be expressed by the formulas

$$V_n^{(1)}(x) = \frac{1}{(n-2)S_n(1)} \int_S \tau(y) \frac{\partial}{\partial n_y} \frac{1}{|x-y|^{n-2}}\, dS_y\,, \qquad n \geq 3\,, \tag{18a}$$

$$V_2^{(1)}(n) = \frac{1}{2\pi} \int_S \tau(y) \frac{\partial}{\partial n_y} \ln \frac{1}{|x-y|}\, dS_y\,. \tag{18b}$$

As in the previous example, we shall prove the case $n \geq 3$. Indeed, for $\phi \in D$, we have

$$\begin{aligned}
-\left\langle \frac{\partial}{\partial n}[\tau(x)\delta(S)] * \frac{1}{|x|^{n-2}}, \phi \right\rangle &= -\left\langle \frac{\partial}{\partial n}[\tau(y)\delta(S)] \otimes \frac{1}{|z|^{n-2}}, \phi(y+z) \right\rangle \\
&= -\left\langle \frac{\partial}{\partial n}[\tau(y)\delta(S)], \left\langle \frac{1}{|z|^{n-2}}, \phi(y+z) \right\rangle \right\rangle \\
&= \int_S \tau(y) \frac{\partial}{\partial n} \left[\int \frac{\phi(y+z)}{|z|^{n-2}} dx \right] dS_y \\
&= \int_S \tau(y) \left[\frac{\partial}{\partial n} \int \frac{\phi(x)}{|x-y|^{n-2}}\, dx \right] dS_y \\
&= \int_S \tau(y) \left[\int \phi(x) \frac{\partial}{\partial n_y} \frac{1}{|x-y|^{n-2}} dx \right] dS_y \\
&= \int \phi(x) \left[\int_S \tau(y) \frac{\partial}{\partial n} \frac{1}{|x-y|^{n-2}} dS_y \right] dx\,,
\end{aligned}$$

from which the required formula follows.

Example 7. We have come across many generalized functions that vanish from some point to infinity such as $x_+^\lambda = x^\lambda H(x)$. Formulas for their convolution take a simple form. For instance, the convolution of f and g that vanish on the negative axis is

$$\begin{aligned}
f * g &= \int_{-\infty}^{\infty} f(y)g(x-y)dy = \int_0^\infty f(y)g(x-y)dy \\
&= \begin{cases} \int_0^x f(y)g(x-y)dy\,, & x \geq 0\,, \\ 0\,, & x < 0\,, \end{cases}
\end{aligned}$$

because $g(x-y) = 0$ for $y < x$.

Let us denote the distribution

$$x_+^{\lambda-1}/\Gamma(\lambda)\,, \tag{19}$$

which we studied in Example 5 of Section 4.4, as Φ_λ and show that it satisfies the relation

$$\Phi_\lambda * \Phi_\mu = \Phi_{\lambda+\mu}\,. \tag{20}$$

For Re $\lambda > 0$, Re $\mu > 0$ we have

$$\Phi_\lambda * \Phi_\mu = \frac{x_+^{\lambda-1}}{\Gamma(\lambda)} * \frac{x_+^{\mu-1}}{\Gamma(\mu)} = \int_0^x \frac{t^{\lambda-1}(x-t)^{\mu-1}}{\Gamma(\lambda)\Gamma(\mu)}\, dt$$

and

$$\Phi_{\lambda+\mu} = x_+^{\lambda+\mu-1}/\Gamma(\lambda+\mu)\,.$$

Thus (20) will hold for Re $\lambda > 0$, Re $\mu > 0$ if we can prove that

$$\int_0^x t^{\lambda-1}(x-t)^{\mu-1}dt = \frac{\Gamma(\lambda)\Gamma(\mu)}{\Gamma(\lambda+\mu)}x_+^{\lambda+\mu-1}\,, \qquad x \geq 0\,. \tag{21}$$

But this follows by setting $t = x\tau$ and using the identity involving the beta and the gamma functions,

$$\beta(\lambda,\mu) = \int_0^1 \tau^{\lambda-1}(1-\tau)^{\mu-1}dt = \Gamma(\lambda)\Gamma(\mu)/\Gamma(\lambda+\mu)\,.$$

By the principle of analytic continuation, (20) holds for all other complex values of λ and μ.

Example 8. Consider an integrable function $f(x)$ such that $f(x) = 0$ for $x < 0$. Its convolution with $\Phi_n(x)$ ($n = 1, 2, \ldots$), mentioned in Example 7, yields

$$\Phi_n(x) * f(x) = \frac{1}{(n-1)!}\int_0^x (x-y)^{n-1}f(y)\,dy\,. \tag{22}$$

It is known [26] that

$$\begin{aligned} I_n f &= \int_0^x \int_0^{y_{n-1}} \cdots \int_0^{y_2}\int_0^{y_1} f(y)dy\,dy_1\cdots dy_{n-1} \\ &= \frac{1}{(n-1)!}\int_0^x (x-y)^{n-1}f(y)\,dy\,. \end{aligned} \tag{23}$$

Combining (22) and (23), we have

$$I_n f = \Phi_n(x) * f(x)\,. \tag{24}$$

By virtue of (4.4.50), we have

$$\Phi_{-n}(x) = \delta^{(n)}(x)\,.$$

Thus

$$I_{-n}f = \Phi_{-n}(x) * f(x) = \delta^{(n)}(x) * f(x) = (d^n/dx^n)f(x)\,. \tag{25}$$

This means that the convolution of f with $\Phi_n(x)$ gives an n-fold integration, and with $\Phi_{-n}(x)$ the convolution gives an n-fold differentiation.

The foregoing relations are readily generalized to the case that λ is an arbitrary real or complex number and, instead of f, we have a distribution t that is concentrated in $x \geq 0$. Relations (24) and (25) then become

$$I_\lambda t = (x_+^{\lambda-1}/\Gamma(\lambda)) * t \tag{26}$$

and

$$I_{-\lambda}t = d^\lambda t/dx^\lambda\,. \tag{27}$$

Because

$$\Phi_{\lambda+\mu} * t = (\Phi_\lambda * \Phi_\mu) * t = \Phi_\lambda * (\Phi_\mu * t)\,, \tag{28}$$

we find, by putting $\lambda = -\mu$, that differentiation and integration of the same order are inverse operations with respect to one another. Equation (28) further implies that

$$\frac{d^\lambda}{dx^\lambda}\left(\frac{d^\mu t}{dx^\mu}\right) = \frac{d^{\lambda+m} t}{dx^{\lambda+\mu}} \tag{29}$$

is true for all λ and μ.

When λ is not an integer, the quantities I_λ and $I_{-\lambda}$ are called the fractional integral and fractional derivative, respectively, of order λ.

Example 9. With the help of the previous two examples we can prove the following interesting results in the theory of ordinary differential equations and integral equations.

(a) The only solution of the ordinary differential equation

$$d^\lambda s/dx^\lambda = t(x)\,, \qquad s, t \in D'\,, \tag{30}$$

is

$$s(x) = \Phi_\lambda * t\,. \tag{31}$$

This follows on writing (30) as $(\Phi_{-\lambda} * s) = t$ and then taking the convolution of both sides by Φ_λ.

(b) By virtue of the notation of Example 8, we can write the Abel integral equation

$$t(x) = \frac{1}{\Gamma(1-\alpha)} \int_0^x \frac{s(y)}{(x-y)^\alpha}\,dy\,, \qquad x \geq 0\,, \tag{32}$$

as

$$t(x) = \Phi_{-\alpha+1} * s\,, \tag{33}$$

which can be inverted to give

$$s(x) = \Phi_{\alpha-1} * t\,. \tag{34}$$

Two remarks are in order here. The condition $\alpha < 1$, which guarantees the convergence of (32), has been omitted because of the previous discussion. Second, to guarantee the existence of convolution (33), we assume that $s(x)$ and $t(x)$ vanish for $x < 0$.

(c) Let us follow Schwartz [1] and define

$${}_a\Phi_\lambda(x) = e^{ax}\Phi_\lambda(x)\,. \tag{35}$$

For instance,

$${}_a\Phi_{-1}(x) = e^{ax}\Phi_{-1}(x) = e^{ax}\delta'(x) = \delta'(x) - a\delta(x)\,. \tag{36}$$

Because $e^{ax}s(x) * e^{ax}t(x) = e^{ax}\{s(x) * t(x)\}$, it follows that

$${}_a\Phi_\lambda(x) *_a \Phi_\mu^{(x)} = {}_a\Phi_{\lambda+\mu}(x)\,. \tag{37}$$

Using these results in case (a), we find that the only solution of the ordinary differential equation

$$(d/dx - a)^{\lambda} s(x) = t(x)\,, \tag{38}$$

or ${}_a\Phi_{-\lambda} * s(x) = t(x)$, is

$$s(x) = {}_a\Phi_{\lambda} * t(x)\,. \tag{39}$$

Example 10. Convolutions arise naturally in many physical problems. We demonstrate this by considering the initial value problem

$$\partial^2 u/\partial t^2 - \partial^2 u/\partial x^2 = 0\,, \tag{40}$$

$$u(x,0) = \phi(x)\,, \qquad (\partial u/\partial t)(x,0) = \psi(x)\,. \tag{41}$$

This is the famous problem of the motion of a string which is given initial displacement $\phi(x)$ and initial velocity $\psi(x)$.

The general solution of (40) is clearly

$$u(x,t) = f(x+t) + g(x-t)\,, \tag{42}$$

as can be verified by direct substitution. Then from (41) we get

$$u(x,0) = f(x) + g(x) = \phi(x)\,, \tag{43}$$

and

$$\partial u(x,0)/\partial t = f'(x) - g'(x) = \psi(x)\,. \tag{44}$$

Integration of (44) with respect to x yields

$$f(x) - g(x) = \int_a^x \psi(s)ds\,, \tag{45}$$

where the constant of integration has been incorporated into the lower limit. Relations (43) and (45) enable us to solve for $f(x)$ and $g(x)$, so that

$$f(x) = \frac{1}{2}\phi(x) + \frac{1}{2}\int_a^x \psi(s)ds\,, \qquad g(x) = \frac{1}{2}\phi(x) - \frac{1}{2}\int_a^x \psi(s)ds\,.$$

When these expressions are substituted into (42), we obtain

$$u(x,t) = \frac{1}{2}[\phi(x+t) + \phi(x-t)] + \frac{1}{2}\int_{x-t}^{x+t} \psi(s)ds\,, \tag{46}$$

which is the well-known d'Alembert formula.

Let us interpret this formula in the light of the present theory. For this purpose, let us first take the special case $\phi(x) = 0$. Then (46) becomes

$$u(x,t) = \frac{1}{2}\int_{x-t}^{x+t} \psi(s)\,ds\,. \tag{47}$$

We can express it as a distribution with the help of the generalized function $E(x, t)$ defined as

$$E(x,t) = \begin{cases} \frac{1}{2}, & |x| < t, \\ 0, & |.x| > t, \end{cases}$$
$$= \frac{1}{2} H(t+x) H(t-x) = \frac{1}{2} H(t - |x|), \tag{48}$$

so that (47) takes the form

$$u(x,t) = \int_{-\infty}^{\infty} E(x-s,t)\psi(s)ds = E * \psi. \tag{49}$$

In Chapter 10 we shall show that $E(x, t)$ is the fundamental solution of the wave equation (40).

As a second special case, let us set $\psi(x) = 0$. Then (46) reduces to

$$u(x,t) = \frac{1}{2}[\phi(x+t) + \phi(x-t)]$$
$$= \frac{1}{2}\int_{-\infty}^{\infty} [\delta(t+x-s) + \delta(t-x+s)]\phi(s)ds. \tag{50}$$

Now from (48) it follows that

$$\partial E(x-s,t)/\partial t = \frac{1}{2}[\delta(t+x-s)H(t-x+s) + \delta(t-x+s)H(t+x-s)],$$

which, for $t > 0$, becomes

$$\partial E(x-s,t)/\partial t = \frac{1}{2}[\delta(t+x-s) + \delta(t-x+s)]. \tag{51}$$

Combining (50) and (51) we obtain

$$u(x,t) = \int_{-\infty}^{\infty} \frac{\partial E(x-s,t)}{\partial t} \phi(s)ds = \frac{\partial}{\partial t}\int_{-\infty}^{\infty} E(x-s,t)\phi(s)ds = \frac{\partial}{\partial t}(E * \phi), \quad t > 0, \tag{52}$$

the so-called Stokes rule. The general formula now follows by adding (49) and (52)

$$u(x,t) = E * \psi + (\partial E/\partial t) * \phi.$$

7.8. The Fourier Transform of a Convolution

Let f and g be two locally integrable functions. Then the Fourier transform of the convolution $f * g$ is

$$
\begin{aligned}
[f * g]^\wedge(u) &= \int_{-\infty}^{\infty} f * g e^{iux}\, dx \\
&= \int_{-\infty}^{\infty} e^{iux} dx \int_{-\infty}^{\infty} f(x-y) g(y)\, dy \\
&= \int_{-\infty}^{\infty} g(y) e^{iuy} dy \int_{-\infty}^{\infty} f(x-y) e^{iu(x-y)} d(x-y) \\
&= \widehat{g}(u)\widehat{f}(u) = \widehat{f}(u)\widehat{g}(u)\,. \qquad (1)
\end{aligned}
$$

Thus, the Fourier transform of the convolution of f and g is the product of their Fourier transforms. A similar result holds for inverse Fourier transforms.

Relation (1) also holds for two singular distributions s and t under certain restrictions. For instance, if at least one of these distributions has compact support, then relation (1) holds. For further study of this subject the reader should see reference [27]. We illustrate this concept with a few elementary examples.

Example 1. In view of (7.6.6) and (6.4.7) we have

$$[t^{(k)}(x)]^\wedge(u) = [\delta^{(k)} * t]^\wedge(u) = (-iu)^k \widehat{t}(u)\,,$$

which agrees with (6.3.21a).

Example 2. Using relation (7.6.2b) we obtain

$$[t(x-y)]^\wedge u = [t(x) * \delta(x-y)]^\wedge = \widehat{t}(u) e^{iuy}\,,$$

which agrees with (6.3.25a).

Example 3. Consider the relation

$$\delta'(x-a) * \delta''(x-b) = \int_{-\infty}^{\infty} \delta'(x-a-y)\delta''(y-b) dy = \delta'''(x-a-b)\,,$$

which follows as a special case from Exercise 10. Taking the Fourier transform of both sides and using (6.4.7) and (7.6.2b), we find that

$$(-iu)e^{iua}(-iu)^2 e^{iub} = (-iu)^3 e^{iu(a+b)}\,,$$

which is an identity.

Example 4. We end this section with a physical example in which the Fourier transform, the direct product, the convolution, and the Fourier transform of convolution are all used. In optics, $F(u, v)$ is the Fourier transform of a function $f(x, y)$; it gives the far-field behavior due to a source of light whose strength (or intensity) is $f(x, y)$ per unit area in the x, y plane. Let us consider the special case when an infinitely narrow slit parallel to the y axis is a light source with unit intensity, so that $f(x, y) = \delta(x) \otimes 1(y)$. Accordingly, the far-field behavior is $F(u, v) = 1(u) \otimes 2\pi\delta(v)$, which is a light line perpendicular to the slit.

Next consider two slits with intensity $f_1(x, y)$ and $f_2(x, y)$. Then the far-field behavior of the slit with intensity $f_1(x, y) * f_2(x, y)$ is $F_1(u, t)F_2(u, t)$.

By extending these concepts one can study the interference phenomena due to a system of slits [21].

7.9. Distributional Solutions of Integral Equations

Our aim in this section is to solve the integral equations that involve generalized functions. We shall limit our study to the convolution-type integral equations because generalized functions are very suitable for them. Indeed, from the sifting property, we find that the solution of the integral equation

$$\int_a^x k(x-t)f(t)dt = k(x)$$

is $f(t) = \delta(t)$. As another example, it is easily verified that the solution of the integral equation

$$\int_0^x \sin(x-t)f(t)dt = 1 + x\,,$$

is $f(t) = 1 + t + \delta(t) + \delta'(t)$. Often, it is helpful to take the Fourier transform of the convolution type integral equation, because then, we get a simple algebraic equation to solve for the transformed functions [26] as we demonstrate in the solution of the Cauchy type integral equation given below.

By setting $h(x) = \delta(x) + k(x)$, we can write the Volterra integral equation

$$f(x) + \int_0^x k(x-t)f(t)dt = g(x)\,, \qquad x \geq 0\,, \tag{1}$$

as

$$h * f = g\,, \tag{2}$$

where we have extended the functions f and g for $x < 0$. Incidentally, by this substitution the distinction between the integral equations of the first and second kind has disappeared. In order to solve this equation it suffices to find the distribution h_1 such that

$$h * h_1 = \delta\,. \tag{3}$$

Then the solution of (2) is

$$f = h_1 * g\,. \tag{4}$$

We found in Example 1 of Section 7.4 that $\delta(x)$ plays the role of the unit element in the convolution algebra. Accordingly, we observe from (3) that h_1 is the inverse of $h = \delta + k$. Taking the inverse of $\delta + k$, however, is like taking the inverse of $1 + \lambda$, so we set

$$h_1 = \delta + \sum_{n=1}^{\infty}(-1)^n(k^*)^n\,, \tag{5}$$

where $(k^*)^n$ is the convolution product of k by itself taken n times. When the kernel is suitably smooth, the series (5) can be shown to converge and thus can be substituted in (4) to yield the solution

$$f(x) = g(x) + \int_0^x k_1(t-x)g(t)dt\,,$$

where

$$k_1(t-x) = \sum_{n=1}^{\infty}(-1)^n (k^*)^n\,. \tag{6}$$

We now discuss a few important singular integral equations.

Cauchy-type integral equation

Recall that the function $1/x$ defines the distribution

$$\left\langle \frac{1}{x}\,,\,\phi(x)\right\rangle = \lim_{\varepsilon\to 0}\int_{|x|>\varepsilon}\frac{\phi(x)}{x}\,dx\,,$$

for a test function $\phi(x)$. From formula (6.4.29a) we find that $[1/x]^\wedge = \pi i \operatorname{sgn} u$. Let us use this information to solve the famous singular Cauchy-type integral equation

$$\lambda f(x) + \frac{1}{\pi i}\int_{-\infty}^{*\infty}\frac{f(t)}{x-t}\,dt = g(x)\,, \tag{7}$$

where the $*$ means that the integral is the Cauchy principal value and where the functions f and g are tempered distributions. Taking the Fourier transform of both sides of this equation and using the convolution formula

$$[(1/x) * f]^\wedge = [1/x]^\wedge \widehat{f} = \pi i \widehat{f}(u)\operatorname{sgn} u\,,$$

we obtain

$$(\lambda + \operatorname{sgn} u)\widehat{f}(u) = \widehat{g}(u)\,, \tag{8}$$

The result of multiplying both sides of (8) by $\lambda - \operatorname{sgn} u$ is

$$(\lambda^2 - 1)\widehat{f} = \lambda\widehat{g} - (\operatorname{sgn} u)\widehat{g}\,,$$

so that

$$\widehat{f} = \frac{\lambda}{\lambda^2-1}\,\widehat{g} - \frac{\operatorname{sgn} u}{\lambda^2-1}\,\widehat{g}\,, \tag{9}$$

where we have assumed that $\lambda^2 \neq 1$. Finally, we take the inverse Fourier transform of (9) and obtain

$$\begin{aligned} f(x) &= \frac{\lambda}{\lambda^2-1}\,g(x) - \frac{1}{\lambda^2-1}F^{-1}[\operatorname{sgn} u\,\widehat{g}] \\ &= \frac{\lambda}{\lambda^2-1}\,g(x) - \frac{1}{\lambda^2-1}\,\frac{1}{\pi i}\int_{-\infty}^{*\infty}\frac{g(t)}{x-t}\,dt\,, \end{aligned} \tag{10}$$

which is the required solution.

Abel-type integral equation

We found in Example 9 of Section 7.7 that the solution of the Abel integral equation

$$\frac{1}{\Gamma(1-\alpha)}\int_0^x \frac{g(t)dt}{(x-t)^\alpha} = f(x) \tag{11}$$

is $g(x) = f(x) * \Phi_{-\alpha+1}(x)$ where $\Phi_\lambda(x) = x_+^{\lambda-1}/\Gamma(\lambda)$. Thus

$$\begin{aligned} g(x) &= f(x) * \frac{d}{dx}\Phi_\alpha = \frac{d}{dx}f(x) * \Phi_\alpha \\ &= \frac{1}{\Gamma(\alpha)}\int_0^x \frac{f'(t)dt}{(x-t)^{1-\alpha}} \,. \end{aligned} \tag{12}$$

For $0 < \alpha < 1$, the integral on the left side (23) is convergent, and therefore this integral equation can be solved by classical means [26]. We have solved it here by the distributional approach for all values of α.

We can solve various integral equations related to Abel's equation by the help of generalized functions. We present two examples.

Example 1. First we solve the integral equation

$$s\int_0^s \frac{g(t)dt}{(s^2-t^2)^{1/2}} = f(s) = \begin{cases} f_1(s)\,, & 0 \le s < a\,, \\ f_2(s)\,, & a < s < \infty\,, \end{cases} \tag{13}$$

where

$$\lim_{s\to a} f_1(s) \neq \lim_{s\to a} f_2(s)\,. \tag{14}$$

The formal solution of (13) is [26]

$$g(t) = \frac{2}{\pi}\frac{d}{dt}\int_0^t \frac{f(u)du}{(t^2-u^2)^{1/2}} \,. \tag{15}$$

Now set $u = t\sin\vartheta$, so that

$$(t^2-u^2)^{1/2} = t\cos\vartheta \qquad du = t\cos\vartheta\, d\vartheta\,.$$

and (15) becomes

$$g(t) = \frac{2}{\pi}\frac{d}{dt}\int_0^{\pi/2} f(t\sin\vartheta)d\vartheta = \frac{2}{\pi}\int_0^{\pi/2} f'(t\sin\vartheta)\sin\vartheta\, d\vartheta\,. \tag{16}$$

To make use of the generalized functions we write

$$f(s) = f_1(s)[1-H(s-a)] + f_2(s)H(s-a)\,, \tag{17}$$

which when differentiated becomes

$$\begin{aligned} f'(s) = {} & f_1'(s)[1-H(s-a)] + f_2'(s)H(s-a) \\ & +\delta(s-a)[f_2(s) - f_1(s)]\,. \end{aligned}$$

Furthermore, we let $t \sin\alpha = a$. Substituting these values in (16) we obtain

$$g(t) = \frac{2}{\pi}\int_0^{\pi/2} \{f_1'(t\sin\vartheta)[1 - H(t\sin\vartheta - t\sin\alpha)] + f_2'(t\sin\vartheta)H(t\sin\vartheta - t\sin\alpha) + \delta(t\sin\vartheta - t\sin\alpha)[f_2(t\sin\vartheta)]\}\sin\vartheta\, d\vartheta\,. \tag{18}$$

Now, in view of the (3.1.1), namely,

$$\delta[f(x)] = \sum_{m=1} \frac{\delta(x - x_m)}{|f'(x_m)|}\,,$$

where the x_m are the simple zeros of $f(x)$, we find that

$$\begin{aligned}\int_0^{\pi/2} &\delta(t\sin\vartheta - t\sin\alpha)[f_2(t\sin\vartheta) - f_1(t\sin\vartheta)]\sin\vartheta d\vartheta\\ &= [f_2(t\sin\alpha) - f_1(t\sin\alpha)](\sin\alpha)/(t\cos\alpha)\\ &= [f_2(a) - f_1(a)]a/t\sqrt{t^2 - a^2}\,.\end{aligned}$$

Substituting this value in (18) and using the definition of the Heaviside function, we obtain the required solution:

$$g(t) = \frac{2}{\pi}\left\{\int_0^{\alpha} f_1'(t\sin\vartheta)\sin\vartheta d\vartheta + \int_\alpha^{\pi/2} f_2'(t\sin\vartheta)\sin d\vartheta + [f_2(a) - f_1(a)]a/t\sqrt{t^2 - a^2}\right\}. \tag{19}$$

Example 2.

$$s\int_s^\infty \frac{g(t)dt}{(t^2 - s^2)^{1/2}} = f(s) = \begin{cases} f_1(s)\,, & 0 \le s \le s\,,\\ f_2(s)\,, & a < s < \infty\,,\end{cases} \tag{20}$$

with condition (14). Its formal solution is [26]

$$g(t) = -\frac{2}{\pi}\frac{d}{dt}\int_t^\infty \frac{f(u)du}{(u^2 - t^2)^{1/2}}\,. \tag{21}$$

In this case we set $u = t\cosh\vartheta$, $du = t\sinh\vartheta d\vartheta$, so that (19) takes the form

$$g(t) = -\frac{2}{\pi}\int_0^\infty f'(t\cosh\vartheta)\cosh\vartheta\, d\vartheta\,. \tag{22}$$

Writing the function $f(s)$ as in (17), setting $t\cosh\alpha = a$, and proceeding as in the first example, we obtain

$$g(t) = \frac{2}{\pi}\left\{\int_0^{\alpha} f_1'(t\cosh\vartheta)\cosh\vartheta d\vartheta + \int_\alpha^{\infty} f_2'(t\cosh\vartheta)\cosh d\vartheta + [f_2(a) - f_1(a)]a/t\sqrt{a^2 - t^2}\right\}.$$

For the unified analysis for the distributional solutions of various singular integral equations, the reader is referred to Estrada and Kanwal [29–31].

We shall discuss the convolution type integral equations in Chapter 13 also. There we shall use the theory of moments.

Wiener-Hopf integal equation

The equation of the type

$$\int_0^\infty K(s-t)g(t)dt = \lambda g(s) + f(s)\,, \quad 0 \le s, t < \infty \tag{23}$$

is called the Wiener-Hopf integral equation [28]. Its distinctive features are the difference kernel and the semiinfinite interval. Observe that although both s and t are nonnegative the argument $s-t$ in the kernel can be negative so that we need the kernel $K(s)$ for $-\infty < s < \infty$. Secondly, although the kernel $K(s-t)$ is the difference kernel we can not apply the Fourier transform because the interval of integration is semiinfinite. We remedy these difficulties as follows.

We set

$$\int_{-\infty}^\infty K(s-t)g(t)dt = \begin{cases} \lambda g(s) + f(s)\,, & 0 < s < \infty\,, \\ u(s)\,, & -\infty < s < 0\,, \end{cases} \tag{24}$$

which introduces a new function $u(s)$. Finally, we extend the definition of the functions $f(s)$, $g(s)$ and $u(s)$ so that

$$f(s) < 0\,, \quad s < 0\,, \quad g(s) = 0\,, \quad s < 0\,, \quad u(s) = 0\,, \quad s \ge 0\,.$$

This enables us to write equation (23) as

$$\int_{-\infty}^\infty K(s-t)g_+(t)dt = \lambda g_+(s) + f_+(s) + u_-(s)\,, \quad -\infty < s, t < \infty\,, \tag{25}$$

where the subscripts $\pm$ indicate the $\pm$ half-line on which the function is non-vanishing.

Let us now demonstrate how the generalized functions enter the picture. The Wiener-Hopf integral equation

$$\int_0^\infty e^{-(s-t)} g(t)dt = 1\,, \tag{26}$$

does not have a classical solution. It is easily verified that the distributional solution is [29]

$$g(s) = \frac{1}{2}H(s) - \frac{1}{2}\delta'(s) + C(\delta(s) + \delta'(s))\,, \tag{27}$$

where C is an arbitrary constant. For the general equation

$$\int_0^\infty e^{-(s-t)} g(t) = f(s)\,, \tag{28}$$

the solution is

$$g(s) = \frac{1}{2}(f_+(s) - f_+''(s) + C(\delta(s) + \delta'(s)) . \tag{29}$$

When $f(s) = 1$, $f_+(s) = 1_+ = H(s)$ and relation (27) follows. For full details see Estrada and Kanwal [29].

Exercises

1. Show that

$$D_x^k D_y^q [s(x) \otimes t(y)] = [D_x^k s(x)] \otimes [D_y^q t(y)] .$$

2. Show that the direct product is linear in the following sense. If α and β are arbitrary numbers and if s_1, s_2, t_1, and t_2 are arbitrary distributions defined over R_n, then

$$s_1 \otimes (\alpha t_1 + \beta t_2) = \alpha(s_1 \otimes t_1) + \beta(s_1 \otimes t_2) ,$$
$$(\alpha s_1 + \beta s_2) \otimes t_1 = \alpha(s_1 \otimes t_1) + \beta(s_2 \otimes t_1) .$$

3. Prove that $D_y^k(s(x) \otimes 1(y)) = 0$ for $|k| \neq 0$.

4. Prove the equations

(a) $f_\alpha * f_\beta = f_{\alpha+\beta}$, $\quad f_\alpha(x) = \dfrac{H(x)x^{\alpha-1}}{\Gamma(\alpha)} e^{\alpha x}$, $\quad \alpha > 0$;

(b) $f_\alpha * f_\beta = f_{\sqrt{\alpha^2+\beta^2}}$, $\quad f_\alpha(x) = \dfrac{1}{\alpha\sqrt{2\pi}} \exp\left(-\dfrac{x^2}{2\alpha^2}\right)$, $\quad \alpha > 0$;

(c) $f_\alpha * f_\beta = f_{\alpha+\beta}$, $\quad f_\alpha(x) = \dfrac{1}{\sqrt{\pi}} \dfrac{\alpha}{\alpha^2 + x^2}$, $\quad \alpha > 0$.

5. If $s(x, y) \in S'_{m+n}$, show that

$$D_x^p D_y^k F_x[s] = F_x[(ix)^p D_y^k s] ,$$

and

$$F_x[D_x^p D_y^p s] = (-iv)^p D_y^k F_x[s] ,$$

where F_x denotes Fourier transform in R_n, and the x space.

6. Establish the following distributional convolution formulas:

(a) $[xH(x) * e^x H(x)] = (e^x - x - 1)H(x)$,

(b) $(H(x) \sin x) * (H(x) \cos x) = \frac{1}{2}H(x)x \sin x$,

(c) $\delta'(x) * \text{Pf}\,(1/x) = \text{Pf}\,(H(-x)/x^2) - \text{Pf}\,(H(x)/x^2)$.

7. Evaluate

(a) $e^{-|x|} * e^{-|x|}$,

(b) $e^{-ax^2} * xe^{-ax^2}$,

(c) $xe^{-ax^2} * xe^{-ax^2}$.

8. By replacing λ with $-\lambda$ in Example 7 and proceeding as in Example 8 of Section 7.6, show that

$$\frac{d^\lambda}{dx^\lambda}\left[\frac{x_+^{\mu-1}}{\Gamma(\mu)}\right] = \frac{x_+^{\mu-\lambda-1}}{\Gamma(\mu-\lambda)}.$$

Since $1_+ = H(x)$, on setting $\mu = 1$ this formula becomes

$$d^\lambda H(x)/dx^\lambda = x_+^{-\lambda}/\Gamma(1-\lambda).$$

Also deduce that

$$D^\lambda \delta^{(l)}(x) = \frac{x_+^{-l-\lambda-1}}{\Gamma(-l-\lambda)} \quad \text{and} \quad D^\lambda \frac{x_+^{\lambda-l-1}}{\Gamma(\lambda-l)} = \delta^{(l)}(x), \qquad l = 0, 1, 2, \dots.$$

9. (a) From relation (7.6.21) deduce the following:

$$\int_{-\infty}^{\infty} y^\lambda H(y)(x-y)^\mu H(x-y)dy = \frac{\lambda!\mu!}{(\lambda+\mu+1)!}\, x^{\lambda+\mu+1} H(x).$$

Discuss its validity. (For instance, when $\lambda + \mu + 1$ is a negative integer, the right-hand side is to be replaced.)

(b) For $\lambda = 0$, (a) reduces to

$$\int_{-\infty}^{\infty} H(y)(x-y)^\mu H(x-y)dy = \frac{\mu! x^{\mu+1}}{(\mu+1)!}\, H(x).$$

From this relation derive

$$\int_0^x \delta^{(m)}(y)(x-y)^\mu H(x-y)dy = \frac{\mu! x^{\mu-m}}{(\mu-m)!}.$$

(c) Similarly, for $\lambda = \mu = 0$, we have $\int_{-\infty}^{\infty} H(y)H(x-y)dy = xH(x)$. Show that

$$\int_0^\infty \delta^{(m)}(y)dy = \int_0^x \delta^{(m-r)}(y)\delta^{(r-1)}(x-y)dy = \delta^{(m-1)}(x).$$

10. Show that

$$\int_{-\infty}^{\infty} \delta^{(m)}(y)\delta^{(n)}(x-y)dy = \delta^{(m+n)}(x),$$

$$\int_{-\infty}^{\infty} \delta(\alpha-y)\delta(x-\alpha)dy = \delta(x-y),$$

$$\int_{-\infty}^{\infty} \delta^{(m)}(\alpha-y)\delta^{(n)}(x-\alpha)dy = \delta^{(m+n)}(x-y).$$

Note that the first of these relations implies that $\delta^{(m)} * \delta^{(n)} = \delta^{(m+n)}$.

11. Prove that, in the notation of Section 7.5,

$$\frac{d^k}{dx^k}(s * \psi) = \left\langle s(y), \frac{\partial^k}{\partial x^k}(\psi(x-y))\right\rangle.$$

12. With the help of the analysis of Section 7.5 regularize the pseudofunction $f(x) = \text{Pf}\,(H(x)/x) = \text{Pf}\,(1/x_+)$.
Hint: Split R_1 into three intervals:
(i) In the interval $\varepsilon < 0$, $f(x) = 0$.
(ii) In $-\varepsilon < x < \varepsilon$, we can appeal to the analysis of Chapter 4 and take the Hadamard finite part of $\langle f, \phi \rangle$.
(iii) In $x \geq \varepsilon$, the function is regular.
Sketch the regularization for $\varepsilon = \frac{1}{4}$ and compare with the graph of $f(x)$.

13. Show that
(a) $H(x) * \text{Pf}\,(H(x)/x) = H(x) \ln x$,
(b) $\text{Pf}\left(\frac{H(x)}{x}\right) * \text{Pf}\left(\frac{H(x)}{x}\right) = \frac{d^2}{dx^2} \int_0^x \ln \xi ln(x - \xi) d\xi$.

14. Show that

$$\text{Pf}\left(\frac{1}{(x-\xi)^n}\right) = \left\{\frac{(-1)^n}{n!}\, \delta^{(n)}(x-\xi)\right\} * \text{Pf}\left(\frac{1}{x}\right).$$

15. Let $f(x)$ be a continuous function in R_1 that is periodic with period 2π. Show that $\widehat{f}(u) = \sum_{n=-\infty}^{\infty} b_n \delta_n (u - n)$, and relate b_n to the coefficients of the Fourier series of $f(x)$.

16. Show that

$$\text{Pf}\,(1/x) * \text{Pf}\,(1/x) = -\pi^2 \delta(x).$$

CHAPTER 8

The Laplace Transform

8.1. A Brief Discussion of the Classical Results

The main applications of the Laplace transform are directed toward problems in which the time t is the independent variable. We shall therefore use this variable in this chapter. Let $f(t)$ be a complex-valued function of the real variable t such that $f(t)e^{-ct}$ is abolutely integrable over $0 < t < \infty$, where c is a real number. Then the Laplace transform of $f(t)$, $t \geq 0$, is defined as

$$\tilde{f}(s) = \mathcal{L}\{f(t)\} = \int_0^\infty f(t)e^{-st}dt\,, \qquad \operatorname{Re} s > c\,, \tag{1}$$

where $s = \sigma + i\omega$. The Laplace transform defined by (1) has the following basic properties.

(1) Linearity. Let the Laplace transforms of the functions $f(t)$ and $g(t)$ be $\tilde{f}(s)$ and $\tilde{g}(s)$, respectively, and let α and β be any constants. Then the Laplace transform of the function $h(t)$ defined by $h(t) = \alpha f(t) + \beta g(t)$ is $\tilde{h}(s) = \alpha \tilde{f}(s)\beta\tilde{g}(s)$.

(2) The uniqueness theorem. If $\tilde{f}(s) = \tilde{g}(s)$ on some vertical line in their region of convergence, then $f(t) = g(t)$.

(3) The transform of the nth derivative. If the function $f(t)$ is n-times continuously differentiable, then

$$\mathcal{L}\{f^{(n)}(t)\} = [f^{(n)}(t)]^{\sim} = s^n\tilde{f}(s) - s^n\tilde{f}(s) - s^{n-1}f(0) - s^{n-2}f'(0) - \cdots - f^{(n-1)}(0)\,. \tag{2}$$

(4) The convolution theorem. Let $\tilde{f}(s)$ and $\tilde{g}(s)$ be the Laplace transforms of the function $f(t)$ and $g(t)$. Then the Laplace transform of the convolution $h(t) = \int_0^t f(\tau)g(t-\tau)d\tau$, $t \geq 0$, is

$$\tilde{h}(s) = \tilde{f}(s)\tilde{g}(s)\,. \tag{3}$$

(5) The inverse transform. The formula for the inverse of the Laplace transform is

$$f(t) = \mathcal{L}^{-1}\{\tilde{f}(s)\} = \frac{1}{2\pi i}\int_{\sigma-i\infty}^{\sigma+i\infty} \tilde{f}(s)e^{st}ds\,, \tag{4}$$

where σ is $\operatorname{Re} s$. The usual method of evaluating this complex integral is by analytically continuing $\tilde{f}(s)$ into the complex plane for $\operatorname{Re} s < c$, converting the line integral into a contour integral, and then applying the residue theorem.

(6) $$\mathcal{L}^{-1}\{\tilde{f}(as+b)\} = (1/a)\exp(-bt/a)f(t/a)\,. \tag{5}$$

For the special case when $a = 1$, $b = -\alpha$, (5) becomes

$$\mathcal{L}^{-1}\{\tilde{f}(s-\alpha)\} = e^{\alpha t}f(t)\,. \tag{6}$$

(7) If $\mathcal{L}^{-1}\{\tilde{f}(s)\} = f(t)$ is valid for $c \geq 0$, and if $f(t)$ is assigned arbitrary values for $-c \leq t < 0$, then

$$\mathcal{L}^{-1}\{e^{-cs}\tilde{f}(s)\} = f(t-c)H(t-c)\,. \tag{7}$$

where $H(t-c)$ is the Heaviside function.

All these results are discussed in many classical texts on the subject.

8.2. The Laplace Transform Distributions

Recall that when we attempted to define the Fourier transform of a distribution by the classical formula, in Chapter 6, we got into difficulty, which we resolved by defining a new class of test functions. The same difficulty arises in the present situation. Indeed, if we formally extend definition (8.1.1) to a distribution $f(t)$ *whose support is bounded on the left at* 0, we have

$$\tilde{f}(s) = \int_0^\infty f(t)e^{-st}dt = \langle f(t), e^{-st}\rangle\,. \tag{1}$$

We examine this relation assuming that there exists a real number c such that $e^{-ct}f(t)$ is a distribution belonging to S' (the class of tempered distributions). Then we can rewrite (1)

$$\tilde{f}(s) = \langle e^{-ct}f(t),\ H(t)e^{-(s-c)t}\rangle\,, \tag{2}$$

where $H(t)$ is the Heaviside function. For Re $s > c$, the function $H(t)e^{-(s-c)t}$ is a test function in S, and definition (2) makes sense. However, this is not the case, as is clear from the distribution $f(t) = H(t)e^{t^2}$; $f(t)$ is a member of D', but there is no value of c for which $H(t)e^{t^2-ct} \in S$. Accordingly, we define a new class of test functions and its dual class [32].

Definition 1. The space L of *test functions of exponential decay* is the space of the complex-valued functions $\phi(t)$ satisfying the following properties:

(1) $\phi(t)$ is infinitely differentiable; i.e., $\phi(t) \in C^\infty(R_n)$.
(2) $\phi(t)$ and its derivatives of all orders vanish at infinity faster than the reciprocal of the exponential of order c; that is, $|e^{ct}D^k\phi(t)| < M, \forall c, k$.

Definition 2. A function $f(t)$ is of *exponential growth* if and only if $f(t)$ together with all its derivatives grows at infinity more slowly than the exponential function of order c; i.e., there exist real constants c and M such that $|D^k f(t)| \leq Me^{ct}$.

Definition 3. A linear continuous functional over the space L of test functions is called a *distribution of exponential growth*. This dual space of L is denoted L'.

In view of these definitions we find that all the distributions belonging to L' have the Laplace transform based on the definition

$$\tilde{f}(s) = \int_0^\infty f(t)e^{-st}dt = \langle f(t), e^{-st}\rangle\,, \tag{3}$$

for real $s \in (c, \infty)$. This follows from the relation

$$\left| \int_0^\infty f(t)e^{-st}dt \right| \leq M \int_0^\infty e^{-(s-c)t}dt\,, \tag{4}$$

the right side of which is finite for Re $s > c$, and from the fact that $e^{-st} \in L$.

We shall come across various other spaces of this nature in Chapter 13.

Since this definition agrees with (1) of the classical transform, most of the known formulas remain the same. For example, all the properties of Section 8.1 hold for a distribution. However, Property 3 of Section 8.1 will be examined carefully in the next section.

Example 1. The Laplace transform of the Heaviside function is

$$[H(t)]^{\sim} = \int_0^\infty e^{-st}dt = \frac{1}{s}\,. \tag{5}$$

Example 2. The delta function and its derivatives

$$(a) \quad [\delta(t-a)]^{\sim} = \int_0^\infty e^{-st}\delta(t-a)dt = e^{-sa} \tag{6}$$

$$(b) \quad [\delta'(t-a)]^{\sim} = \int_0^\infty e^{-st}\delta'(t-a)dt = -\{(d/dt)(e^{-st})\}_{t=0} = se^{-sa}\,. \tag{7}$$

Continuing this process, we obtain

$$(c) \quad [\delta^{(n)}(t-a)]^{\sim} = s^n e^{-sa}\,. \tag{8}$$

By *formally* setting $a = 0$ in (6) and (8), we obtain

$$(d) \quad [\delta(t)]^{\sim} = 1\,, \tag{9}$$

$$(e) \quad [\delta^{(n)}(t)]^{\sim} = s^n\,. \tag{10}$$

We have emphasized the word "formally" because the integral $\int_0^\infty e^{-st}\delta(t)dt$ is not defined. Indeed, $\int_0^\infty e^{-st}\delta(t)dt = \int_{-\infty}^\infty e^{-st}H(t)\delta(t)dt$. This means that $\widetilde{\delta}(t) = H(0)$, which is usually taken to be $\frac{1}{2}$. However, definition (9) is consistent with the fact that $\delta(t)$ is the derivative of the Heaviside function, as explained in the next section.

8.3. The Laplace Transform of the Distributional Derivatives and Vice Versa

By definition (8.2.3), we have

$$\mathcal{L}\{\widetilde{f'}(t)\} = \langle \overline{f}\,'(t), e^{-st}\rangle\,. \tag{1}$$

In order to reconcile this result for distributions with Property 3 of Section 8.1, we should first evaluate $\overline{f}\,'(t)$ and then apply the right side of (1). For this purpose, let us discuss the function

$$f(t) = \begin{cases} g_1(t), & t < a, \\ g_2(t), & t > a, \end{cases}$$
$$= g_1(t)H(a-t) + g_2(t)H(t-a), \tag{2}$$

where $a > 0$ and $g_1(t)$ and $g_2(t)$ are continuously differentiable functions. The classical derivative $f'(t)$ of (2) is

$$f'(t) = g_1'(t)H(a-t) + g_2'(t)H(t-a), \tag{3}$$

for all $t \neq a$, while the distributional derivative is

$$\overline{f}\,'(t) = f'(t) + [f]\delta(t-a), \tag{4}$$

where we have used the notation, as in Chapter 5, $[f] = (f(a+) - f(a-)$.

The Laplace transform of (3) is

$$\int_0^\infty e^{-st} f'(t)dt = \int_0^a g_1'(t)e^{-st}dt + \int_a^\infty g_2'(t)e^{-st}dt$$
$$= [e^{-st}g_1(t0]_0^a + s\int_0^a g_1(t)e^{-st} + [e^{-st}g_2(t)dt]_a^\infty$$
$$+ s\int_a^\infty g_2(t)e^{-st}dt$$
$$= s\left\{\int_0^a g_1(t)e^{-st}dt + \int_a^\infty g_2(t)e^{-st}dt\right\} - e^{-as}[f] - g_1(0)$$
$$= s\widetilde{f}(s) - f(0) - [f]e^{-as}, \tag{5}$$

where we have written $f(0)$ for $g_1(0)$.

On the other hand,

$$\mathcal{L}\{\overline{f}\,'(t)\} = \langle \overline{f}\,'(t),\, e^{-st}\rangle = \langle f' + [f]\delta(t-a), e^{-st}\rangle$$
$$= \int_0^\infty f'(t)e^{-st}dt + [f]e^{-as} = s\widetilde{f}(s) - f(0), \tag{6}$$

where we have used (5). Thus Property 3 of Section 8.1 holds.

Relation (5) makes sense even when we allow a to tend to zero, because in that case we have

$$\mathcal{L}\{f'(t)\} = \int_0^\infty g_2'(t)e^{-st}dt = s\int_0^\infty g_2(t)e^{-st}dt - g_2(0)$$
$$= s\widetilde{f}(s) - f(0+) = s\widetilde{f}(s) - f(0-) - [f(0+) - f(0-)]e^{-os}, \tag{7}$$

which is consistent with (5).

To examine the situation in a different way, we write (2) with $a = 0$: $f(t) = g_1(t)H(-t) + g_2(t)H(t)$. Then for $t > 0$,

$$\overline{f}'(t) = g_1'(t)H(-t) + g_2'H(t) + [g_2(0) - g_1(0)]\delta(t) = g_2'(t) + [f]\delta(t)\,,$$

where $[f] = [f(0+) - f(0-)]$. thus

$$\mathcal{L}\{f'(t)\} = \mathcal{L}\{g_2'(t)\} + [f]\widetilde{\delta}(t) = s\widetilde{f}(s) - f(0+) + [f]\widetilde{\delta}(t)\,. \tag{8}$$

For (6) and (8) to agree we should have $\widetilde{\delta}(t) = 1$, as stipulated in (8.2.9).

Finally, we check consistency by using (8) for $f(t) = H(t)$:

$$\mathcal{L}\{\overline{H}'(t)\} = s(1/s) - H(0+) + [H(0+) - H(0-)] = 1\,,$$

where we have used (8.2.5).

For functions $f(t)$ that have support only for positive values of t, so that $f(0-) = 0$, (8) becomes

$$\mathcal{L}\{\overline{f}'(t)\} = s\widetilde{f}(s)\,. \tag{9}$$

Continuing in this fashion, we find that

$$\mathcal{L}\{\overline{f}^{(n)}(t)\} = s^n\widetilde{f}(s)\,. \tag{10}$$

Let us now prove, by induction, the result

$$(d^k/ds^k)[\widetilde{f}(s)] = \langle f(t), (-t)^k e^{-st}\rangle\,, \tag{11}$$

which gives the formula for the derivative of the Laplace transform. It is true for $k = 0$, by the definition of the transform. Now let us assume (11) to be true for k replaced by $k - 1$; i.e.,

$$(d^{k-1}/ds^{k-1})\{\widetilde{f}(s)\} = \langle f(t), (-t)^{k-1} e^{-st}\rangle\,.$$

Then

$$\begin{aligned}\frac{d^k}{ds^k}\{\widetilde{f}(s)\} &= \frac{d}{ds}\left[\frac{d^{k-1}}{ds^{k-1}}\{\widetilde{f}(s)\}\right] = \frac{d}{ds}[\langle f(t), (-t)^{k-1}e^{-st}\rangle] \\ &= \left[\left\langle f(t), (-t)^{k-1}\frac{d}{ds}e^{-st}\right\rangle\right] = \langle f(t), (-t)^k e^{-st}\rangle\,,\end{aligned}$$

as desired.

8.4. Examples

Example 1. Since $H(t)\ln t$ is a regular distribution, we have

$$\mathcal{L}\{H(t)\ln t\} = \int_0^\infty \ln t e^{-st} dt .$$

On our setting $st = \eta$, this integral becomes

$$\frac{1}{s}\int_0^\infty e^{-\eta}(\ln\eta - \ln s)d\eta = -\frac{1}{s}(\gamma + \ln s) ,$$

where $\gamma = -\int_0^\infty e^{-\eta}\ln\eta d\eta = 0.5772\cdots$ is Euler's constant. Thus

$$\mathcal{L}\{H(t)\ln t\} = [H(t)\ln t]^{\sim} = -(1/s)(\gamma + \ln s) . \tag{1}$$

Example 2. We can use the result of the previous example to evaluate the Laplace transform of the singular distribution Pf $[H(t)/t]$. Because

$$\text{Pf}\,[H(t)/t] = (\overline{d}/dt)[H(t)\ln t] ,$$

we obtain, from (1) and (8.3.9),

$$\mathcal{L}\{\text{Pf}\,[H(t)/t]\} = \mathcal{L}\{(\overline{d}/dt)[H(t)\ln t]\} = s[-(1/s)(\gamma + \ln s)] = -(\gamma + \ln s) . \tag{2}$$

Similarly, because [see (4.2.3)]

$$\text{Pf}\left(\frac{H(t)}{t^2}\right) = -\frac{\overline{d}}{dt}\left[\text{Pf}\left(\frac{H(t)}{t}\right)\right] - \delta'(t) ,$$

we find that

$$\mathcal{L}\{\text{Pf}\,[H(t)/t^2]\} = s(\ln s + \gamma) - s = s(\ln s + \gamma - 1) . \tag{3}$$

Example 3. Let us find the Laplace transform of the function $t_+^\lambda = H(t)t^\lambda$, where $\lambda \neq -1, -2, -3, \ldots$. Since $t_+^\lambda \in L'$, we have $\mathcal{L}\{t_+^\lambda\} = \int_0^\infty e^{-st}t^\lambda dt$. Letting $u = st$ for $s > 0$, it follows that

$$\mathcal{L}\{t_+^\lambda\} = \frac{1}{s^{\lambda+1}}\int_0^\infty e^{-u}u^\lambda du = \frac{\Gamma(\lambda+1)}{s^{\lambda+1}} . \tag{4}$$

In particular, for $\lambda = 0$ we recover formula (8.2.5).

Example 4. The Laplace transform of a periodic function. Let $f(t)$ be a function that vnaishes identically outside the finite interval $(0, T)$. The periodic extension of $f(t)$ of period T is the function obtained by summing the translates $f(t - kT)$, for $k = 0, \pm1, \pm2, \ldots$ (as shown in Figure 8.1):

$$f_T(t) = \sum_{k=-\infty}^{\infty} f(t - kT) . \tag{5}$$

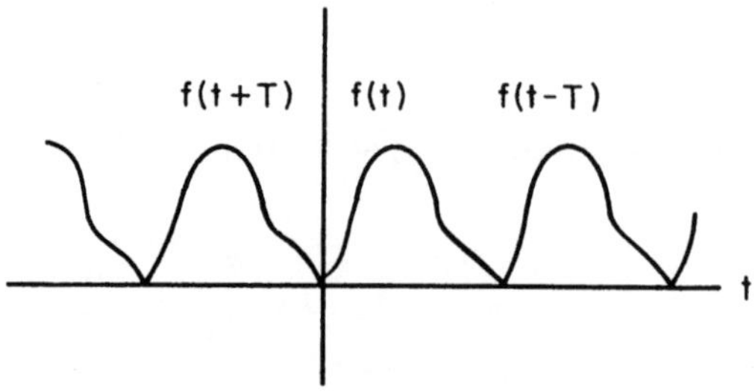

Figure 8.1. The periodic extension of $f(t)$.

Then we can show that

$$\mathcal{L}\{f_T(t)\} = \frac{1}{1-e^{-sT}} \int_0^T e^{-st} f(t)dt\,. \tag{6}$$

The proof follows on writing (5) as the convolution

$$f_T(t) = \sum_{k=-\infty}^{\infty} \{f(t) * \delta(t-kT)\} = f(t) * \sum_{k=-\infty}^{\infty} \delta(t-kT)\,. \tag{7}$$

Now

$$\mathcal{L}\left\{\sum_{k=-\infty}^{\infty} (t-kT)\right\} = \mathcal{L}\left\{\sum_{k=0}^{\infty} \delta(t-kT)\right\}$$

$$= 1 + e^{-sT} + e^{-2sT} + e^{-3sT} = \frac{1}{1-e^{-sT}}\,. \tag{8}$$

This summation will be valid if

$$|e^{-st}| = |e^{-(\sigma+i\omega)T}| = e^{-\sigma T} < 1\,,$$

which is true for all $\sigma > 0$.

Now we apply the convolution theorem (8.1.3) to (7), so that

$$\mathcal{L}\{f_T(t)\} = \widetilde{f}(s)\mathcal{L}\left\{\sum_{k=0}^{\infty} \delta(t-kT)\right\}\,, \tag{9}$$

and (6) follows.

Example 5. In many physical applications we encounter the convolution equation

$$f(t) * x(t) = g(t)\,. \tag{10}$$

Applying the Laplace transform to this, we obtain $\widetilde{f}(s)\widetilde{x}(s) = \widetilde{g}(s)$, so $\widetilde{x}(s) = (\widetilde{f})^{-1}g$. By inversion we obtain $x(t) = (\widetilde{f})^{-1} * g$, where $\mathcal{L}\{(f^*)\} = (\widetilde{f}(s))^{-1}$, and we assume that $(\widetilde{f}(s))^{-1}$ is the Laplace transform of a right-sided distribution.

For example, when

$$f = \sum_{k=0}^{\infty} a_k D^k \delta(t)\,, \tag{11}$$

so that (10) is an ordinary differential equation, we have

$$\mathcal{L}\{(\tilde{f})^{-1}\} = 1\Big/\sum_{k=0}^{n} a_k s^k\,. \tag{12}$$

The method of partial fractions now helps us in solving the problem.

Example 6. Let us find

$$\mathcal{L}^{-1}\left\{\frac{1}{1+s^2}\right\}^{\nu+1/2}, \qquad s>0\,,\quad \nu > -\frac{1}{2}\,.$$

The factors of $\{1/(1+s^2)\}^{\nu+1/2}$ are $\{1/(s+i)\}^{\nu+1/2}$ and $\{1/(s-i)\}^{\nu+1/2}$. The relation

$$\mathcal{L}\{H(t)e^{\lambda t}t^{\nu-1}/\Gamma(\nu)\} = 1/(s-\lambda)^{\nu}\,, \qquad \nu>0\,, \tag{13}$$

which is easily proved, shows that we can set

$$f(t) = \left\{H(t)e^{it}t^{\nu-1/2}\Big/\left(\nu+\frac{1}{2}\right)\right\}, \quad \text{and} \quad g(t) = \left\{H(t)e^{-it}t^{\nu-1/2}/\Gamma\left(\nu+\frac{1}{2}\right)\right\},$$

so that

$$\tilde{f}(s) = \left(\frac{1}{s-i}\right)^{\nu+1/2}, \quad \text{and} \quad \tilde{g}(s) = \left(\frac{1}{s+i}\right)^{\nu+1/2}.$$

Next, we use the convolution theorem to obtain $h_\nu(t) = f(t) * g(t)$; and find that the required value is

$$\begin{aligned} h_\nu(t) &= \frac{H(t)\exp(it)t^{2\nu}}{[\Gamma(\nu+\frac{1}{2})]^2}\int_{-1}^{1}\exp[-it(1+u)]\left(\frac{1-u^2}{4}\right)^{\nu-1/2}\frac{du}{2} \\ &= H(t)\left[\sqrt{\pi}/\Gamma\left(\nu+\frac{1}{2}\right)\right](t/2)^{\nu}J_\nu(t)\,. \end{aligned}$$

Example 7. Let us solve the initial-boundary value problem

$$\partial^2 v/\partial t^2 - \partial^2 v/\partial x^2 = 0\,, \qquad 0<x<\infty\,,\quad t>0\,, \tag{14a}$$

$$v(x,0) = 0\,, \tag{14b}$$

$$(\partial v/\partial t)(x,0) = 0\,, \tag{14c}$$

$$v(0,t) = h(t)\,. \tag{14d}$$

We multiply both sides of (14a) by e^{-st} and integrate from $t=0$ to $t=\infty$. Integrating the first term by parts twice and using the initial conditions and the fact that e^{-st} is exponentially small at $t = \infty$ for Re $s > 0$, we obtain the boundary value problem

$$-d^2\tilde{v}/dx^2 + s^2\tilde{v} = 0, \qquad 0 < x < \infty, \tag{15a}$$

$$\tilde{v}(0, s) = \widetilde{h}(s), \tag{15b}$$

where $\widetilde{v}(x, s)$ and $\widetilde{h}(s)$ are the Laplace transforms of $v(x, t)$ and $h(t)$, respectively. To make this two-point boundary value problem complete, we add the requirement that

$$\lim_{x\to\infty} \tilde{v}(x, s) = 0, \qquad \text{Re } s > 0. \tag{15c}$$

The solution of system (15) is

$$\widetilde{v}(x, s) = \widetilde{h}(s)e^{-sx}. \tag{16}$$

Because

$$\mathcal{L}^{-1}\{\widetilde{h}(s)\} = h(t), \qquad \mathcal{L}^{-1}\{e^{-sx}\} = \delta(t - x),$$

we find that by inverting (16) and using the convolution theorem

$$\begin{aligned} v(x, t) &= \int_0^t h(\tau)\delta(t - x - \tau)d\tau = \begin{cases} h(t - x), & t > x, \\ 0, & t < x, \end{cases} \\ &= h(t - x)H(t - x). \end{aligned}$$

Example 8. Let us attempt to solve the initial-boundary value problem

$$\partial^2 u/\partial x^2 - \partial^2 u/\partial t^2 = 0, \qquad x > 0, \quad t > 0, \tag{17a}$$

$$u(x, 0) = 0, \tag{17b}$$

$$u(0, t) = e^{ct}, \tag{17c}$$

$$u_x(0, t) = 0, \tag{17d}$$

with the help of generalized functions and Laplace transform. We set

$$\begin{aligned} v(x, t) &= H(x)H(t)u(x, t), \qquad (18) \\ \bar{\partial} v/\partial x &= \delta(x)H(t)u(x, t) + H(x)H(t)u_x(x, t) \\ &= \delta(x)H(t)u(0, t) + H(x)H(t)u_x(x, t), \\ \bar{\partial}^2 v/\partial x^2 &= \delta'(x)H(t)u(0, t) + \delta(x)H(t)u_x(x, t) + H(x)H(t)u_{xx}(x, t) \\ &= \delta'(x)H(t)u(0, t) + \delta(x)H(t)u_x(0, t) + H(x)H(t)u_{xx}(x, t) \\ &= \delta'(x)H(t)e^{ct} + H(x)H(t)u_{xx}(x, t), \end{aligned}$$

where we have used conditions (17c) and (17d). Similarly,

$$\begin{aligned} \frac{\bar{\partial}^2 v}{\partial t^2} &= H(x)\delta'(t)u(x, 0) + H(x)H(t)u_{tt}(x, t) \\ &= H(x)\delta(t)u_t(x, 0) + H(x)H(t)u_{tt}(x, t). \end{aligned}$$

Consequently, system (17) is equivalent to the inhomogeneous equation

$$\overline{\partial}^2 v/\partial x^2 - \overline{\partial}^2 v/\partial t^2 = e^{ct} H(t)\delta'(x) - u_t(x,0)H(x)\delta(t)\,, \tag{19}$$

which we solve by taking the Laplace transform of both sides with respect to x. Then

$$\left(\frac{\overline{d}^2}{dt^2} - s^2\right) V(s,t) = -se^{ct}H(t) + A(s)\delta(t)\,, \tag{20}$$

where $V(s,t)$ and $A(s)$ are the Laplace transforms of $v(x,t)$ and $u_t(x,0)$, respectively.
The solution of (20) is derived by setting

$$V(s,t) = G(s,t)H(t)\,, \tag{21}$$

so that

$$\frac{\overline{d}}{dt} V(s,t) = \delta(t)G(s,0) + H(t)\frac{dG(s,t)}{dt}\,,$$
$$\frac{\overline{d}^2}{dt^2} V(s,t) = \delta'(t)G(s,0) + \delta(t)\frac{dG(s,0)}{dt} + H(t)\frac{d^2G(s,t)}{dt^2}\,.$$

Then the inhomogeneous equation (20) is equivalent to the initial value problem

$$d^2G(s,t)/dt^2 - s^2G(s,t) = -se^{ct}\,, \tag{22a}$$
$$G(s,0) = 0\,, \tag{22b}$$
$$dG(s,0)/dt = A(s)\,. \tag{22c}$$

The solution of this system is easily found to be

$$G(s,t) = \frac{se^{ct}}{s^2-c^2} + \frac{1}{2}\frac{e^{st}}{c-s} - \frac{1}{2}\frac{e^{-st}}{s+c} + \frac{1}{s}A(s)\sinh st\,,$$

so from (21) we obtain

$$V(s,t) = \left\{\frac{se^{ct}}{s^2-c^2} + \frac{1}{2}\frac{e^{st}}{c-s} - \frac{1}{2}\frac{e^{-st}}{s+c} + \frac{1}{s}A(s)\sinh st\right\} H(t)\,. \tag{23}$$

Because the value of $A(s)$ is unspecified, we can eliminate the undesirable term containing e^{st}. We set $A(s) = s/(s-c)$, and (23) then reduces to

$$V(s,t) = [s/(s^2-c^2)](e^{ct} - e^{-st})H(t)\,. \tag{24}$$

Now we use the relations

$$\mathcal{L}^{-1}[s/(s^2-c^2)] = \cosh cx H(x) \quad \text{and} \quad \mathcal{L}^{-1}(e^{-ts}F(s)) = f(x-t)H(x-t);$$

with these, inversion of (24) yields

$$v(x,t) = \begin{cases} \frac{1}{2}(e^{c(x+t)} - e^{c(x-t)})H(t)H(x)\,, & t < x\,, \\ \frac{1}{2}(e^{c(x+t)} - e^{c(t-x)})H(t)H(x)\,, & t > x\,. \end{cases}$$

Finally, appealing to (18), we obtain

$$u(x,t) = \begin{cases} \frac{1}{2}(e^{c(x+t)} - e^{c(x-t)}), & t < x, \\ \frac{1}{2}(e^{c(x+t)} - e^{c(t-x)}), & t > x. \end{cases}$$

Exercises

1. Recall that $\sigma = \operatorname{Re} s$ and prove that for $\sigma > 0$,

   (a) $\mathcal{L}\{H(t)e^{\pm iwt}\} = 1/(s \mp iw)$,
   (b) $\mathcal{L}\{H(t)\cos wt\} = s/(s^2 + w^2)$,
   (c) $\mathcal{L}\{H(t)\sin wt\} = w/(s^2 + w^2)$,
   (d) $\mathcal{L}\{H(t)J_0(t)\} = (s^2 + 1)^{-1/2}$,
   (e) $\mathcal{L}\{H(t)I_0(t)\} = (s^2 - 1)^{-1/2}$.

2. Establish the following relations

   (a) $\mathcal{L}\{t - \tau\}H(t - \tau) = e^{-st}\tilde{f}(s)$,
   (b) $\mathcal{L}\{e^{-\alpha t}f(t)\} = \tilde{f}(s + \alpha)$,
   (c) $\mathcal{L}\{f(at)\} = (1/a)\tilde{f}(s/a)$.

Find the appropriate value of Re s for which each holds.

3. Show that

   (a) $\mathcal{L}\{\delta(t + \beta)\} = (1/|\alpha|)e^{s\beta/\alpha}$,
   (b) $\mathcal{L}\{\text{Pf}\,[H(t)\sin t/t^2]\} = (i/2)\{s + i\alpha)\ln(s + i\alpha)$
   $-(s - i\alpha)\ln(s - i\alpha) + \alpha(1 - \gamma)$, where γ is Euler's constant;
   (c) $\mathcal{L}\{\text{Pf}\,[H(t)\cosh \alpha t]\} = \frac{1}{2}\ln(s^2 - \alpha^2) - \gamma$.

4. The *square wave* function $f(t)$ is defined by

$$f(t) = \begin{cases} 1, & 0 < t < T, \\ -1, & T < t < 2T, \end{cases} \qquad \text{and} \qquad f(t + 2T) = f(T).$$

Show that its Laplace transform is $\tilde{f}(s) = (1/s)\tanh\frac{1}{2}sT$.

CHAPTER 9

Applications to Ordinary Differential Equations

9.1. Ordinary Differential Operators

In Section 2.6 we defined the differential operator L,

$$\begin{aligned} Lt &= \left(a_n(x)\frac{d^n}{dx^n} + a_{n-1}\frac{d^{n-1}}{dx^{n-1}} + \cdots + a_1\frac{d}{dx} + a_0 \right) t \\ &= \sum_{m=0}^{n} a_m(x)\frac{d^m t}{dx^m}\,, \end{aligned} \tag{1}$$

and its formal adjoint L^*,

$$L^*\phi = \sum_{m=0}^{n} (-1)^m d^m(a_m(x)\phi)/dx^m\,, \tag{2}$$

where the coefficients $a_m(x)$ are infinitely differentiable functions, t is a distribution, and ϕ is a test function. These operators are related by the equation

$$\langle Lt, \phi\rangle = \langle t, L^*\phi\rangle\,. \tag{3}$$

This means that the action of Lt on ϕ is equivalent to the action of t on the test function $\psi = L^*\phi$.

Our aim is to find the solution of the ordinary differential equation

$$Lt = \sum_{m=0}^{\infty} a_m(x) d^m t/dx^m = \tau\,, \tag{4}$$

where τ is an arbitrary known distribution. As defined in Section 2.6, the fundamental solution is the solution for $\tau = \delta(x)$. A distribution t is a solution of (4) if for every test function ϕ we have

$$\langle Lt, \phi\rangle = \langle \tau, \phi\rangle\,, \tag{5a}$$

or equivalently,

$$\langle t, L^*\phi\rangle = \langle \tau, \phi\rangle\,. \tag{5b}$$

In searching for a solution t of differential equation (4) we may have the following situations:

(1) The solution t is a sufficiently smooth function, so that the operation in (4) can be performed in the classical sense and the resulting equation is an identity. Then t is the classical solution.

(2) The solution t is not sufficiently smooth, so that the operation in (4) cannot be performed, but it satisfies (5) as a distribution. It is then a weak solution.
(3) The solution t is a singular distribution and satisfies (5). It is then a distributional solution.

All these solutions are called *generalized solutions*.

We really gain no new insight into the solutions of the classical problems in ordinary differential equations by using the theory of distributions. However, the theory does enhance our knowledge if discontinuities are present in these equations or if we want to find the fundamental solutions. This will become clear in the next sections.

9.2. Homogeneous Differential Equations

Let us start with the simplest differential equation,

$$dt/dx = 0\,. \tag{1}$$

We know that the only classical solution to this equation is $t = c$. We now prove the following theorem:

Theorem. *The only generalized solution to (1) is $t = c$.*

To prove this theorem we need the following lemma:

Lemma. *Any test function $\psi(x)$ can be represented as the derivative of another test function $\phi(x)$ if and only if*

$$\int_{-\infty}^{\infty} \psi(x)dx = 0\,. \tag{2}$$

Proof. When $\psi(x) = \phi'(x)$, $\phi(x) \in D$ we have

$$\int_{-\infty}^{\infty} \psi(x)dx = [\phi(x)]_{-\infty}^{\infty} = 0\,.$$

On the other hand, $\phi(x) = \int_{-\infty}^{x} \psi(u)du$ is an infinitely differentiable function, and $\psi(x)$ and $\phi'(x)$ vanish outside the same interval.

Proof of the Theorem. Equation (1) implies that $\langle t', \phi\rangle = -\langle t, \phi'\rangle = 0$. Thus

$$\langle t, \psi\rangle = 0 \tag{3}$$

for every test function that is the dervative of another test function, provided (1) is satisfied.

Let us now take a fixed test function $\phi_0(x)$ that is normalized such that

$$\int_{-\infty}^{\infty} \phi_0(x)dx = 1\,, \tag{4}$$

and write

$$\phi(x) = \phi_0(x) \int_{-\infty}^{\infty} \phi(u)du + \phi(x) - \phi_0(x) \int_{-\infty}^{\infty} \phi(u)du\,, \qquad \phi \in D\,. \tag{5}$$

Let

$$\psi(x) = \phi(x) - \phi_0(x) \int_{-\infty}^{\infty} \phi(u)du\,. \tag{6}$$

Then

$$\int_{-\infty}^{\infty} \psi(x)dx = \int_{-\infty}^{\infty} \phi(x)dx - \int_{-\infty}^{\infty} \phi_0(x)dx \int_{-\infty}^{\infty} \phi(u)du = 0\,,$$

where we have used (4). Then, in view of the lemma, $\psi(x)$ is the derivative of some test function, and from (3) it follows that $\langle t, \psi \rangle = 0$. Accordingly, (5) becomes

$$\langle t, \phi \rangle = \langle t, \phi_0 \rangle \int_{-\infty}^{\infty} \phi(u)du\,.$$

But $\langle t, \phi_0 \rangle$ is a constant, say, c, so the preceding relation becomes

$$\langle t, \phi \rangle = \int_{-\infty}^{\infty} c\phi(u)du = \langle c, \phi \rangle\,, \tag{7}$$

or $t = c$. From this theorem we immediately have the following two corollaries.

Corollary 1. If two generalized functions s and t have the same derivative, then $s = t + c$.

Corollary 2. Let $T = (t_1, t_2, \ldots, t_n)$ be an n-dimensional vector distribution. Then the solution of the differential equation

$$dT/dx = 0 \tag{8}$$

is $T = C$, $C = (c_1, c_2, \ldots, c_n)$. These ideas can be extended to the nth-order ordinary differential equation

$$Lt = a_n t^{(n)} + a_{n-1} t^{(n-1)} + \cdots + a_1 t' + a_0 t = 0\,, \tag{9}$$

where a_j, $j = 0, 1, \ldots, n$, are infinitely differentiable functions and $a_n \neq 0$. Indeed, this equation can be transformed to a system of linear first-order ordinary differential equations, as will be explained in Section 9.10. The solution in this case has also the same form as the classical solution.

9.3. Inhomogeneous Differential Equations: The Integral of a Distribution

The simplest inhomogeneous ordinary differential equation is

$$dt/dx = f\,, \tag{1}$$

where f is a continuous function. A distribution t is a solution of this equation if

$$\langle t', \phi \rangle = \langle f, \phi \rangle\,, \tag{2a}$$

or

$$\langle t, \phi' \rangle = \langle f, -\phi \rangle\,, \tag{2b}$$

for every $\phi \in D$. To solve (2b) we appeal to decomposition (9.2.5),

$$\phi(x) = \phi_0(x) \int_{-\infty}^{\infty} \phi(u)du + \phi(x) - \phi_0(x) \int_{-\infty}^{\infty} \phi(u)du\,.$$

Then we have

$$\langle t, \phi \rangle = \left\langle t, \phi_0 \int_{-\infty}^{\infty} \phi(u)du \right\rangle + \langle t, \psi \rangle\,, \tag{3}$$

where $\psi(x)$ is defined in (9.2.6). Since $\psi(x)$ is the derivative of the function

$$\phi_1(x) = \int_{-\infty}^{x} \phi(u)du - \int_{-\infty}^{x} \phi_0(v)dv \int_{-\infty}^{\infty} \phi(u)du\,, \tag{4}$$

we can use $\psi(x)$ for $\phi'(x)$ and $\phi_1(x)$ for $\phi(x)$ in (2). Thus $\langle t, \psi \rangle = \langle f, -\phi_1 \rangle$, which means that $\langle t, \psi \rangle$ is known, and we can set $\langle t, \psi \rangle = \langle t_0, \phi \rangle$. Also, $\langle t, \phi_0 \rangle$ is merely an arbitrary constant. Accordingly, we can write (3) as $\langle t, \phi \rangle = \langle c, \phi \rangle + \langle t_0, \phi \rangle$, so that

$$t = c + t_0\,, \tag{5}$$

which agrees with the classical solution; c is the solution of the homogeneous equation and t_0 is the particular solution.

As a corollary, we find that for a vector equation

$$dT/dx = F\,, \tag{6}$$

where $T = (t_1, t_2, \ldots, t_n)$ and $F = (f_1, f_2, \ldots, f_n)$, the solution is

$$T = C + T_0\,, \tag{7}$$

where C is a constant vector distribution and T_0 is the particular solution which arises in a manner similar to (4).

The generalized solution of the inhomogeneous ordinary differential equation

$$a_n t^{(n)} + a_{n-1} t^{(n-1)} + \cdots + a_0 t = f\,, \tag{8}$$

where $a_0, \ldots, a_n$ are infinitely differentiable functions, t is a scalar distribution, $a_n \neq 0$, and f is a continuous function (or the generalized solution of its equivalent system of the first order equations), is identical to the classical solution.

We shall discuss in Section 9.10 the equation obtained from (8) by substituting $\delta(x)$ for $f(x)$ on the right side.

9.4. Examples

Example 1. To find the general solution of the equation

$$x^m dt/dx = 0, \qquad m \geq 1, \tag{1}$$

we appeal to the relation $dH/dx = \delta(x)$ and use the derivatives of $\delta(x)$. Indeed, we assume that

$$t(x) = c_1 + c_2 H(x) + c_3\delta(x) + c_4\delta'(x) + \cdots + c_{m+1}\delta^{m-2}(x), \tag{2}$$

so that

$$t'(x) = c_2\delta(x) + c_3\delta'(x) + \cdots + c_{m+1}\delta^{m-1}(x).$$

Thus

$$\begin{aligned}\langle x^m t'(x), \phi\rangle =& \langle c_2 x^m \delta(x), \phi(x)\rangle + \langle c_3 x^m \delta'(x), \phi(x)\rangle \\ &+ \cdots + \langle c_{m+1} x^m \delta^{(m-1)}(x), \phi(x)\rangle = 0,\end{aligned}$$

and we have $x^m t'(x) = 0$ as required. Hence, (2) is the general solution of (1).

For $m = 1$ the solution reduces to

$$t(x) = c_1 + c_2 H(x). \tag{3}$$

The Heaviside function $H(x)$, although an ordinary function, is not differentiable. Therefore (3) is a weak solution.

For $m \geq 2$, (2) is the distributional solution.

Example 2. We show that the general solution of $x\, dt/dx = 1$ is $t = c_1 + c_2 H(x) + \ln|x|$. The complementary part $c_1 + c_2 H(x)$ follows from Example 1. To prove the rest, we differentiate t, obtaining

$$dt/dx = c_2\delta(x) + \text{Pf}\,(1/x).$$

Thus

$$\begin{aligned}\langle x dt/dx, \phi(x)\rangle &= c_2\langle x\delta(x), \phi(x)\rangle + \langle x\text{Pf}\,(1/x), \phi(x)\rangle \\ &= \lim_{\varepsilon\to 0}\left[\int_{-\infty}^{-\varepsilon} \frac{x\phi(x)}{x} dx + \int_{\varepsilon}^{\infty} \frac{x\phi(x)}{x} dx\right] \\ &= \lim_{\varepsilon\to 0}\left[\int_{-\infty}^{-\varepsilon} \phi(x)dx + \int_{\varepsilon}^{\infty} \phi(x)dx\right] \\ &= \int_{-\infty}^{\infty} \phi(x)dx = \langle 1, \phi\rangle,\end{aligned}$$

as required.

Example 3. We show that the general solution of $x^2 dt/dx = \text{Pf}\,(1/x)$ is

$$c_1 + c_2 H(x) + c_2\delta(x) - \frac{1}{2}\,\text{Pf}\,(1/x^2).$$

From Example 1, the complementary function is $c_1 + c_2 H(x) + c_3\delta(x)$. The particular solution is obtained by noting that

$$x^2 \frac{d}{dx}\left(-\frac{1}{2}\,\mathrm{Pf}\left(\frac{1}{x^2}\right)\right) = \mathrm{Pf}\left(\frac{1}{x}\right).$$

9.5. Fundamental Solutions and Green's Functions

Recall that the fundamental solution for the differential operator L is defined as $LE(x) = \delta(x)$, where L is defined in (9.1.1). Let us start with the single ordinary differential equation

$$LE(x) = (a_1 d/dx + a_0)E = \delta(x).$$

Since there is no loss of generality in taking $a_1 = 1$, we discuss this differential equation in its customary form,

$$LE = dE/dx + aE = \delta(x), \tag{1}$$

where we have set $a_0 = a$. If $a = 0$, the solution is clearly $t(x) = H(x)$+const. Fortunately, upon setting $E(x) = e^{-ax}t(x)$, (1) takes the same form, namely,

$$dt/dx = \delta(x)e^{ax} = \delta(x), \tag{2}$$

because

$$\langle \delta(x)e^{ax}, \phi(x)\rangle = \phi(0) = \langle \delta(x), \phi(x)\rangle, \qquad \text{for all} \quad \phi \in D.$$

Thus

$$E(x) = e^{-ax}t(x) = e^{-ax}(H(x) + C) = H(x)e^{-ax} + Ce^{-ax}, \tag{3}$$

where C is a constant.

Inasmuch as $U = e^{-ax}$ is the unique solution of the initial value problem

$$dU/dx + aU = 0, \qquad U\big|_{x=0} = 1, \tag{4}$$

we can write (3)

$$E(x) = H(x)U(x) + Ce^{-ax}. \tag{5}$$

Just as in the analysis of Sections 9.2 and 9.3, these arguments extend to a system of first-order ordinary differential equations

$$LE = dE/dx + AE = \delta(x)I \tag{6}$$

where the fundamental solution E is now an $n \times n$ matrix, A is a given $n \times n$ matrix (whose entries are infinitely differentiable functions), and I is the identity matrix. In this case the initial value problem that corresponds to (4) is

$$dU/dx + AU = 0, \qquad U\big|_{x=0} = I. \tag{7}$$

Its solution is $U = e^{-xA}$. To find the solution of (6) we let B denote an arbitrary $n \times n$ matrix whose entries are distributions. Then, applying Leibniz's formula,

$$L(UB) = (LU)B + UB' = UB',$$

we find that the solution of (6) is equal to UB, where B satisfies the differential equation

$$dB/dx = U^{-1}\delta(x) = e^{xA}\delta(x) = I\delta(x). \tag{8}$$

Its solution is $B = H(x)I + C$, where C now is a constant $n \times n$ matrix and where $H(x)I$ is the diagonal matrix with all its diagonal entries equal to the Heaviside function. Since $E = UB = e^{-xA}B$, we have

$$E = H(x)e^{-xA} + e^{-xA}C. \tag{9}$$

This solution helps us in solving the inhomogeneous equation for a distribution vector t,

$$dt/dx + At = T, \tag{10}$$

where T is a distribution vector with a compact support. Indeed, the particular solution is the convolution

$$t = E(x) * T, \tag{11}$$

while the complementary solution is the same as the classical one. The same is true for the single differential equation (1).

Incidentally, (9) is called the right fundamental solution of the operator L. The left fundamental solution is the distribution that satisfies the equation

$$dE/dx + EA = \delta(x)I. \tag{12}$$

Its solution is

$$E = H(x)e^{-xA} + Ce^{-xA}. \tag{13}$$

The next natural step would be to consider the general nth-order ordinary differential operator. We shall, of course, attend to it, but first we want to discuss second-order ordinary differential equations, because they are the cornerstone of mathematical physics and theoretical mechanics.

9.6. Second-Order Differential Equations with Constant Coefficients

The simplest equation of this type is

$$d^2E/dx^2 = \delta(x - \xi). \tag{1}$$

Since $H'(x) = \delta(x)$, integration of (1) gives

$$dE(x, \xi)/dx = H(x - \xi) + \alpha(\xi), \tag{2}$$

where $\alpha(\xi)$ is an arbitrary function. Next, we integrate (2) and obtain

$$\begin{aligned} E(x,\xi) &= \int_{-\infty}^{x} H(x-\xi)dx + x\alpha(\xi) + \beta(\xi) \\ &= (x-\xi)H(x-\xi) + x\alpha(\xi) + \beta(\xi)\,, \end{aligned} \tag{3}$$

where $\beta(\xi)$ is another arbitrary function. It is easily verified that (3) satisfies (1). Equation (3) shows that $E(x,\xi)$ is a continuous and piecewise differentiable function.

This solution helps us in solving the inhomogeneous equation

$$d^2t/dx^2 = \tau(x)\,, \tag{4}$$

where $\tau(x)$ is a distribution with compact support. Indeed,

$$t(x) = E * \tau\,, \tag{5}$$

as is easily verified by operating on both sides of (5) with d^2/dx^2.

Let us see if we can solve, with the help of the above analysis, an initial or boundary value problem in the classical theory of ordinary differential equations. Let us consider the boundary value problem

$$d^2u/dx^2 = f(x)\,, \qquad 0 \le x \le 1\,, \qquad u(0) = u(1) = 0\,, \tag{6}$$

where $f(x)$ is an integrable function with compact support. Appealing to relations (3) and (5), we find that

$$\begin{aligned} u(x) &= \int_{-\infty}^{\infty} (x-\xi)H(x-\xi)f(\xi)d\xi + x\int_{-\infty}^{\infty} \alpha(\xi)f(\xi)d\xi \\ &\quad + \int_{-\infty}^{\infty} \beta(\xi)f(\xi)d\xi \\ &= \int_{-\infty}^{x} (x-\xi)f(\xi)d\xi + x\int_{-\infty}^{\infty} \alpha(\xi)f(\xi)d\xi + \int_{-\infty}^{\infty} \beta(\xi)f(\xi)d\xi\,. \end{aligned}$$

Boundary conditions (6) then give

$$0 = -\int_{-\infty}^{0} \xi f(\xi)d\xi + \int_{-\infty}^{\infty} \beta(\xi)f(\xi)d\xi \quad \text{and}, \tag{7}$$

$$0 = \int_{-\infty}^{1} (1-\xi)f(\xi)d\xi + \int_{-\infty}^{\infty} \alpha(\xi)f(\xi)d\xi + \int_{-\infty}^{\infty} \beta(\xi)f(\xi)d\xi\,. \tag{8}$$

From (7) we find that $\beta(\xi) = \xi H(-\xi)$. Then (8) yields

$$\alpha(\xi) = \begin{cases} -1 + \xi H(\xi)\,, & -\infty \le \xi \le 1\,, \\ 0\,, & \xi > 1\,. \end{cases}$$

Thus the required solution is

$$\begin{aligned} u(x) &= \int_0^x (x-\xi)f(\xi)d\xi - x\int_0^1 (1-\xi)f(\xi)d\xi \\ &= \int_0^1 (x-\xi)H(x-\xi)f(\xi)d\xi - \int_0^1 x(1-\xi)f(\xi)d\xi \\ &= \int_0^1 [(x-\xi)H(x-\xi) - x(1-\xi)]f(\xi)d\xi\,, \qquad (9) \\ &= \int_0^x G(x,\xi)f(\xi)d\xi\,, \qquad (10) \end{aligned}$$

where

$$\begin{aligned} G(x,\xi) &= (x-\xi)H(x-\xi) - x(1-\xi) \\ &= \begin{cases} -x(1-\xi)\,, & x<\xi\,, \\ -\xi(1-x)\,, & x>\xi\,. \end{cases} \qquad (11) \end{aligned}$$

The function $G(x,\xi)$ is called Green's function. It satisfies differential equation (1) and the same boundary conditions as does $u(x)$, namely,

$$G(0,\xi) = G(1,\xi) = 0\,. \qquad (12)$$

We can go a step further and examine the case of inhomogeneous boundary values, that is, $u(0) = a$, $u(1) = b$. For this purpose we split the function u into two parts u_1 and u_2 such that $u = u_1 + u_2$. The function u_1 satisfies the same differential equation and has the same boundary values as the function u in (6). Accordingly, its value, from (9), is

$$u_1 = \int_0^x (x-\xi)f(\xi)d\xi - x\int_0^1 (1-\xi)f(\xi)d\xi\,.$$

The function u_2 satisfies the boundary value problem

$$d^2u_2/dx^2 = 0\,, \qquad u_2(0) = a\,, \qquad u_2(1) = b\,.$$

We can look at the quantity a as the strength of the jump discontinuity at 0 and $-b$ as the strength of the jump discontinuity at $x = 1$ (so that the function rises at 0 and falls back at $x = 1$). Then we can appeal to (5.1.5) and obtain

$$\overline{d}^2u_2/dx^2 = a\delta'(x) - b\delta'(x-1)\,, \qquad u_2(0) = 0\,, \qquad u_2(1) = 0\,.$$

Its solution follows from (9) and is

$$u_2 = bx - a(x-1).$$

Thus

$$u(x) = \int_0^x (x-\xi)f(\xi)d\xi - x\int_0^1 (1-\xi)f(\xi)d\xi + bx - a(x-1)\,. \qquad (13)$$

Let us now consider the initial value problem

$$d^2u/dx^2 + a^2u = \delta(x)\,, \qquad u\big|_{x=0+} = 0\,, \qquad u'\big|_{x=0+} = 1\,. \tag{14}$$

Here u is defined for $x \geq 0$. Accordingly, the solution is such that u is zero for $x < 0$ and satisfies the differential equation $d^2u/dx^2 + a^2u = 0$ for $x > 0$. The solution that satisfies the conditions $u(0) = 0$, and $u'(0) = 1$, is $\sin(ax)/a$. Thus the solution of the initial value problem (14) is

$$u(x) = (1/a)H(x)\sin(ax)\,. \tag{15}$$

It is the fundamental solution E (also called the Causal solution) of the operator $d^2/dx^2 + a^2$.

Let us use this fundamental solution to construct the solution of the initial value problem

$$d^2u/dx^2 + a^2u = f(x)\,, \qquad u\big|_{x=0+} = u_0\,, \qquad u'\big|_{x=0+} = u_1\,, \tag{16}$$

where f is a continuous function for $x \geq 0$. For this purpose we continue the functions u and f in the following way:

$$v(x) = \begin{cases} 0\,, & x < 0\,, \\ u(x)\,, & x \geq 0\,, \end{cases}$$

$$g(x) = \begin{cases} 0\,, & x < 0\,, \\ f(x)\,, & x \geq 0\,. \end{cases}$$

We then find from (5.1.3) and (5.1.5) that

$$\overline{v}'(x) = v' + u_0\delta(x)\,, \qquad \overline{v}''(x) = v'' + u_0\delta'(x) + u_1\delta(x)\,.$$

Accordingly, the function v satisfies the differential equation

$$\frac{\overline{d}^2 v}{dx^2} + a^2 v = g(x) + u_0\delta'(x) + u_1\delta(x)\,, \tag{17}$$

whose solution is

$$\begin{aligned} v(x) &= E * (g + u_0\delta' + u_1\delta) = E * g + u_0E' + u_1E \\ &= \frac{1}{a}\int_0^x f(y)\sin a(x-y)dy + u_0\cos(ax) + u_1\frac{\sin(ax)}{a}\,. \end{aligned} \tag{18}$$

An Alternative approach

We can obtain (3) and various other interesting results by an alternative approach. For this purpose let us first recall (2.7.12),

$$\frac{\overline{d}^2}{dx^2}\left(\frac{1}{2}|x-\xi|\right) = \delta(x-\xi)\,. \tag{19}$$

Accordingly, the solution (Green's function) for the boundary value problem

$$\frac{d^2}{dx^2}G(x,\xi) = \delta(x-\xi)\,, \tag{20}$$

$$G(0,\xi) = 0\,, \qquad G(1,\xi) = 0\,, \tag{21}$$

is

$$G(x,\xi) = \frac{1}{2}|x-\xi| + xA(\xi) + B(\xi)\,. \tag{22}$$

Applying the boundary conditions (21), we find that

$$A(\xi) = \xi - \frac{1}{2}\,, \qquad B(\xi) = -\frac{1}{2}\xi\,. \tag{23}$$

Substituting these values in (22), we recover (11).

At this stage it is useful to introduce the symbols $x_<$ and $x_>$. They stand for the values

$$x_< = \min(x,\xi) = \begin{cases} x\,, & a \le x \le \xi\,, \\ \xi\,, & \xi \le x \le b\,, \end{cases}$$

and

$$x_> = \max(x,\xi) = \begin{cases} \xi\,, & a \le x \le \xi\,, \\ x\,, & \xi \le x \le b\,, \end{cases}$$

as shown in Figure 9.1. The corresponding quantities $G_<$ and $G_>$ stand for the values of G in the $x_<$ and $x_>$ regions, respectively.

Figure 9.1.

In the present case, $a = 0$, $b = 1$, so that $G_< = x(\xi - 1)$ and $G_> = \xi(x-1)$. The functions $G_<$ and $G_>$ satisfy differential equation (1) in the $x_<$ and $x_>$ regions respectively. The function $G_<$ satisfies the boundary condition $G(0,\xi) = 0$ while $G_>$ satisfies the condition $G(1,\xi) = 0$. At $x = \xi$ these two solutions are equal. However, there is a jump in their derivatives; that is

$$\left[dG_>/dx - dG_</dx\right]_{x=\xi} = 1\,. \tag{24}$$

The preceding remarks are valid for a general Sturm-Liouville problem,

$$-\frac{d}{dx}\left[p(x)\frac{dG(x,\xi)}{dx}\right] + q(x)G(x,\xi) = \delta(x-\xi)\,, \qquad a \le x\,, \quad \xi \le b\,, \tag{25}$$

$$G(a,\xi)=0\,, \qquad G(b,\xi)=0\,, \tag{26}$$

where p and q are real-valued functions on $a \le x \le b$, p, q are continuous in this interval and p is positive. In this case,

$$G(x,\xi)\big|_{x=\xi_+} = G(x,\xi)\big|_{x=\xi_-} \tag{27}$$

and

$$\left.\frac{dG(x,\xi)}{dx}\right|_{x=\xi_+} - \left.\frac{dG(x,\xi)}{dx}\right|_{x=\xi_-} = -\frac{1}{p(\xi)}\,, \tag{28}$$

where (28) follows by integrating (25).

Example 1. In quantum mechanics there is a very interesting phenomenon in which a particle passes through a potential barrier that classical mechanics predicts is impenetrable. Because this phenomenon implicitly involves motion, let us begin with the time-dependent Schrödinger wave equation

$$\frac{1}{i}\frac{\partial}{\partial t}\,\psi(x,t) = \left[-\varepsilon^2\frac{\partial^2}{\partial x^2} + V(x)\right]\psi(x,t)\,, \tag{29}$$

where $\psi(x,t)$ is the wave function, V is the potential, and ε is a suitable parameter.

To illustrate the phenomenon of tunneling, we make two very simple choices. First, we assume that the time dependence of $\psi(x,t)$ is purely oscillatory, so that

$$\psi(x,t) = y(x)e^{iEt}\,. \tag{30}$$

Second, we choose $V(x) = \delta(x)$, the Dirac delta function. Then (29) reduces to

$$\varepsilon^2 y'' + Ey = y(x)\delta(x) = y(0)\delta(x)\,. \tag{31}$$

We solve this equation by the procedure outlined in this section. The solutions of (31) are

$$y(x) = ae^{-ix\sqrt{E}/\varepsilon} + be^{ix\sqrt{E}/\varepsilon}\,, \qquad x<0\,, \tag{32}$$

and

$$y(x) = ce^{-ix\sqrt{E}/\varepsilon} + de^{ix\sqrt{E}/\varepsilon}\,, \qquad x>0\,. \tag{33}$$

There are four constants a, b, c, d in these solutions. Two can be determined by physical considerations. Suppose we aim a monoenergetic unit-amplitude incident beam of particles moving toward $x = 0$ from the left. Then $a = 1$. The term $be^{ix\sqrt{E}/\varepsilon}$ in (32) is then a reflected wave, moving left, for $x \le 0$. In the region $x > 0$ there will be only a transmitted wave, moving right. Accordingly, $d = 0$ in (33) and we have the following solutions:

$$y(x) = \begin{cases} e^{-ix\sqrt{E}/\varepsilon} + be^{ix\sqrt{E}/\varepsilon}\,, & x<0\,, \qquad (34)\\ ce^{-ix\sqrt{E}/\varepsilon}\,, & x>0\,. \qquad (35)\end{cases}$$

The remaining two constants are determined by the following considerations:

(1) $y(x)$ is continuous at $x = 0$;

(2) $[dy/dx]_{0+} - [dy/dx]_{0-} = y(0)/\varepsilon^2$.

We thereby obtain

$$b = \frac{2\varepsilon\sqrt{E}i - 1}{4\varepsilon^2 E + 1}, \qquad c = \frac{2\varepsilon\sqrt{E}\left(2\varepsilon\sqrt{E} + 1\right)}{4\varepsilon^2 E + 1}. \tag{36}$$

The quantities $R = |b|^2$ and $T = |c|^2$ are called the *reflection* and *transmission coefficients*, respectively. Their values are

$$R = \frac{1}{r\varepsilon^2 E + 1}, \qquad T = \frac{4\varepsilon^2 E}{4\varepsilon^2 E + 1}. \tag{37}$$

In the language of quantum mechanics, R is the probability that an incident particle of energy E will be reflected, and T is the probability that the incident particle will be transmitted. Note that the total probability that a particle will be reflected or transmitted is unity because $T + R = 1$. When $E \to \infty$, $R \to 0$ and $T \to 1$, and when $E \to 0$, $T \to 0$ and $r \to 1$. For these two limits the classical and the quantum-mechanical predictions agree.

Example 2. Let us find the Green's function that satisfies the Sturm-Liouville problem

$$\frac{\overline{d}^2 G(x,\xi)}{dx^2} + G(x,\xi) = \delta(x-\xi), \qquad 0 \le x \le \pi, \tag{38}$$

$$G(0) - G'(0) = 0, \qquad G(\pi) + G'(\pi) = 0. \tag{39}$$

To find the complete solution of this system, we first find the solution in the region $x_< \ (0 \le x \le \xi)$ and $x_> \ (\xi < x \le 1)$.

In the $x_<$ region, the solution of the homogeneous part of (38) is

$$G(x,\xi) = \widetilde{A}\sin x + \widetilde{B}\cos x, \tag{40}$$

so that

$$G'(x,\xi) = \widetilde{A}\cos x - \widetilde{B}\sin x. \tag{41}$$

In view of the boundary condition at $x = 0$, we obtain $\widetilde{A} = \widetilde{B}$. Thus the solution in the $x_<$ region is

$$G(x,\xi) = \widetilde{A}(\sin x + \cos x) = \sqrt{2}\widetilde{A}[(1/\sqrt{2})\sin x + (1/\sqrt{2})\cos x].$$

$$= A\sin(x + \pi/4), \qquad A = \sqrt{2}\widetilde{A}. \tag{42}$$

In the $x_>$ region, we take the solution of the homogeneous part of equation (38) to be

$$G(x,\xi) = \widetilde{C}\sin x + \widetilde{D}\sin x, \tag{43}$$

so that

$$G'(x,\xi) = \widetilde{C}\cos x - \widetilde{D}\sin x. \tag{44}$$

Then the boundary condition (39) yields $\widetilde{C} = -\widetilde{D}$. Thus

$$G(x,\xi) = C\sin(x - \pi/4), \qquad C = \sqrt{2}\widetilde{C}. \tag{45}$$

The condition of continuity is

$$A\sin(x+\pi/4)\big|_{x=\xi} = C\sin(x-\pi/4)\big|_{x=\xi},$$

or

$$\frac{A}{\sin(\xi-\pi/4)} = \frac{C}{\sin(\xi+\pi/4)} = B = \text{const}\,. \tag{46}$$

This leaves only B undetermined, which we find by applying the jump condition

$$dG/dx\big|_{x\xi+} - dG\big|_{x=xi-} = 1$$

or

$$B[\sin(\xi+\pi/4)\cos(\xi-\pi/4) - \sin(\xi-\pi/4)\cos(\xi+\pi)] = 1\,,$$

which yields $B = 1$. Thus

$$G(x,\xi) = \begin{cases} \sin(\xi-\pi/4)\sin(x+\pi/4)\,, & x<\xi\,, \\ \sin(\xi+\pi/4)\sin(x-\pi/4)\,, & x>\xi\,. \end{cases} \tag{47}$$

Example 3. Let us use the information of the previous example to solve the boundary value problem

$$u''(x) + u(x) = f(x)\,, \qquad 0 \le x \le \pi\,, \tag{48}$$

$$f(x) = \begin{cases} f_1(x) = 0\,, & 0 \le x < \pi/2\,, \\ f_2(x) = 2x/\pi\,, & \pi/2 < x \le \pi\,. \end{cases}$$

$$u(0) - u'(0) = 0\,, \qquad u(\pi) + u'(\pi) = 0\,. \tag{49}$$

Thus, the jump discontinuity is at the point $\xi = \pi/2$. In view of the jump in the function f and the Green's function at $x = \pi/2$, we expect the solution u and its derivative u' to have jump discontinuities at $\pi/2$. As in Chapter 5, let the square brackets label the jump of a function at $x = \pi/2$. We thus set

$$[u] = u\left(\frac{\pi}{2}+\right) - u\left(\frac{\pi}{2}-\right) = \alpha \quad [u'] = u'\left(\frac{\pi}{2}+\right) - u'\left(\frac{\pi}{2}-\right) = \beta\,.$$

Now recall the values of the distributional derivatives, namely,

$$\overline{u}' = u' + \alpha\delta(x-\pi/2)\,, \qquad \overline{u}'' = u'' + \alpha\delta'(x-\pi/2) + \beta\delta(x-\pi/2)\,.$$

Thus (48) can be written as

$$\overline{u}'' + u = f(x) + \beta\delta(x-\pi/2) + \alpha\delta'(x-\pi/2)\,,$$

so that

$$\begin{aligned} u(x) &= \int_0^\pi G(x,t)\left[f(t) + \beta\delta\left(t-\frac{\pi}{2}\right) + \alpha\delta'\left(t-\frac{\pi}{2}\right)\right]dt \\ &= \frac{2}{\pi}\int_{\pi/2}^\pi G(x,t)f(t)dt + \beta G\left(x,\frac{\pi}{2}\right) - \alpha\frac{\partial G}{\partial t}\left(x,\frac{\pi}{2}\right) \\ &= \frac{1}{\sqrt{2}}\sin\left(x-\frac{\pi}{4}\right)\left(1-\frac{4}{\pi}\right) + \beta\left(x,\frac{\pi}{2}\right) - \alpha\frac{\partial G}{\partial t}\left(x,\frac{\pi}{2}\right), \end{aligned}$$

where we have used the value for G from (47).

As in the previous example, we find the values of u in the $x_<$ and $x_>$ regions. In the $x_<$ region,

$$u_< = \int_0^{\pi} [G_<(x,t) f(t)] dt + \beta G_< \left(x, \frac{\pi}{2}\right) - \alpha \frac{\partial G_<}{\partial t} \left(x, \frac{\pi}{2}\right) .$$

Next, we substitute the required values from the previous example, namely,

$$G_<(x,t)\Big|_{t=\pi/2} = [\sin(t - \pi/4)\sin(x + \pi/4)]_{\pi/2} = (\sqrt{2}/2)\sin(x + \pi/4) ,$$
$$\partial G_</\partial t\Big|_{t=\pi/2} = [\cos(t - \pi/4)\sin(x + \pi/4)]_{\pi/2} = (\sqrt{2}/2)\sin(x + \pi/4) ,$$

and obtain

$$u_< = (1/\sqrt{2})[\sin(x - \pi/4)(1 - 4/\pi) + \sin(x + \pi/4)(\beta - \alpha)] . \tag{50}$$

In the $x_>$ region,

$$u_> = \frac{1}{\sqrt{2}} \sin\left(x + \frac{\pi}{4}\right)\left(1 - \frac{4}{\pi}\right) + \beta G_> \left(x + \frac{\pi}{2}\right) - \alpha \left(x + \frac{\partial G_>}{\partial t}\right) \left(x, \frac{\pi}{2}\right) . \tag{51}$$

Because $G_>(x,t) = \sin(t + \pi/4)\sin(x - \pi/4)$, we have

$$G_> \left(x + \frac{\pi}{2}\right) = \frac{1}{\sqrt{2}} \sin\left(x - \frac{4}{\pi}\right) , \qquad \frac{\partial}{\partial t} G_> \left(x + \frac{\pi}{2}\right) = -\frac{1}{\sqrt{2}} \sin\left(x - \frac{\pi}{4}\right) .$$

Thus (51) yields

$$u_> = (1/\sqrt{2})\sin(x - \pi/4)[\beta + \alpha + 1 - 4/\pi] . \tag{52}$$

Combining (50) and (52), we have

$$u(x) = \frac{1}{\sqrt{2}} \begin{cases} \sin(x - \pi/4)(1 - 4/\pi) + \sin(x + \pi/4)(\beta - \alpha) , & 0 \le x \le \pi/2 , \\ \sin(x - \pi/4)[\beta + \alpha + 1 - 4/\pi] , & \pi/2 \le x \le \pi . \end{cases}$$

9.7. Eigenvalue Problems

Let us now give a brief description of the spectral theory of the Sturm-Liouville problem with general boundary conditions. For this purpose we find the Green's function $G(x, \xi, \lambda)$ that satisfies the system [33,34]

$$LG - \lambda r G = -(pG')' + qG - \lambda r G = \delta(x - \xi) , \qquad a \le x \le b , \tag{1a}$$
$$B_a G = (\cos\alpha) G(a) - (\sin\alpha) G'(a) = 0 , \tag{1b}$$
$$B_b G = (\cos\beta) G(b) + (\sin\beta) G'(b) = 0 , \tag{1c}$$

where the quantities p, q, a, and b have been defined in Section 9.6. The function $r(x)$ is a continuous real-valued function that is positive in $[a, b]$, λ is the eigenvalue, and α and

β are given real numbers, $0 \le \alpha < \pi, 0 \le \beta < \pi$. The signs in the boundary conditions (1b) and (1c) have been chosen so that the eigenvalues decrease as α or β increases.

In the notation of Figure 9.1, system (1) can be split into two simpler systems. In the $x_<$ region, $G(x, \xi, \lambda)$ is a solution of the homogeneous equation

$$LG - \lambda r G = 0, \qquad B_a G = 0. \tag{2}$$

We can connect the solution of this system to the solution $\phi_\lambda(x)$ of the initial value problem

$$L\phi_\lambda(x) - \lambda r \phi_\lambda(x) = 0, \tag{3a}$$

$$\phi_\lambda(a) = \sin\alpha, \tag{3b}$$

$$\phi'_\lambda(a) = \cos\alpha, \tag{3c}$$

so that

$$B_a\phi_\lambda(a) \equiv 0. \tag{3d}$$

Thus $G(x, \xi, \lambda)$ is a constant multiple of $\phi_\lambda(x)$ in the $x_<$ region. In the $x_>$ region, by the same token, let $\chi_\lambda(x)$ be the unique solution of the initial value problem

$$L\chi_\lambda(x) - \lambda r \chi_\lambda(x) = 0, \tag{4a}$$

$$\chi_\lambda(b) = \sin\beta, \tag{4b}$$

$$\chi'_\lambda(b) = -\cos\beta, \tag{4c}$$

so that

$$B_b\chi_\lambda(b) \equiv 0. \tag{4d}$$

Thus $G(x, \xi, \lambda)$ is a constant multiple of $\chi_\lambda(x)$ in the $x_>$ region. From this discussion it follows that

$$G(x, \xi, \lambda) = A\phi_\lambda(x_<)\chi_\lambda(x_>). \tag{5}$$

where A is a constant. G is clearly continuous at $x = \xi$. To find the constant, we appeal to the jump condition (9.6.28), which holds for (1a) also. Thus

$$A[\phi_\lambda(\xi)\chi'_\lambda(\xi) - \phi'_\lambda(\xi)\chi_\lambda(\xi)] = -1/p(\xi).$$

The quantity inside the brackets is the Wronskian $W_\xi(\phi_\lambda, \chi_\lambda)$ of ϕ_λ and χ_λ evaluated at $x = \xi$. Thus

$$AW_\xi(\phi_\lambda, \chi_\lambda) = -1/p(\xi). \tag{6}$$

Now we appeal to Abel's formula for the Wronskian, namely,

$$W(\phi_\lambda, \chi_\lambda) = \phi_\lambda(x)\chi'_\lambda(x) - \chi_\lambda(x)\phi'_\lambda(x) = \omega(\lambda)/p(x), \tag{7}$$

where $\omega(\lambda)$ is independent of x but may depend on λ. From (6) and (7) we find that $A = -1/\omega(\lambda)$, and from (5) we have

$$G(x, \xi, \lambda) = -\phi_\lambda(x_<)\chi_\lambda(x_>)/\omega(\lambda). \tag{8}$$

It follows from (7) that the zeros of $\omega(\lambda)$ coincide with the eigenvalues of system (1). Indeed, let $\lambda = \mu$ be a zero of $\omega(\lambda)$, so that $\omega(\mu) = 0$. From (7) it follows that the Wronskian of $\phi_\mu(x)$ and $\chi_\mu(x)$ is zero, and therefore these functions are linearly dependent. However, neither of these functions can vanish identically because of their initial values. Accordingly, $\phi_\mu(x) = k\chi_\mu(x)$, where k is a nonzero constant and both these functions satisfy the two boundary conditions of system (1). Thus, μ is an eigenvalue of (1), while $\phi_\mu(x)$ is the corresponding eigenfunction. Furthermore, we shall soon find [see (14)] that at an eigenvalue, Green's function has a singularity and therefore $\omega(\lambda)$ must vanish. Thus, the zeros of $\omega(\lambda)$ coincide with the eigenvalue of system (1). Let us label them as

$$\lambda_1 < \lambda_2 < \cdots \lambda_n < \cdots, \qquad \lambda_n \to \infty,$$

and set

$$\phi_{\lambda_n}(x) = k_n \chi_{\lambda_n}(x). \tag{9}$$

To normalize these eigenfunctions we appeal to the residue of G at $\lambda = \lambda_n$, which is

$$-\frac{\phi_{\lambda_n}(x_<)\chi_{\lambda_n}(x_>)}{\omega'(\lambda_n)} = -k_n \frac{\chi_{\lambda_n}(x_<)\chi_{\lambda_n}(x_>)}{\omega'(\lambda_n)} = -\frac{\phi_{\lambda_n}(x_<)\phi_{\lambda_n}(x_>)}{k_n\omega'(\lambda_n)}$$

Note that $\chi_{\lambda_n}(x_<)\chi_{\lambda_n}(x_>) = \chi_{\lambda_n}(x_<)\chi_{\lambda_n}(\xi)$ in both the $x_<$ and $x_>$ regions. Similarly, $\phi_{\lambda_n}(x_<)\phi_{\lambda_n}(x_<) = \phi_{\lambda_n}(x)\phi_{\lambda_n}(\xi)$, and the required value of the residue of G at $\lambda = \lambda_n$ is

$$-k_n\chi_{\lambda_n}(x)\chi_{\lambda_n}(\xi)/\omega'(\lambda_n) = -\phi_{\lambda_n}(x)\phi_{\lambda_n}(\xi)/k_n\omega'(\lambda_n) = u_n(x)\overline{u}_n(\xi),$$

where $u_n(x)$ are the orthonormalized eigenfunctions of system (1) and $\overline{u}_n(x)$ are their complex conjugates. From this relation it follows that

$$u_n(x) = \pm\phi_{\lambda_n}(x)/[k_n\omega'(\lambda_n)]^{1/2} = \pm[k_n/\omega'(\lambda_n)]^{1/2}\chi_{\lambda_n}(x). \tag{10}$$

Next we expand Green's function in terms of the orthonormalized eigenfunctions $u_n(x)$, as obtained in the foregoing, so that

$$G(x, \xi, \lambda) = \sum_{n=1}^{\infty} g_n(\xi, \lambda)u_n(x), \tag{11}$$

where

$$g_n = \langle G, u_n\rangle = \int_a^b r(x)G(x, \xi, \lambda)\overline{u}_n(x)\,dx. \tag{12}$$

To find g_n we multiply (10) by $\overline{u}_n(x)$ and integrate from a to b. We then use (12) and the sifting property. We thus obtain $\langle LG, u_n\rangle - \lambda g_n = \overline{u}_n(\xi)$. Because $\langle LG, u_n\rangle = \langle G, Lu_n\rangle = \lambda_n g_n$, we have

$$g_n(\xi, \lambda_n) = \overline{u}_n(\xi)/(\lambda_n - \lambda). \tag{13}$$

From (11) it then follows that

$$G(x, \xi, \lambda) = \sum_{n=1}^{\infty} \frac{u_n(x)\overline{u}_n(\xi)}{\lambda_n - \lambda}. \tag{14}$$

Two interesting results follow from this discussion. First, we find that the solution $w(x, \lambda)$ of the inhomogeneous system

$$-(pw')' + qw - \lambda rw = f(x), \qquad a \leq x \leq b, \tag{15a}$$

$$B_a w = B_b w = 0, \tag{15b}$$

is

$$w(x, \lambda) = \int_a^b G(x, \xi, \lambda) f(\xi) d\xi = \sum_{n=1}^{\infty} u_n(x) \frac{\int_a^b f(\xi) \overline{u}_n(\xi) d\xi}{(\lambda_n - \lambda)}$$
$$= \sum_{n=1}^{\infty} \frac{\langle f, u_n \rangle u_n(x)}{\lambda_n - \lambda}. \tag{16}$$

Second, when we expand $\delta(x)/r(x)$, we obtain

$$\frac{\delta(x-\xi)}{r(x)} = \sum_{n=1}^{\infty} \left\langle \frac{\delta(y-\xi)}{r(y)}, \overline{u}_n(y) \right\rangle u_n(x)$$
$$= \sum_{n=1}^{\infty} \left(\int_a^b \frac{\delta(y-\xi)}{r(y)} r(y) \overline{u}_n(y) dy \right) u_n(x),$$
$$= \sum_{n=1}^{\infty} u_n(x) \overline{u}_n(\xi). \tag{17}$$

We end this section with an example.

Example. Let us consider the system

$$(xu')' + \lambda xu = 0, \qquad 0 \leq x \leq 1, \tag{18a}$$

$$\lim_{x \to 0} xu' = 0, \tag{18b}$$

$$u(1) = 0. \tag{18c}$$

The only solution that satisfies (18a) and (18b) is $J_0\left(\sqrt{\lambda}x\right)$. The boundary condition (18c) implies that $J_0\left(\sqrt{\lambda_n}\right) = 0$, which gives the eigenvalues of system (18). The corresponding eigenfunctions are

$$u_n(x) = \sqrt{2} J_0\left(\sqrt{\lambda_n}x\right) / J_0\left(\sqrt{\lambda_n}\right).$$

Then from relation (17) it follows that

$$\frac{\delta(x-\xi)}{x} = \sum_{n=1}^{\infty} \frac{2}{\left[J_0\left(\sqrt{\lambda_n}\right)\right]^2} J_0\left(\sqrt{\lambda_n}x\right) J_0\left(\sqrt{\lambda_n}\xi\right). \tag{19}$$

9.8. Second-Order Differential Equations with Variable Coefficients

Let us consider the differential equation

$$a(x)\frac{d^2y}{dx^2}+b(x)\frac{dy}{dx}+c(x)y=\sum_{n=0}^{N}\beta_n\delta^{(n)}(x-\xi),$$

where $\delta^{(n)}(x)$ is the nth derivative of $\delta(x)$. To fix the ideas, we shall take $N=1$ so that we have only to solve

$$Ly=[a(x)d^2/dx^2+b(x)d/dx+c(x)]y=\beta_0\delta(x-\xi)+\beta_1\delta'(x-\xi). \tag{1}$$

Since this is a nonhomogeneous equation, its solution is expressed as

$$y(x)=y_h(x)+y_p(x), \tag{2}$$

where $y_h(x)$ is the solution of the homogeneous equation $Ly=0$ and $y_p(x)$ is a particular solution. The part $y_h(x)$ is the same as we study in the classical analysis, we therefore attend only to the particular solution, which we assume to be

$$y_p(x)=G(x)H(x-\xi), \tag{3}$$

where $H(x)$ is the Heaviside function and $G(x)$ is an unknown function. Then we appeal to (2.6.25), obtaining

$$y_p'(x)=G'(x)H(x-\xi)+G(\xi)\delta(x-\xi), \tag{4a}$$
$$y_p''(x)=G''(x)H(x-\xi)+G'(\xi)\delta(x-\xi)+G(\xi)\delta'(x-\xi). \tag{4b}$$

Thus

$$b(x)y_p'(x)=b(x)G'(x)H(x-\xi)+b(\xi)G(\xi)\delta(x-\xi), \tag{5a}$$
$$\begin{aligned}a(x)y_p''(x)&=a(x)G''(x)H(x-\xi)+a(x)G'(\xi)\delta(x-\xi)+a(x)G(\xi)\delta'(x-\xi)\\&=a(x)G''(x)H(x-\xi)+a(\xi)G'(\xi)\delta(x-\xi)+G(\xi)[-a'(\xi)\delta(x-\xi)\\&\quad+a(\xi)G(\xi)\delta'(x-\xi)],\\&=a(x)G''(x)H(x-\xi)+[a(\xi)G'(\xi)-a'(\xi)G(\xi)]\delta(x-\xi)\\&\quad+a(\xi)G(\xi)\delta'(x-\xi).\end{aligned} \tag{5b}$$

Substituting (3) and (5) in (1), we get

$$\begin{aligned}&a(x)G''(x)H(x-\xi)+[a(\xi)G'(\xi)-a'(\xi)G(\xi)]\delta(x-\xi)\\&\qquad+a(\xi)G(\xi)\delta'(x-\xi)+b(x)G'(x)H(x-\xi)\\&\qquad+b(\xi)G(\xi)\delta(x-\xi)+c(x)G(x)H(x-\xi)\\&\quad=\beta_0\delta(x-\xi)+\beta_1\delta'(x-\xi),\end{aligned}$$

or

$$\begin{aligned}(LG(x))H(x-\xi)+[a(\xi)G'(\xi)-a'(\xi)G(\xi)+b(\xi)G(\xi)]\delta(x-\xi)\\ a(\xi)G(\xi\delta'(x-\xi)\\ =\beta_0\delta(x-\xi)+\beta_1\delta'(x-\xi)\,.\end{aligned} \tag{6}$$

Equating the coefficients of $H(x-\xi)$, $\delta(x-\xi)$ and $\delta'(x-\xi)$ on both sides of this equation we have

$$LG=a(x)G''(x)+b(x)G'(x)+c(x)G(x)=0\,, \tag{7a}$$

$$a(\xi)G'(x)+(b(\xi)-a'(\xi))G(\xi)=\beta_0\,, \tag{7b}$$

$$a(\xi)G(\xi)=\beta_1\,. \tag{7c}$$

System (7) is the initial value problem for $G(x)$ and can be solved by the classical method, thereby solving the original problem completely.

For the special case when $\beta_1=0$, we have the fundamental solution $E(x-\xi)$,

$$E(x-\xi)=E(x,\xi)=G(x)H(x-\xi)\,. \tag{8}$$

Let us illustrate these concepts with the following examples.

Example 1. Let us find Green's function for the harmonic oscillator, that is,

$$d^2y/dx^2+k^2y=\delta(x-\xi)\,. \tag{9}$$

Comparing (9) with (1), we find that

$$a(x)=1\,,\qquad b(x)=0\,,\qquad c(x)=k^2\,,$$
$$\beta_0=1\,,\qquad \beta_1=0\,.$$

Thus system (7) reduces to

$$(d^2/dx^2+k^2)G(x)=0\,, \tag{10a}$$

$$G'(\xi)=1\,, \tag{10b}$$

$$G(\xi)=0\,. \tag{10c}$$

The general solution of (10a) is

$$G(x)=A\sin(kx+b)\,. \tag{11}$$

Application of initial conditions (10b) and (10c) yields $G(\xi)=A\sin(k\xi+B)=0$, so that $B=-k\xi$ and

$$Ak\cos(k\xi+B)=Ak\cos 0=1\,.$$

Thus $A=1/k$. Substituting these values in (11), we obtain

$$G(x)=[(1/k)\sin k(x-\xi)]\,,$$

and the general solution of (9) and the fundamental solution are

$$y(x) = \alpha \sin(kx + B) + [(1/k)\sin k(x - \xi)]H(x - \xi)$$

and

$$E(x - \xi) = [(1/k)\sin k(x - \xi)]H(x - \xi)\,,$$

respectively. This fundamental solution agrees with (9.6.15).

Example 2. Consider

$$(1 - x^2)\frac{d^2y}{dx^2} - 2x\frac{dy}{dx} = \delta(x - \xi)\,. \tag{12}$$

Here $a(x) = 1 - x^2$, $b(x) = -2x$, $c(x) = 0$, $\beta_0 = 1$, $\beta_1 = 0$. System (7) becomes

$$\left[(1 - x^2)\frac{d^2}{dx^2} - 2x\frac{d}{dx}\right]G(x) = 0\,, \tag{13a}$$

$$(1 - \xi^2)G'(\xi) = 1, \tag{13b}$$

$$(1 - \xi^2)G(\xi) = 0\,. \tag{13c}$$

The general solution of (13a) is

$$G(x) = A\ln\frac{1 + x}{1 - x} + B\,. \tag{14}$$

Initial condition (13c) yields

$$A\ln\frac{1 + \xi}{1 - \xi} + B = 0\,. \tag{15}$$

Since

$$G'(x) = A\left(\frac{1}{1 + x} + \frac{1}{1 - x}\right) = \frac{2A}{1 - x^2}\,,$$

from condition (13b) we find that $(1 - \xi^2)2A/(1 - \xi^2) = 1$, so $A = \frac{1}{2}$. Then (15) gives the value of B as

$$B = -\frac{1}{2}\ln\frac{1 + \xi}{1 - \xi} = \frac{1}{2}\ln\frac{1 - \xi}{1 + \xi}\,.$$

Substituting these values of A and B in (14), we obtain

$$G(x) = \frac{1}{2}\left(\ln\frac{1 - \xi}{1 + \xi}\frac{1 + x}{1 - x}\right)\,, \tag{16}$$

so the general solution of (12) is

$$y = \alpha\ln\frac{1 + x}{1 - x} + \beta + \frac{1}{2}\left(\ln\frac{1 - \xi}{1 + \xi}\frac{1 + x}{1 - x}\right)H(x - \xi)\,. \tag{17}$$

The fundamental solution, $E(x, \xi)$, is

$$E(x - \xi) = E(x, \xi) = \frac{1}{2}\left(\ln\frac{1 - \xi}{1 + \xi}\frac{1 + x}{1 - x}\right)H(x - \xi)\,.$$

Example 3. Let us extend the method of the previous examples to solve the differential equation

$$d^2y/dx^2 + 3dy/dx + 2y = \delta'''(x)\,. \tag{18}$$

It is clear that the terms $\delta(x)$ and $\delta'(x)$ will appear in the expression for $y(x)$. However, in view of (2.6.15) we need only assume that the coefficients of $\delta(x)$ and $\delta'(x)$ are constants. Accordingly, we assume

$$y(x) = G(x)H(x) + a\delta(x) + b\delta'(x)\,. \tag{19}$$

Differentiating twice we obtain

$$y'(x) = G'(x)H(x) + G(0)\delta(x) + a\delta'(x) + b\delta''(x)\,,$$
$$y''(x) = G''(x)H(x) + G'(0)\delta(x) + G(0)\delta'(x) + a\delta''(x) + b\delta'''(x)\,.$$

Substitution of these in (18) yields the following system:

$$\begin{aligned} G''(x) + 3G'(x) + 2G(x) &= 0\,, \\ G'(0) + 3G(0) + 2a &= 0\,, \\ G(0) + 3a + 2b &= 0\,, \\ a + 3b &= 0\,, \\ b &= 1\,. \end{aligned}$$

Solving this system, we obtain $G(x) = -e^{-x} + 8e^{-2x}$, $a = -3$, $b = 1$, so from (19) the particular solution is

$$y(x) = (-e^{-x} + 8e^{-2x})H(x) - 3\delta(x) + \delta'(x)\,. \tag{20}$$

Example 4. Let us find the fundamental solution $E(x)$ of the differential operator $d^2/dx^2 - a^2$ with the help of the Fourier transform. Accordingly, we take the Fourier transform of the equation $d^2E/dx^2 - a^2E = \delta(x)$ and obtain $-(u^2 + a^2)\widehat{E} = 1$ or $\widehat{E}(u) = -1/(a^2 + u^2)$. Taking the inverse transform, we find that

$$E(x) = F^{-1}\left(-\frac{1}{a^2 + u^2}\right) = -\frac{1}{2a}e^{-a|x|}\,. \tag{21}$$

9.9. Fourth-Order Differential Equations

Fourth-order ordinary differential equations arise in various steady state and vibration problems in applied mechanics. We present the Green's function theory of such an equation in this section. Let us consider the differential equation [35]

$$\begin{aligned} Ly &= a(x)\frac{d^4y}{dx^4} + b(x)\frac{d^3y}{dx^3} + c(x)\frac{d^2y}{dx^2} + d(x)\frac{dy}{dx} + e(x)y \\ &= \beta_0\delta(x-\xi) + \beta_1\delta'(x-\xi) + \beta_2\delta''(x-\xi) + \beta_3\delta'''(x-\xi)\,, \end{aligned} \tag{1}$$

where we have stopped at the term $\delta'''(x)$ for the sake of simplicity; the following analysis can be easily extended to include $\delta^{(n)}(x-\xi)$ for $n \geq 4$. Again, the general solution $y(x)$ is

$$y(x) = y_h(x) + y_p(x), \tag{2}$$

where $y_h(x)$ is the solution of the homogeneous equation and $y_p(x)$ is the particular solution. Our aim is to find the particular solution, which we again assume to be

$$y_p(x) = G(x)H(x-\xi). \tag{3}$$

Differentiating this relation four times we obtain

$$y_p'(x) = G'(x)H(x-\xi) + G(\xi)\delta(x-\xi), \tag{4a}$$

$$y_p''(x) = G''(x)H(x-\xi) + G'(\xi)\delta(x-\xi) + G(\xi)\delta'(x-\xi), \tag{4b}$$

$$\begin{aligned} y_p'''(x) = {} & G'''(x)H(x-\xi) + G''(\xi)\delta(x-\xi) + G'(\xi)\delta'(x-\xi) \\ & + G(\xi)\delta''(x-\xi), \end{aligned} \tag{4c}$$

$$\begin{aligned} y_p^{\text{iv}}(x) = {} & G^{\text{iv}}(x)H(x-\xi) + G'''(\xi)\delta(x-\xi) + G''(\xi)\delta'(x-\xi) \\ & + G'(\xi)\delta''(x-\xi) + G(\xi)\delta'''(x-\xi), \end{aligned} \tag{4d}$$

where we have used (2.6.25). Similarly, we obtain

$$d(x)y_p'(x) = d(x)G'(x)H(x-\xi) + d(\xi)G(\xi)\delta(x-\xi), \tag{5a}$$

$$\begin{aligned} c(x)y_p''(x) = {} & c(x)G''(x)H(x-\xi) + [c(\xi)G'(\xi) - c'(\xi)G(\xi)]\delta(x-\xi) \\ & + c(\xi)G(\xi)\delta'(x-\xi), \end{aligned} \tag{5b}$$

$$\begin{aligned} b(x)y_p'''(x) = {} & b(x)G'''(x)H(x-\xi) \\ & + [b(\xi)G''(\xi) - b'(\xi)G'(\xi) + b''(\xi)G(\xi)]\delta(x-\xi) \\ & + [b(\xi)G'(\xi) - 2b'(\xi)G(\xi)]\delta'(x-\xi) + b(\xi)G(\xi)\delta''(x-\xi), \end{aligned} \tag{5c}$$

$$\begin{aligned} a(x)y_P^{\text{iv}}(x) = {} & a(x)G^{\text{iv}}(x)H(x-\xi) \\ & + [a(\xi)G'''(\xi) - a[(\xi)G''(\xi) + a''(\xi)G'(\xi) - a'''(\xi)G(\xi)]\delta(x-\xi) \\ & + [a(\xi)G''(\xi) - 2a'(\xi)G'(\xi) + 3a''(\xi)G(\xi)]\delta'(x-\xi) \\ & + [a(\xi)G'(\xi) - 3a'(\xi)G(\xi)]\delta''(x-\xi) + a(\xi)G(\xi)\delta'''(x-\xi). \end{aligned} \tag{5d}$$

The next step is to substitute these values in (1), so that

$$\begin{aligned} & (LG)H(x-\xi) + \{a(\xi)G'''(\xi) - [a'(\xi) - b(\xi)]G''(\xi) \\ & \quad + [a''(\xi) - b'(\xi) + c(\xi)]G'(\xi) \\ & \quad - [a'''(\xi) - b''(\xi) + c'(\xi) - d(\xi)]G(\xi)\}\delta(x-\xi) \\ & \quad + \{a(\xi)G''(\xi) - [2a'(\xi) - b(\xi)]G'(\xi) \\ & \quad + [3a''(\xi) - 2b'(\xi) + c(\xi)]G(\xi)\}\delta'(x-\xi) \\ & \quad + \{a(\xi)G'(\xi) - [3a'(\xi) - b(\xi)]G(\xi)\}\delta''(x-\xi) + a(\xi)G(\xi)\delta'''(x-\xi) \\ & = \beta_0\delta(x-\xi) + \beta_1\delta'(x-\xi) + \beta_2\delta''(x-\xi) + \beta_3\delta'''(x-\xi). \end{aligned} \tag{6}$$

We equate the coefficients of the generalized functions on both sides of (6) to obtain the initial value problem

$$LG = a(x)G^{\text{iv}}(x) + b(x)G'''(x) + c(x)G''(x) + d(x)G'(x) + e(x)G(x) = 0\,, \tag{7a}$$

$$\begin{aligned} a(\xi)G'''(\xi) - [a'(\xi) - b(\xi)]G''(\xi) + [a''(\xi) - b'(\xi) + c(\xi)]G'(\xi) \\ - [a'''(\xi) - b''(\xi) + c'(\xi) - d(\xi)]G(\xi) = \beta_0\,, \end{aligned} \tag{7b}$$

$$a(\xi)G''(\xi) - [2a'(\xi) - b(\xi)]G'(\xi) + [3a''(\xi) - 2b'(\xi) + c(\xi)]G(\xi) = \beta_1\,, \tag{7c}$$

$$a(\xi)G'(\xi) - [3a'(\xi) - b(\xi)]G(\xi) = \beta_2\,, \tag{7d}$$

$$a(\xi)G(\xi) = \beta_3\,. \tag{7e}$$

Its solution yields the value of $G(x)$ and then equation (3) yields the value of $y_p(x)$.

To find the fundamental solution of the operator L, we set $\beta_0 = 1$ and $\beta_1 = \beta_2 = \beta_3 = 0$.

Example. Let us solve the differential equation

$$(d^2/dx^2)[EI_0(1+mx)^3 d^2y/dx^2] = \beta_1\delta'\left(x - \frac{1}{2}l\right)\,, \tag{8}$$

which embodies the static problem of finding the deflection y of a linearly tapered beam that is free at $x = 0$, clamped at $x = l$, and subjected to a couple β_1 at its midlength (see Figure 9.2). The quantity D is Young's modulus, $I = I_0(1+mx^3)$ is the moment of inertia of the cross section, and I_0 and m are constants. Written explicitly, (8) becomes

$$\begin{aligned} &EI_0(1+mx)^3y^{\text{iv}} + 6EI_0m(1+mx)^2y''' + 6EI_0m^2(1+mx)^2y''(x)\,, \\ &= \beta_1\delta'\left(x - \frac{1}{2}l\right)\,. \end{aligned} \tag{9}$$

Comparing (9) with (1), we have

$$\begin{aligned} &a(x) = EI_0(1+mx)^3\,, \qquad b(x) = 6EI_0m(1+mx)^2\,, \\ &c(x) = 6EI_0m^2(1+mx)\,,\; d(x) = e(x) = 0\,. \end{aligned} \tag{10}$$

To find the particular solution we start with (3) with $\xi = \frac{1}{2}l$, so that (7a) becomes

$$G^{\text{iv}} + \frac{6m}{1+mx}G''' + \frac{6m^2}{(1+mx)^2}G'' = 0\,. \tag{11}$$

Its solution is

$$G(x) = c_1 + c_2x + c_2(1+mx) + c_4\ln(1+mx)\,, \tag{12}$$

where the c_1, c_2, c_3, and c_4 are constants. To find these constants we substitute (12) in the

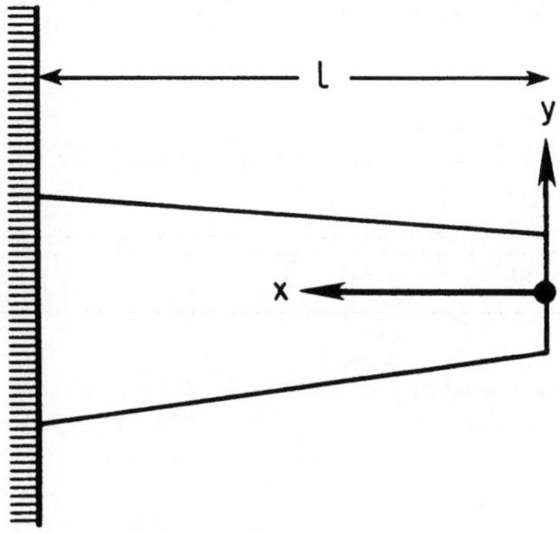

Figure 9.2. The deflection of a tempered beam.

initial conditions (7b)–(7e). After some algebraic work we obtain

$$
\begin{aligned}
c_1 &= -\frac{\beta_1 l}{4mEI_0(1+\frac{1}{2}ml)^2} - \frac{\beta_1}{2m^2EI_0(1+\frac{1}{2}ml)}\,, \\
c_2 &= \frac{\beta_1}{2mEI_0(1+\frac{1}{2}ml)^2}\,, \qquad c_3 = \frac{\beta_1}{2m^2EI_0}\,, \qquad c_4 = 0\,.
\end{aligned}
\tag{13}
$$

From (3), (12), and (13) we find that the particular solution of (9) is

$$
\begin{aligned}
y_p(x) &= G(x)H\left(x - \frac{1}{2}l\right) \\
&= \frac{\beta_1}{EI_0}\left[-\frac{1}{2m^2\left(1+\frac{1}{2}ml\right)} - \frac{l}{4m\left(1+\frac{1}{2}ml\right)^2}\right. \\
&\quad \left. + \frac{x}{2m\left(1+\frac{1}{2}ml\right)^2} + \frac{1}{2m^2(1+mx)}\right] H\left(x - \frac{l}{2}\right).
\end{aligned}
\tag{14}
$$

The general solution can now be readily obtained by adding to (14) the solution to the homogeneous equation, namely,

$$
y_h(x) = b_1 + b_2 x + b_3(1+mx) + b_4 \ln(1+mx)\,, \tag{15}
$$

where the b_1, b_2, b_3 and b_4 are constants. These constants are obtained with the help of the boundary conditions that the beam is free at $x = 0$ and clamped at $x = l$, that is,

$$
y(l) = 0\,, \qquad y'(l) = 0\,, \qquad y''(l) = 0\,, \qquad y'''(l) = 0\,. \tag{16}
$$

We leave the algebraic details to the reader (Exercise 6). The solution is

$$\begin{aligned}\frac{2mEI_0}{\beta_1}y(x) = &-\left(\frac{l}{(1+ml)^2} - \frac{l}{2(1+\frac{1}{2}ml)^2} - \frac{1}{m(1+ml)} - \frac{1}{m(1+\frac{1}{2}ml)}\right)\\ &+\left(\frac{1}{(1+ml)^2} - \frac{1}{(1+\frac{1}{2}ml)^2}\right)x\\ &+\left(-\frac{1}{m(1+\frac{1}{2}ml)} - \frac{l}{2(1+\frac{1}{2}ml)^2} + \frac{x}{(1+\frac{1}{2}ml)^2}\right.\\ &\left.+\frac{1}{m(1+mx)}\right)H\left(x-\frac{1}{2}l\right). \end{aligned} \tag{17}$$

9.10. Differential Equations of nth Order

Finally, we consider the fundamental solution of the nth-order differential operator (9.1.1) with $a_n = 1$. This means that we have to solve the differential equation

$$\frac{\bar{d}^n E}{dx^n} + a_{n-1}(x)\frac{\bar{d}^{n-1}E}{dx^{n-1}} + \cdots + a_0(x)E = \delta(x). \tag{1}$$

Using the standard transformation

$$u_1 = E, \qquad u_2 = u_1', \qquad \ldots, \qquad u_n = u_{n-1}', \tag{2}$$

we can write (1) as a first order system,

$$\bar{d}U/dx + AU = F, \tag{3}$$

where

$$U = \begin{bmatrix} u_1 \\ \vdots \\ u_n \end{bmatrix}, \qquad A = \begin{bmatrix} 0 & -1 & 0 & \ldots & 0 \\ 0 & 0 & -1 & \ldots & 0 \\ \vdots & & & & \vdots \\ 0 & & & & -1 \\ a_0 & a_1 & a_2 & \ldots & a_{n-1} \end{bmatrix}, \qquad F = \begin{bmatrix} 0 \\ 0 \\ \vdots \\ \delta(x) \end{bmatrix}. \tag{4}$$

Let us now recall a few well-known results from the theory of ordinary differential equations. First, the general solution of the equation

$$LU = dU/dx + A(x)U = 0,$$

where $A(x)$ is a given $n \times n$ matrix of differentiable functions, can be written as $U(x) = K(x)C$, where C is a constant vector and $K(x)$ is a fundamental matrix of the differential equation, namely, a solution that satisfies the condition $\det K(x) \neq 0$. From the matrix

K we form the Green's function (matrix) for the initial value problems associated with the operator L by setting $G(x, y) = K(x)[K(y)]^{-1}$.

Second, $G(x, y)$ is independent of the particular fundamental matrix K that we use. Moreover, if A is a constant, then $G(x, y)$ depends only on $x - y$ i.e., $G(x, y) = g(x - y)$, where $g(x)$ satisfies the initial value problem

$$dg/dx + Ag = 0\,, \qquad g(0) = I\,.$$

The Green's function $G(x, y)$ enables us to solve the inhomogeneous equation (3) where F is a vector whose components are continuous functions. Indeed,

$$U(x) = \int_0^x G(x, y)F(y)dy + G(x, 0)C\,, \tag{5}$$

where C is a constant vector, is the solution of (3) and satisfies the initial condition $U(0) = C$.

We are interested only in evaluating the fundamental solution E. Accordingly, in view of (2) we take the first component of (5) and obtain

$$\begin{aligned} E(x) &= \int_0^x G_{1n}(x, y)\delta(y)dy + \sum_{i=1}^{n} c_i G_{1i}(x, 0) \\ &= G_{1n}(x, 0)H(x) + \sum_{i=1}^{n} c_i G_{1i}(x, 0)\,, \end{aligned} \tag{6}$$

where G_{ij} and c_i are the components of G and C respectively.

Suppose that $u_1(x), \ldots, u_n(x)$ are n linearly independent solutions of the homogeneous equation associated with (1). Then

$$K(x) = \begin{bmatrix} u_1(x) & \cdots & u_n(x) \\ u_1'(x) & \cdots & u_n'(x) \\ \vdots & & \vdots \\ u_1^{(n-1)}(x) & \cdots & u_n^{(n-1)}(x) \end{bmatrix} \tag{7}$$

is a fundamental matrix of (3). Thus

$$G_{1n}(x, y) = \sum_{i=1}^{n} u_i(x)A_{ni}(y)\,, \tag{8}$$

where the quantity A_{ni} is the (n, i) element of the inverse of K, i.e.,

$$\begin{aligned} A_{ni} =& (-1)^{n+i} \begin{vmatrix} u_1 & \cdots & u_{i-1} & u_{i+1} & \cdots & u_n \\ \vdots & & \vdots & \vdots & & \vdots \\ u_1^{(n-2)} & \cdots u_i^{(n-2)} & u_{i+1}^{(n-2)} & \cdots & & u_n^{(n-2)} \end{vmatrix} \\ & \times \begin{vmatrix} u_1 & \cdots & u_n \\ \vdots & \vdots & \vdots \\ u_1^{(n-1)} & \cdots & u_n^{(n-1)} \end{vmatrix}^{-1}\,. \end{aligned} \tag{9}$$

Comparing (8) and (9) we obtain

$$G_{1n}(x,y) = \begin{vmatrix} u_1(y) & \dots & u_n(y) \\ \vdots & & \vdots \\ u_1^{(n-2)}(y) & & u_n^{(n-2)}(y) \\ u_1(x) & \dots & u_n(x) \end{vmatrix} \times \begin{vmatrix} u_1(y) & \dots & u_n(y) \\ \vdots & & \vdots \\ u_1^{(n-1)}(y) & \dots & u_n^{(n-1)}(y) \end{vmatrix}^{-1} . \tag{10}$$

Then from (6) we find that the fundamental solution is

$$E(x) = G_{1n}(x,0)H(x) + \sum_{i=1}^{n} c_i u_i(x) . \tag{11}$$

In view of this discussion, (11) can be written as

$$E(x) = g(x)H(x) + \sum_{i=1}^{n} c_i u_i(x) , \tag{12}$$

where g satisfies the initial value problem

$$\frac{d^n g}{dx^n} + a_{n-1}\frac{d^{n-1} g}{dx^{n-1}} + \cdots + a_0 g = 0 , \tag{13a}$$

$$g(0) = g(0) = \cdots = g^{(n-2)}(0) = 0 , \tag{13b}$$

$$g^{(n-1)}(0) = 1 . \tag{13c}$$

If the polynomial

$$p(x) = x^n + a_{n-1}x^{n-1} + \cdots\cdots + a_1 x + a_0 , \tag{14}$$

has roots $\lambda_1, \dots, \lambda_n$ that are simple, then it is easy to show that

$$g(x) = \sum_{i=1}^{n} e^{\lambda_i x}/p'(\lambda_i) , \tag{15}$$

and so

$$E(x) = H(x)\sum_{i=1}^{n} e^{\lambda_i x}/p'(\lambda_i) + \sum_{i=1}^{n} c_i e^{\lambda_i x} . \tag{16}$$

We can readily extend the analysis of this section to the case in which the right side of (1) contains terms such as $\delta'(x), \delta''(x), \dots, \delta^{(k)}(x)$ on the same lines as presented in the previous sections.

9.11. Ordinary Differential Equations with Singular Coefficients

Linear homogeneous systems of differential equations with infinitely smooth coefficients have no solutions in the space of generalized functions other than the classical solutions.

However, for equations with singularities in the coefficients, there may appear new solutions in generalized functions. For instance, $y(x) = \delta(x)$ is a solution of the confluent hypergeometric equation

$$xy'' + (2 - x)y' - y = 0, \tag{1}$$

the hypergeomtric equation

$$x(1 - x)y'' + (2 - 4x)y' - 4y = 0, \tag{2}$$

and the Bessel equation

$$x^2y'' + xy' + (x^2 - 1)y = 0. \tag{3}$$

The proof follows by appealing to formula (2.6.26):

$$x^n\delta^{(m)}(x) = \begin{cases} 0, & m < n, \\ (-1)^n \frac{m!}{(m-n)!}\delta^{(m-n)}(x), & m \geq n. \end{cases} \tag{4}$$

Let us verify it for the equation (3). In this case we find that $x^2\delta''(x) = 2\delta(x)$, $x\delta'(x) = -\delta(x)$, $x^2\delta(x) = 0$, and we have an identity.

Analysis of this section is based on the works of Littlejohn and Kanwal [36] and Krall, Kanwal and Littlejohn [37]. For the detailed study of this subject the reader is referred to Wiener [38]. We start our discussion with the confluent hypergeometric equation.

Confluent hypergeometric equation

$$xy'' + (m - x)y' - py = 0. \tag{5}$$

Its classical solutions are

$$y_1(x) = x^{1-m}{}_1F_1(1 - m + p; 2 - m : x) = \sum_{r=0}^{\infty} \frac{(1 - m + p)_r\, x^{1+r-m}}{(2 - m)_r\, r!}, \tag{6}$$

$m \neq 0, -1, \ldots,$ and

$$y_2(x) = x^{1-m}{}_1F_1(1 - m - p; 2 - m : x) = \sum_{r=0}^{\infty} \frac{(1 - m + p)\, x^{1+r-m}}{(2 - m)_r r!}, \tag{7}$$

$m \neq 2, 3, \ldots.$ Here ${}_1F_1$ is the hypergeometric function and $(p)_r$ is the Pochhammer's symbol,

$$p_0 = 1, \qquad (p)_r = p(p+1)\cdots(p+r-1), \qquad r = 1, \cdots.$$

Our aim is to present solutions of the form

$$y(x) = \sum_{n=1}^{\infty} a_n\delta^{(n)}(x). \tag{8}$$

We proceed as is done in deriving the power series solution except that now we have to appeal to the basic concept of distribution theory, namely, use the test function $\phi(x) \in D(x)$. Accordingly, we have to examine the quantity

$$\langle xy'' + (m-x)y' - py\,, \phi\rangle = \langle xy'', \phi\rangle + \langle (m-x)y\,\phi\rangle + \langle -py, \phi\rangle\,. \tag{9}$$

Now

$$\langle xy'', \phi\rangle = \langle y'', x\phi\rangle = \langle y, (x\phi)''\rangle = \langle y, x\phi'' + 2\phi'\rangle\,,$$

$$\langle (m-x)y', \phi\rangle = \langle y, -(m-x)\phi' + \phi\rangle\,,$$

and $\langle py, \phi\rangle = \langle y, p\phi\rangle$. Substituting these values in (7) we get

$$\langle xy'' - (m-x)y' - py, \phi\rangle = \langle y, x\phi'' + x\phi' + (2-m)\phi' + (1-p)\phi\rangle\,. \tag{10}$$

Next, we substitute series (8) in the right side of (10) so that we have to evaluate the relation

$$\left\langle \sum_{n=0}^{\infty} a_n \delta^{(n)}(x),\ x\phi'' + x\phi' + (2-m)\phi' + (1-p)\phi \right\rangle, \tag{11}$$

which is equal to

$$\sum_{n=0}^{\infty} a_n \langle -n\delta^{(n+1)}(x) + n\delta^{(n)}(x) + (2-m)\delta^{(n-1)}(x) - (1-p)\delta^{n}(x), \phi\rangle$$

$$= \left\langle \sum_{r=1}^{\infty} [a_{r-1}(m-r-1) - a_r(r-p+1)]\delta^{(r)}(x) - (1-p)a_0\delta(x), \phi \right\rangle. \tag{12}$$

From (10) and (12) it follows that if $y(x)$ is a solution of equation (5) then $(1-p)a_0 = 0$ and for $r = 1, 2, \ldots,$ we have the recurrence relation

$$a_{r-1}(m-r-1) - a_r(p-r-1) = 0\,. \tag{13}$$

For finding the coefficients a_r we must consider four cases.

Case 1. If p is not a positive integer, then $p - r - 1 \neq 0$ so that

$$a_r = \frac{a_{r-1}(m-r-1)}{(p-r-1)}\,.$$

Because $a_0 = 0$, we find that $a_r = 0, r = 1, \cdots$. This yields $y(x) = 0$.

Case 2. If p is a positive integer then $p - r - 1 = 0$ so that $r = p - 1$. When m is not a positive integer we have $a_0 = 0, a_1 = 0, \ldots, a_{p-2} = 0$. But a_{p-1} is arbitrary and we set it equal to 1 so that

$$a_{p+n} = \frac{(p+1-m)(p+2-m)\cdots(p+n+1-m)}{(n+1)!}\,.$$

Thereby, we have the solution

$$
\begin{aligned}
y(x) &= \sum_{n=-1}^{\infty} \frac{(p+1-m)_{n+1}}{(n+1)!} \, \delta^{(p+n)}(x) \\
&= \sum_{r=0}^{\infty} \frac{(p+1-m)_r}{r!} \, \delta^{(p+r-1)}(x) \,. \qquad (14)
\end{aligned}
$$

Case 3. If p is a positive integer, m is a positive integer and $m \leq p$, then $a_0 = 0$, $a_1 = 0, \ldots, a_{p-2} = 0$. Again a_{p-1} is arbitrary and we set it equal to 1. Thus

$$
y(x) = \sum_{r=0}^{\infty} \frac{(p+1-m)_r}{r!} \, \delta^{p+r-1}(x) \,. \qquad (15)
$$

Case 4. If p is a positive integer, m is a positive integer and $m > p$, then $a_0 = 0$, $a_1 = 0, \ldots, a_{p-2} = 0$ and a_{p-1} is arbitrary and we set it equal to 1. This yields

$$
a_p = (p+1-m), \ldots, a_{p+n} = \frac{(p-m+1) \cdots \quad (p+n+1-m)}{(n+1)!} \,,
$$

$n = 1, \cdots, m-2$. But then $a_{m-1} = 0$ and $a_n = 0$, $n = m, m+1, \ldots$. Thus, the solution in this case is

$$
y(x) = \sum_{n=-1}^{n-p-2} \frac{(p+1-m)_{n+1}}{(n-1)!} \, \delta^{(p+n)}(x) \,. \qquad (16)
$$

We have already found that $\delta(x)$ is a solution of the confluent hypergeometric equation (1). A few more examples are:

1. Equation:

$$
xy + (10-x)y' - 3y = 0 \,.
$$

Solution:

$$
\begin{aligned}
y(x) &= \sum_{n=2}^{8} (-1)^{n-2} \binom{6}{n-2} \delta^n(x) \\
&= \overset{(2)}{\delta}(x) - 6\overset{(3)}{\delta}(x) + 15\overset{(4)}{\delta}(x) - 20\overset{(5)}{\delta}(x) + 15\overset{(6)}{\delta}(x) - 6\overset{(7)}{\delta}(x) + \overset{(8)}{\delta}(x) \,.
\end{aligned}
$$

2. Equation:

$$
xy'' + (4-x)y' - 7y = 0 \,.
$$

Solution:

$$
y(x) = \sum_{n=0}^{\infty} \binom{n-3}{n-6} \delta^{(n)}(x) = \sum_{r=0}^{\infty} \frac{(4)_r}{r!} \, \delta^{(r+6)}(x) \,.
$$

3. Equation:

$$xy'' + (3 - x)y' - y = 0\,.$$

Solution:

$$y(x) = \delta(x) - \delta'(x)\,.$$

4. Equation:

$$xy'' + \left(\frac{3}{2} - x\right)y' - y = 0\,.$$

Solution:

$$y(x) = \sum_{n=0}^{\infty}\left(\left(\frac{1}{2}\right)_n \frac{1}{n!}\,\delta^{(n)}(x)\right).$$

The Hypergeometric Equation. Recall that one classical solution of the hypergeometric equation

$$x(1-x)y''(x) + [\gamma - (\alpha + \beta + 1)x]y'(x) - \alpha\beta y(x) = 0 \tag{17}$$

for the case when γ is neither a negative integer nor zero, is

$$y = {}_2F_1(\alpha, \beta, \gamma : x) = \sum \frac{(\alpha)_n(\beta)_n}{(\gamma)_n n!}\,x^n\,. \tag{18}$$

When γ is not a positive integer, we get the second solution of equation (17)

$$\begin{aligned} y &= x^{1-c}{}_2F_1(1-\gamma+\alpha,\ 1-\gamma+\beta, 2-\gamma : x) \\ &= \sum_{n=0}^{\infty} \frac{(1-\gamma+\alpha)_n(1-\gamma+\beta)_n}{(2-c)_n\, n!}\,x^{1-\gamma+n}\,. \end{aligned} \tag{19}$$

We want the solution in the form of the delta series (8). When we proceed in the same fashion as we have done for the confluent hypergeometric equation we get

$$a_0(1-\alpha)(1-\beta) = 0\,, \tag{20}$$

$$a_n(n+1-\alpha)(n+1-\beta) + a_{n-1}(n+1-\gamma) = 0\,, \qquad n = 1, 2, \ldots\,. \tag{21}$$

We must examine the following cases.

Case 1. If neither α or β is a positive integer, then all the coefficients are zero and $y(x) \equiv 0$.

Case 2. When α is a positive integer, but β is not, then $a_0 = 0, \ldots, a_{\alpha-2} = 0$, but $a_{\alpha-1}$ is arbitrary. With $a_{\alpha-1} = 0$, the recurrence relation yields

$$a_{\alpha+m-1} = \frac{(-1)^m(\alpha-\gamma+1)_m}{m!\,(\alpha-\beta+1)_m}$$

so that the solution is

$$y(x) = \sum_{m=0}^{\infty} \frac{(-1)^m (\alpha - \gamma + 1)_m}{m!\, (\alpha - \beta + 1)_m}\, \delta^{(\alpha+m-1)}(x)\,. \tag{22}$$

When γ is a positive integer, $\alpha < \gamma$, the above series terminates and we have

$$y(x) = \sum_{m=0}^{\gamma-\alpha-1} \frac{(-1)^m (\alpha - \gamma + 1)_m}{m!\, (\alpha - \beta + 1)_m}\, \delta^{(\alpha+m-1)}(x)\,. \tag{23}$$

Case 3. If both α and β are positive integers, we may assume that $\alpha \geq \beta$. If γ is not an integer we get the series (22) as a solution. When γ is an integer, $\alpha \geq \beta \geq \gamma$, we again get the same solution.

When γ is an integer, $\alpha \geq \gamma > \beta$, $a_0, \ldots, a_{\beta-2}$ are zero while $a_{\beta-1}$ is arbitrary and determines $a_\beta, \ldots, a_{\gamma-2}$. The coefficients $a_{\gamma-1} \cdots a_{\alpha-2}$ are zero while $a_{\alpha-1}$ is arbitrary and determines $a_\alpha, a_{\alpha-1}, \ldots$. Therefore we have two solution

$$y_1(x) = \sum_{m=0}^{\gamma-\beta-1} \frac{(-1)^m (\beta - \gamma + 1)_m}{m!\, (\beta - \alpha + 1)_m}\, \delta^{(\beta+m-1)}(x)\,, \tag{24}$$

and

$$y_2(x) = \sum_{m=0}^{\infty} \frac{(-1)^m (\alpha - \gamma + 1)_m}{m!\, (\alpha - \beta + 1)_m}\, \delta^{(\beta+m-1)}(x)\,. \tag{25}$$

If γ is a positive integer, $\gamma > \alpha \geq \beta$, then

$$y(x) = \sum_{m=0}^{\gamma-\beta-1} \frac{(-1)^m (\alpha - \gamma + 1)_m}{m!\, (\alpha - \beta + 1)_m}\, \delta^{(\alpha+m-1)}(x)\,. \tag{26}$$

We have already found that $\delta(x)$ is a solution of the hypergeometric differential equation (2). A few other examples are:

1. Equation ($\alpha = \beta = \gamma = 1$):

$$x(1-x)y''(x) + [1 - 3x]y'(x) - y(x) = 0\,. \tag{27}$$

Solution:

$$y(x) = \sum_{n=0}^{\infty} (-1)^n \frac{\delta^{(n)}(x)}{n!} = \delta(x-1)\,. \tag{28}$$

It is interesting to observe that the classical solution is

$$y(x) = {}_2F_1(x) = \sum_{n=0}^{\infty} \frac{(1)_n (1)_n x^n}{(1)_n n!} = \sum_{n=0}^{\infty} x^n = \frac{1}{1-x}\,. \tag{29}$$

Comparing (28) and (29) we observe the intriguing similarity. The function $1/1 - x$ has a pole of order 1 and the delta function $\delta(x-1)$ also has a simple pole.

2. Equation ($\alpha = \beta = 2, \gamma = 1$)

$$x(1-x)y'' + [1+5x]y' - 4y = 0. \tag{30}$$

Solution:

$$y(x) = \delta'(x) + \sum_{n=2}^{\infty} \frac{(-1)^{n+1} n \delta^{(n)}(x)}{(n-1)!} = \delta'(x-1) - \delta''(x-1). \tag{31}$$

The classical solution of (30) is

$$y(x) = {}_2F_1(x) = \sum_{n=0}^{\infty} \frac{(2)_n (2)_n x^n}{(1)_n n!} = \sum_{n=0}^{\infty} (n+1)^2 x^n = \frac{1+3x}{(1-x)^3}. \tag{32}$$

Examining the solutions (31) and (32) we observe that both the delta series and the classical solutions have a pole of order 3.

The Bessel Equation. The classical solutions of the Bessel equation

$$x^2 y'' + xy' + (x^2 - \nu^2)y = 0, \tag{33}$$

are $J_\nu(x)$ and $J_{-\nu}(x)$ when ν is not an integer and $J_\nu(x)$ and $Y_\upsilon(x)$ when υ is an integer. Our present aim is to derive the delta series solution. For this purpose we proceed as in the two previous cases and find the coefficients a_n of the delta series (8) satisfying the recurrence relation

$$a_{n+2} = -a_n \frac{(n+1-\nu)(n+1+\nu)}{(n+1)(n+2)}.$$

Thus, we obtain the two solutions

$$y_1(x) = a_0 \sum_{n=0}^{\infty} (-1)^n \frac{\prod_{j=1} (2j-1-\nu) \prod_{j=1}^{n} (2j-1+\nu)}{(2n)!} \delta^{(2n)}(x), \tag{34}$$

$$y_2(x) = a_1 \sum_{n=0}^{\infty} \frac{(-1)^n \prod_{j=1}^{n} (2j-\nu) \prod_{j=1}^{n} (2j+\nu)}{(2n+1)!} \delta^{(2n+1)}(x). \tag{35}$$

If $\nu = 2N+1 > 0$, is an odd integer, then $a_{2N-2} = 0$, and the first series terminates. When $\nu = 2N > 0$ is an even integer, then $a_{2N+1} = 0$ and the second series terminates. Accordingly, the singular solution of the Bessel equation

$$x^2 y'' + xy' + (x^2 - (m+1)^2)y = 0, \tag{36}$$

for m an integer, is

$$y(x) = C \sum_{k=0}^{[m/2]} \frac{(m-k)!}{4^k k!(m-2k)!} \delta^{(m-2k)}(x). \tag{37}$$

We have already found that $y_0 = a_0\delta(x)$ satisfies the Bessel equation (36) for $m = 0$. Similarly, when

$$
\begin{aligned}
m &= 1, & y_1 &= a_0\delta^{(2)}(x),\\
m &= 2, & y_2(x) &= a_0(\delta(x) + 4\delta^{(2)}(x)),\\
m &= 3, & y_3(x) &= a_1(\delta'(x) + 2\delta^{(3)}(x)),\\
m &= 4, & y_4(x) &= a_0(\delta(x) + 12\delta^{(2)}(x)) + 16\delta^{(4)}(x)),\\
m &= 5, & y_5(x) &= a_1(\delta'(x) + \frac{16}{3}\delta^{(3)}(x)) + \frac{16}{3}\delta^{(5)}(x)).
\end{aligned}
$$

Let us observe that the Laplace transform of $y_1(x)$, as given by (34), for $\nu = 0$, is

$$\mathcal{L}y_1 = a_0 \sum_{n=0}^{\infty} \frac{(-1)^n (2n)!}{a^{2n}(n!)^2} s^{2n} = a_0(1+s^2)^{-1/2},$$

where we have used formula (8.2.8). Because $(1+s^2)^{-1/2}$ is the Laplace transform of $J_0(x)$ (see Exercise 1.d of Chapter 8) we find that $y_1 = a_0 J_0(x)$.

Similarly, if $v = 0$, the Laplace transform of $y_2(x)$, as given by (35), is

$$\mathcal{L}(y_2) = a_1 \sum_{n=0}^{\infty} \frac{(-1)^n 2^{2n} (n!)^2}{(2n+1)!} s^{2n+1} = a_1 \frac{\sinh^{-1} s}{(1+s^2)^{1/2}}.$$

Because $(\sinh^{-1} s)|(1+s^2)^{1/2}$ is the Laplace transform of $-(\pi/2)Y_0(x)$, the Bessel function of the second kind, we have $y_2(x) = -a_1(\pi/2)Y_0(x)$.

For arbitrary $\nu > -1$

$$\mathcal{L}(J_\nu(x)) = \frac{((1+s^2)^{1/2} - s)^\nu}{(1+s^2)^{1/2}}. \tag{38}$$

If $a_0 = 1, a_1 = -\nu$ and $y = y_1 + y_2$, then

$$\mathcal{L}(y) = \frac{((1+s^2)^{1/2} - s)^\nu}{(1+s^2)^{1/2}}, \tag{39}$$

as in (38). Thus

$$
\begin{aligned}
J_\nu(x) &= \sum_{n=0}^{\infty} \frac{(-1)^n \prod_{j=1}^{n}(2j-1-\nu)\prod_{j=1}^{n}(2j-1+\nu)\overset{(2n)}{\delta(x)}}{2n!}\\
&= \sum_{n=0}^{\infty} \frac{(-1)^n \prod_{j=1}^{n}(2j-\nu)\prod_{j=1}^{n}(2j+\nu)\,\overset{(2n+1)}{\delta(x)}}{(2n+1)}.
\end{aligned} \tag{40}
$$

The Bessel function $J_{-\nu}(x)$ is represented by the same formula as (40) except that we have to change the sign in the second part.

When $\nu > 0$ is an integer, Y_ν does not possess a Laplace transform. Instead the finite series (37) is found.

Another interesting thing to observe is that if we take the Fourier transforms of equations (34) and (35) and set $\widehat{y}(x) = y(u)$, we obtain

$$(1-u^2)y'' - 3uy' + m(m+2)y = 0\,, \tag{41}$$

and

$$y(u) = A \sum_{k=0}^{[(m-1)/2]} (-1)^k \binom{m-1-k}{m} (2x)^{(m-1-2k)}\,, \tag{42}$$

respectively, where A is a constant Equation (41) is the Chebyshev equation and (42) is the $(m-1)$st Chebyshev polynomial of the second kind.

Exercises

1. Solve Example 1 of Section 9.8 by the method of Section 9.6.

2. Show that the solution of

$$x\frac{d^2G(x,\xi)}{dx^2} + \frac{dG(x,\xi)}{dx} = \delta(x-\xi)\,, \qquad 0 \le x \le 1\,,$$

for $G(x,\xi)$ bounded as $x \to 0$ and $G(1,\xi) = 0$, is

$$G(x,\xi) = \begin{cases} \ln \xi\,, & 0 < x \le \xi\,, \\ \ln x\,, & \xi \le x \le 1\,. \end{cases}$$

3. Solve the boundary value problem of Example 3 of Section 9.6 for the special case

$$f(x) = \begin{cases} 0\,, & 0 \le x \le \pi/2\,, \\ x^2/\pi\,, & \pi/2 \le x < \pi\,. \end{cases}$$

4. Solve the boundary value problem of Example 3 of Section 9.6 with $f(x)$ replaced by the derivative $f'(x)$ on the right side of (9.6.48).

5. Solve Exercise 4 for the special value of the function $f(x)$ given in Exercise 3.

6. Derive (9.9.17), completing the algebraic details.

7. The governing differential equation that describes the response of a viscoelastic incomplete ring to a suddenly applied force at its free end is of the form

$$a\frac{d^3y}{dt^3} + b\frac{dy}{dt} = \alpha H(t) + \beta_0\delta(t) + \beta_1\delta'(t) + \beta_2\delta''(t)\,,$$

where $a, b, \alpha, \beta_0, \beta_1, \beta_2$ are appropriate constants related to the problem. Solve this differential equation with $a = b = \alpha = \beta_0 = \beta_1 = \beta_2 = 1$.

8. Show that the infinite series

$$y = \sum_{n=0}^{\infty} \frac{2^{n+1}\delta^{(n)}(x)}{n!(n+1)!}$$

formally satisfies the first-order ordinary differential equation

$$x^2 dy/dx - 2y = 0\,.$$

It is interesting to observe that $x = 0$ is an essential singularity for this differential equation.

9. Prove that the general solution of the differential equation

$$x^n \frac{d^m y}{dx^m} = 0\,,$$

is

$$y(x) = \sum_{k=0}^{m-1} a_k H(x) x^{m-k-1} + \sum_{k=m}^{n-1} b_k \delta^{(k-m)}(x) + \sum_{k=0}^{m-1} c_k x^k\,.$$

Hint: Take the Fourier transform of the differential equation and solve the transformed equation.

CHAPTER 10

Applications to Partial Differential Equations

10.1. Introduction

Recall from Chapter 2 that the differential operator L of order p in n independent variables $x_1, x_2, \ldots, x_n$, is

$$Lu = \sum_{|k| \leq p} a_k(x) D^k u \,, \tag{1}$$

where the coefficients a_k have partial derivatives of all orders. Its formal adjoint L^* is defined as

$$L^* v = \sum_{|k| \leq p} (-1)^k D^k (a_k v) \,. \tag{2}$$

Here u and v are functions having derivatives of order p in R_n.

An interesting relation involving the operators L and L^* is

$$vLu - uL^* v = \text{div } J(u, v) \,, \tag{3}$$

where J is a vectorial bilinear form in u, v, and derivatives of u and v of order $p - 1$ or less. We prove this result for $p = 2$. The generalization to higher dimensions offers no extra difficulty.

For $p = 2$ we have

$$Lu = \left(\sum_{i=1}^{n} \sum_{j=1}^{n} A_{ij} \frac{\partial^2}{\partial x_i \partial x_j} + \sum_{j=1}^{n} B_j \frac{\partial}{\partial x_j} + C \right) u \,, \tag{4}$$

where we assume that $A_{ij} = A_{ji}$, so that

$$\begin{aligned} vA_{ij} \frac{\partial^2 u}{\partial x_j \partial x_j} &= \frac{\partial}{\partial x_j} \left[(vA_{ij} \frac{\partial u}{\partial x_i} \right] - \left[\frac{\partial}{\partial x_i} (vA_{ij}) \frac{\partial u}{\partial x_j} \right] \\ &= \frac{\partial}{\partial x_j} \left[(vA_{ij}) \frac{\partial u_i}{\partial x_j} \right] - \frac{\partial}{\partial x_j} \left[\frac{\partial}{\partial x_j} (vA_{ij}) u \right] + \left[\frac{\partial^2}{\partial x_i \partial x_j} (vA_{ij}) \right] u \,. \end{aligned}$$

Also

$$vB_j \frac{\partial u}{\partial x_j} = \frac{\partial}{\partial x_j} (vB_j u) - u \frac{\partial}{\partial x_j} (vB_j) \,.$$

Thus

$$
\begin{aligned}
vLu &= u\left\{\sum_{i=1}^{n}\sum_{j=1}^{n}\frac{\partial^2}{\partial x_i\partial x_j}(vA_{ij}) - \sum_{i=1}^{n}\frac{\partial}{\partial x_i}(vB_j) + vC\right\} \\
&+ \sum_{I=1}^{n}\frac{\partial}{\partial x_i}\left\{v\sum_{j=1}^{n}(A_{ij})\frac{\partial u}{\partial x_j} - \sum_{j=1}^{n}\frac{\partial}{\partial x_j}(vA_{ij})u + vB_i u\right\} \\
&= uL^*v + \operatorname{div} J(u, v)\,, \qquad (5)
\end{aligned}
$$

where $J = (J_1, J_2, \ldots, J_n)$ and

$$
J_i = \left\{v\sum_{j=1}^{n}A_{ij}\frac{\partial u}{\partial x_j} - \sum_{j=1}^{n}\frac{\partial}{\partial x_j}(vA_{ij})u + vB_i u\right\}\,, \tag{6}
$$

and we have proved (3).

Finally, we apply the divergence theorem and obtain

$$
\int_R (vLu - uL^*v)dx = \int_S \widehat{n}\cdot J\; dS\,, \tag{7}
$$

where R is a bounded region in R_n with boundary S whose exterior unit normal is denoted $\widehat{n} = (n_1, n_2, \ldots n_n)$.

Example 1. For the Laplace operator

$$
L = -\nabla^2 = -\left(\frac{\partial^2}{\partial x_1^2} + \frac{\partial^2}{\partial x_2^2} + \cdots + \frac{\partial^2}{\partial x_n^2}\right)\,, \tag{8}
$$

it follows that $L^* = L$, so it is a self-adjoint operator. Then (5) becomes

$$
(v\nabla^2 u - u\nabla^2 v)dx = \operatorname{div}\,(v\operatorname{grad} u - u\operatorname{grad} v)\,,
$$

so that (7) yields Green's second identity,

$$
\int_R (v\nabla^2 u - u\nabla^2 v)dx = \int_S \left(v\frac{\partial u}{\partial n} - u\frac{\partial v}{\partial n}\right) dS\,. \tag{9}
$$

Example 2. For the heat operator

$$
L = \partial/\partial t - \nabla^2\,, \tag{10}
$$

we have

$$
L^* = -\partial/\partial t - \nabla^2 \tag{11}
$$

and

$$
vLu - uL^*v = \frac{\partial}{\partial t}(uv) + \sum_{i=1}^{n}\frac{\partial}{\partial x_i}\left[u\frac{\partial v}{\partial x_i} - v\frac{\partial u}{\partial x_i}\right]\,.
$$

Next, we set

$$J = \left(vu, \frac{\partial v}{\partial x_1}u - \frac{\partial u}{\partial x_1}v, \ldots, \frac{\partial v}{\partial x_n}u - v\frac{\partial u}{\partial x_n} \right),$$

and let e_t denote a unit vector in the t direction, so that we can write J as

$$J = e_t(vu) + u\text{grad}_x v - v\text{grad}_x u,$$

where the subscript x indicates that the differential operator is with respect to x. Accordingly,

$$\int_t \int_R [vLu - uL^*v]dx\, dt = \int_S [e_t vu + u\text{grad}_x v - v\text{grad}_x u] \cdot \hat{n} \in dS. \qquad (12)$$

Example 3. The wave operator

$$L = \Box^2 = \partial^2/\partial t^2 - \nabla^2, \qquad (13)$$

where $\Box^2$ is called the D'Alembertian operator, is clearly self-adjoint. In this case (7) becomes

$$\int_t \int_R [vL(u) - uL(v)]dx\, dt$$
$$= \int_S \left[e_t \left(v\frac{\partial u}{\partial t} - u\frac{\partial v}{\partial t} \right) + u\text{grad}_x v - v\text{grad}_x u \right] \cdot \hat{n} \in dS. \qquad (14)$$

10.2. Classical and Generalized Solutions

The classification of classical and generalized solutions made in the previous chapter is valid in the present case as well. However, there is one important difference. In the case of the ordinary differential equation $Lu = 0$ with constant coefficients, every solution is the classical solution. The matter is quite different for partial differential equations. The solutions in a similar situation may now include generalized functions. For instance, $\partial u/\partial x_1 = 0$, in R_2, has among its solutions the generalized function $\delta(x_2)$,

$$\langle \delta(x_2), \phi(x_1, x_2) \rangle = \int_{-\infty}^{\infty} \phi(x_1, 0)dx_1.$$

A function $u(x)$ is called a *classical solution* of the partial differential equation

$$Lu = s, \qquad (1)$$

where $s(x)$ is a given smooth function in R_n, if $u(x)$ has continuous derivatives of order p and satisfies (1) identically. If $s(x)$ is a distribution, then $u(x)$ is said to be a *generalized solution* of (1) if

$$\langle Lu, \phi \rangle = \langle s, \phi \rangle, \qquad (2)$$

that is, if $u(x)$ satisfies (1) in the distributional sense. In view of relation (10.1.3) we have

$$\langle u, L^*\phi\rangle = \langle s, \phi\rangle\,. \tag{3}$$

Accordingly, we have only to verify that, for each $\phi \in D$, the action of u on $L^*\phi$ is the same as the action of s on ϕ. In case we are interested in the generalized solution of (1) in an open region R and not the entire R_n, then we seek a distribution u that satisfies (3) for all $\phi \in D(R)$, that is, for all test functions with support in R.

For the homogeneous part of (1), namely,

$$Lu = 0\,, \tag{4}$$

we have the following result:

Theorem.
(a) If u is a classical solution of (4) in R, then it is also a generalized solution in R.

(b) Let U be a function with continuous derivatives of order p that is a generalized solution of (4) in R. Then u is a classical solution in R.

Proof.
(a) Let ϕ be an arbitrary test function with support in the interior of R. Then, in view of (10.1.7),

$$\langle u, L^*\phi\rangle = \int_R uL^*\phi dx = \int_R \phi Lu dx + \int_S n\cdot J\, dS = \int_S n \cdot J\, dS.$$

However, ϕ vanishes on S, so that $J = 0$, and we have $\langle u, L^*\phi\rangle = 0$, so u is a generalized solution.

(b) As in (a) we find that $\langle u, L^*\phi\rangle = \int_R \phi Lu\, dx$. Since we have assumed that $\langle u, L^*\phi\rangle = 0$, we have $\int_R \phi Lu\, dx = 0$. Now, if $Lu \neq 0$ in the neighborhood of any point x_0 in R, say, it is positive there, then we choose ϕ such that $\phi > 0$ in a sphere with center at x_0 and zero elsewhere, so that $\int_R \phi Lu\; dx \neq 0$, which is a contradiction. Of course, as pointed out earlier, we have generalized solutions that are not classical solutions.

10.3. Fundamental Solutions

We shall mainly be interested in the equations wherein the coefficients are constants. The theory of partial differential equations stems from the intensive and extensive study of a few basic equations of mathematical physics, and the coefficients in all of these are constants. Such equations arise in the study of gravitation, electromagnetism, perfect fluids, elasticity, heat transfer, and quantum mechanics. Of great importance in the study of these equations are their fundamental solutions. Recall that a fundamental solution $E(x)$ is a generalized function that satisfies the equation

$$LE(x) = \delta(x)\,. \tag{1}$$

This solution is not unique, because we can add to it any solution of the homogeneous equation. This understood, in the sequel we shall select the fundamental solution from among the particular solutions according to its behavior at infinity or other appropriate criteria. The next few sections are devoted to the fundamental solutions of various important partial differential equations of physics, mechanics, and engineering.

In the study of these solutions the following interesting concept is helpful. It is called *Hadamard's method of descent*:

Given the solution of a partial differential equation in R_{n+1}, we can find its solution in R_n or in a still lower dimension. In doing so, we *descend* from the higher-dimensional problem to a lower-dimensional one. For instance, the solution of the initial value problem for the wave equation in two dimensions can be obtained from that in three dimensions. Specifically, let us consider a linear partial differential equation

$$L\left(D, \frac{\partial}{\partial x_{n+1}}\right) u = f(x) \otimes \delta(x_{n+1}) , \tag{2}$$

in the space R_{n+1} of variables (x, x_{n+1}), where $x = (x_1, \ldots, x_n)$, D is $\partial/\partial x_j$, $j = 1, \ldots, n$, $f \in D'(R_n)$,

$$L\left(D, \frac{\partial}{\partial x_{n+1}}\right) = \sum_{q=1}^{p} \frac{\partial^q}{\partial x_{n+1}} L_q(D) + L_0(D) , \tag{3}$$

and $L_q(D)$ are partial differential operators involving the variables $x_1, \ldots, x_n$.

When we say that the generalized function $g \in D'(R_{n+1})$ allows the continuation over functions of the form $\phi(x)1(x_{n+1})$ where $\phi \in D(R_n)$, we mean the following: Given an arbitrary sequence of functions $\{\psi_m(x_{n+1})\}$, $m = 1, 2, \ldots$, belonging to $D(R_1)$, where R_1 is the space with variables X_{n+1} and converging to 1 in R_1 [i.e., $1(x_{n+1})$], then there is the limit

$$\lim_{m\to\infty} \langle g, \phi(x)\psi_m(x_{n+1})\rangle = \langle g, \phi(x)1(x_{n+1})\rangle = \langle g_0, \phi\rangle , \qquad \phi \in D(R_n) . \tag{4}$$

In view of the completeness of D', we find that $g_0 \in D'(R_n)$.

Specifically, for $g(x)$ such that $g(x) = f(x) \otimes \delta(x_{n+1})$, the inhomogeneous term in (2), we have

$$\begin{aligned} \langle g_0, \phi\rangle &= \lim_{m\to\infty} \langle g(x), \phi(x)\psi_m(x_{n+1})\rangle \\ &= \lim_{m\to\infty} \langle f(x) \otimes \delta(x_{n+1}), \phi(x)\psi_m(x_{n+1})\rangle \\ &= \lim_{m\to\infty} \langle f(x), \delta(x_{n+1})\phi(x)\psi_m(x_{n+1})\rangle \\ &= \lim_{m\to\infty} \langle f(x), \phi(x)\psi_m(0)\rangle \\ &= \langle f(x), \phi(x)\rangle , \qquad \phi \in D . \end{aligned} \tag{5}$$

Accordingly, the method of descent can be stated as follows: If the solution $u \in D'(R_{n+1})$ of (1) allows the continuation (4), then the distribution $u_0 \in D'(R_n)$ is the solution of the equation

$$L_0(D)u_0 = f(x). \tag{6}$$

For instance, if the locally integrable function $E(x,t)$ is the fundamental solution of the operator $L(D, \partial/\partial t)$, then the distribution

$$E_0(x) = \int_{-\infty}^{\infty} E(x,t)dt, \tag{7}$$

is the fundamental solution of the operator L_0. Indeed, in view of the Lebesgue theorem on the passage of the limit under the integral sign, we have

$$\begin{aligned}\lim_{m\to\infty}\langle E(x,t), \phi(x)\psi_m(t)\rangle &= \lim_{m\to\infty}\int E(x,t)\phi(x)\psi_m(t)dx\,dt \\ &= \int E(x,t)\phi(x)dx\,dt \\ &= \int \phi(x)\left(\int_{-\infty}^{\infty} E(x,t)dt\right)dx \\ &= \langle E_0(x), \phi(x)\rangle,\end{aligned} \tag{8}$$

where E_0 is defined in (7) and $\phi \in D$. Moreover, this limit does not depend on the sequence $\{\psi_m(t)\}$. Hence $E_0(x)$ is the fundamental solution of the operator L_0, as required.

10.4. The Cauchy-Riemann Operator

For two-dimensional space we can use the complex variable $z = x + iy$. Then $\bar{z} = x - iy$, $dz = dx + i\,dy$, and the Cauchy-Riemann operator is

$$\frac{\partial}{\partial \bar{z}} = \frac{1}{2}\left(\frac{\partial}{\partial x} + i\frac{\partial}{\partial y}\right). \tag{1}$$

Accordingly, we proceed to solve the partial differential equation

$$\frac{\partial E(x,y)}{\partial x} + i\frac{\partial E(x,y)}{\partial y} = \delta(x,y) = \delta(x)\delta(y), \tag{2}$$

to derive the value of the fundamental solution $E(x,y)$. We take the Fourier transform of (2) with respect to y.

$$d\widehat{E}(x,u)/dx + u\widehat{E}(x,u) = \delta(x), \tag{3}$$

where we have used the notation of Chapter 6. We have studied this equation in Section 9.5. All its fundamental solutions are given by (9.5.5), and in the present case the required solution is

$$\widehat{E}(x,u) = [H(x) + C(u)]e^{-ux}. \tag{4}$$

Now we encounter a difficulty. The function e^{-ux} is not tempered when $x > 0$ and $u \to -\infty$ or when $x < 0$ and $u \to \infty$. We can remedy this by defining $C(u)$ as follows:

$$C(u) = -H(u) = \begin{cases} -1\,, & u > 0\,, \\ 0\,, & u < 0\,. \end{cases} \tag{5}$$

Then (4) becomes

$$\widehat{E}(x, u) = \begin{cases} -H(-x)e^{-ux}\,, & u > 0\,, \\ H(x)e^{-ux}\,, & U < 0\,, \end{cases} \tag{6}$$

and we have a tempered distribution. Its inverse is

$$\begin{aligned} E(x, y) &= \frac{1}{2\pi}\left[H(x)\int_{-\infty}^{0} e^{-(x+iy)u}du - H(-x)\int_{0}^{\infty} e^{-(x+iy)u}du\right] \\ &= \frac{1}{2\pi}\,\frac{1}{x+iy} = \frac{1}{2\pi z}\,. \end{aligned} \tag{7}$$

From (1), (2), and (7) we find that $1/\pi z$ is the fundamental solution of the Cauchy-Riemann operator $\partial/\partial\bar{z}$. Accordingly, we can write relation (2) succinctly as

$$\frac{\partial}{\partial\bar{z}}\,\frac{1}{z} = \pi\delta(x, y)\,. \tag{8}$$

10.5. The Transport Operator

We now study the transport equation

$$\frac{1}{\nu}\,\frac{\partial E(x, t)}{\partial t} + (v \cdot \text{grad}_{\,x} E) + \alpha E = \delta(x, t)\,, \qquad |v| = 1\,, \tag{1}$$

where $E(x, t)$ is the velocity distribution function, v is the velocity, and ν and α are some suitable parameters. Applying the Fourier transform to this equation with respect to x, we obtain

$$\frac{1}{\nu}\,\frac{\partial \widehat{E}(u, t)}{\partial t} + [\alpha - i(v \cdot u)]\widehat{E}(u, t) = \delta(t)\,. \tag{2}$$

Its solution is

$$\widehat{E}(u, t) = \nu H(t)\exp\{[iv \cdot u - \alpha]\nu t\}\,, \tag{3}$$

which is tempered. Accordingly, we can apply the inverse Fourier transform, obtaining

$$E(x, t) = F_u^{-1}[\widehat{E}(u, t)] = \nu H(t)e^{-\alpha\nu t}F^{-1}[\exp((iv \cdot u)\nu t)]\,.$$

Because from (6.4.12), $F^{-1}[\exp((iv \cdot u)\nu t)] = \delta(x - \nu t v)$, we find that the fundamental solution of the transport operator is

$$E(x, t) = \nu H(t)e^{-\alpha\nu t}\delta(x - \nu t v)\,. \tag{4}$$

Here we encounter our first opportunity to use the method of descent, explained in Section 10.3. We determine the fundamental solutions $E_0(x, t)$ of the steady state operator

$$v \cdot \text{grad } E_0(x) + \alpha E_0(x) = \delta(x) . \tag{5}$$

We have

$$\begin{aligned}\langle E(x, t), \phi(x)1(t)\rangle &= \nu \int_0^\infty e^{-\alpha vt} \langle \delta(x - vtv), \phi(x)\rangle dt \\ &= \nu \int_0^\infty e^{-\alpha vt} \phi(vtv)dt = \int_0^\infty e^{-\alpha s} \phi(sv)ds \\ &= \left\langle \frac{e^{-\alpha|x|}}{|x|^2} \delta\left(v - \frac{x}{|x|}\right), \phi\right\rangle .\end{aligned}$$

Comparing this with (10.3.8), we obtain

$$E_0(x_1) = \left(e^{-\alpha|x|}/|x|^2\right) \delta(v - x/|x|). \tag{6}$$

10.6. The Laplace Operator

We have already derived the fundamental solution of the Laplace operator in different ways (in Chapters 4 and 6). Let us rederive and examine this solution starting with the n-dimensional Laplace operator

$$\nabla^2 = \frac{\partial^2}{\partial x_1^2} + \frac{\partial^2}{\partial x_2^2} + \cdots + \frac{\partial^2}{\partial x_n^2} . \tag{1}$$

When $f(x)$ depends only on $r = |x| = (x_1^2 + x_2^2 + \cdots + x_n^2)^{1/2}$, then the Laplacian of this function takes a rather simple form

$$\nabla^2 f(r) = r^{1-n} \frac{d}{dr}\left(r^{n-1} \frac{df}{dr}\right) = \frac{d^2 f}{dr^2} + \frac{n-1}{r} \frac{df}{dr} , \tag{2}$$

from which it follows that if $r \neq 0$, then r^{2-n} is a solution of $\nabla^2 f = 0$. Since the function r^{2-n} is a locally integrable function in the neighborhood of $x = 0$, we can define a distribution r^{2-n} through the relation

$$\langle r^{2-n}, \phi\rangle = \int r^{2-n} \phi(x)dx , \qquad \phi \in D . \tag{3}$$

Since the differential operator ∇^2 is self-adjoint, we have

$$\begin{aligned}\langle \nabla^2 r^{2-n}, \phi\rangle = \langle r^{2-n}, \nabla^2 \phi\rangle &= \int r^{2-n} \nabla^2 \phi \, dx \\ &= \lim_{\varepsilon \to 0} \int r^{2-n} H(r - \varepsilon) \nabla^2 \phi \, dx .\end{aligned} \tag{4}$$

When we apply Green's second identity to the region

$$R = \{x \in R_n - (|x| < \varepsilon)\},$$

we obtain

$$\int_R r^{2-n}\nabla^2\phi\, dx = \int_S \left[r^{2-n}\frac{\partial\phi}{\partial r} - \phi\frac{\partial}{\partial r}(r^{2-n})\right] dS, \tag{5}$$

where S is the surface of the sphere of radius ε with the center at the origin. We have seen in Section 3.3 that the surface area of S is $(2\pi^{n/2}\varepsilon^{n-1})/\Gamma(n/2)$. Accordingly, the right side of (5) has the value

$$-\frac{2(2-n)\pi^{n/2}}{\Gamma(n/2)}\phi(0) + o(\varepsilon),$$

where little o is the usual symbol for the order of magnitude of a quantity. Letting $\varepsilon \to 0$, we find from (4) and (5) that

$$\nabla^2 r^{2-n} = \frac{2(2-n)\pi^{n/2}}{\Gamma(n/2)}\delta(x) = -(n-2)S_n(1)\delta(x), \tag{6}$$

which agrees with (4.4.73), found by a different method. Recalling the definition of a fundamental solution for the differential operator given in (10.3.1), we find from (6) that if $n > 2$,

$$E(r) = \frac{\pi^{-n/2}\Gamma(n/2)r^{2-n}}{2(n-2)} = \frac{1}{(n-2)S_n(1)r^{n-2}}, \tag{7}$$

is the fundamental solution of the Laplace operator $-\nabla^2$. In particular,

$$E(r) = 1/4\pi r \tag{8}$$

is the fundamental solution in R_3.

The case $n = 2$ requires special consideration. In this case we replace r^{2-n} by $\ln r$. The result, which corresponds to (5), as explained in Example 2 of Section 5.9, is

$$\int_R \ln r\nabla^2\phi\, dx = -\int_S \left[\ln r\frac{\partial\phi}{\partial r} - \frac{1}{r}\phi\right] dS = -2\pi\phi(0) + o(\varepsilon),$$

which yields

$$-\nabla^2\left(\frac{1}{2\pi}\ln r\right) = \delta(x). \tag{9}$$

Thus

$$E(r) = (1/2\pi)\ln r \tag{10}$$

is the fundamental solution for this case. Solutions (8) and (10) are the *Newtonian* and *logarithmic potentials*, respectively.

Alternatively, we can derive the fundamental solution (7) by taking the Fourier transform of both sides of the equation

$$-\nabla^2 E(x) = \delta(x), \qquad x = (x_1, x_2, \ldots, x_n) \in R_n. \tag{11}$$

The result, in the notation of Chapter 6, is

$$|u|^2\widehat{E}(u) = 1 \qquad \text{or} \qquad \widehat{E}(u) = 1/|u|^2 .$$

Taking the inverse transform and using (6.4.65), we recover (7).

Single- and double-layer potentials. We now derive the formulas for the discontinuities across a surface element of the second-order partial derivatives of a harmonic function due to single- and double-layer potentials. These formulas follow readily from the analysis of Sections 5.5 and 5.6.

Let $F(x)$ be a singular function with respect to the surface S. The Laplacian of F takes the form

$$\begin{aligned}\overline{\nabla}^2 F &= \nabla^2 F + (B - 2\Omega A)\delta(S) + A\delta'(S)\\ &= \nabla^2 F + (B - 2\Omega A)\delta(S) + 2A\Omega\delta(S) + Ad_n\delta(S)\\ &= \nabla^2 F + B\delta(S) + Ad_n\delta(S) .\end{aligned} \tag{12}$$

If F is harmonic in the complement of S, then

$$\overline{\nabla}^2 F = B\delta(S) + Ad_n\delta(S) . \tag{13}$$

When F vanishes at infinity, (11) and (13) yield, by convolution,

$$F(x) = -\int_S \left[B(\xi)E(x-\xi) - A(\xi)\frac{dE(x-\xi)}{dn}\right] dS_\xi , \tag{14}$$

where $E(x)$ is the fundamental solution already derived, namely,

$$E(x) = \begin{cases} \frac{1}{2\pi}\ln\frac{1}{|x|} , & \text{for} \quad n = 2 , \\ \frac{1}{(n-2)S_n(1)|x|^{n-2}} , & \text{for} \quad n \geq 3 . \end{cases} \tag{15}$$

If we let $F \in R_3$ such that

$$\nabla^2 F = 0, \qquad \text{outside } S , \tag{16}$$

$$[F] = 0 , \tag{17}$$

$$[dF/dn] = B , \tag{18}$$

then F is the *single-layer potential*. From (14) we find that the solution is

$$F(x) = -\int_S B(\xi)E(x-\xi)dS_\xi . \tag{19}$$

Furthermore, from the analysis of Sections 5.5 and 5.6 we find that $[\partial F/\partial x_i] = Bn_i$, and

$$0 = [\nabla^2 F] = C - 2\Omega B , \tag{20}$$

$$[d^2F/dn^2] = 2\Omega B\,; \tag{21}$$

$$\left[\frac{\partial^2 F}{\partial x_i \partial x_j}\right] = Cn_in_j + \frac{\partial B}{\partial x_i}n_j + \frac{\partial B}{\partial x_j}n_i + B\frac{\delta n_i}{\delta x_j}$$
$$= \left(2\Omega n_in_j + \frac{\delta n_i}{\delta x_j}\right)B + \frac{\delta B}{\delta x_i}n_j + \frac{\delta B}{\delta x_j}n_i\,; \tag{22}$$

$$\left[\frac{\partial^2 F}{\partial x_i^2}\right] = \left(2\Omega n_i^2 + \frac{\delta n_i}{\delta x_i}\right)B + 2\frac{\delta B}{\delta x_i}n_i\,, \qquad i \text{ not summed} \tag{23}$$

$$\left[\frac{\partial^2 F}{\partial n \partial x_i}\right] = 2\left(\Omega B n_i + \frac{\delta B}{\delta x_i}\right)\,. \tag{24}$$

If $F \in E$ satisfies the system

$$\nabla^2 F = 0\,, \qquad \text{outside } \Sigma\,, \tag{25}$$

$$[F] = A\,, \tag{26}$$

$$\left[\frac{dF}{dn}\right] = 0\,, \tag{27}$$

then (14) reduces to

$$F(x) = \int_S A(\xi)\frac{dE(x-\xi)}{dn}\,dS_\xi\,, \tag{28}$$

which is the *double-layer potential.*

The required jump relations are

$$\left[\frac{\partial F}{\partial x_i}\right] = \frac{\delta A}{\delta x_i} \tag{29}$$

$$0 = [\nabla^2 F] = C + \nabla^2 A\,,$$
$$[d^2F/dn^2] = -\nabla^2 A \tag{30}$$

$$\left[\frac{\partial^2 F}{\partial x_i \partial x_j}\right] = Cn_in_j + \frac{\delta^2 A}{\delta x_i \delta x_j} - \mu_{jk}n_i\frac{\delta A}{\delta x_k}$$
$$= \frac{\delta^2 A}{\partial x_i \delta x_j} - \nabla^2 A n_in_j - \mu_{jk}n_i\frac{\delta A}{\delta x_k} \tag{31}$$

$$\left[\frac{\partial^2 F}{\partial x_i^2}\right] = \frac{\delta^2 A}{\delta x_i^2} - \nabla^2 A n_i^2 - \mu_{ik}n_i\frac{\delta A}{\delta x_k}\,, \qquad i \text{ not summed} \tag{32}$$

$$\left[\frac{\partial^2 F}{\partial n \partial x_i}\right] = -\mu_{ik}\frac{\delta A}{\delta x_k} - \nabla^2 A n_i\,. \tag{33}$$

Poisson's integral formula. We can obtain the well-known Poisson's integral formula from our analysis in a very simple manner. Our aim is to solve the Dirichlet problem

$$\nabla^2 u(r, \vartheta, \varphi) = 0\,, \qquad r < a\,, \tag{34}$$

$$u(a, \vartheta, \varphi) = f(\vartheta, \varphi)\,, \tag{35}$$

where (r, ϑ, φ) are spherical polar coordinates and $r = a$ is the sphere on which the boundary value $f(\vartheta, \varphi)$ is prescribed. In order to use our formula, we must introduce the function that has a jump discontinuity across the sphere. This is achieved by defining the function $\widetilde{u}(r, \vartheta, \varphi)$

$$\widetilde{u}(r, \vartheta, \varphi) = \begin{cases} u(r, \vartheta, \varphi)\,, & r < a\,, \\ -(a/r)u(a^2/r, \vartheta, \varphi)\,, & r > a\,, \end{cases} \tag{36}$$

so that

$$\overline{\nabla}^2 \widetilde{u} = 0 \qquad \text{in the entire space.} \tag{37}$$

Accordingly,

$$A = [\widetilde{u}] = -2f(\vartheta, \varphi)\,, \qquad B = \left[\frac{d\widetilde{u}}{dn}\right] = \left[\frac{\partial \widetilde{u}}{\partial r}\right] = \frac{1}{a} f(\vartheta, \varphi)\,. \tag{38}$$

Substituting these values in (14) leads to the integral

$$u(r, \vartheta, \varphi) = -\frac{1}{4\pi} \int_0^{\pi} \int_0^{2\pi} \left[\frac{1}{a} f(\vartheta, \varphi) \frac{1}{|x - \xi|} + 2f(\vartheta, \varphi) \frac{\partial}{\partial r'} \frac{1}{|x - \xi|}\right]_{r'=a} \times a^2 \sin\vartheta' d\vartheta' d\varphi'\,. \tag{39}$$

Since $x = (r, \vartheta, \varphi)$, and $\xi = (r', \vartheta', \varphi)$, we have

$$|x - \xi| = (r^2 + r'^2 - 2rr' \cos\Theta)^{1/2}\,,$$

where

$$\cos\Theta = \cos\vartheta \cos\vartheta' + \sin\vartheta \sin\vartheta' \cos(\varphi - \varphi')\,.$$

Substituting this value in (39), we obtain

$$\begin{aligned} u(r, \vartheta, \varphi) =& -\frac{1}{4\pi} \int_0^{\pi} \int_0^{2\pi} f(\vartheta', \varphi') \\ & \times \left\{\frac{1}{a} \frac{1}{(r^2 + r'^2 - 2rr' \cos\Theta)^{1/2}} - \frac{2(r - r\cos\Theta)}{(r^2 + r'^2 - 2rr' \cos\Theta)^{3/2}}\right\}_{r'=a} \\ & \times a^2 \sin\vartheta' d\vartheta' d\varphi' \\ =& \frac{1}{4\pi} a(a^2 - r^2) \int_0^{\pi} \int_0^{2\pi} \frac{f(\vartheta', \varphi') \sin\vartheta' d\vartheta' d\varphi'}{(r^2 + a^2 - 2ar\cos\Theta)^{3/2}}\,, \end{aligned} \tag{40}$$

which is Poisson's integral formula.

10.7. The Heat Operator

Let us now study the heat operator

$$\partial/\partial t - \kappa \nabla^2\,, \tag{1}$$

where t is time and κ is a positive constant. For the sake of simplicity let us take $\kappa = 1$, subsequently we shall restore it. Accordingly, our aim is to solve the differential equation

$$\partial E(x,t)/\partial t - \nabla^2 E(x,t) = \delta(x)\delta(t)\,, \qquad t \geq 0\,, \qquad x \in R_n\,. \tag{2}$$

By taking the Fourier transform of this equation with respect to x, we obtain

$$d\widehat{E}(u,t)/dt + |u|^2 \widehat{E}(u,t) = \delta(t)\,. \tag{3}$$

Its solution is

$$\widehat{E}(u,t) = H(t)\exp(-t|u|^2)\,. \tag{4}$$

From relation (6.3.18b) its inverse transform is

$$E(x,t) = \left(2\sqrt{\pi t}\right)^{-n} \exp(-|x|^2/4t)\,. \tag{5}$$

Let us observe in passing that for $t \geq 0$

$$\begin{aligned} \int E(x,t)dx &= \left(2\sqrt{\pi t}\right)^{-n} \int \exp\left(-\frac{|x|^2}{4t}\right) dx \\ &= \prod_{i=1}^{n} \frac{1}{\sqrt{\pi}} \int_{-\infty}^{\infty} e^{-u_i^2} du_i = 1\,. \end{aligned} \tag{6}$$

For heat equation (2) the concept of the *causal* fundamental solution $C(x,t)$ is of interest. It is defined as the particular solution of (2) that vanishes for $t < 0$. That is

$$\partial C/\partial t - \nabla^2 C = \delta(x)\delta(t)\,, \qquad C = 0\,, \qquad t < 0\,. \tag{7}$$

Accordingly,

$$C(x,t) = (2\sqrt{\pi t})^{-n} H(t)\exp(-|x|^2/4t)\,. \tag{8}$$

It is a C^∞ function except at $t = 0$. Furthermore, it is invariant under the space rotations because it depends only on $|x|^2$. This solution is called *causal* (or *nonanticipative*) because no output signal can ever precede the input singal that gives rise to it.

Furthermore, the causal fundamental solution $C(x,t)$ coincides for $t > 0$ with the solution of the initial value problem

$$\partial u/\partial t - \nabla^2 u = 0\,, \qquad t > 0\,, \tag{9}$$

$$u(x, 0+) = \delta(x)\,. \tag{10}$$

Indeed, if we define $C(x,t) = H(t)u(x,t)$, then $\nabla^2 C = H(t)\nabla^2 u$, and

$$\begin{aligned} \partial C/\partial t &= H(t)\partial u/\partial t + u(x,t)dH/dt \\ &= H(t)\partial u/\partial t + u(x,0+)\delta(t) \\ &= H(t)\partial u/\partial t + \delta(x)\delta(t)\,, \end{aligned}$$

so that

$$\partial C/\partial t - \nabla^2 C = H(t)[\partial u/\partial t - \nabla^2 u] + \delta(x)\delta(t) = \delta(x)\delta(t)$$

as required.

Let us now restore the constant κ in (2) and obtain the results corresponding to those derived above. For instance, (5) becomes

$$E(x,t) = \left(2\sqrt{\kappa\pi t}\right)^{-n} \exp(-|x|^2/4\kappa t)\,. \tag{11}$$

To verify this solution, we observe that for all $\phi \in D$

$$\begin{aligned}
&\langle \partial E/\partial t - \kappa\nabla^2 E, \phi\rangle - \langle E, \partial\phi/\partial t + \kappa\nabla^2\phi\rangle \\
&\qquad = -\iint_0^\infty E(x,t)\left(\frac{\partial\phi}{\partial t} + \kappa\nabla^2\phi\right) dx\,dt \\
&\qquad = -\lim_{\varepsilon\to 0}\iint_\varepsilon^\infty E(x,t)\left(\frac{\partial\phi}{\partial t} + \kappa\nabla^2\phi\right) dx\,dt\,.
\end{aligned} \tag{12}$$

Next, we integrate (12) by parts, using $\partial E/\partial t - \kappa\nabla^2 E = 0,\ t > 0$, and obtain

$$\begin{aligned}
&\left\langle \frac{\partial E}{\partial t} - \kappa\nabla^2 E, \phi\right\rangle \\
&\qquad = \lim_{\varepsilon\to 0}\left[\int E(x,\varepsilon)\phi(x,\varepsilon)dx + \iint_\varepsilon^\infty \left(\frac{\partial E}{\partial t} - \kappa\nabla^2 E\right)\phi\,dx\,dt\right] \\
&\qquad = \lim_{\varepsilon\to 0}\int E(x,\varepsilon)\phi(x,\varepsilon)dx = \lim_{\varepsilon\to 0}\int E(x,\varepsilon)\phi(x,0)dx\,,
\end{aligned} \tag{13}$$

because, in view of (6),

$$\left|\int E(x,t)[\phi(x,\varepsilon) - \phi(x,0)]dx\right| \le M\varepsilon \int E(x,\varepsilon)dx = M\varepsilon\,,$$

where M stands for the maximum of $\phi(x,\varepsilon) - \phi(x,0)$. Moreover,

$$\begin{aligned}
\left|\int E(x,t)[\phi(x) - \phi(0)]dx\right| &\le M(2\sqrt{\kappa\pi t})^{-\pi}\int \exp\left(-\frac{|x|^2}{4t}\right)|x|dx \\
&= M(\sqrt{2\kappa\pi t})^{-n}S_n(1)\int_0^\infty r^n \exp\left(-\frac{r^2}{4t}\right)dr \\
&= M'\sqrt{t}\int_0^\infty y^n e^{-y^2}\,dy = C\sqrt{t}\,,
\end{aligned}$$

where $S_n(1)$ is the area of the n-dimensional sphere of unit radius and M' and c are appropriate constants. Thus it follows that when $t \to 0+$ we have

$$\begin{aligned}
\langle E,\phi\rangle &= \int E(x,t)\phi(x)dx = \phi(0)\int E(x,t)dx + \int E(x,t)[\phi(x) - \phi(0)]dx \\
&= \phi(0) = \langle\delta,\phi\rangle\,.
\end{aligned} \tag{14}$$

Combining (13) and (14), we have the required verificiation.

Let us use the method of descent to obain the fundamental solution for the Laplace operator for $n \geq 3$ from the fundamental solution of the heat equation. From (5) we have

$$E(x) = \int_{-\infty}^{\infty} E(x,t)dt = \int_{0}^{\infty} (2\sqrt{\pi t})^{-n} \exp\left(-\frac{|x|^2}{4t}\right) dt$$
$$= \frac{|x|^{-n+2}}{4\pi^{n/2}} \int_0^\infty e^{-s} s^{(n/2)-2} ds = \Gamma\left(\frac{n}{2}-1\right) \frac{|x|^{-n+2}}{4\pi^{n/2}}$$
$$= \frac{1}{(n-2)S_n(1)} |x|^{-n+2}, \qquad n \geq 3,$$

which is precisely the value of the fundamental solution for the Laplacian.

10.8. The Schrödinger Operator

We now solve the partial differential equation

$$-\left(\nabla^2 E(x,t) - \frac{1}{i}\frac{\partial E(x,t)}{\partial t}\right) = \delta(x,t) = \delta(x)\delta(t), \qquad x \in R_n, \qquad t \geq 0. \tag{1}$$

This is the free Schrödinger equation. Additional terms are included if there is a potential, or other physical interactions are present. There is also a constant coefficient to the ∇^2 term, just as in the heat equation.

Taking the Fourier transform of (1) with respect to x, we find

$$\frac{1}{i}\frac{d\widehat{E}(u,t)}{dt} + |u|^2\, \widehat{E}(u,t) = \delta(t), \tag{2}$$

whose solution is

$$\widehat{E}(u,t) = iH(t)\exp(-it|u|^2). \tag{3}$$

To find its inverse Fourier transform we can apply the relation (6.3.18) or observe that

$$\widehat{E}(u,t,\varepsilon) = iH(t)\exp(-(\varepsilon+it)|u|^2), \qquad \varepsilon > 0,$$

converges in the distributional sense to $\widehat{E}(u,t)$ as $\varepsilon \to 0$. Then the inverse Fourier transform of $\widehat{E}(t,\varepsilon)$ converges to that of $\widehat{E}(t)$. Accordingly, we find that

$$E(x,t,\varepsilon) = (2\pi)^{-n}\left\{\int_{R_n} \exp(-ix\cdot u - (\varepsilon+it)|u|^2)du\right\}$$
$$= (2\pi)^{-n}\left\{\int_{R_n} \exp(-(\varepsilon+it)|u|^2)du\right\}\exp\left(-\frac{|x|^2}{4(\varepsilon+it)}\right).$$

The limit of this quantity as $\varepsilon \to 0$ such that $\varepsilon > 0$, $t > 0$, is

$$E(x,t) = (2\pi)^{-n} t^{-n/2} J^n \exp(-|x|^2/4it), \tag{4}$$

where J is the Fresnel integral $\int_{-\infty}^{\infty} e^{-iy^2} dy = (1-i)\pi/\sqrt{2} = \pi e^{-i\pi/4}$. Thus

$$E(x,t) = \exp[-i(n-2)\pi/4](4\pi t)^{-n/2}\exp(-|x|^2/4it) \tag{5}$$

and the corresponding causal fundamental solution $C(x,t)$ is

$$C(x,t) = H(t)\exp[-i(n-2)\pi/4](4\pi t)^{-n/2}\exp(-|x|^2/4it)\,. \tag{6}$$

It is easier to solve the initial value problem for the wave function $\psi(x,t)$,

$$\frac{1}{i}\frac{\partial\psi(x,t)}{\partial t} - \nabla^2\psi(x,t) = 0\,, \tag{7}$$

$$\psi(x,0) = f(x)\,. \tag{8}$$

The Fourier transforms of these equations are

$$\frac{1}{i}\frac{\partial}{\partial t}\widehat{\psi}(u,t) + |u|^2\widehat{\psi}(u,t) = 0\,, \tag{9}$$

$$\widehat{\psi}(u,0) = \widehat{f}(u)\,, \tag{10}$$

respectively. The solution of (9) and (10) is

$$\widehat{\psi}(u,t) = e^{-it|u|^2}\widehat{f}(u)\,. \tag{11}$$

Taking the inverse transform, we obtain

$$\psi(x,t) = F^{-1}(e^{-it|u|^2}\widehat{f}(u)) = \left(\frac{\pi}{t}\right)^{3/4} e^{\pm 3\pi i/4}\int e^{i|x-y|^2/4t} f(y)\,dy\,. \tag{12}$$

It is interesting to observe from (11) that $|\psi(u,t)| = |\widehat{f}(u)|$ is independent of t. Thus

$$\int |\psi(x,t)|^2 dx = \frac{1}{(2\pi)^3}\int |\psi(u,t)|^2 du$$

is also independent of t. Accordingly, we find that if the wave function is normalized at $t=0$, it remains normalized.

10.9 The Helmholtz Operator

Our contention is that the function [39]

$$E(x) = e^{ik|x|}/4\pi|x|\,, \tag{1}$$

where $k \in (0,\infty)$, satisfies the Helmholtz equation in R_3,

$$-(\nabla^2 + k^2)E(x) = \delta(x)\,. \tag{2}$$

Indeed, in view of the relations

$$\frac{\partial}{\partial x_j}(|x|) = \frac{x_j}{|x|}, \quad \frac{\partial}{\partial x_j}\left(\frac{1}{|x|}\right) = -\frac{x}{|x|^3}, \qquad j = 1, 2, 3, \ldots,$$

and

$$\frac{\partial}{\partial x_j}(e^{ik|x|}) = \frac{ikx_j}{|x|}\, e^{ik|x|},$$

we have

$$\begin{aligned}\nabla^2 e^{ik|x|} &= \frac{\partial^2}{\partial x_j \partial x_j} e^{ik|x|} = \frac{\partial}{\partial x_j}\left[\frac{ikx_j}{|x|} e^{ik|x|}\right] \\ &= \left(\frac{2ik}{|x|} - k^2\right) e^{ik|x|},\end{aligned}$$

where we have used the summation convention. Thus

$$\begin{aligned}\nabla^2 \frac{e^{ik|x|}}{|x|} &= \frac{\partial^2}{\partial x_j \partial x_j}\left[\frac{e^{ik|x|}}{|x|}\right] \\ &= \frac{\partial}{\partial x_j}\left[\frac{1}{|x|}\frac{\partial}{\partial x_j} e^{ik|x|} + e^{ik|x|}\frac{\partial}{\partial x_j}\frac{1}{|x|}\right] \\ &= e^{ik|x|}\frac{\partial^2}{\partial x_j \partial x_j}\frac{1}{|x|} + 2\frac{\partial}{\partial x_j} e^{ik|x|}\frac{\partial}{\partial x_j}\frac{1}{|x|} + \frac{1}{|x|}\frac{\partial^2}{\partial x_j \partial x_j} e^{ik|x|}. \end{aligned} \tag{3}$$

However,

$$\frac{\partial^2}{\partial x_j \partial x_j}\frac{1}{|x|} = \nabla^2 \frac{1}{|x|} = -4\pi\delta(x),$$

so that (3) reduces to

$$-(\nabla^2 + k^2) e^{-ik|x|}/4\pi|x| = \delta(x). \tag{4}$$

Similarly, we can prove that the complex conjugate

$$\overline{E} = e^{-ik|x|}/4\pi|x| \tag{5}$$

also satisfies the Helmholtz equation. Setting $k = -im$ in (2), we find that

$$-(\nabla^2 - m^2) e^{-m|x|}/4\pi|x| = \delta(x), \tag{6}$$

so that the fundamental solution of the operator $-(\nabla^2 - m^2)$ is

$$E = e^{-m|x|}/4\pi|x|. \tag{7}$$

To extend these results to the n-dimensional case, we appeal to (6.4.73),

$$\widehat{f}(|u|) = (2\pi)^{n/2}\int_0^\infty f(r) r^{n/2}\frac{J_{(n-2)/2}(|u|r)}{|u|^{(n-2)/2}}\, dr, \tag{8}$$

which gives the Fourier transform of a radial function $f(|x|)$, $|x| = r$. Then we consider the general differential equation [40]

$$-(\nabla_n^2 - 1)^l E(x) = \delta(x)\,, \tag{9}$$

where l is a positive integer. We take its Fourier transform, obtaining

$$\widehat{E}(|u|) = -(-1)^l/(1+|u|^2)^l\,. \tag{10}$$

Next, we use the identity [22]

$$\int_0^\infty \frac{z^{\nu+1}J_\nu(az)dz}{(z^2+1)^\mu} = \frac{a^{\mu-1}K_{\nu-\mu+1}(a)}{(\mu-1)!2^{\mu-1}}\,, \tag{11}$$

where K is the modified Bessel function, $a > 0$, and $-1 < \mathrm{Re}\,\nu < 2\mathrm{Re}\,\mu - \frac{1}{2}$. Hence, from (8), (10), and (11), and considering the inverse Fourier transform, we get the fundamental solution for the operator $-(\nabla_n^2 - 1)^\ell$ in S,

$$E(|x|) = \frac{(-1)^{l+1}x^{l-(n/2)}}{(\pi)^{n/2}2^{(n/2)+l-1}(l-1)!}\,K_{(n/2)-l}(|x|)\,. \tag{12}$$

From this relation we readily find that the fundamental solution for the operator $-(\nabla_n^2 - m^2)^l$ is

$$E(|x|) = \frac{m^{2l}(-1)^{l+1}|mx|^{l-(n/2)}}{(\pi)^{n/2}2^{(n/2)+l-1}(l-1)!}\,K_{(n/2)-l}(m|x|)\,. \tag{13}$$

Similarly, the fundamental solution of the operator $-(\nabla_n^2 + k^2)^l$ follows by setting $m = -ik$ in (13);

$$E(|x|) = \frac{(-ik)^{2l}(-1)^{l+1}|-ikx|^{l-(n/2)}}{(\pi)^{n/2}2^{(n/2)+l-1}(l-1)!}\,K_{(n/2)-l}(-ik|x|)\,.$$

Because $K_\nu(-iz) = \frac{1}{2}i\pi e^{i\pi\nu/2}H_\nu^{(1)}(z)$, the Hankel function of the first kind, the preceding formula becomes

$$E(|x|) = \frac{(-ik)^{2l}(-1)^{l+1}|-ik|x||^{l-(n/2)}}{\pi^{n/2}2^{(n/2)+l-1}(l-1)!}\,\frac{1}{2}i\pi e^{(i\pi/2)(n/2-1)}H_{(n/2)-l}^{(1)}(k|x|)\,. \tag{14}$$

When $l = 1$, this reduces to

$$E(|x|) = \frac{i}{4}\left(\frac{1}{2\pi k|x|}\right)^{(n-2)/2} H_{(n/2)-1}^{(1)}(k|x|)\,. \tag{15}$$

For $n = 3$, we recover (1) through the relation $H_{1/2}^{(1)}(z) = -i(\frac{1}{2}\pi z)^{-1/2}e^{iz}$. A direct derivation of the formula for $n = 3$ is outlined in Exercise 4.

For $n = 2$, we find from equations (13) and (14) that the fundamental solutions of the operator $-(\nabla^2 - m^2)$ and $-(\nabla^2 + k^2)$ are $(1/2\pi)K_0(mr)$ and $(i/4)H_0^{(1)}(kr)$ respectively.

Following the analysis of this section we can obtain the fundamental solution of the differential operator $(D^k + \lambda^2)$. For instance the fundamental solution of the operator $(\partial^2/\partial x_1\partial x_2 + \lambda^2)$ is $H(x_1)H(x_2)J_0(2\lambda(x_1x_2)^{1/2})$. Recall that the fundamental solution of the operator D^k is given in Exercise 10(b).

10.10. The Wave Operator

To find the fundamental solution for the wave operator we solve the wave equation

$$\Box^2 E(x,t) = -\left(\nabla^2 E(x,t) - \frac{\partial^2 E(x,t)}{\partial t^2}\right) = \delta(x,t)$$
$$= \delta(x)\delta(t), \qquad x \in R_n, \quad t \in R_1, \tag{1}$$

where $\Box^2$ is the D'Alembertian operator. We have taken the wave speed c to be unity and the source time and source point to be $t = 0$ and $x = 0$, respectively, to simplify the algebra. Subsequently, we shall reintroduce c and take the source time and point to be τ and y, respectively. Let us understand some of the basic concepts about this operator by first considering the case $n = 1$, that is,

$$\partial^2 E/\partial t^2 - \partial^2 E/\partial x^2 = \delta(x)\delta(t), \qquad x, t \in R_1. \tag{2}$$

This equation takes a simple form if we introduce the variables (s, y),

$$s = t - x, \quad y = t + x, \qquad \text{or} \qquad x = \frac{1}{2}(y - s), \quad t = \frac{1}{2}(y + s). \tag{3}$$

Then a test function $\phi(x, y) \in D$ becomes $\widetilde{\phi}(s, y)$ such that

$$\widetilde{\phi}(s,y) = \phi\left(\frac{y+s}{2}, \frac{y-s}{2}\right)$$

and

$$4\left(\frac{\partial}{\partial s}\right)\left(\frac{\partial}{\partial y}\right)\widetilde{\phi}(s,y) = \left[\frac{\partial^2}{\partial t^2} - \frac{\partial^2}{\partial x^2}\right]\phi\left(\frac{y+s}{2}, \frac{y-s}{2}\right).$$

Because

$$\langle E, \phi\rangle = \iint E(x,t)\phi(x,t)dx\,dt = \frac{1}{2}\iint E\left(\frac{y+s}{2} \frac{y-s}{2}\right)\widetilde{\phi}(s,y)\,ds\,dy,$$

it follows that the value $\widetilde{E}(s, y)$ of the distribution $E(x, t)$ in (s, y) coordinates is $\widetilde{E}(s, y) = \frac{1}{2}E[\frac{1}{2}(y+s), \frac{1}{2}(y-s)]$.

Next we take a new test function $\psi(x, t)$ and let

$$\phi = (\partial^2/\partial t^2 - \partial^2/\partial x^2)\psi.$$

Then, in view of (2) we have

$$\widetilde{\psi}(0,0) = \psi(0,0) = \langle\psi(x,t), \delta(x)\delta(t)\rangle = \langle\psi, \Box^2 E\rangle = \langle E, \Box^2\psi\rangle$$
$$= \iint E(s,y)\Box^2\psi(s,y)ds\,dy = 4\iint E(s,y)\frac{\partial^2\widetilde{\psi}(s,y)}{\partial s\,\partial y}ds\,dy$$
$$= 4\iint \frac{\partial^2 E(s,y)}{\partial s\,\partial y}\widetilde{\psi}(s,y)ds\,dy.$$

Accordingly, $\widetilde{E}(s, y)$ satisfies the differential equation

$$4\, \partial^2 \widetilde{E}(s, y)/\partial s\, \partial y = \delta(s)\delta(y)\,. \tag{4}$$

Its solution is found by taking the appropriate linear combinations of

$$\begin{aligned} \widetilde{E}^{(1)}(s, y) &= \frac{1}{4} H(s)H(y)\,, & \widetilde{E}^{(2)}(s, y) &= -\frac{1}{4} H(s)H(-y)\,, \\ \widetilde{E}^{(3)}(s, y) &= -\frac{1}{4} H(-s)H(y)\,, & \widetilde{E}^{(4)}(s, y) &= \frac{1}{4} H(-s)H(-y)\,. \end{aligned} \tag{5}$$

Finally, we can go back to (x, t) coordinates and evaluate $E(x, t)$ by using the relation

$$E(x, t) = 2E(t - x, t + x)\,. \tag{6}$$

Thus

$$\begin{aligned} E_1^{(1)}(x, t) &= \frac{1}{2} H(t - x)H(t + x)\,, & E_1^{(2)}(x, t) &= -\frac{1}{2} H(t - x)H(-t - x)\,, \\ E_1^{(3)}(x, t) &= \frac{1}{2} H(-t + x)H(t + x)\,, & E_1^{(4)}(x, t) &= \frac{1}{2} H(-t + x)H(-t - x)\,, \end{aligned} \tag{7}$$

where the subscript 1 indicates that we are in R_1. Because

$$E_1^{(1)}(x, t) = \frac{1}{2} H(t - x)H(t + x) = \frac{1}{2} H(t)H(t^2 - x^2) = \frac{1}{2} H(t - |x|)\,, \tag{8}$$

its support is the sector $t + x \geq 0$, $t - x \geq 0$ in the (x, t) plane (see Figure 10.1). This sector is called the forward light cone. Recall from Section 2.8 that the singular support of a distribution is the smallest set outside of which it is a C^∞ function. Now, the distribution $E(x, t)$ is equal to $\frac{1}{2}$ in the interior of the forward light cone and zero outside it. Thus, the singular support of $E_1(x, t)$ is precisely the boundary of this cone, that is, the union of two rays

$$x + t = 0\,, \qquad x - t = 0\,, \qquad t \geq 0\,. \tag{9}$$

The other three distributions in (7) have similar interpretations.

From a physical point of view, we find that an initial impulse at the origin creates a zone of displacement $E(x, t) = \frac{1}{2}$ whose edges propagate along the lines $x \pm t = 0$. These lines are called the *characteristic lines*.

Let us now return to the case of R_3 [41]. Taking the Fourier transform of (1) with respect to x, we obtain

$$d^2 \widehat{E}(u, t)/dt^2 + |u|^2 \widehat{E}(u, t) = \delta(t)\,, \tag{10}$$

whose solution, which follows from (9.6.15), is

$$\widehat{E}(u, t) = H(t) \sin(|u|t)/|u|\,. \tag{11}$$

Accordingly, the solution of (10), with support in $t \geq 0$, is

$$\widehat{E}_+(u, t) = H(t) \sin(|u|t)/|u|\,, \tag{12}$$

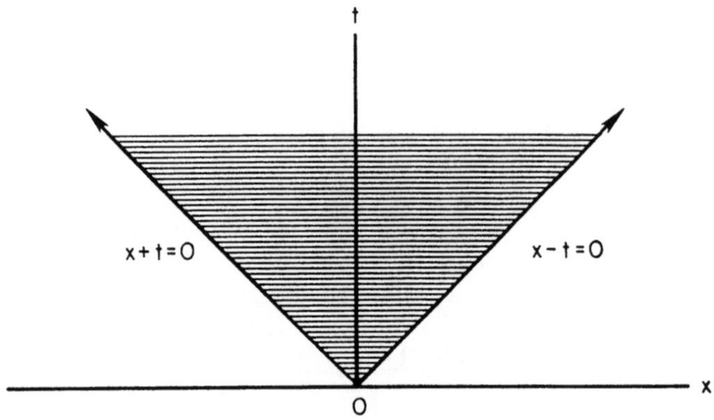

Figure 10.1. The support of the fundamental solution to the wave operator in R_1.

and that for $t \leq 0$ is

$$\widehat{E}_-(u, t) = -H(-t)\sin(|u|t)/|u|. \tag{13}$$

The inverse Fourier transform of (11) follows from (6.4.79);

$$E(x, t) = H(t)F^{-1}\sin(|u|t)/|u| = H(t)(1/4\pi t)\delta(S), \tag{14}$$

where $\delta(S)$ is single layer on the spherical surface of radius t. Thus,

$$E(x, t) = \frac{H(t)}{4\pi t}\delta(S) = \frac{H(t)}{2\pi}\delta(t^2 - |x|^2),$$

where we have used the relation $\delta(S) = 2R\delta(R^2 - |x|^2)$ when the radius of the sphere S is R, or

$$E(x, t) = \frac{H(t)}{2\pi}\left\{\frac{\delta(t - |x|)}{2|x|} + \frac{\delta(t + |x|)}{2|x|}\right\} = \frac{H(t)}{4\pi|x|}\delta(t - |x|),$$

where we have used the fact that $\delta(t + |x|) = 0$ for $t < 0$. Thus, setting $|x| = r$, we have

$$E(x, t) = H(t)\delta(t - r)/4\pi r. \tag{15}$$

From this relation it follows that the fundamental solution of the wave equation for $n = 3$ has as support the surface of the forward light cone with vertex at the source point, namely, $t^2 - |x|^2 = 0$, $t \geq 0$, (see Figure 10.2). The upper part is called the *forward sheet* and extends to the future ($t > 0$). The lower sheet is called the *backward* or *retrograde sheet* and extends to the past ($t < 0$).

Next, we derive the fundamental solution for the case $n = 2$, using the method of descent, that is

$$E_2(x_1, x_2, t) = \int_{-\infty}^{\infty} \frac{H(t)\delta[t - \{(x_1^2 + x_2^2) + x_3^2\}^{1/2}]}{4\pi[x_1^2 + x_1^2) + x_3^2]^{1/2}}\, dx_3$$
$$= \frac{H(t)}{2\pi} \int_0^{\infty} \frac{\delta[t - \{(x_1^2 + x_2^2) + x_3^2\}^{1/2}]}{[(x_1^2 + x_2^2) + x_3^2]^{1/2}}\, dx_3\,. \tag{16}$$

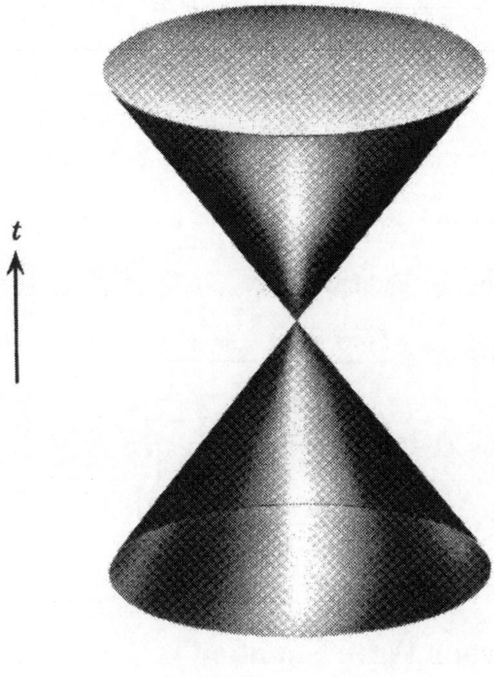

Figure 10.2. The support of the fundamental solution to the wave equation in R_3.

We set $x_3^2 = v^2 - (x_1^2 + x_2^2)$, so that

$$dx_3 = v\,dv/x_3 = v\,dv/[v^2 - (x_1^2 + x_2^2)]^{1/2}\,.$$

Then from (16) we have

$$E_2(x_1, x_2, t) = \frac{H(t)}{2\pi} \int_{(x_1^2+x_2^2)^{1/2}}^{\infty} \frac{\delta(t - v)}{v} \frac{v\,dv}{[v^2 - (x_1^2 + x_2^2)]^{1/2}}$$
$$= \frac{H(t)}{2\pi} \int_{|x|}^{\infty} \frac{\delta(t - v)}{[v^2 - |x|^2]^{1/2}}\, dv = H(t) \frac{H(t - |x|)}{2\pi[t^2 - |x|^2]^{1/2}}\,. \tag{17}$$

This, of course, can be derived directly by the Fourier transform technique (see Exercise 5).

From (17) we can recover the corresponding value for $n = 1$ by a similar approach. Indeed,

$$E_1(x_1, t) = H(t) \int_{-\infty}^{\infty} \frac{H(t - (x_1^2 + x_2^2)^{1/2})}{2\pi [t^2 - x_1^2 - x_2^2]^{1/2}} \, dx_2$$
$$= H(t) \frac{H(t - |x_1|)}{\pi} \int_0^{(t^2 - x_1^2)^{1/2}} \frac{dx_2}{[t^2 - x_1^2 - x_2^2]^{1/2}}$$
$$= H(t) \frac{H(t - |x_1|)}{\pi} \int_0^1 \frac{du}{(1 - u^2)^{1/2}} = \frac{1}{2} H(t) H(t - |x_1|) , \tag{18}$$

which agrees with (8) for $t > 0$.

Finally, we introduce the wave speed c and take the source point and the source time to be y and τ, respectively, so that the differential equation becomes

$$\partial^2 E(x, t)/\partial t^2 - c^2 \nabla^2 E = \delta(x - y)\delta(t - \tau) ; \tag{19}$$

and the corresponding values of E_3, E_2, and E_1 are

$$E_3(x, t) = \frac{\delta(c(t - \tau) - |x - y|)}{4\pi c |x - y|} , \tag{20}$$

$$E_2(x, t) = \frac{H(c(t - \tau) - |x - y|)}{2\pi [c^2 t^2 - |x - y|^2]^{1/2}} , \tag{21}$$

$$E_1(x, t) = \frac{H(c(t - \tau) - |x - y|)}{2c} . \tag{22}$$

10.11. The Inhomogeneous Wave Equation

We now present two forms of the solution to the inhomogeneous wave equation in three dimensions, and some related results [14, 42, 43],

$$\Box^2 \phi(x, t) = \frac{\partial^2 \phi(x, t)}{\partial t^2} - c^2 \nabla^2 \phi(x, t) = f(x, t) . \tag{1}$$

In view of the fundamental solution (10.10.20), we obtain by convolution

$$\phi(x, t) = \int_{-\infty}^{t} \int_{-\infty}^{\infty} \frac{f(y, \tau)\delta(g)}{4\pi R} \, dy \, d\tau , \tag{2}$$

where

$$g = \tau - t + R/c , \qquad |x - y| = R . \tag{3}$$

To obtain the *first form*, we substitute

$$d\tau = \frac{dg}{[\partial g/\partial \tau]_y} = dg$$

in (2) and use the sifting property, so that (2) becomes

$$\phi(x,t) = \int_{-\infty}^{\infty} \left[\frac{f(y,\tau)}{4\pi R} \right]_{g=0} dy \,. \tag{4}$$

When $g = 0$, we have $\tau = t - R/c$, so (4) yields the well-known *retarded potential*,

$$\phi(x,t) = \int_{-\infty}^{\infty} \frac{f(y, t - R/c)}{4\pi R} \, dy \,. \tag{5}$$

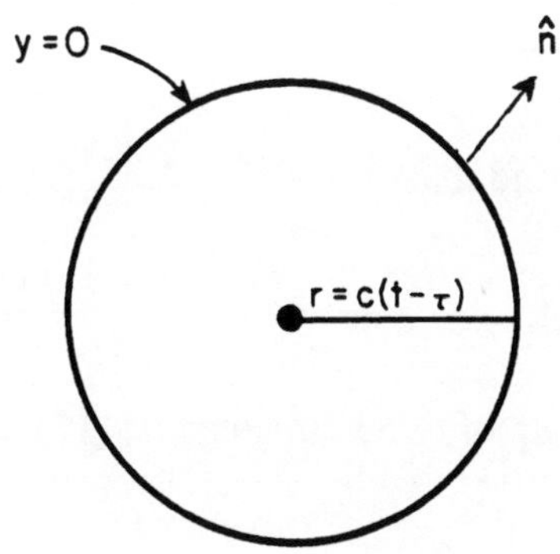

Figure 10.3. Surface area element of the sphere $g = 0$.

The *second form* follows from the observation that

$$dy = \frac{dy_1\, dy_2\, dg}{|\partial g/\partial y_3|} \,, \tag{6}$$

where $\partial g/\partial y_3$ is evaluated for τ and (y_1, y_2) fixed. Thus

$$dy = \frac{dy_1 dy_2}{(\partial g/\partial y_3)/|\nabla g|} \, \frac{dg}{|\nabla g|} = \frac{d\Omega\, dg}{|\nabla g|} \,, \tag{7}$$

where $d\Omega$ is the element of surface area of the sphere $g = 0$ with center at x and radius $r = c(t - \tau)$ (see Figure 10.3) and we have used

$$\frac{dy_1 dy_2}{|\partial g/\partial y_3|/|\nabla g|} = \frac{dy_1 dy_2}{|n_3|} = d\Omega \,, \qquad |\nabla g| = \frac{1}{c} \,, \tag{8}$$

while n_3 is the third component of the unit normal $\widehat{n}$ on the surface of the sphere $g = 0$. Combining (6)–(8), we have

$$dy = c\, d\Omega\, dg \,, \tag{9}$$

so that (2) becomes

$$\phi(x,t) = \frac{c}{4\pi}\int \frac{f(y,t)}{R}\delta(g)\,dg\,d\Omega\,d\tau = \frac{c}{4\pi}\int_{-\infty}^{\infty}\int_{g=0}\frac{f(y,\tau)}{R}\,d\Omega\,d\tau$$
$$= \frac{1}{4\pi}\int_{-\infty}^{t}\frac{d\tau}{t-\tau}\int_{g=0} f(y,\tau)d\Omega\,. \tag{10}$$

The two-dimensional case. Now we use the method of descent and obtain the solution of the two-dimensional inhomogeneous wave equation

$$\Box_2^2\phi = f(x_1, x_2, \tau)\,, \tag{11}$$

where $\phi(x_1, x_2, t)$. This method applied to (10) yields

$$\phi(x_1, x_2, t) = \frac{1}{4\pi}\int_{-\infty}^{t}\frac{d\tau}{t-\tau}\int_{g=0} f(y_1, y_2, \tau)d\Omega\,, \tag{12}$$

where from the Figure 10.4 we find that

$$d\Omega = \frac{dy_1dy_2}{\cos\theta}\,, \qquad \cos\theta = \frac{[c^2(t-\tau)^2 - (x_1-y_1)^2 - (x_2-y_2)^2]^{1/2}}{c(t-\tau)}\,.$$

Since the sphere $g = 0$ is made up of two hemispheres, (12) becomes

$$\phi(x_1, x_2, t) = \frac{1}{2\pi}\int_{-\infty}^{t} c\,d\tau\int_{r\le c(t-\tau)}\frac{f(y_1, y_2, \tau)}{[c^2(t-\tau)^2 - R^2]^{1/2}}\,dy_1dy_2\,, \tag{13}$$

where, as before, $R^2 = |x-y|^2 = (x_1-y_1)^2 + (x_2-y_2)^2$.

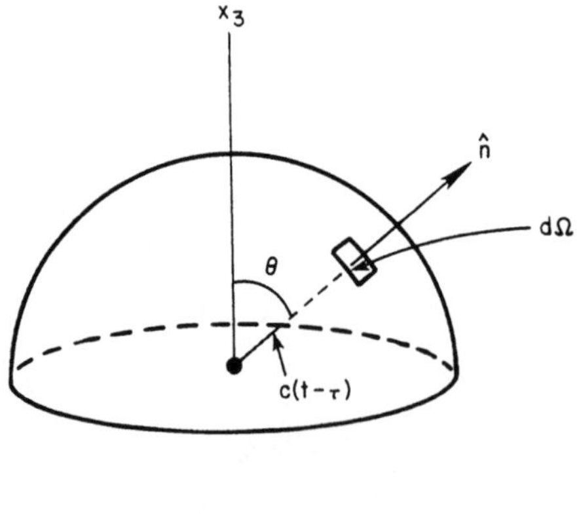

Figure 10.4.

An initial value problem for the wave equation. We now use the information gathered above to solve the following initial value problem in three dimensions [42]

$$\Box^2\phi(x,t) = f(x,t)\,, \qquad t \geq 0\,, \tag{14}$$

$$\phi(x,0) = g_0(x)\,, \tag{15}$$

$$\partial\phi(x,0)\partial t = g_1(x)\,. \tag{16}$$

Let us solve for $\phi(x,t)H(t)$ instead. Distributional differentiation yields

$$\overline{\Box}^2[H(t)\phi(x,t)] = H(t)f(x,t) + \frac{1}{c^2}g_1(x)\delta(t) + \frac{1}{c^2}g_0(x)\delta'(t)\,. \tag{17}$$

This problem can be split into three parts $\phi = \phi_1 + \phi_2 + \phi_3$, where these functions are the solutions of the following equations:

$$\Box^2[H\phi_1] = H(t)f(x,t)\,, \qquad \Box^2[\phi_2] = \frac{1}{c^2}g_1(x)\delta(t)\,,$$

$$\Box^2[H\phi_3] = \frac{1}{c^2}g_0(x)\delta'(t)\,.$$

The solutions of these equations are obtained from the previous discussion;

$$H(t)\phi_1(x,t) = \frac{1}{4\pi}\int_0^t \frac{d\tau}{t-\tau}\int_{g=0} f(y,\tau)d\Omega\,,$$

$$\begin{aligned} H(t)\phi_2(x,t) &= \frac{1}{4\pi c^2}\int \frac{g_1(y)}{R}\delta(-t+R/c)\delta(\tau)d\tau\, dy \\ &= \frac{1}{4\pi c}\int \frac{g_1(y)}{R}\delta(R-ct)dy\,, \end{aligned}$$

where we have used $\delta(R/c - t) = c\delta(R-ct)$. Now we observe that $dy = R^2 dr\, d\omega$, where ω is the solid angle seen from $y = 0$. Then the previous relation becomes

$$\begin{aligned} H(t)\phi_2(x,t) &= \frac{1}{4\pi c}\int Rg_1(y)\delta(R-ct)dr\, d\omega = \frac{1}{4\pi}\int_{r=ct} tg_1(y)d\omega \\ &= \frac{1}{4\pi c^2 t}\int_{R=ct} g_1(y)c^2t^2 d\omega \\ &= t\frac{1}{4\pi c^2t^2}\int_{R=ct} g_1(y)dS = tM_t[g_1(y)]\,, \end{aligned}$$

where we have used the relation $c^2t^2 d\omega = dS$, which is the element of surface on the sphere $r = ct$ with center at x and $M_t[g_1(y)]$ is the mean of $g_1(y)$ on this sphere.

Similarly,

$$\begin{aligned} H(t)\phi_3(x,t) &= \frac{1}{4\pi c^2}\frac{\partial}{\partial t}\int \frac{g_0(y)}{R}\delta(\tau - t + R/c)\delta(\tau)d\tau\, dy \\ &= \frac{\partial}{\partial t}\{t[M_t g_0(y)]\}\,, \end{aligned}$$

because the integrand is identical to that used in the evaluation of ϕ_2. Combining these results, we obtain

$$H(t)\phi(x,t) = \frac{1}{4\pi}\int_0^t \frac{d\tau}{t-\tau}\int_{g=0} f(y,\tau)d\Omega + tM_t[g_1(y)] + \frac{\partial}{\partial t}\{tM_t[g_0(y)]\}. \quad (18)$$

Example 1. Let us use (18) to find $\phi(x,t)$ for $|x| > a$ that satisfies the system

$$\Box^2\phi = 0, \quad (19)$$

$$\phi(x,0) = \begin{cases} 1 & \text{for} \quad |x| \le a, \\ 0 & \text{for} \quad |x| > a. \end{cases} \quad (20)$$

The solution is spherically symmetric, so the observer can be placed on one of the axes. We have $\phi = 0$ for $t < (R-a)/c$ and $t > (R+a)/c$, whereas $\phi \neq 0$ for $(R-a)/c \le t \le (R+a)/c$. Because $R \ge a$, the intersection of the sphere of radius ct and the source region can be approximated by a circle (see Figure 10.5) whose area is $\pi[a^2 - (R-ct)^2]$, so

$$M_t[\phi(x,0)] = \frac{1}{4\pi c^2 t^2}\pi[a^2 - (R-ct)^2] = \frac{a^2-(R-ct)^2}{4c^2t^2}.$$

Now we apply (18), obtaining

$$\phi(R,t) = \begin{cases} \frac{1}{4c^2}\frac{\partial}{\partial t}\left[\frac{a^2-(R-ct)^2}{t}\right], & \text{for} \quad \frac{R-a}{c} \le t \le \frac{R+a}{c}, \\ 0, & \text{otherwise}. \end{cases} \quad (21)$$

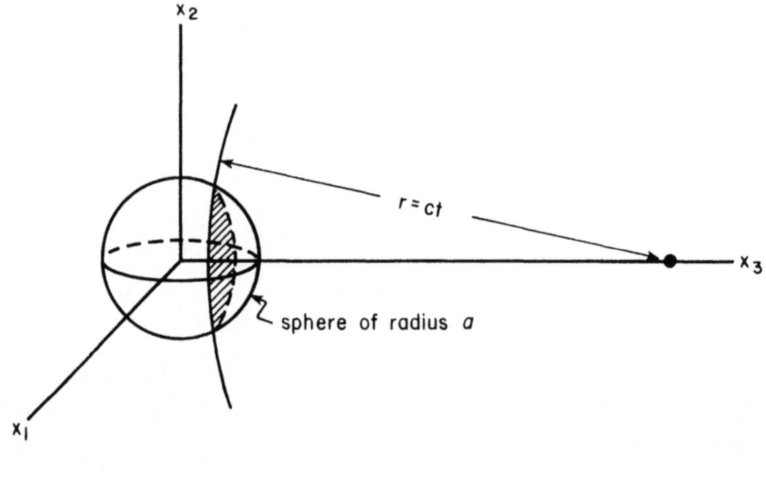

Figure 10.5. The intersection of the source region and a sphere of radius ct.

Moving Sources.

Example 2. A moving point source, $n = 3$. In radiation problems, in acoustics, and in electromagnetism we encounter moving sources. Consider a point source moving with velocity v through an infinite medium that is at rest, so that the volume source density is [44]

$$q(y, \tau) = q_0(\tau)\delta(y - v\tau), \tag{22}$$

where $y = (y_1, y_2, y_3)$ and τ is a scalar. Accordingly, we have to solve the inhomogeneous wave equation

$$(\partial^2/\partial t^2 - c^2\nabla^2)\phi = q_0(\tau)\delta(y - v\tau). \tag{23}$$

In Section 10.10 we observed that the solution of the equation

$$(\partial^2/\partial t^2 - c^2\nabla^2)E(x, t; y, \tau) = \delta(x - y)\delta(t - \tau), \tag{24}$$

is

$$E(x, t; y, \tau) = \delta(t - \tau - |x - y|/c)/4\pi|x - y|. \tag{25}$$

thus by convolution we have

$$\begin{aligned}\phi(x, t) &= \frac{1}{4\pi}\int_{-\infty}^{\infty}\int \frac{q_0(\tau)}{|x - y|}\delta\left(t - \tau - \frac{|x - y|}{c}\right)\delta(y - v\tau)dy\,d\tau \\ &= \frac{1}{4\pi}\int_{-\infty}^{\infty} \frac{q_0(\tau)}{|x - v\tau|}\delta\left(t - \tau - \frac{|x - v\tau|}{c}\right)d\tau.\end{aligned} \tag{26}$$

Next we use (3.1.1), namely,

$$\delta[f(\tau)] = \sum_{i=1}^{n} \frac{\delta(\tau - \tau_i)}{|f'(\tau_i)|},$$

where the τ_i are the simple zeros of $f(\tau)$. In the integral of (26) we have

$$f(\tau) = \frac{|x - v\tau|}{c} + \tau - t,$$

so that

$$\frac{df}{d\tau} = \frac{v^2\tau - v \cdot x}{c|x - v\tau|} + 1,$$

and we have to find the roots of the equation

$$|x - v\tau|/c + \tau - t = 0. \tag{27}$$

This is a quadratic in τ and has at most two roots τ_1 and τ_2. Accordingly, (26) takes the form

$$\phi(x, t) = \frac{1}{4\pi}\sum_{i=1}^{2} \frac{q_0(\tau_i)}{|(v^2\tau_i - v \cdot x)/c + |x - v\tau_i||}. \tag{28}$$

Let us denote τ_1 and τ_2 as $t^{\pm}$, respectively, and introduce the Mach number (vector)

$$M = v/c\,, \tag{29}$$

and the separation vectors

$$R^{\pm} = x - v\tau^{\pm}\,, \tag{30}$$

in (28). Then we have

$$\phi(x,t) = \frac{1}{4\pi}\left[\frac{q_0(\tau^+)}{R^+|1 - M\cos\phi^+|} + \frac{q_0(\tau^-)}{R^-|1 - M\cos\phi^-|}\right], \tag{31}$$

where

$$\cos\theta^{\pm} = M\cdot R^{\pm}/|M||R^{\pm}|\,, \tag{32}$$

is the cosine of the angle between the vectors $R^{\pm}$ and M. Note that (27) takes the simple form

$$\tau^{\pm} = t - R^{\pm}/c\,. \tag{33}$$

Example 3. A moving line source, $n = 2$. Let us now take the radiation field from an infinitely long line source moving perpendicular to its own axis with uniform velocity v through a fluid at rest, so that its strength can be expressed [45] $q(\tau)\delta(y - v\tau)$, where τ is time and $y = (y_1, y_2, y_3)$. Then the velocity potential, $\phi(y,\tau)$, for this field satisfies the wave equation

$$\partial^2\phi/\partial\tau^2 - c^2\nabla^2\phi = q(\tau)\delta(y - v\tau)\,. \tag{34}$$

In Section 10.10 we found that the solution of the equation

$$(\partial^2/\partial t^2 - c^2\nabla^2)E(x,y;t,\tau) = \delta(x-y)\delta(t-\tau)\,, \tag{35}$$

is given by (10.10.21), that is,

$$E(x,y;t,\tau) = \frac{H(c(t-\tau) - |x-y|)}{2\pi c[c^2(t-\tau)^2 - |x-y|^2]^{1/2}}\,. \tag{36}$$

Comparing (34) and (35), we find that the value of the velocity potential $\phi(x,t)$ is given by the convolution integral

$$\phi(x,t) = \int q(\tau)\delta(y - v\tau)E(x,y;t,\tau)dy\,d\tau\,. \tag{37}$$

For the spherical case of a source moving along the x_1 axis with a constant speed V such that $V < c$, we take $q(\tau) = q_0 e^{-i\omega\tau}$, where q_0 is a constant. Then (37) reduces to

$$\phi(x,t) = \frac{q_0}{2\pi c}\int \frac{e^{-i\omega\tau}H(c(t-\tau) - [(x_1 - V\tau)^2 + x_2^2]^{1/2})}{\{c^2(t-\tau)^2 - [(x_1 - V\tau)^2 + x_2^2]\}^{1/2}}\,d\tau\,. \tag{38}$$

Now, introduce a new variable ξ,

$$\xi^2 = \frac{(1-M^2)^2}{(x_1/c - Mt)^2 + (1-M^2)(x_2/c)^2}\left[\tau + \frac{(x_1M/c) - t}{1-M^2}\right]^2, \tag{39}$$

where $M = V/c$ is the Mach number. Then (38) takes the form

$$\phi(x,t) = \frac{q_0}{2\pi c^2(1-M^2)^{1/2}} \exp\left[-i\frac{\omega}{1-M^2}\left(t - \frac{Mx_1}{c}\right)\right] \int_1^\infty \frac{1}{(\xi^2-1)^{1/2}}$$
$$\times \exp\left\{i\frac{\omega}{1-M^2}\left[\left(\frac{x_1}{c} - Mt\right)^2 + (1-M^2)\left(\frac{x_2}{c}\right)^2\right]\xi\right\} d\xi\,. \tag{40}$$

This formula takes an elegant form if we use the integral representation of the Hankel function $H_0^{(1)}(x)$,

$$H_0^{(1)}(x) = \frac{2}{i\pi}\int_0^\infty \frac{ixt}{(t^2-1)^{1/2}}\,dt\,. \tag{41}$$

Then (40) becomes

$$\phi(x,t) = \frac{iq_0}{4c^2(1-M^2)^{1/2}} \exp\left[-i\frac{\omega}{1-M^2}\left(t - \frac{Mx_1}{c}\right)\right]$$
$$\times H_0^{(1)}\left(\frac{\omega[(x_1 - Vt)^2 + (1-M^2)x_2^2]^{1/2}}{c(1-M^2)}\right)\,. \tag{42}$$

Example 4. Moving surface sources. In this case the inhomogeneous wave equation is [14]

$$\Box^2\phi(x,t) = q(x,t)\delta(S) = q(x,t)|\nabla f|\delta(f)\,, \tag{43}$$

where $f = f(x,t)$. The surface $f(x,t) = 0$ can expand and move. Again by convolution, we find from (10.10.20) that

$$4\pi\phi(x,t) = \int \frac{q|\nabla f|}{R}\delta(f)\delta(g)dy\,d\tau = \int \frac{1}{R}[q|\nabla f|\delta(f)]_{ret}dy\,, \tag{44}$$

where for a function ψ, $[\psi(y;x,t]_{ret} = \psi(y, t - R/c)$ and the subscript ret stands for retarded time. Let $F(y;x,t) = [f(y,\tau)]_{ret}$. Then if the surface Σ is described by $F(y;x,t) = 0$, with x and t fixed, we find that

$$dy = \frac{dy_1 dy_2\,dF}{|\partial F/\partial y_3|} = \frac{dy_1 dy_2\,dF}{|\partial F/\partial y_3|/|\nabla F|}\,\frac{1}{|\nabla F|} = \frac{dF\,d\Sigma}{|\nabla F|}\,. \tag{45}$$

Because $F = [f(y,\tau)]_{ret} = f(y, \tau - R/c)$, we have

$$\frac{\partial F}{\partial y_i} = \left[\frac{\partial f}{\partial y_i} + \frac{\widehat{R}_i}{c} + \frac{\partial f}{\partial t}\right]_{ret} = \left[|\nabla f|\left(n_i + \frac{\widehat{R}_i}{c}\frac{\partial f/\partial t}{|\nabla f|}\right)\right]_{ret} = [|\nabla f|(n_i - M_n\widehat{R}_i)]_{ret}\,,$$

where $M_n = -(\partial f/\partial t)/(c|\nabla f|) = v_n/c$ is the Mach number based on the local normal velocity $v_n = -(\partial f/\partial t)/|\nabla f|$ of the surface $f = 0$, and $R_i = (x_i - y_i)/R$. Thus

$$|\nabla F|^2 = \frac{\partial F}{\partial y_i}\frac{\partial F}{\partial y_i} = [|\nabla f|^2(n_i - M_n\widehat{R}_i)^2]_{ret}$$
$$= [|\nabla f|^2(1 + M_n^2 - 2M_n\cos\theta)]_{ret} = [|\nabla f|^2\Lambda^2]_{ret}\,, \tag{46}$$

where $\Lambda^2 = 1 + M_n^2 - 2M_n \cos\theta$ and the quantity θ is the angle between the normal to $f = 0$ and the radial direction $x - y$. Combining (44)–(46), we finally obtain

$$4\pi\phi(x,t) = \int \frac{1}{R}[q|\nabla f|]_{ret}\delta(F)\frac{dF\,d\Sigma}{[|\nabla f|\Lambda]_{ret}} = \int_{F=0} \frac{1}{R}\left[\frac{q}{\Lambda}\right]_{ret} d\Sigma\,.$$

10.12. The Klein-Gordon Operator

Let $\rho = (t^2 - x_1^2 - x_2^2 - \cdots - x_n^2)^{1/2}$. Then for the d'Alembert operator $\Box^2 = \partial^2/\partial t^2 - \partial^2/\partial x_1^2 - \cdots - \partial^2/\partial x_n^2$, we find that

$$\Box^2 u(t, x_1, \ldots, x_n) = \frac{1}{\rho^n}\frac{\partial}{\partial\rho}\left(\rho^n \frac{\partial u}{\partial\rho}\right), \tag{1}$$

for a function $u(x,t)$. The proof is straightforward and is left as an exercise (see Exercise 6). This formula is similar to (10.6.2) for the Laplacian operator, whose fundamental solution we found to be $-(1/(n-2)S_n(1)r^{n-2})$, which depends only on r. It is therefore natural to look for the fundamental solution of the d'Alembert operator that is a function of ρ only. For this purpose we redefine ρ as follows:

$$\begin{aligned}\rho &= \rho(t, x_1, \ldots, x_n)\\ &= \begin{cases} (t^2 - x_1^2 - \cdots - x_n^2)^{1/2}, & t > r = (x_1^2 + x_2^2 + \cdots + x_n^2)^{1/2},\\ 0, & \text{otherwise}.\end{cases}\end{aligned} \tag{2}$$

(cf. Example 6 of Section 4.4). Let us first define the distribution ρ^λ, where λ is a complex number. This presents no problem in the half-plane Re $\lambda > -1$, since ρ^λ is locally integrable for these values of λ. Indeed,

$$\langle \rho^\lambda, \phi\rangle = \int_{t>0} \rho^\lambda \phi(t,x)dx\,dt\,, \tag{3}$$

where ϕ is a test function, defines an analytic generalized function for Re $\lambda > -2$. Using analytic continuation, one can extend (3) to a meromorphic function in the whole complex plane. This is achieved with the help of (1). Indeed, because

$$\Box^2\rho^\lambda = \lambda(\lambda + n - 1)\rho^{\lambda-2} \tag{4}$$

is valid for Re $\lambda > -2$, it follows that for all λ and $\phi \in D$ we have

$$\langle \Box^2\rho^\lambda, \phi\rangle = \lambda(\lambda+n-1)\langle \rho^{\lambda-2}, \phi\rangle\,. \tag{5}$$

By iteration of (5) we obtain more generally that

$$\langle \rho^\lambda, \phi\rangle = \frac{\langle \Box^{2k}\rho^{\lambda+2k}, \phi\rangle}{(\lambda+2)(\lambda+4)\cdots(\lambda+2k)(\lambda+n+1)\cdots(\lambda+n+2k+1)}, \tag{6}$$

which gives the values of ρ^λ in the strip $-2k-2 < \lambda \le -2k$ in terms of the values of ρ^λ in $\operatorname{Re}\lambda > -2$. From this relation it also follows that the singular points of $\rho\lambda$ are $\lambda = -2k$, $k = 1, 2, 3, \ldots$, and $\lambda = -n-2k-1$, $k = 0, 1, 2, 3, \ldots$. Thus for n even all the singular points are simple poles, and for n odd the points $-2, -4, \ldots, -n+1$ are simple while the points $-n-1, -n-3, \ldots$ are double poles.

We now normalize the function ρ^λ by putting

$$Z_\mu = \rho^{\mu-n-1}/\pi^{(n-1)/2}2^{\mu-1}\Gamma\left(\frac{\mu}{2}\right)\Gamma\left(\frac{\mu+1-n}{2}\right), \tag{7}$$

thereby removing the singularities of ρ^λ and producing an entire generalized function. To prove this, we observe that the points $\lambda = -2, -4, \ldots$ for $\lambda > -n-1$ are simple poles of ρ^λ; then the residues at these points follow from (6):

$$\operatorname*{Res}_{\lambda=-2k}\langle\rho^\lambda,\phi\rangle = \lim_{\lambda\to-2k}(\lambda+2k)\langle\rho^\lambda,\phi\rangle = \frac{(-1)^{k-1}\langle\Box^{2k}H,\phi\rangle}{2^{k-1}\Gamma(k)(n-2)\cdots(n-2k)}, \tag{8}$$

where $H = H(t-(x_1^2+\cdots+x_n^2)^{1/2}) = \rho^0$ is the Heaviside function. Accordingly,

$$\begin{aligned}
\lim_{\mu\to-2k+n+1} Z_\mu &= \lim_{\lambda\to-2k}\left[\rho^\lambda/\pi^{(n-1)/2}2^{\lambda+n}\Gamma\left(\frac{\lambda+n+1}{2}\right)\Gamma\left(\frac{\lambda+2}{2}\right)\right] \\
&= \left[\pi^{(n-1)/2}2^{-2k+n}\Gamma\left(\frac{n+1}{2}-k\right)\right]^{-1}\lim_{\lambda\to-2k}\left[\rho^\lambda/\Gamma\left(\frac{\lambda+2}{2}\right)\right] \\
&= \left[\pi^{(n-1)/2}2^{-2k+n}\Gamma\left(\frac{n+1}{2}-k\right)\right]^{-1} \\
&\quad\times\lim_{\lambda\to-2k}\frac{\frac{1}{2}(\lambda/2+1)\cdots(\lambda/2+k-1)[(\lambda+2)\rho^\lambda]}{\Gamma(\lambda/2+k+1)} \\
&= \left[\pi^{(n-1)/2}2^{-2k+n}\Gamma\left(\frac{n+1}{2}-k\right)\right]^{-1}\left(\frac{1}{2}(-k+1)\cdots(-1)\right) \\
&\quad\times\frac{(-1)^k\Box^{2k}H}{2^{k-1}(k-1)!(n-2)\cdots(n-2k)} \\
&= \pi\left(\frac{n}{2}-k\right)\Box^{2k}H/\pi^{(n-1)/2}2^{n-2}\Gamma\left(\frac{n+1}{2}-k\right)\Gamma\left(\frac{n}{2}\right).
\end{aligned} \tag{9}$$

Next suppose that n is even so that $\lambda = n-1$ is a simple pole of ρ^λ. In order to compute the residue at this point, we set $x_i = r\omega_i$, $i = 1, \ldots, n$. Then

$$\begin{aligned}
\langle\rho^\lambda,\phi\rangle &= \int_{t>r}(t^2-r^2)^{\lambda/2}\phi(t,r_1,\ldots,r_n)r^{n-1}\,dt\,dr\,d\Omega_n \\
&= \int_0^\infty\int_0^t(t^2-r^2)^{\lambda/2}r^{n-1}\widehat{\Phi}(t,r)\,dr\,dt = \int_0^\infty t^{\lambda+n}\Phi(t,\lambda)dt,
\end{aligned} \tag{10}$$

where

$$\widehat{\Phi}(t,r) = \int_{\Omega_n}\phi(t,r\omega_1,\ldots,r\omega_n)d\Omega_n \tag{11}$$

and

$$\Phi(t,\lambda) = \int_0^1 (1-r^2)^{\lambda/2} r^{n-1} \widehat{\Phi}(t, tr)\, dr\,. \tag{12}$$

Thus

$$\begin{aligned} \operatorname*{Res}_{\lambda=-n-1} \langle \rho^\lambda, \phi\rangle &= \lim_{\lambda\to -n-1} \Phi(0,\lambda) = \widehat{\Phi}(0,0) \int_0^1 (1-r^2)^{\lambda/2} r^{n-1}\, dr \\ &= \frac{(-1)^{n/2}\pi^{(n_1)/2}}{\Gamma((n+1)/2)} \langle \delta(x), \phi(x)\rangle\,. \end{aligned} \tag{13}$$

Subsequently,

$$\begin{aligned} Z_0 = \lim_{\mu\to 0} Z_\mu &= \lim_{\lambda\to -n-1} \left[\rho^\lambda / \pi^{(n-1)/2} 2^{\lambda_n} \Gamma\left(\frac{\lambda+n+1}{2}\right)\Gamma\left(\frac{\lambda+2}{2}\right)\right] \\ &= \lim_{\lambda\to -n-1} \left[\pi^{(n-1)/2} 2^{\lambda+n} \Gamma\left(\frac{\lambda+2}{2}\right)\right]^{-1} \left[(\lambda+n+1)\rho^\lambda / \Gamma\left(\frac{\lambda+n+3}{2}\right)\right] \\ &= \delta(x)\,. \end{aligned} \tag{14}$$

With the help of (6), the analysis can be generalized to give

$$\operatorname*{Res}_{\lambda=-n+1+2k} \langle \rho^\lambda, \phi\rangle = \frac{(-1)^{n/2}\pi^{(n+1)/2}}{4^k k!\,\Gamma((n+1)/2+k)} \langle \Box^{2k}\delta, \phi\rangle\,, \tag{15}$$

so that

$$Z_{-2k} = \Box^{2k}\delta\,. \tag{16}$$

Finally, for odd n it follows from a similar calculation that the coefficient $c_{-2}^{(k)}$ of the expansion of $\langle \rho^\lambda, \phi\rangle$ about $-n-1-2k$ $(k = 0, 1, 2, \ldots)$ is

$$c_{-2}^{(k)} = \frac{2(-1)^{(n-1)/2}\pi^{(n-1)/2}}{4^k k!\,\Gamma((n+1)/2+k)} \langle \Box^{2k}\delta, \phi\rangle\,,$$

so that (16) is valid for this case as well.

Now that from (6) it follows that

$$\Box^{2k} Z_\mu = Z_{\mu-2k}\,. \tag{17}$$

Furthermore, we can convolute Z_μ and Z_ν since their supports lie in the forward light cone. We then have

$$Z_\mu * Z_\nu = Z_{\mu+\nu}\,.$$

The distributions Z_μ are called the *Riesz distributions*.

Example 1. Let us consider the set $D'_+(\Gamma)$ formed by all the distributions with support in the forward light cone Γ. The convolutions of the members of $D'_+(\Gamma)$ are also in $D'_+(\Gamma)$. Moreover, it can be shown that this convolution algebra has no zero divisors. Hence Z_μ has the unique inverse $Z_{-\mu}$ in $D'_+(\Gamma)$.

Consider now the differential equation

$$\Box^{2k} f = g\,, \tag{18}$$

where the distributions f and g vanish for $t < 0$. Then

$$Z_{-2k} * f = (\Box^{2k}\delta) * f = \delta * \Box^{2k} f = g\,,$$

so that $f = Z_{2k} * g$ is the unique distributional solution. Because the initial values can be added to the differential equation, it follows that such an initial value problem has a unique solution.

The Distribution $_mZ_{2k}$.
Our aim in this section is to find the solution of the Klein-Gordon operator

$$\Box^2 + m^2\,. \tag{19}$$

For this purpose we let m^2 be any complex number and define

$${}_mZ_{-2k} = [(\Box^2 + m^2)\delta]^{*k}\,, \tag{20}$$

where $*k$ in the exponent means that we have k-fold convolution. Since these distributions are in $D'_+(\Gamma)$, they have unique inverses ${}_mZ_{2k}$ such that

$${}_mZ_{2k} *_m Z_{-2k} = \delta\,. \tag{21}$$

Formally, we have

$$\begin{aligned} {}_mZ_{2k} &= (\Box^2 + m^2)^{*(-k)} = (\Box^2\delta * [\delta + m^2(\Box^2\delta)^{*(-1)}])^{*(-k)} \\ &= (\Box^2\delta)^{*(-k)}[\delta + m^2(\Box^2\delta)^{*(-1)}]^{*(-k)} \\ &= \sum_{p=0}^{\infty} \binom{k+p-1}{p} (-1)^p m^{2p} Z_{2p+2k}\,. \end{aligned} \tag{22}$$

From this discussion it follows that the iterated Klein-Gordon equation

$$(\Box^2 + m^2)^k f = g \tag{23}$$

has a unique solution for the distributions f and g vanishing for $t < 0$, and the same is true for the initial value problem. Indeed, the solution is

$$f =_m Z_{2k} * g\,. \tag{24}$$

Finally, we observe that on substituting (7) in (22) we obtain

$${}_mZ_{2k} = \frac{1}{\pi^{(n-1)/2}\Gamma(k)2^{k+(n+1)/2-1}} \left(\frac{\rho}{m}\right)^{k-(n+1)/2} J_{k-(n+1)/2}(m\rho)\,, \tag{25}$$

where $J_{k-(n+1)/2}$ is the Bessel function of order $k - (n+1)/2$.

Example 2. Let us find the fundamental solution $E(x,t)$ of the Klein-Gordon operator, i.e., the solution of the equation

$$(\Box^2 + m^2)E(x,t) = \delta(x,t) \tag{26}$$

in R_4, where $E(x,t) = 0$ for $t < 0$. From the foregoing analysis we know that the solution is $E(x,t) =_m Z_2$, so from (25) we have

$$\begin{aligned} E(x,t) &= \frac{1}{4\pi}\frac{m}{\rho} J_{-1}(m\rho) = \frac{1}{4\pi}\frac{1}{\rho^2}\sum_{q=0}^{\infty}\frac{(-1)^q (m\rho/2)^{2q}}{q!\Gamma(q)} \\ &= \frac{1}{2\pi}\lim_{q\to 0}\frac{\rho^{2q}}{\Gamma(q)} + \frac{1}{2\pi}\rho^{-2}\sum_{q=1}^{\infty}(-1)^4\frac{(m\rho/2)^{2q}}{q!\Gamma(q)} \\ &= Z_2 - (m/4\pi)J_1(m\rho)/\rho\,. \end{aligned} \tag{27}$$

Because $Z_2 = (1/4\pi)\text{Res}_{\lambda=-2}\rho^\lambda = (1/2\pi)\delta(t-(x_1^2+x_2^2+x_3^2)^{1/2})$, (27) becomes

$$E(x,t) = \frac{1}{2\pi}\delta(t-(x_1^2+x_2^2+x_3^2)^{1/2}) - \frac{m}{4\pi}\frac{J_1(m\rho)}{\rho}\,. \tag{28}$$

Example 3. Let us solve the equation

$$(\Box^2 + m_1^2)(\Box^2 + m_2^2)E(x,t) = \delta(x,t) \tag{29}$$

in R_4, where $E(x,t)$ vanishes for $t < 0$.

We can write (29) as ${}_{m_1}Z_{-2} *_{m_2} Z_{-2}E(x,t) = \delta$, so that

$$\begin{aligned} E(x,t) &= ({}_{m_1}Z_{-2})^{*(-1)} * ({}_{m_2}Z_{-2})^{*(-1)} = \frac{1}{m_2^2 - m_1^2}\left[{}_{m_1}Z_2 - {}_{m_2}Z_2\right] \\ &= \frac{1}{m_2^2 - m_1^2}\left[\frac{m_2}{4\pi}\frac{J_1(m_2\rho)}{\rho} - \frac{m_1}{4\pi}\frac{J_1(m_1\rho)}{\rho}\right]. \end{aligned} \tag{30}$$

Many results of Sections 10.10 and 10.11 can be deduced from the present formulas and are left as exercises for the reader.

Exercises

1. Derive (10.10.16) with the help of the Fourier transform.

2. Use the method of Section 10.6 and obtain the Green's function for the half-space problem, that is, solve
(a) the Dirichlet problem

$$\nabla^2 u = f(x), \quad x_3 > 0, \quad u(x_1, x_2, 0) = g(x_1, x_2),$$

(b) the Neumann problem

$$\nabla^2 u = f(x), \quad x_3 > 0, \quad (\partial u/\partial x_3)(x_1, x_2; 0) = g(x_1, x_2).$$

3. Derive the Poisson integral formula for a circle by following the steps given in Section 10.6 in the derivation of the corresponding formula for a sphere.

4. Find the fundamental solution of the Helmholtz equation in three dimensions,

$$-(\nabla^2 + k^2)E(x) = \delta(x),$$

by the following steps:

(a) Take the Fourier transform of both sides of this equation and obtain $-(k^2 - |u|^2)\widehat{E}(|u|) = 1$.

(b) Show that the solution of the latter equation is

$$E(u) = \delta(k^2 - |u|^2) - \mathrm{Pf}\left(\frac{1}{k^2 - |u|^2}\right).$$

(c) Expand the terms of this equation.

(d) Use the relation $-\widehat{t}(x - a) = e^{ia\cdot u}\widehat{t}(u)$.

(e) Use the fact that the Fourier transform of the step function $H(x)$ is

$$\widehat{H}(\pm x) = \pi\delta(x) \pm i\,\mathrm{Pf}\,(1/x),$$

as well as exercise 18 of Chapter 6.

5. Derive the two-dimensional fundamental solution $E(x, t)$ given by (10.10.17) directly by the Fourier transform method (recall Exercise 26 of Chapter 6).

6. Derive (10.12.1).

7. Show that the fundamental solution of the dissipative wave equation

$$\left(\overline{\nabla}^2 - a^2\frac{\partial}{\partial t} - \frac{1}{c^2}\frac{\partial^2}{\partial t^2}\right)E(x, x', t, t') = \delta(x - x')\delta(t - t')$$

is

$$\begin{aligned}E(x, x', t, t') =& \frac{c\exp[-\frac{1}{2}a^2c^2(t - t')]}{4\pi|x - x'|}\\ &\times\left\{\delta[c(t - t) - |x - x'|] + \frac{a^2c|x - x'|}{2[|x - x'|^2 - c^2(t - t')^2]^{1/2}}\right.\\ &\times J_1\left[\frac{1}{2}a^2c[|x - x'| - c^2(t - t')^2]^{1/2}\right]\\ &\left.\times H[(c(t - t') - |x - x'|)]\right\}H(t - t').\end{aligned}$$

As a special case derive the fundamental solution of the telegraph operator

$$\frac{\partial^2}{\partial x^2} - 2\frac{\partial}{\partial t} - \frac{1}{c^2}\frac{\partial^2}{\partial t^2}.$$

8. Show that the value of the fundamental solution $E(x, t)$ with support in the region $t \geq 0$ for the n-dimensional wave operator $\partial^2/\partial t^2 - \nabla_n^2 = -\Box_n^2$ is

$$E(x, t) = \frac{1}{2^n \pi^{(n-1)/2}((n-1)/2)!}\left(\frac{\partial^2}{\partial t^2} - \nabla_n^2\right)^{(n-1)/2} [H(t^2 - |x|^2)H(t)]$$

when n is odd and

$$E(x, t) = \frac{1}{2^n \pi^{n/2}}\left(\frac{\partial^2}{\partial t^2} - \nabla_n^2\right)^{n/2} [t^2 - |x|^2]^{1/2} H(t^2 - |x|^2)H(t)$$

when n is even.

9. By taking the Fourier transforms with respect to x_n of the fundamental solutions as given in the Exercise 8, replacing u_n (the Fourier transform of x_n) by k and then changing $n-1$ to n, show that the fundamental solution of the operator $-(\Box_n^2 + k^2)$ with support in the region $t \geq 0$ is

$$E(x, t) = [2^n(\pi)^{n/2}(n/2)!k]^{-1}(\Box_n^2 + k^2)^{n/2}[H(t)H(t^2 - |x|^2)\sin(k(t^2 - |x|^2)^{1/2})],$$

when n is even and

$$\begin{aligned} E(x, t) =& [2^{n+1}\pi^{(n-1)/2}((n+1)/2))!k]^{-1}(\Box_n^2 + k^2)^{(n+1)/2} \\ &\times [H(t)H(t^2 - |x|^2)(t^2 - |x|^2)^{1/2}]J_1(k(t^2 - |x|^2)^{1/2}), \end{aligned}$$

when n is odd.

10. Prove that the fundamental solution $E(x, t)$ of the differential operator

$$(\partial^2/\partial t^2 - \alpha\nabla^2)(\partial^2/\partial t^2 - \beta\nabla^2)$$

in the three dimensional case with support $t \geq 0$, is

$$\begin{aligned} E(x, t) =& \frac{1}{4\pi(\alpha^2 - \beta^2)}\left\{\left(t - \frac{|x|}{t}\right) H(\alpha t - |x|)\right. \\ &\left. - \left(t - \frac{|x|}{\beta}\right) H(\beta t - |x|)\right\}. \end{aligned}$$

For a detailed study of the fundamental solutions of the products of hyperbolic differential operators see Ortner [46].

11. Let $P(D) = P(\partial/\partial t, \partial/\partial x_1, \partial/\partial x_3, \ldots, \partial/\partial x_m)$ be the homogeneous hyperbolic differential operator of order m. Find the distributional differential equation that is equivalent

to the Cauchy problem

$$P(D)u = f\,, \qquad t \geq 0\,,$$

$$\partial^k u/\partial t^k = g_k(x_1, \ldots, x_n)\,, \qquad k = 0, 1, \ldots, m-1, \qquad t = 0\,.$$

12. Show that the fundamental solution $E(x, \alpha)$ of the Stokes equations

$$\nabla E(x, \alpha) = 0\,, \qquad \nabla p = \mu \nabla^2 E(x, \alpha) + 8\pi \mu \alpha \delta(x)\,,$$

where α is a given vector, is $E(x, \alpha) = \alpha/r + (\alpha \cdot x)x/r^3$, $r = |x|$.

13. For a homogeneous, linearly elastic, isotropic medium, the displacement equations of equilibrium are

$$\mu[(1-2\nu)^{-1}\nabla\, \nabla \cdot u(x) + \nabla^2 u(x)] + f(x) = 0\,,$$

where $u = (u_1, u_2, u_3)$ denotes the displacement field, μ is the shear modulus, v is the Poisson ratio of the material, and f is the body force per unit of volume. Show that for $f(x) = 16\pi\mu(1-\nu)\delta(x)\alpha$, where α is a given vector, the solution is

$$u(x, \alpha) = (3 - 4v)\alpha/r + (a \cdot x)x/r\,, \qquad r = |x|\,.$$

14. The equations of motion for steady state elastodynamics are

$$\mu \left[\frac{1}{1-2\nu} \nabla\, \nabla \cdot u + \nabla^2 u + m^2 u \right] + f = 0\,.$$

where m is a suitable parameter and the other quantities are as defined in the previous Exercise. Show that for $f(x) = 4\pi\mu\delta(x)\alpha$ the solution u is

$$u(x, \alpha) = \left[\frac{e^{-imr}}{r}\alpha + \frac{1}{m^2}\nabla(\alpha \cdot \nabla)\left(\frac{e^{-imr}}{r} - \frac{e^{-imtr}}{r} \right) \right],$$

where $r = |x|$ and $\tau^2 = (1-2\nu)/2(1-\nu)$.

Derive the result of Exercise 13 from this value of $u(x)$ as $m \to 0$.

15. Show that the Green's tensor satisfying the vector wave equation

$$\frac{1}{c^2} \frac{\partial^2 G_{ij}(x, ct)}{\partial t^2} + \nabla \times [\nabla \times G_{ij}(x, ct)] = \delta_{ij}\delta(t)\delta(x)\,, \qquad x = (x_1, x_2, x_3)\,,$$

is

$$G_{ij} = \frac{1}{4\pi r}\delta\left(t - \frac{r}{c}\right)\left(\delta_{ij} - \frac{x_i x_j}{r^3}\right) + \frac{1}{4\pi r^3} c^2 t H\left(t - \frac{r}{c}\right)\left(\delta_{ij} - \frac{3x_i x_j}{r^t}\right)$$
$$+ \frac{1}{3} c^2 t H(t)\delta(x)\delta_{ij}\,.$$

16. With the help of Exercise 15, derive the formal solution of the electromagnetic wave equation,

$$\frac{1}{c^2}\frac{\partial^2 E}{\partial t^2} + \nabla \times (\nabla \times E) = -\frac{\partial J}{\partial t},$$

where E is the electric field and J is the current density. This is achieved by first writing the left side of this equation in distributional derivatives. The initial conditions and boundary conditions thereby appear as sources on the left side of the equation. Convoluting the sum of these sources with G_{ij} then yields the required solution. The result is simplified when the initial and boundary conditions are specified explicitly.

17. Show that the solution of the heat equation when initially a mass M is distributed uniformly on the surface S of a sphere of radius r_0, i.e., the solution of the initial value problem

$$u_t - \nabla^2 u = 0, \qquad x \in R_n, \qquad u(x, 0) = M\delta(S)/S_n(1)r_0^{n-1}$$

is

$$u(x, t) = c_n \frac{M}{t} \frac{e^{-(|x|^2+r_0^2)/4t}}{(|x|r_0^{(n/2)-1})} I_{(n/2)-1}\left(\frac{|x|r_0}{2t}\right),$$

where

$$c_n = \begin{cases} \dfrac{1}{2\sqrt{\pi}} \cdot \dfrac{(n-2)!}{((n-3)/2)!}, & \text{for } n \text{ odd}, \\ \dfrac{1}{\pi} 2^{3n/2-5}(n-2)(n-2)! & \text{for } n \text{ even}. \end{cases}$$

Deduce that the value of $u(0, t)$ is the same as for the case when the initial value $u(x, 0)$ is replaced by $M\delta(x - a)$, where $|x| = r_0$. Thus an observer at the origin cannot tell the difference between the effect of a mass M at a single point at a distance r_0 or the mass M distributed uniformly over a sphere of radius r_0. *Hints*: (i) In view of the equivalence of the initial value problem (10.7.10) and (10.7.11) with the inhomogeneous equation (10.7.2), show, by convolution and the sifting property of $\delta(S)$, that

$$u(x, t) = \frac{1}{2(\pi t)^{n/2}} \frac{M}{\omega_n r_0^{n-1}} \int_S e^{-(x-\xi)^2/4t}\, dS_\xi, \qquad \xi \in S.$$

(ii) Cut the surface S into infinitely small rings of thickness dh situated at a distance h from the origin so that $|x - \xi|^2$ is constant on each ring. Use the fact that the area of this ring is $dS = S_{n-1}(1)r_0(r_0^2 - h^2)^{(n-3)/2}dh$. Then deduce that

$$u(x, t) = \frac{M}{2(\pi t)^{n/2}} \frac{S_{n-1}(1)}{S_n(1)} e^{-(|x|^2+r_0^2)/4t} \int_{-1}^{1} (1 - s^2)^{(n-3)/2} e^{(|x|r_0|2t)s} ds,$$

where $s = h/r_0$.

18. Consider Oseen's flow equations

$$\operatorname{div} v = 0, \qquad \partial v/\partial x + \operatorname{grad} p - \nabla^2 v/4\sigma = t\delta(S),$$

where $v = (v_1, v_2, v_3)$ is the velocity vector, $x_1 = x, x_2 = y, x_3 = z, t = fe_x + ge_y + he_z$, with e_x, e_y, e_z being unit vectors along the x, y, z directions, respectively, p is the (scalar) pressure; 4σ is the Reynold's number; and $t\delta(S)$ describes the action of the body S on the fluid motion. Take the Fourier transform of these equations and deduce that

$$\widehat{p} = (iu/\rho^2)[t\delta(S)]^\wedge ,$$

$$\widehat{v} = -\frac{1}{iu_1}\left\{\frac{1}{\rho_2} - \frac{1}{\rho^2 - 4\sigma i u_1}\right\}\{-iu(-iu\cdot[t\delta(S)]^\wedge) + \rho^2[t\delta(S)]^\wedge\},$$

where $\rho^2 = u_1^2 + u_2^2 + u_3^2$. Determine their inverse transforms by using the identities

$$F^{-1}\left(\frac{1}{\rho^2 - 4\sigma i u_1}\right) = \frac{\exp\{2\sigma(x - |x|)\}}{4\pi|x|},$$

$$\int_{-\infty}^{x}\frac{\partial}{\partial y}\frac{\exp\{2\sigma(x-|x|)}{4\pi|x|}\,dx = -\frac{1}{4\pi}\frac{y}{y^2+z^2}\left(1+\frac{x}{|x|}\right)\exp\{2\sigma(x-|x|)\}$$

and the convolution theorem.

19. Using the notation and the method of Exercise 18, show that the solution of the Oseen's equations

$$\operatorname{div} v = a\delta(x, y, z),$$

$$\frac{\partial v}{\partial x} + \operatorname{grad} p - \frac{1}{4\sigma}\nabla^2 v = b\delta(x, y, z)e_x ,$$

where a, b are constants, is

$$v = b\left\{\operatorname{grad}\left(\frac{1 - e^{2\sigma(x-|x|)}}{4\pi|x|}\right) + \frac{\sigma}{\pi|x|}e^{2\sigma(x-|x|)}e_x\right\} - a\operatorname{grad}\left(\frac{1}{4\pi|x|}\right).$$

Derive the value of the pressure p.

20. In the previous two exercises we have considered the nondimensionalized and steady Oseen's equations. Let us now consider the unsteady Oseen's equations in their physical dimensions, that is

$$\operatorname{div} v = 0,$$

$$\partial v/\partial t + U\partial v/\partial x + 1/\rho \operatorname{grad} p - \nu\nabla^2 v = b\delta(x)\delta(y)\delta(z)\delta(t)\widehat{e}_y ,$$

where U and b are constants, ν is the kinematic viscosity. Find the values of p and v with the help of following two hints.

(i) Take the divergence of the second equation and show that p satisfies Poisson's equation

$$\nabla^2 p = \rho b\delta(x)\delta'(y)\delta(z)\delta(t).$$

(ii) The solution of the homogeneous equation

$$\frac{\partial f}{\partial t} + U\frac{\partial f}{\partial x} - \nu\,\nabla^2 f = 0,$$

is

$$f = \frac{1}{R}\left\{1 - \operatorname{erfc}\left(\frac{R}{2\sqrt{\nu t}}\right)\right\},$$

where $R^2 = (x - Ut)^2 + y^2 + z^2$ and erfc is the complementary error function

$$\operatorname{erfc}\eta = \frac{1}{\sqrt{\pi}}\int_\eta^\infty \exp(-\xi^2)d\xi .$$

CHAPTER 11

Applications to Boundary Value Problems

In this chapter we present solutions to boundary value problems, arising in various disciplines of mathematical physics for axially symmetric bodies, including dumbbells, elongated rods, and prolate bodies, of which spheres and spheroids are special cases. The methods rest on exploring the fundamental solutions of partial differential equations, as presented in the previous chapters, and then taking a suitable axial distribution of the Dirac delta function and its derivatives on a segment of the axis of symmetry of the body. This idea is extended to include distribution of these functions on arbitrary straight lines, curves, and disks [47–55].

11.1. Poisson's Equation

Let us consider Poisson's equation in n dimensions;

$$\nabla^2 u(x) = -\frac{2(n-2)\pi^{n/2}}{\Gamma(n/2)} F(x), \qquad n \geq 3, \tag{1}$$

where $u(x)$ is the generalized n-dimensional potential, $F(x)$ is some forcing function, and $x = (x_1, x_2, \ldots, x_n)$ is the position vector. In Section 10.4 we found that for the special case $F(x) = \delta(x)$ we obtain the fundamental solution $E(x)$

$$u(x) = E(x) = 1/r^{n-2}, \qquad r = |x|. \tag{2}$$

This value of $u(x)$ gives the generalized electrostatic potential due to an n-dimensional sphere of radius 1. For a distribution $F(x)$ of delta functions over a volume V, the corresponding value of $u(x)$ is

$$u(x) = E * F = \int_V \frac{F(x')dV}{|x-x'|^{n-2}}, \tag{3}$$

where $*$ denotes the convolution and the variable of integration is x'.

Our aim is to use (3) to obtain the generalized electrostatic potential of a large class of axially symmetric bodies for which a sphere, a dumbbell, and a spheroid are special cases. For this purpose, we prescribe an axial distribution of sources along the axis of symmetry, which we take to be the x_1 axis. Accordingly,

$$F(x) = f(x_1)\delta(x_2)\delta(x_3)\cdots\delta(x_n) \qquad (-c_1 \leq x_1 \leq c_2), \tag{4}$$

where $f(x_1)$ is the strength of the distribution and c_1 and c_2 are two positive constants. For bodies with fore-and-aft symmetry we have $c_1 = c_2 = c$. In terms of cylindrical polar coordinates, the shape of an axially symmetric body can be written

$$\rho = \rho_0(x), \qquad -a \le x \le a, \qquad \rho_0(\pm a) = 0, \tag{5}$$

where we let x denote x_1 and where $\rho^2 = x_2^2 + x_3^2 + \cdots + x_n^2$. In view of the symmetry, $\rho_0(x) = \rho_0(-x)$. The geometry of this configuration is explained in Figure 11.1.

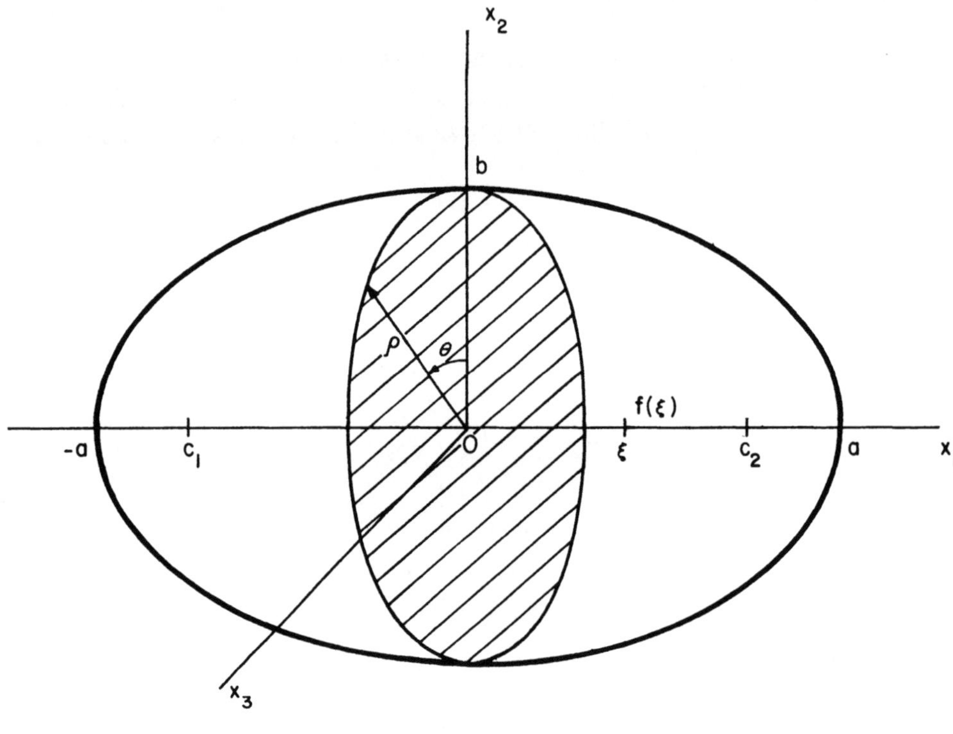

Figure 11.1. Line distribution of delta functions along the x_1 axis in $c_1 \le x_1 \le c_2$ of an axially symmetric body.

In the next stage, we substitute F given by (4) in (3) and obtain

$$u(\rho, x) = \int_{-c_1}^{c_2} \frac{f(\xi)d\xi}{[(x-\xi)^2 + \rho^2]^{(n-2)/2}}, \tag{6}$$

where we have used the sifting property of the delta function. If the potential u is prescribed on the surface S to be $g(x)$, then (6) becomes

$$g(x) = \int_{-c_1}^{c_2} \frac{f(\xi)d\xi}{[(x-\xi)^2 + \rho^2]^{(n-2)/2}} \qquad \text{on} \quad S : \rho = \rho_0(x). \tag{7}$$

This is a Fredholm integral equation of the first kind for $f(x)$. Using this technique we can solve the potential problems for a wide class of axially symmetric configurations and in various fields of mechanics. Furthermore, we can solve both the direct and inverse problems. For the direct problem, the body profile $\rho = \rho_0(x)$, as well as $g(x)$, is prescribed, and we evaluate the distribution $f(x)$ and the parameters c_1 and c_2 from (7). For the inverse problem we are given $f(x)$, $g(x)$ and the parameters c_1 and c_2, then (7) provides a body profile.

The value $g(x) = 1$ is of special interest because it leads to the evaluation of the n-dimensional capacity of S. This is defined as

$$C(n) = (n-2)\int_v F(x)dx\,, \tag{8}$$

where $dx = dx^1 dx^2 \cdots dx^n = dv$. This relation, in view of (4), becomes

$$C(n) = (n-2)\int_{-c_1}^{c_2} f(\xi)d\xi\,. \tag{9}$$

Let us now consider an axial distribution of dipoles along the axis of symmetry such that

$$F(x) = h(x)\delta'(y)\delta(x_3)\cdots\delta(x_n)\,, \qquad c_1 \le x \le c_2\,, \tag{10}$$

where $x_1 = x$, $x_2 = y$, and $h(x)$ is the strength of the distribution. Thus, the dipoles are directed along the y axis. Substituting (10) in (3), we obtain

$$u(x,\rho) = -(n-2)y\int_{-c_1}^{c_2} \frac{h(\xi)d\xi}{[(x-\xi)^2+\rho^2]^{n/2}}\,. \tag{11}$$

Let $u(x,\rho) = yl(x)$ on the surface S. Then (11) becomes

$$l(x) = -(n-2)\int_{-c_1}^{c_2} \frac{h(\xi)d\xi}{[(x-\xi)^2+\rho^2]^{n/2}}\,, \qquad (x,\rho)\in S\,, \tag{12}$$

which like (7), is a Fredholm integral equation of the first kind for $h(x)$, provided that the equation $\rho = \rho_0(x)$, of the surface S, is known. Alternately, if $h(x)$, $l(x)$, c_1, and c_2 are prescribed, then (12) yields us the body profile.

In the sequel we take $c_1 = c_2 = c$.

11.2. Dumbbell-Shaped Bodies

Let us take two isolated delta distributions of equal strength such that

$$f(x) = A_0\delta(x+c) + A_0\delta(x-c)\,,$$

where A_0 is a constant. This gives a dumbbell-shaped body, as shown in Figure 11.2. We substitute in (11.1.7), and find that for $g(x) = 1$ we have

$$1 = A_0(1/4R_1^{n-2} + 1/R_2^{n-2})\,, \tag{1}$$

where

$$R_1 = [(x+c)^2 + \rho^2]^{1/2}, \qquad R_2 = [(x-c)^2 + \rho^2]^{1/2}. \tag{2}$$

To obtain the relationship between the geometric parameters c/a and d/a (see Figure 11.2), we evaluate (1) at the terminal points $x = \pm a$, $\rho = 0$ and at the dumbbell neck $x = 0$, $\rho = \pm d$. The result is

$$\frac{1}{A_0} = \frac{1}{(a+c)^{n-2}} + \frac{1}{(a-c)^{n-2}}, \qquad \frac{1}{A_0} = \frac{2}{(c^2+d^2)^{(n-2)/2}}. \tag{3}$$

Elimination of A_0 yields the desired relation

$$\frac{1}{(a+c)^{n-2}} + \frac{1}{(a-c)^{n-2}} = \frac{2}{(c^2+d^2)^{(n-2)/2}}. \tag{4}$$

Furthermore, from (1) and (3) we derive the equation for the n-dimensional dumbbell

$$\frac{1}{[(x+c)^2+\rho^2]^{(n-2)/2}} + \frac{1}{[(x-c)^2+\rho^2]^{(n-2)/2}} = \frac{1}{(a+c)^{n-2}} + \frac{1}{(a-c)^{n-2}}. \tag{5}$$

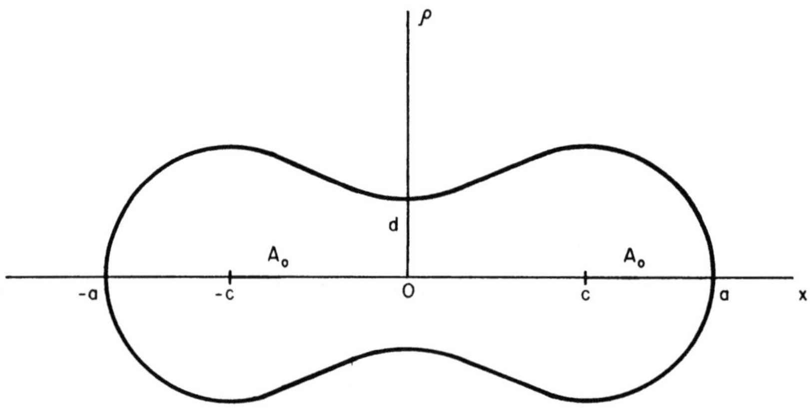

Figure 11.2. Dumbbell-shaped bodies that can be represented by two delta functions.

The capacity of this configuration follows from (11.1.9):

$$\begin{aligned} C(n) &= 2(n-2)A_0 = 2(n-2)\left[\frac{1}{(a+c)^{n-2}} + \frac{1}{(a-c)^{n-2}}\right]^{-1} \\ &= (n-2)(c^2+d^2)^{(n-2)/2}. \end{aligned} \tag{6}$$

Electrostatics

This corresponds to the case $n = 3$. In this case (11.1.6) and (4)–(6) become, respectively,

$$u(x) = \frac{a^2 - c^2}{2a}\left[\frac{1}{[(x+c)^2+\rho^2]^{1/2}} + \frac{1}{[(x-c)^2+\rho^2]^{1/2}}\right], \tag{7}$$

$$\frac{a}{a^2 - c^2} = \frac{1}{(c^2+d^2)^{1/2}}, \tag{8}$$

$$\frac{2a}{a^2 - c^2} = \frac{1}{[(x+c)^2+\rho^2]^{1/2}} + \frac{1}{[(x-c)^2+\rho^2]^{1/2}}, \tag{9}$$

$$C(3) = (a^2 - c^2)/a = (c^2 + d^2)^{1/2}. \tag{10}$$

Thus, the capacity of a dumbbell in three dimensions is equal to that of a sphere of radius $(c^2 + d^2)^{1/2}$.

Axially symmetric configurations are represented in Figure 11.3 for different values of d/a.

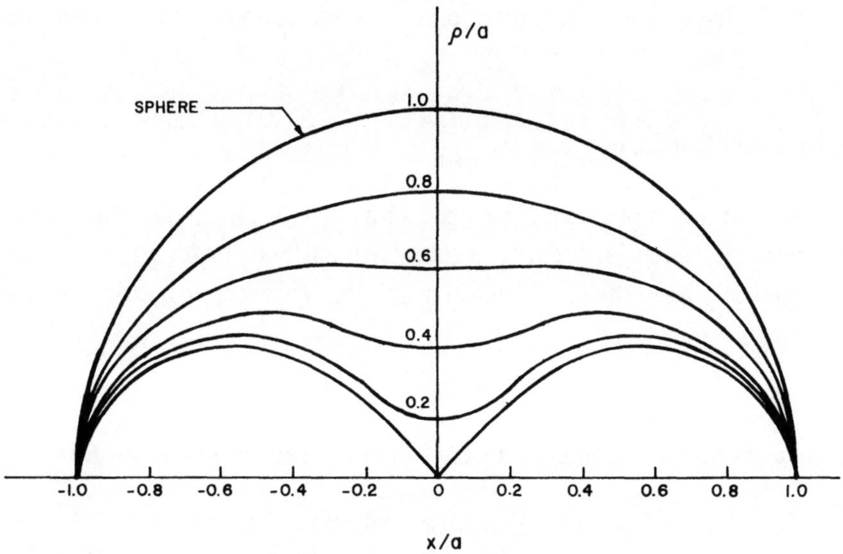

Figure 11.3. A class of axially symmetric bodies generated by a pair of delta functions. The numbers along the ρ/a axis designate values of d/a.

Perfect fluids

The relation

$$\psi(n-2) = U\rho^{n-3}(n-3)^{-1}[1 - u(n)] \tag{11}$$

relates the stream function $\psi(n-2)$ for an $(n-2)$-dimensional body of revolution moving with a uniform speed U along its axis of symmetry in a perfect fluid (or, equivalently, the fluid streaming past the body with speed U) with potential $u(n)$ in a space of n dimensions. This potential function assumes the value unity on the surface of the body and vanishes at infinity. Thus, for the flow of a perfect fluid in three dimensions, we need the value of $u(5)$.

Now, from relations (11.1.6), (11.2.3), and (11.2.4), for $n = 5$ it follows that

$$u(5) = \frac{(a^2 - c^2)^3}{a(a^2 + 3c^2)} \left[\frac{1}{R_1^3} + \frac{1}{R_2^3} \right], \tag{12}$$

$$C(5) = \frac{6(a^2 - c^2)^3}{a(a^2 + 3c^2)} = 3(c^2 + d)^{3/2}. \tag{13}$$

Then (11) yields

$$\psi(3) = \frac{1}{2} U \rho^2 \left[1 - \frac{(a^2 - c^2)^3}{a(a^2 + 3c^2)} \left\{ \frac{1}{R_1^3} + \frac{1}{R_2^3} \right\} \right]. \tag{14}$$

This gives the value of the stream function for a wide class of dumbbell-shaped bodies.

Cavities in elastic medium

The solution of a torsion problem for a shaft with a cavity can be obtained from the solution of the electrostatic potential for a seven-dimensional body, with $g(x) = 1$, provided the cavity and the body have the same meridian profile. This isoperimetric equality is

$$\psi(5) = \frac{1}{4} \rho^4 [1 - u(7)], \tag{15}$$

where ψ now stands for the elastic stream function. The strain energy E is

$$E = (2G/\pi)[(2\pi^2/15)C(7) - V(5)], \tag{16}$$

where $C(7)$ is the capacity, $V(5)$ is the volume of the body, and G is the shear modulus.

The present method is suitable for solving some of these problems in a very simple fashion. For instance, for a spherical cavity we take $f(x) = a^5 \delta(x)$. Then (11.1.6) and (11.1.9) yield

$$u(7) = a^5/(x^2 + \rho^2)^{5/2}, \qquad C(7) = 5a^5.$$

Substituting these values in (15) and (16) and using (3.3.3) for $V(n)$ for $n = 5$, we obtain

$$\psi(5) = \frac{1}{4} \rho^4 [1 - a^5/(x^2 + \rho^2)^{5/2}], \qquad E = \frac{4}{15} G \pi a^5. \tag{17}$$

Similarly, for a dumbbell-shaped cavity,

$$u(7) = \frac{(a^2 - c^2)^5}{2a(a^4 + 10a^2c^2 + 5c^4)} \left[\frac{1}{R_1^5} + \frac{1}{R_2^5} \right],$$

$$C(7) = \frac{5(a^2 - c^2)^5}{a(a^4 + 10a^2c^2 + 5c^4)},$$

$$V(5) = 2\pi^2 \int_{-a}^{a} \int_{0}^{\rho_0(x)} \rho^3 \, d\rho \, dx = \frac{1}{2}\pi^2 \int_{-a}^{a} \rho_0^4(x) dx .$$

Accordingly,

$$\psi(5) = \frac{1}{4}\rho^4 \left[1 - \frac{(a^2 - c^2)^5}{2a(a^4 + 10a^2c^2 + 5c^4)} \left(\frac{1}{R_1^5} + \frac{1}{R_2^5} \right) \right], \tag{18}$$

and

$$E = \pi G \left[\frac{4(a^2 - c^2)^5}{3a(a^4 + 10a^2c^2 + 5c^4)} = \int_{-a}^{a} \rho_0^4(x) \, dx \right] . \tag{19}$$

11.3. Uniform Axial Distributions

Capacity ($n = 3$)

Let the uniform distribution be

$$f(x) = A , \tag{1}$$

A constant. Then for $g(x) = 1$, (11.1.7) yields

$$\frac{1}{A} = \ln \frac{R_2 - (x - c)}{R_1 - (x + c)}, \tag{2}$$

where R_1 and R_2 are defined in (11.2.2). This equation is equivalent to

$$\frac{(\rho^2 + (x - c)^2)^{1/2} - (x - c)}{(\rho^2 + (x + c)^2)^{1/2} - (x + c)} = \lambda ,$$

where λ is a constant, or

$$\frac{x^2}{[c^2(4\lambda^3 + 8\lambda^2 + 4\lambda)]/4\lambda(1 - \lambda)^2} + \frac{\rho^2}{[c^2(4\lambda^3 + 8\lambda^2 + 4\lambda)]/(1 - \lambda^2)^2} = 1 ,$$

which is the equation of a spheroid. The equation of a prolate spheroid is

$$x^2/a^2 + \rho^2/b^2 = 1 , \qquad b^2 = (1 - e^2)a^2 , \quad c = ae , \tag{3}$$

where e is the eccentricity. Accordingly (2) becomes

$$\frac{1}{A} = \ln \frac{1 + e}{1 - e} = 2\tanh^{-1} e . \tag{4}$$

The value of the potential function $u(x, \rho)$, from (11.1.6) is

$$u(x, \rho) = (2\tanh^{-1} e)^{-1} \ln \frac{R_2 - (x - c)}{R_1 - (x + c)}, \tag{5}$$

and applying (11.1.9), we obtain the value of the capacity

$$C = ae(\tanh^{-1} e)^{-1}. \tag{6}$$

An interesting limiting case is that of a slender body, which is obtained by setting $b/a \ll 1$. In this case (6) reduces to

$$C = \frac{a}{\ln(2a/b)} (1 + O(\varepsilon^2)),$$

where $\varepsilon = b/a = (1 - e^2)^{1/2}$ is called the slenderness parameter. Equation (5) can be put in familiar form by using prolate-spheroidal coordinates ξ, η, ϕ:

$$x = c\xi\eta, \qquad y + iz = c[(\xi^2 - 1)(1 - \eta^2)]^{1/2} e^{i\phi}, \qquad e = 1/\xi_0. \tag{7}$$

Then (5) takes the form

$$u(\xi) = \frac{\ln[(\xi + 1)/(\xi - 1)]}{\ln[(\xi_0 + 1)/(\xi_0 - 1)]} = \frac{Q_0(\xi)}{Q_0(\xi_0)}, \tag{8}$$

where $Q_0(\xi)$ is the Legendre function of the second kind.

The corresponding results for an oblate spheroid can be derived by replacing c by ic and e by $ie(1 - e^2)^{-1/2}$ in (5) and (6) and by replacing ξ by $i\xi$ in (8). The corresponding formulas are

$$C(3) = c/\sin^{-1} e = be/\sin^{-1} e, \tag{9}$$

and

$$u(x, \rho) = Q_0(i\xi)/Q_0(i\xi_0) = \cot^{-1} \xi/\cot^{-1} \xi_0. \tag{10}$$

As $e \to 1$ or $\xi_0 \to 0$, the oblate spheroid reduces to a thin circular disk of radius b and (9) and (10) reduce to

$$C(3) = 2b/\pi, \qquad u(x, \rho) = 2/\pi \cot^{-1} \xi.$$

The results for a sphere follows upon setting $e = 0$.

Perfect fluids (n = 5)

In this case, if we set $f(x) = A$ and $g(x) = 1$ in (11.1.7), we obtain

$$\frac{1}{A} = \frac{1}{\rho^2} \left[\frac{x + c}{R_1} - \frac{x - c}{R_2} \right], \qquad \rho = \rho_0(x), \quad |x| \le a, \tag{11}$$

where R_1 and R_2 are given by (11.2.2). For the semimajor axis ($x = a,\ \rho = 0$) and the semiminor axis ($x = 0,\ \rho = b$), (11) yields the relations

$$2ac(a^2 - c^2)^{-2} = 1/A\,, \qquad 2cb^{-2}(b^2 + c^2)^{1/2} = 1/A\,. \tag{12}$$

Hence

$$(a^2 - c^2)^2 = ab^2(b^2 + c^2)^{1/2}\,, \tag{13}$$

which relates the parameters b/a and c/a. The shape function $\rho = \rho_0(x; a; c)$ is now known from (11)–(13). These bodies are prolate shaped.

The value of the potential $u(5)$ follows from relations (11.1.6)

$$u(5) = \frac{A}{\rho^2}\left[\frac{x + c}{R_1} - \frac{x - c}{R_2}\right]. \tag{14}$$

Next, we use (11.2.11) to obtain the stream function for the motion of these configurations in a perfect fluid.

$$\psi(3) = \frac{1}{2}U\rho^2\left[1 - \frac{A}{\rho^2}\left(\frac{x + c}{R_1} - \frac{x - c}{R_2}\right)\right], \tag{15}$$

where A is given by (12). The capacity $C(5)$ is

$$C(5) = 6Ac = \frac{3(a^2 - c^2)^2}{a} = 3b^2(b^2 + c^2)^{1/2}\,. \tag{16}$$

Two limiting cases are of special interest.

(1) $c \to 0$, $A \to \infty$ such that $cA \to \frac{1}{2}a^3$. In this limit we recover the results for a sphere.

(2) $b/a \ll 1$ leads to an elongated rod. In this case (13)–(16) reduce to

$$\frac{c}{a} = -\frac{b}{2a} - \frac{7}{6}\left(\frac{b}{a}\right)^3 + O\left(\frac{b}{a}\right)^4, \tag{17}$$

$$u(5) = \frac{b^2}{\rho^2}\left(1 + \frac{b^2}{c^2}\right)^{1/2}\left[1 - \frac{\rho^2}{2}\,\frac{x^2 + x^2}{(c^2 - x^2)^2} + O\left(\frac{c\rho}{c^2 - x^2}\right)^4\right], \tag{18}$$

$$C(5) = 3ab^2\left[1 - \frac{b}{2a} + \frac{1}{2}\left(\frac{b}{a}\right)^2 + O\left(\frac{b}{a}\right)^3\right], \tag{19}$$

$$\psi(3) = \frac{1}{2}U\rho^2\left[1 - \frac{b^2}{\rho^2}\left(1 + \frac{b^2}{c^2}\right)^{1/2}\left\{1 - \frac{\rho^2}{2}\,\frac{x^2 + c^2}{(x^2 - x^2)^2} + O\left(\frac{c\rho}{c^2 - x^2}\right)^4\right\}\right], \tag{20}$$

respectively. The body shape in this limit, from (11), is

$$\frac{\rho_0}{b} = 1 - \left(\frac{b}{c}\right)^2\frac{x^2(3c^2 - x^2)}{4(c^2 - x^2)^4} + O\left(\frac{b}{c}\right)^4, \qquad (c^2 - x^2) \gg b^2\,. \tag{21}$$

Cavities in an elastic medium (n = 7)

The relation corresponding to (2) and (11) in this case is

$$\frac{1}{A} = \frac{1}{\rho^4}\left[\frac{x+c}{R_1} - \frac{x-c}{R_2} - \frac{1}{3}\left\{\frac{(x+c)^3}{R_1^3} - \frac{(x-c)^3}{R_2^3}\right\}\right], \tag{22}$$

while those corresponding to (12) are

$$\frac{2ac(a^2+c^2)}{(a^2-c^2)^4} = \frac{1}{A}, \qquad \frac{2c(2c^2+3b^2)}{3b^4(c^2+b^2)^{3/2}} = \frac{1}{A}. \tag{23}$$

Hence

$$\frac{a(a^2+c^2)}{(a^2-c^2)^4} = \frac{2c^2+3b^2}{3b^4(c^2+b^2)^{3/2}}, \tag{24}$$

which relates the parameters b/a and c/a. Furthermore, (22)–(24) provide the body shape $\rho = \rho_0(x; a; c)$.

The quantities $u(7)$ and $C(7)$ are

$$u(7) = \frac{3b^4(c^2+b^2)^{3/2}}{2c(2c^2+3b^2)\rho^4}\left[\frac{x+c}{R_1} - \frac{x-c}{R_2} - \frac{1}{3}\left\{\frac{(x+c)^3}{R_1^3} - \frac{(x-c)^3}{R_2^3}\right\}\right], \tag{25}$$

$$C(7) = 5(a^2-c^2)^4/a(a^2+c^2) = 15b^4(c^2+b^2)^{3/2}/(2c^2+3b^2). \tag{26}$$

We can then appeal to relations (11.2.15) and (11.2.16) to obtain

$$\psi(5) = \frac{1}{4}\rho^4\left[1 - \frac{3b^4(c^2+b^2)^{3/2}}{2c(2c^2+3b^2)\rho^4} \times\left\{\frac{x+c}{R_1} - \frac{x-c}{R_2} - \frac{1}{3}\frac{(x+c)^3}{R_1^3} + \frac{1}{3}\frac{(x-c)^3}{R_2^3}\right\}\right], \tag{27}$$

$$E = \frac{2G}{\pi}\left[\frac{\pi^2 b^4(c^2+b^2)^{3/2}}{2c^2+3b^2} - \frac{\pi^2}{2}\int_{-a}^{a}\rho_0^4(x)dx\right], \tag{28}$$

where we have used $V(5) = \frac{1}{2}\pi^2\int_{-a}^{a}\rho_0^4(x)dx$. As $c \to 0$, $A \to \infty$ such that $cA \to 1/2a^5$, we recover the results for a spherical cavity as given in Section 11.2.

For a cavity with the shape of an elongated rod, that is, for $b/a \ll 1$, we have

$$\frac{c}{a} = 1 - \frac{(3)^{1/4}}{2}\frac{b}{a} - \frac{(3)^{1/2}}{8}\left(\frac{b}{a}\right)^2 + O\left(\frac{b}{a}\right)^3,$$

$$C(7) = \left[\frac{15}{2}\left(\frac{b}{a}\right)^4 + O\left(\frac{b}{a}\right)^5\right]a^5,$$

$$u(7) = \frac{b^4}{\rho^4}\left(1 + \frac{b^2}{c^2}\right)^{3/2}\left(1 + \frac{3}{2}\frac{b^2}{c^2}\right)^{-1}\left[1 + \frac{3}{8}\rho^4\left(\frac{1}{(c+x)^4} + \frac{1}{(c-x)^4}\right)\right.$$
$$\left. + O\left(\frac{c\rho}{c^2 - x^2}\right)^6\right], \qquad (c^2 - x^2) \geq b^2,$$

$$E = \frac{G\pi ab^4}{4}\left[\frac{c}{a} - 8 - 3\left(\frac{b}{c}\right)^4\left(\frac{7}{2} + \frac{2c^4(a^2 + 3c^2)}{3(a^2 - c^2)^3}\right) + O\left(\frac{b}{c}\right)^6\right].$$

A class of configurations is sketched in Figure 11.4 for various values of b/a.

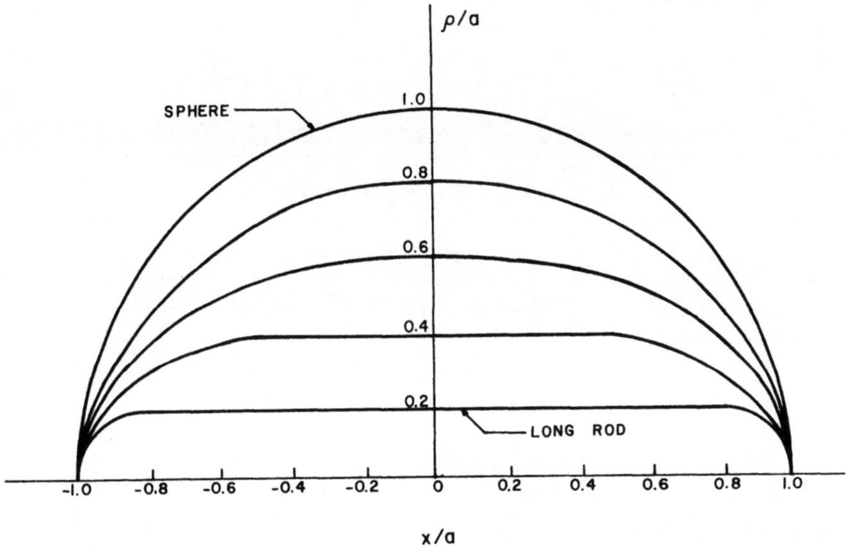

Figure 11.4. A class of axially symmetric cavities of uniform distribution of delta functions. The numbers along the ρ/a axis designate the values of b/a.

11.4. Linear Axial Distributions

In this section we study the polarization potential, so that $g(x) = -E_0 x$, where E_0 is the uniform electrostatic field in which a conductor is lying. Furthermore, we take

$$f(x) = Bx. \qquad (1)$$

Then (11.1.7) yields

$$-E_0 x = B\left[x \ln\left\{\frac{R_2-(x-c)}{R_1-(x+c)}\right\} - R_1 + R_2\right]. \tag{2}$$

Proceeding in the same manner as we did in the analysis following (11.3.2), we find that the right side of (2) is a multiple of x only if the shape of the body is a spheroid. For a prolate spheroid, as defined by (11.3.3), we find from (2) that

$$B = E_0\left[2e - \ln\left(\frac{1+e}{1-e}\right)\right]^{-1}.$$

The potential $u(3)$ is obtained from (11.1.6),

$$u(x,\rho) = \frac{E_0}{2e - \ln[(1+e)/(1-e)]}\left[x \ln\left\{\frac{R_2-(x-c)}{R_1-(x+c)}\right\} - R_1 + R_2\right]. \tag{3}$$

The polarization potential $P(3)$ is

$$\begin{aligned} P(3) &= -\int_S u\frac{\partial u}{\partial n}dS \\ &= -\frac{4}{3}\pi E_0^2 a^3(1-e^2)^{1/2}\left[\ln\left(\frac{1+e}{1-e}\right) - \frac{2e}{1-e^2}\right]\left[\ln\left(\frac{1+e}{1-e}\right) - 2e\right]^{-1}. \end{aligned} \tag{4}$$

In terms of the prolate spheroidal coordinates defined by (11.3.7), (3) and (4) take the forms

$$u(\xi,\eta) = -cE_0\xi\eta Q_1(\xi)/Q_1(\xi_0)$$

and

$$P(3) = -\frac{4}{3}\pi E_0^2 c^3 \xi_0^2(\xi_0^2-1)^{1/2}[Q_1^1(\xi)/Q_1(\xi_0)],$$

respectively.

11.5. Parabolic Axial Distributions, n = 5

Let us now take

$$f(x) = A + Bx^2. \tag{1}$$

Setting $g(x) = 1$ and substituting this value of $f(x)$ in (11.1.7) yields

$$\begin{aligned} &\frac{A}{\rho^2}\left(\frac{x+c}{R_1} - \frac{x-c}{R_2}\right) + B\left\{\frac{x-c}{R_1} - \frac{x+c}{R_2} + \frac{x^2}{\rho^2}\left(\frac{x+c}{R_1} - \frac{x-c}{R_2}\right)\right. \\ &\qquad \left. + \ln\frac{R_2-(x-c)}{R_1-(x+c)}\right\} = 1, \qquad |x| \le a. \end{aligned} \tag{2}$$

This relation is again valid only for a spheroid. For a prolate spheroid, as defined by (11.3.3), we find that

$$B = -\frac{A}{c^2}, \qquad A\left[\frac{2e}{1-e^2} - \ln\frac{1+e}{1-e}\right] = a^2e^2, \tag{3}$$

which determines the values of both A and B. Appealing to (11.1.6), we obtain

$$u(x,\rho) = \frac{A}{c^2\rho^2}\left[(c-x)R_1 + (c+x)R_2 + \rho^2 \ln\left\{\frac{R_1-(x+c)}{R_2-(x-c)}\right\}\right]. \tag{4}$$

This in turn enables us to use (11.2.11) to obtain the stream function for perfect fluids,

$$\psi(3) = \frac{1}{2}U\rho^2\left[1 - \frac{A}{c^2\rho^2}\left((c-x)R_1 + (c+x)R_2 + \rho^2\ln\left\{\frac{R_1-(x+c)}{R_2-(x-c)}\right\}\right)\right]. \tag{5}$$

In prolate spheroidal coordinates (11.3.7), (4) and (5) become

$$u(x,\rho) = \frac{A}{c^2}\left[\frac{2\xi}{\xi^2-1} - \ln\frac{\xi+1}{\xi-1}\right], \tag{6}$$

and

$$\psi(3) = \frac{1}{2}Uc^2(\xi^2-1)(1-\eta^2) + U\,A(\xi^2-1)(1-\eta^2)\left[\frac{1}{2}\ln\frac{\xi+1}{\xi-1} - \frac{\xi}{\xi^2-1}\right], \tag{7}$$

respectively. Furthermore, from (1) and (11.1.9) we obtain

$$C(5) = \frac{4a^3e^3}{2e/(1-e^2) - \ln[(1+e)/(1-e)]}. \tag{8}$$

The values of these quantities for an oblate spheroid can be obtained by using the limiting process, as explained for the $n = 3$ case in Section 11.3. They are

$$\begin{aligned} u(x,\rho) &= \frac{2Ai}{c^2}\left[\frac{\xi}{\xi^2+1} - \cot^{-1}\xi\right] \\ \psi(3) &= \frac{1}{2}Uc^2(1+\xi^2)(1-\eta^2) - iUA(1+\xi^2)(1-\eta^2) \\ &\quad \times [-\cot^{-1}\xi + \xi/(1+\xi^2)], \\ C(t) &= -2b^3e^3[e(1-e^2)^{1/2} - \sin^{-1}e]^{-1}, \end{aligned}$$

where in this case $A = \frac{1}{2}b^2e^2i[e(1-e^2)^{1/2} - \sin^{-1}e]^{-1}$. As $e \to 1$ we derive the corresponding results for a thin circular disk of radius b,

$$\begin{aligned} u(x,\rho) &= (2/\pi)[\xi/(\xi^2+1) - \cot^{-1}\xi], \\ \psi(3) &= \frac{1}{2}Ub^2(1+\xi^2)(1-\eta^2) - (Ub^2/\pi)(1+\xi^2)(1-\eta^2) \\ &\quad \times [-\cot^{-1}\xi + \xi/(1+\xi^2)], \\ C(5) &= 4b^3/\pi\,. \end{aligned}$$

Similarly if we set $e = 0$, we obtain the corresponding formulas for a sphere.

11.6. The Fourth-Order Polynomial Distribution, n = 7; Spheroidal Cavities

Let us take

$$f(x) = A_0 + A_2 x^2 + A_4 x^4\,, \qquad |x| \le c\,. \tag{1}$$

With this value of $f(x)$ and $g(x) = 1$, (11.1.7) yields

$$A_0 B_{5,0} + A_2 B_{5,2} + A_4 B_{5,4} = 1\,, \qquad |x| \le a\,, \tag{2}$$

where

$$B_{m,n}(x, \rho_0(x); c) = \int_{-c}^{c} \frac{\xi^n d\xi}{[(x-\xi)^2 + \rho^2]^{m/2}}\,, \qquad n = 0, 1, 2, \ldots, \quad m = -1, 1, 3, 5, \ldots\,. \tag{3}$$

These satisfy the recurrence relations

$$B_{m,n} = -\frac{c^{n-1}}{m-2}\left(\frac{1}{R_2^{m-2}} + \frac{(-1)^n}{R_1^{M-2}}\right) + \frac{n-1}{m-2} B_{m-2,n-2} + x B_{m,n-1}\,, \quad n \ge 2\,, \tag{4}$$

$$\partial B_{m,n}/\partial x = -m[x\, B_{m+2,n} - B_{m+2,n+1}]\,, \tag{5}$$

and

$$\partial B_{m,n}/\partial x = -m\rho B_{m+2,n}\,. \tag{6}$$

Thus

$$B_{5,0} = \frac{1}{\rho^4}\left[\frac{x+c}{R_1} - \frac{x-c}{R_2} - \frac{1}{3}\frac{(x+c)^3}{R_1^3} + \frac{1}{3}\frac{(x-c)^3}{R_2^3}\right], \tag{7a}$$

$$B_{5,2} = -\frac{c}{3}\left(\frac{1}{R_2^3} + \frac{1}{R_1^3}\right) + \frac{x}{3}\left(\frac{1}{R_1^3} - \frac{1}{R_2^3}\right) + \frac{1}{3} B_{3,0} + x^2 B_{5,0}\,, \tag{7b}$$

$$B_{5,4} = \frac{c^3}{3}\left(\frac{1}{R_2^3} + \frac{1}{R_1^3}\right) + \frac{c^2 x}{3}\left(\frac{1}{R_1^3} - \frac{1}{R_2^3}\right) + B_{3,2} + \frac{2x}{3} B_{3,1} + x^2 B_{5,2}\,, \tag{7c}$$

$$B_{3,0} = \frac{1}{\rho^2}\left(\frac{x+c}{R_1} - \frac{x-c}{R_2}\right), \tag{7d}$$

$$B_{3,1} = \left(\frac{1}{R_1} - \frac{1}{R_2}\right) + x B_{3,0}\,, \tag{7e}$$

$$B_{3,2} = -c\left(\frac{1}{R_1} + \frac{1}{R_2}\right) + B_{1,0} + x B_{3,1}\,, \tag{7f}$$

$$B_{1,0} = \ln \frac{R_2 - (x-c)}{R_1 - (x+c)}\,. \tag{7g}$$

By evaluating these quantities on the surface of the prolate spheroid (11.3.3), and with a slight algebraic manipulation, it follows that

$$c^4 B_{5,0} - 2c^2 B_{5,2} + B_{5,4} = \frac{4e^3}{3(1-e^2)^2} - \frac{2e}{1-e^2} + \ln\frac{1+e}{1-e} . \tag{8}$$

Comparing (2) and (8) we obtain

$$A_0 = A_4 c^4, \quad A_2 = -2c^2 A_4 , \quad A_4 = \left[\frac{4e^3}{3(1-e^2)^2} - \frac{2e}{1-e^2} + \ln\frac{1+e}{1-e}\right] . \tag{9}$$

Next, we make use of (11.1.6) and (11.1.9) and obtain

$$u(7) = A_4[c^4 B_{5,0} - 2c^2 B_{5,2} + B_{5,4}] , \tag{10}$$

$$C(7) = \frac{16}{3}\, a^5 e^5 A_4 . \tag{11}$$

We substitute these values in (11.2.15) and (11.2.16) and derive

$$\psi(5) = \frac{1}{4}\rho^4[1 - A_4(c^{14} B_{5,0} - 2c^2 B_{5,2} + B_{5,4})] ,$$
$$E = (16G\pi a^5/45)[4e^5 A_4 - 3(1-e^2)^2] ,$$

where we have used

$$V(5) = (8\pi^2 a^5/15)(1-e^2)^2 .$$

The corresponding relations for an oblate spheroid are

$$C(7) = -\frac{8}{3} b^5 e^5 [(2e^3 + e)(1-e^2)^{1/2} - \sin^{-1} e]^{-1}$$
$$E = -(16\pi G/45)[8b^5 e^5 \{(2e^3 + e)(1-e^2)^{1/2} - \sin^{-1} e\}^{-1} + 3b^5(1-e^2)^{1/2}] .$$

When $e \to 1$, these relations reduce to their corresponding values for a penny-shaped crack of radius b,

$$C(7) = 16b^5/3\pi , \qquad E = \frac{64}{45}\, Gb^5 .$$

11.7. The Polarization Tensor for a Spheroid

The polarization potentials u_i, $i = 1, 2, 3$, created by uniform electric fields in the x_i ($x = x_1$, $y = x_2$, $z = x_3$) directions, respectively, are obtained by solving the boundary value problems

$$\nabla^2 u_i = -4\pi F_i(x) ,$$
$$u_i = -x_i \quad \text{on} \quad S , \qquad u_i = O(1/r^2) \quad \text{as} \quad r \to \infty , \tag{1}$$

where $r^2 = x^2 + y^2 + x^2$ and S is the surface of the spheroid, as defined in (11.3.3). The boundary value problem (1) for $i = 1$ can be solved by taking the line distribution of delta functions along the axis of symmetry (x-axis), that is, $F_1(x) = f_1(x)\delta(y)\delta(z)$. Then, following the analysis of Section 11.1, we find that

$$u_1(x, \rho) = \int_{-c}^{c} \frac{f_1(\xi) d\xi}{[(x-\xi)^2 + \rho^2]^{1/2}} . \tag{2}$$

Letting (x, ρ) approach S and using the boundary condition, we have

$$-x = \int_{-c}^{c} \frac{f_1(\xi) d\xi}{[(x-\xi)^2 + \rho^2]^{1/2}}, \qquad (x, \rho) \in S, \tag{3}$$

which is a Fredholm integral equation of the first kind. This equation has been solved in Section 11.4. The solution is

$$f_1(x) = Bx, \qquad B = \left[2e - \ln \frac{1+e}{1-e}\right]^{-1} . \tag{4}$$

Then (2) becomes

$$u_1(x, \rho) = \left[2e - \ln \frac{1+e}{1-e}\right]^{-1} \left[x \ln \left\{\frac{R_2 - (x-c)}{R_1 - (x+c)}\right\} - R_1 + R_2\right], \tag{5}$$

where R_1 and R_2 are defined by (11.2.2). In prolate spheroidal coordinates, (5) takes the form

$$u_1(\xi, \eta) = -c\xi_0 \eta Q_1(\xi)/Q_1(\xi_0) . \tag{6}$$

Solutions of the boundary value problems (1) for $i = 2, 3$ are obtained by taking the line distributions of dipoles, that is, $F_2(x) = f_2(x)\delta'(y)\delta(z)$ and $F_3(x) = f_2(x)\delta(y)\delta'(z)$ along the axis of symmetry. The values of $u_i(x, \rho)$, $i = 2, 3$, follow by appealing to (11.1.11). Indeed,

$$u_i(x, \rho) = -x_i \int_{-c}^{c} \frac{f_i(\xi)\, d\xi}{[(x-\xi)^2 + \rho^2]^{3/2}}, \qquad i = 2, 3 . \tag{7}$$

In view of the boundary conditions, we find that (7) reduces to the Fredholm integral equation

$$1 = \int_{-c}^{c} \frac{f_i(\xi)\, d\xi}{[(x-\xi)^2 + \rho^2]^{3/2}}, \qquad i = 2, 3 . \tag{8}$$

The solution of (8) is achieved by taking

$$f(x) = A + Bx^2, \qquad |x| \leq c . \tag{9}$$

Then (8) yields

$$\begin{aligned} 1 = &\frac{A}{\rho^2}\left(\frac{x+c}{R_1} - \frac{x-c}{R_2}\right) + B\left[\left(\frac{x-c}{R_1} - \frac{x+c}{R_2}\right)\right. \\ &\left. + \frac{x^2}{\rho^2}\left(\frac{x+c}{R_1} - \frac{x-c}{R_2}\right) + \ln\left\{\frac{R_2 - (x-c)}{R_1 - (x+c)}\right\}\right], \qquad |x| \leq a . \end{aligned} \tag{10}$$

This relation is valid for the prolate spheroid (11.3.3) if

$$B = -\frac{A}{c^2}, \qquad A\left[\frac{2e}{1-e^2} - \ln\frac{1+e}{1-e}\right] = a^2e^2. \tag{11}$$

Combining (7), (9) and (11) we obtain the values of u_i, $i = 2, 3$,

$$u_i(x, \rho) = -\frac{x_i A}{c^2\rho^2}\left[(c-x)R_1 + (c+x)R_2 + \rho^2 \ln\left\{\frac{R_1 - (x+c)}{R_2 - (x-c)}\right\}\right]. \tag{12}$$

In terms of prolate spheroidal coordinates, these relations take the forms

$$u_2(\xi, \eta, \varphi) = -c(\cos\varphi)[(\xi_0^2 - 1)(1 - \eta^2)]^{1/2} Q_1^1(\xi)/Q_1^1(\xi_0), \tag{13}$$

and

$$u_3(\xi, \eta, \varphi) = -c(\sin\varphi)[(\xi_0^2 - 1)(1 - \eta^2)]^{1/2} Q_1^1(\xi)/Q_1^1(\xi_0). \tag{14}$$

The formula for the components Q_{ij} of the polarization tensor is

$$\begin{aligned} Q_{ij} &= -\int_S u_i \frac{\partial u_j}{\partial n}\, dS \\ &= -c(\xi_0^2 - 1)\int_0^{2\pi}\int_{-1}^{1} u_i \left.\frac{\partial u_i}{\partial \xi}\right|_{\xi=\xi_0} d\eta\, d\varphi, \qquad i, j = 1, 2, 3. \end{aligned} \tag{15}$$

From the symmetry of the spheroid it is clear that

$$Q_{ij} = 0, \qquad i \neq j. \tag{16}$$

Next, we use the values of u_i, $i = 1, 2, 3$, as given by (6), (13), and (14) in (15) and obtain

$$Q_{11} = -(4\pi c^3/3)\xi_0^2(\xi_0^2 - 1)Q_1^1(\xi_0)/Q_1(\xi_0), \tag{17}$$

$$Q_{22} = Q_{33} = -(4\pi c^3/3)[\xi_0(\xi_0^2 - 1) - 2(\xi_0^2 - 1)^{1/2}/Q_1^1(\xi_0)]. \tag{18}$$

In terms of the eccentricity $e = 1/\xi_0$, these take the forms

$$Q_{11} = -\frac{4\pi a^3(1-e^2)}{3}\left[\ln\frac{1+e}{1-e} - \frac{2e}{1-e^2}\right]\left[\ln\frac{1+e}{1-e} - 2e\right]^{-1}, \tag{19}$$

$$\begin{aligned} Q_{22} = Q_{33} = &-\frac{4\pi a^3(1-e^2)}{3}\left[2e\frac{2e^3}{1-e^2} - \ln\frac{1+e}{1-e}\right] \\ &\times\left[\frac{2e}{1-e^2} - \ln\frac{1+e}{1-e}\right]^{-1}. \end{aligned} \tag{20}$$

Replacing e by $ie(1-e^2)^{-1/2}$ in (19) and (20), we derive the corresponding values for an oblate spheroid;

$$\begin{aligned} Q_{11} = &-\frac{4\pi b^3(1-e^2)}{3}[\sin^{-1} e - e(1-e^2)^{1/2}] \\ &\times [(1-e^2)^{1/2}\sin^{-1} e - e]^{-1}, \end{aligned} \tag{21}$$

$$Q_{22} = Q_{33} = -\frac{4\pi b^3}{3}[e + e^3 - (1-e^2)^{1/2}]\sin^{-1} e \times [e(1-e^2)^{1/2} - \sin^{-1} e]^{-1} . \tag{22}$$

As $e \to 1$, the oblate spheroid reduces to a thin circular disk of radius b, and (21) and (22) become

$$Q_{11} = 0 , \qquad Q_{22} = Q_{33} = 16b^3/3 , \tag{23}$$

respectively. The limiting case of a sphere follows by letting $e \to 0$. We then have

$$Q_{11} = Q_{22} = Q_{33} = 8\pi a^3/3 , \tag{24}$$

as is well known.

The limiting case of a slender body is obtained by taking $b/a = (1-e^2)^{1/2} \ll 1$. Then (19) and (20) reduce, respectively, to

$$Q_{11} = -\frac{4\pi ab^2}{3}\left[\ln\frac{2a}{b} - \frac{a^2}{b^2}\right]\left[\ln\frac{2a}{b} - 1\right]^{-1}\left[1 + O\left(\frac{b}{a}\right)^2\right] , \tag{25}$$

$$Q_{22} = Q_{33} = -\frac{4\pi ab^2}{3}\left[1 - \frac{a^2}{b^2} - \ln\frac{2a}{b}\right]\left[\ln\frac{2a}{b} + \frac{1}{2} - \frac{a^2}{b^2}\right]^{-1}\left[1 + O\left(\frac{a}{b}\right)^2\right] . \tag{26}$$

11.8. The Virtual Mass Tensor for a Spheroid

In this section we consider the potentials $u_i(x)$, $i = 1, 2, 3$, that satisfy the boundary value problems

$$\nabla^2 u_i = -4\pi F_i(x) , \qquad i = 1, 2, 3, \tag{1}$$

$$\widehat{n} \cdot \nabla u_i = -\widehat{n} \cdot e_x \quad \text{on} \quad S , \qquad u_i = O(1/r^2) \quad \text{at} \quad \infty , \tag{2}$$

where $\widehat{n}$ is the unit vector normal to the surface S and the e_x, e_y, e_z, are the unit vectors in the x, y, z, directions, respectively. Then, the u_i, $i = 1, 2, 3$, are the velocity potentials of a body with surface S and moving with unit velocity in a perfect fluid in the x, y, z directions, respectively. For a prolate spheroid, $\widehat{n}$ is

$$\widehat{n} = (a^2 - e^2x^2)^{-1/2}\left(\frac{b}{a}xe_x + \frac{a}{b}ye_y + \frac{a}{b}ze_z\right) . \tag{3}$$

We solve these boundary value problems by the same distribution of singularities as in Section 11.7. That is, we set $F_1(x) = f_1(x)\delta(y)\delta(z)$, $F_2(x) = f_2(x)\delta'(y)\delta(z)$, and $F_3(x) = f_2(x)\delta(y)\delta'(z)$.

The integral representation formulas for u_1, u_2, u_3 are given by (11.7.2) and (11.7.7). Applying the boundary conditions (2) to these formulas, we obtain the integral equations

$$x = \int_{-c}^{c} \frac{(a^2 - x\xi) f_1(\xi) d\xi}{[(x-\xi)^2 + \rho^2]^{3/2}}, \tag{4}$$

$$-\frac{1}{3(1-e^2)} = \int_{-c}^{c} f_i(\xi) \left[\frac{(a^2 - x\xi)}{[(x-\xi)^2 + \rho^2]^{5/2}} - \frac{1}{3(1-e^2)} \frac{1}{[(x-\xi)^2 + \rho^2]^{3/2}} \right] d\xi, \qquad i = 2, 3. \tag{5}$$

The solution of (4) is obtained by setting $f_1(x) = Ax$, $|x| \leq c$, so that

$$x = A[a^2 B_{3,1} - x B_{3,2}],$$

where the quantities $B_{m,n}$ are defined by (11.6.3). On the surface of the spheroid (11.3.3) this relation becomes

$$x = A\left[\frac{2e}{1-e^2} - \ln\frac{1+e}{1-e}\right] x,$$

which yields the value of A. Thus,

$$f_1(x) = \left[\frac{2e}{1-e^2} - \ln\frac{1+e}{1-e}\right]^{-1} x. \tag{6}$$

Equations (5) are solved by setting

$$f_i(x) = A_0 - A_2 x^2, \qquad i = 2, 3, \quad |x| \leq c. \tag{7}$$

Substituting (7) in (5), we obtain, after a slight algebraic manipulation,

$$A_0 = A_2 c^2, \qquad A_2 = \left[\frac{2e(1-2e^2)}{1-e^2} - \ln\frac{1+e}{1-e}\right]^{-1}.$$

Hence,

$$f_2(x) = f_3(x) = \left[\frac{2e(1-2e^2)}{1-e^2} - \ln\frac{1+e}{1-e}\right]^{-1} (c^2 - x^2). \tag{8}$$

Next, we substitute these values of $f_i(x)$ in the integral representation formulas (11.7.2) and (11.7.7) and obtain

$$u_1(x, \rho) = \left[\frac{2e}{1-e^2} - \ln\frac{1+e}{1-e}\right]^{-1} B_{1,1}, \tag{9}$$

$$u_2(x, \rho) = -\left[\frac{2e(1-2e^2)}{1-e^2} - \ln\frac{1+e}{1-e}\right]^{-1} y(c^2 B_{3,0} - B_{3,2}), \tag{10}$$

$$u_3(x, \rho) = -\left[\frac{2e(1-2e^2)}{1-e^2} - \ln\frac{1+e}{1-e}\right]^{-1} z(c^2 B_{3,0} - B_{3,2}). \tag{11}$$

The virtual mass tensor W_{ij} is defined as

$$W_{ij} = -\int_S u_i \frac{\partial u_j}{\partial n}\, dS\,, i, j = 1, 2, 3\,. \tag{12}$$

Substituting (9)–(11) in (12), we obtain

$$W_{11} = \frac{4\pi a^3}{3}(1-e^2)\left[\ln\frac{1+e}{1-e} - 2e\right]\left[\frac{2e}{1-e^2} - \ln\frac{1+e}{1-e}\right]^{-1}, \tag{13}$$

$$\begin{aligned} W_{22} = W_{33} &= \frac{4\pi a^3}{3}(1-e^2)\left[\ln\frac{2e}{1-e^2} - \ln\frac{1+e}{1-e}\right] \\ &\times \left[\ln\frac{1+e}{1-e} - \frac{2e(1-2e^2)}{1-e^2}\right]^{-1}, \end{aligned} \tag{14}$$

and

$$W_{ij} = 0 \qquad \text{for} \qquad i \neq j\,. \tag{15}$$

By letting $e \to 0$ in these relations, we recover the corresponding values for a sphere of radius a:

$$W_{11} = W_{22} = W_{33} = 2\pi a^3/3\,, \tag{16}$$

which are well known.

For a slender body, these relations reduce to

$$W_{11} = \frac{4\pi ab^2}{3}\left[\ln\frac{2a}{b} - 1\right]\left[\frac{a^2}{b^2} - \frac{1}{2} - \ln\frac{2a}{b}\right]^{-1}\left[1 + O\left(\frac{a}{b}\right)^2\right], \tag{17}$$

$$\begin{aligned} W_{22} = W_{33} &= \frac{4\pi ab^2}{3}\left[\frac{a^2}{b^2} - \frac{1}{2} - \frac{2a}{b}\right] \\ &\times \left[\ln\frac{2a}{b} - +\frac{a^2}{b^2}\right]\left[1 + O\left(\frac{b}{a}\right)^{-2}\right]. \end{aligned} \tag{18}$$

The value of W_{ij} for an oblate spheroid follows from (13)–(15) on replacing e by $ie(1-e^2)^{-1/2}$:

$$W_{11} = (4\pi b^3/3)[(1-e^2)^{1/2}\sin^{-1} e - e][e(1-e^2)^{1/2} - \sin^{-1} e]^{-1}\,, \tag{19}$$

$$\begin{aligned} W_{22} = W_{33} &= (4\pi b^3/3)(1-e^2)[e(1-e^2)^{1/2} - \sin^{-1} e] \\ &\times [(1-e^2)^{1/2}\sin^{-1} e - e - e^3]^{-1}\,, \end{aligned} \tag{20}$$

and $W_{ij} = 0$ for $i \neq j$. In the limiting case of a thin circular disk of radius b, (19) and (20) reduce to

$$W_{11} = \frac{8}{3}b^3\,, \qquad W_{22} = W_{33} = 0\,. \tag{21}$$

11.9. The Electric and Magnetic Polarizability Tensors

The electric and magnetic polarizability tensors P_{ij} and M_{ij} are defined in terms of the tensor Q_{ij} and W_{ij} as follows:

$$P_{ij} = Q_{ij} + V I_{ij}\,, \tag{1}$$
$$M_{ij} = W_{ij} + V I_{ij}\,, \tag{2}$$

where I_{ij} is the identity tensor and V is the volume of the body. Accordingly, we can use the results of Sections 11.7 and 11.8 to find the values of P_{ij} and M_{ij}.

For a prolate spheroid,

$$P_{11} = \frac{8\pi a^3 e^3}{3}\left[\ln\frac{1+e}{1-e} - 2e\right]^{-1}, \tag{3}$$
$$P_{22} = P_{33} = \frac{16\pi a^3 e^3}{3}\left[\frac{2e}{1-e^2} - \ln\frac{1+e}{1-e}\right]^{-1}, \tag{4}$$
$$M_{11} = \frac{8\pi a^3 e^3}{3}\left[\frac{2e}{1-e^2} - \ln\frac{1+e}{1-e}\right]^{-1}, \tag{5}$$
$$M_{22} = M_{33} = \frac{16\pi a^3 e^3}{3}\left[\ln\frac{1+e}{1-e} - \frac{2e(1-2e^2)}{1-e^2}\right]^{-1}, \tag{6}$$
$$P_{ij} = M_{ij} = 0\,, \qquad i \neq j\,.$$

For an oblate spheroid,

$$P_{11} = -(4\pi b^3 e^3/3)(1-e^2)^{1/2}[(1-e^2)^{1/2}\sin^{-1} e - e]^{-1}\,, \tag{7}$$
$$P_{22} = P_{33} = -(8\pi b^3 e^3/3)[e(1-e^2)^{1/2} - \sin^{-1} e]^{-1}\,, \tag{8}$$
$$M_{11} = -(4\pi b^3 e^3/3)[e(1-e^2)^{1/2} - \sin^{-1} e]^{-1}\,, \tag{9}$$
$$M_{22} = M_{33} = -(8\pi b^3 e^3/e)(1-e^2)^{1/2}[(1-e^2)^{1/2}\sin^{-1} e - e - e^3]^{-1}\,, \tag{10}$$

$P_{ij} = M_{ij} = 0, i \neq j$.

For a sphere,

$$P_{11} = P_{22} = P_{33} - 4\pi a^3\,, \tag{11}$$
$$M_{11} = M_{22} = M_{33} - 2\pi a^3\,, \tag{12}$$
$$P_{ij} = M_{ij} = 0\,, \qquad i \neq j\,. \tag{13}$$

For a slender body,

$$P_{11} = \frac{4\pi a}{3}(a^2 - b^2)\left(\ln\frac{2a}{b} - 1\right)^{-1}\left[1 + O\left(\frac{b}{a}\right)^2\right], \tag{14}$$

$$P_{22} = P_{33} = \frac{2\pi ab^2}{3}\left(4\ln\frac{2a}{b} - 1\right)\left(\ln\frac{2a}{b} + \frac{1}{2} - \frac{a^2}{b^2}\right)^{-1}\left[1 + O\left(\frac{b}{a}\right)^2\right], \tag{15}$$

$$M_{11} = \frac{2\pi a(a^2 - 3b^2)}{3}\left(\frac{a^2}{b^2} - \frac{1}{2} - \ln\frac{2a}{b}\right)^{-1}\left[1 + O\left(\frac{b}{a}\right)^2\right], \tag{16}$$

$$M_{22} = M_{33} = \frac{2\pi a}{3}(2a^2 - 3b^2)\left(\ln\frac{2a}{b} - 1 + \frac{a^2}{b^2}\right)^{-1}\left[1 + O\left(\frac{b}{a}\right)^2\right], \tag{17}$$

$P_{ij} = M_{ij} = 0, i \neq j$.

For a thin circular disk,

$$P_{11} = 0, \qquad P_{22} = P_{33} = \frac{16}{3}b^3, \tag{18}$$

$$M_{11} = \frac{8}{3}b^3, \qquad M_{22} = M_{33} = 0, \tag{19}$$

$P_{ij} = M_{ij} = 0, i \neq j$.

The values of component M_{11} of the magnetic polarizability tensor and component P_{22} of the electric polarizability tensor for various dumbbell- and prolate-shaped profiles can be found using the isoperimetric equalities

$$M_{11}(3) + V(3) = (2\pi/3)C(5), \tag{20}$$

$$P_{22}(3) + V(3) = 2[M_{11}(3) + V(3)], \tag{21}$$

where $C(5)$ for these profiles is given in Sections 11.2 and 11.3. Indeed, using (11.2.6) with $n = 5$ and (11.3.16), we derive the following results:

For dumbbell-shaped bodies, defined by (11.2.5),

$$M_{11}(3) = \pi\left[2(c^2 + d^2)^{3/2} - \int_{-a}^{a} \rho_0^2(x)\,dx\right], \tag{22}$$

$$P_{22}(3) = \pi\left[4(c^2 + d^2)^{3/2} - \int_{-a}^{a} \rho_0^2(x)\,dx\right], \tag{23}$$

where $\rho = \rho_0(x)$, is the equation of the profile and $V(3) = \pi\int_{-a}^{a}\rho_0^2(x)\,dx$.

Finally, for prolate-shaped bodies, defined by (11.3.11),

$$M_{11}(3) = \pi\left[2b^2(b^2 + c^2)^{1/2} - \int_{-a}^{a} \rho_0^2(x)\,dx\right], \tag{24}$$

$$P_{22}(3) = \pi\left[4b^2(b^2 + c^2)^{1/2} - \int_{-a}^{a} \rho_0^2(x)\,dx\right]. \tag{25}$$

11.10. The Distributional Approach to Scattering Theory

In this section we derive an approximation for the scattered field by distributing delta functions along the axis of symmetry. We discuss only acoustic scattering and start with the inhomogeneous Helmholtz equation

$$(\nabla^2 + k^2)u^s(x) = -4\pi F(x)\,. \tag{1}$$

For the special case $F(x) = \delta(x)$, (1) has the solution (see Section 10.9)

$$u^s(x) = e^{ik|x|}/|x|\,. \tag{2}$$

Let the scatterer S be defined as

$$\rho = \rho_0(x)\,, \qquad -a \le x \le a\,, \qquad \rho_0(\pm a) = 0\,. \tag{3}$$

Then for $F(x) = f(x)\delta(y)\delta(z)$ we find that

$$u^s(x) = \int_{-c}^{c} \frac{e^{ikR}}{R} f(\xi)\,d\xi\,, \tag{4}$$

where $R = [(x-\xi)^2 + \rho^2]^{1/2}$ and $\rho^2 = y^2 + z^2$. We can now discuss the scattering of an acoustic wave by a soft spheroid (the Dirichlet problem) and that by a rigid spheroid (the Neumann problem).

The Dirichlet problem

In the case of a soft spheroid the boundary condition is

$$u^s(x) = -u^i(x) = -e^{ikx} \qquad \text{on} \qquad S\,, \tag{5}$$

where we have taken the incidence to be along the axis of symmetry. From (4) and (5) we derive

$$\int_{-c}^{c} \frac{e^{ikR}}{R} f(\xi)d\xi = -e^{ikx}\,, \qquad (x,\rho) \in S\,. \tag{6}$$

Now for small values of k we expand the functions e^{ikx}, e^{ikR}/R, and $f(x)$ as follows:

$$e^{ikx} = \sum_{n=0}^{\infty} \frac{(ik)^n}{n!} x^n\,, \tag{7}$$

$$\frac{ikR}{R} = \sum_{n=0}^{\infty} \frac{(ik)^n}{n!} R^{n-1}\,, \tag{8}$$

and

$$f(x) = -\sum_{n=0}^{\infty} (ik)^n f_n(x)\,. \tag{9}$$

Substitution of these expressions in (6) yields the integral equations

$$\int_{-c}^{c} \frac{f_0(\xi)}{R}\, d\xi = 1\,, \tag{10}$$

$$\int_{-c}^{c} \frac{f_1(\xi)}{R}\, d\xi = x - \int_{-c}^{c} f_0(\xi)\, d\xi\,, \tag{11}$$

$$\int_{-c}^{c} \frac{f_2(\xi)}{R}\, d\xi = \frac{1}{2}x^2 - \frac{1}{2}\int_{-c}^{c} f_0(\xi) R\, d\xi - \int_{-c}^{c} f_1(\xi)\, d\xi\,, \tag{12}$$

$$\int_{-c}^{c} \frac{f_3(\xi)}{R}\, d\xi = \frac{1}{6}x^3 - \frac{1}{6}\int_{-c}^{c} f_0(\xi) R^2\, d\xi - \frac{1}{2}\int_{-c}^{c} f_1(\xi) R\, d\xi - \int_{-c}^{c} f_2(\xi)\, d\xi\,, \tag{13}$$

$$\begin{aligned}\int_{-c}^{c} \frac{f_4(\xi)}{R}\, d\xi &= \frac{1}{24}x^4 - \frac{1}{24}\int_{-c}^{c} f_0(\xi) R^3 d\xi - \frac{1}{6}\int_{-c}^{c} f_1(\xi) R^2\, d\xi \\ &\quad - \frac{1}{2}\int_{-c}^{c} f_2(\xi) R\, d\xi - \int_{-c}^{c} f_3(\xi)\, d\xi\,, \end{aligned} \tag{14}$$

$$\vdots$$

where $(x, \rho) \in S$. Proceeding as in earlier sections, we find that the solutions of equations (10)–(14) are

$$f_0(x) = [\ln\{(1+e)/(1-e)\}]^{-1} = \alpha^{-1}\,, \tag{15}$$

$$f_1(x) = -2ae\alpha^{-2} + (\alpha_2 e)^{-1} x\,, \tag{16}$$

$$f_2(x) = \frac{2a^2 e}{\alpha^2}\left[\frac{2e}{\alpha} - \frac{\alpha - 2e}{(3-e^2)\alpha - 6e}\right] + \frac{\alpha(3-e^2) - 2e}{2\alpha[(3-e^2)\alpha - 6e]}\, x^2\,, \tag{17}$$

$$\begin{aligned} f_3(x) &= \frac{8a^3 e^2}{\alpha^2}\left[-\frac{e}{\alpha^2} + \frac{1}{3\alpha} - \frac{e}{18}\right] \\ &\quad - \frac{2a^2 e^3[-6e + 2e^3 + \alpha(3-2e^2)]}{(\alpha - 2e)^2[-30e + 8e^3 + 3(5-3e^2)\alpha]}\, x \\ &\quad - \frac{ae}{3\alpha^2}\, x^2 + \frac{[-10e + 4e^3 + (5-3e^2)\alpha]}{2(\alpha - 2e)[-30e + 8e^3 + 3(5-3e^2)\alpha]}\, x^3\,, \end{aligned} \tag{18}$$

$$f_4(x) = D_0 + D_1 x + D_2 x^2 + D_4 x^4\,, \tag{19}$$

where

$$\alpha = \ln[(1+e)/(1-e)]\,, \tag{20}$$

$$
\begin{aligned}
D_0 = & -\frac{a^4}{\alpha}\left[\frac{2e^3}{\alpha^2}\left\{\frac{-8e}{\alpha^2}+\frac{8}{3\alpha}-\frac{5}{9}e\right\}+\frac{e}{2\alpha^2}\{2e+(1-e^2)\alpha\}\right. \\
& \times\left\{\frac{2e}{\alpha}-\frac{\alpha-2e}{(3-e^2)\alpha-6e}\right\} \\
& +\frac{\{\alpha(3-e^2)-2e\}\{2e+2e^3-(1-e^2)^2\alpha\}}{32\alpha\{(3-e^2)\alpha-6e\}}-\frac{2e^2(3-2e^3)}{9\alpha^2} \\
& +\frac{1}{192\alpha}\{10e-6e^3+3(1-e^2)^2\alpha\} \\
& \left.+\frac{2e-(1-e^2)\alpha}{2a^2}D_2+\frac{1}{8}\{-6e+10e^3+3(1-e^2)^2\alpha\}D_4\right], \qquad (21a)
\end{aligned}
$$

$$
D_1 = 2a^3e^3/9(\alpha-2e)^2, \qquad (21b)
$$

$$
\begin{aligned}
D_2 = & -2a^2[(3-e^2)\alpha-6e]^{-1} \\
& \times\left[\frac{e\{2e-(1-e^2)\alpha\}\{2e\alpha(4-e^2)-12e^2-\alpha^2\}}{2\alpha^3\{(3-e^2)\alpha-6e\}}\right. \\
& +\frac{(5-6e^2+e^4)\alpha-2e+6e^3}{32\alpha}-\frac{2e^4}{3\alpha^2} \\
& \left.-\frac{1}{4}\{-30e+26e^3+3(5-e^2)(1-e^2)\alpha\}D_4\right], \qquad (21c)
\end{aligned}
$$

$$
\begin{aligned}
D_4 = & (1/8a)[6e\alpha(42-36e^2+63^4)-(3-e^2)(35--30e^2+3e^4)\alpha^2 \\
& -84e^2+44e^4][210e-110e^3-3(35-30e^3+3e^4)\alpha]^{-2} \\
& \times[(3-e^2)\alpha-6e]^{-1}. \qquad (21d)
\end{aligned}
$$

Thus, we have obtained the value of $f(x)$ up to $O(k^4)$.

To find the far-field behavior of the scattered field $u(x)$ we appeal to (4). In terms of spherical polar coordinates, $x = r\cos\vartheta$, $y = r\sin\vartheta$, we have, for $r \gg 1$,

$$
R = \sqrt{(x-\xi)^2+\rho^2} \simeq r - \xi\cos\vartheta, \qquad (22a)
$$

$$
1/R = 1\sqrt{(x-\xi)^2+\rho^2} \simeq 1/r, \qquad (22b)
$$

and

$$
e^{ikR}\ R \simeq (e^{ikr}\ r)e^{-ik\xi\cos\vartheta}. \qquad (22c)
$$

Substituting this value in (4) we obtain

$$
u^s(r,\vartheta) \simeq (e^{ikr}\ r)j_1(\vartheta), \qquad (23)
$$

where $j_1(\vartheta)$ is the far-field amplitude, given by

$$
j_1(\vartheta) = \int_{-c}^{c} e^{-ik\xi\cos\vartheta} f(\xi)d\xi. \qquad (24)
$$

Next, we substitute in (24) the value of $f(x)$ that we have obtained previously. Then equation (23) yields the far-field value of the scattered fields;

$$u^s \simeq -\frac{e^{ikr}}{r}\left[\frac{2ae}{\alpha} - ik\left(\frac{4a^2e^2}{\alpha^2}\right) - k^2\left\{\frac{8a^3e^3}{\alpha^3} - \frac{4a^3e^2}{3\alpha^2}\right.\right.$$
$$\left. + \frac{a^3e^3}{3\alpha} - \frac{2a^3e^3}{3(\alpha-2e)}\cos\vartheta + \frac{a^3e^3}{3\alpha}\cos^2\vartheta\right\}$$
$$- ik^3\left\{\frac{2a^4e^3}{\alpha^2}\left(-\frac{8e}{\alpha^2} + \frac{8}{3\alpha} - \frac{5}{9}e\right) - \frac{2a^4e^4}{3\alpha^2}\cos^2\vartheta\right\}$$
$$+ k^4\left\{2aeD_0 + \frac{2a^3e^3}{3}D_2 + \frac{2a^5e^5}{5}D_4\right.$$
$$\left. - \frac{2a^5e^5[30e^2 - 14e^4 - 5(25-14e^2)e\alpha + 3(5-3e^2)\alpha^2]}{15(\alpha-2e)^2[-30e + 8e^3 + 3(5-3e^2)\alpha]}\right\}\cos\vartheta$$
$$+ \left\{\frac{4a^5e^6}{3\alpha^3} + \frac{40e - 2\alpha(10+3e^2) + 3e\alpha^2(3-e^2)}{30\alpha^2[(3-e^2)\alpha - 6e]}a^5e^5\right\}\cos^2\vartheta$$
$$\left. - \frac{a^5e^5}{15(\ln(1+e)/(1-e) - 2e)}\cos^3\vartheta + \frac{a^5e^5}{60\alpha}\cos^4\vartheta\right]. \tag{25}$$

The value of the scattering cross-section σ_1 is given by

$$\sigma_1 = 2\pi\int_0^\pi |j_1(\vartheta)|^2 \sin\,\vartheta\, d\vartheta$$
$$\simeq 16\pi a^2\left[\left(\frac{e}{\alpha}\right)^2 + 4\left(\frac{kae}{\alpha}\right)^2\left\{-\left(\frac{e}{\alpha}\right)^2 + \frac{1}{3}\left(\frac{e}{\alpha}\right) - \frac{1}{9}e^2\right\}\right.$$
$$+ (ka)^4\left\{16\left(\frac{e}{\alpha}\right)^6 + \frac{4}{9}\left(\frac{e}{\alpha}\right)^4 + \frac{4}{3}e^2\left(\frac{e}{\alpha}\right)^4 + \frac{197e^4}{2700}\left(\frac{e}{\alpha}\right)^2 + \frac{1}{27}\frac{e^6}{(\alpha-2e)^2}\right.$$
$$- \frac{16}{3}\left(\frac{e}{\alpha}\right)^5 - \frac{8}{27}e^2\left(\frac{e}{\alpha}\right)^3 + \frac{2e}{a^4}\left(\frac{e}{\alpha}\right)\left(D_0 + \frac{a^2e^2D_2}{3} + \frac{a^4e^4}{5}D_4\right)$$
$$+ \frac{e^3}{3}\left(\frac{e}{\alpha}\right)\left[\frac{4}{3}\left(\frac{e}{\alpha}\right)^3 + \frac{40e^2 - 2e\alpha(10+3e^2) + 3e^2\alpha^2(3-e^2)}{30\alpha^2((3-e^2)\alpha - 6e)}\right]$$
$$\left.\left. + \frac{4}{3}\left(\frac{e}{\alpha}\right)^4\left[-8\left(\frac{e}{\alpha}\right)^2 + \frac{8}{3}\left(\frac{e}{\alpha}\right) - \frac{5e^2}{9}\right]\right\}\right]. \tag{26}$$

Neumann problem for a prolate spheroid

In this case the boundary condition is

$$\widehat{n}\cdot\nabla u^s(x) = -\widehat{n}\cdot\nabla u^i = -\widehat{n}\cdot\nabla e^{ikx}, \tag{27}$$

where

$$\widehat{n} = (a^2 - e^2x^2)^{1/2}\left(\frac{b}{a}xe_x + \frac{a}{b}ye_y + \frac{a}{b}ze_z\right). \tag{28}$$

Applying (27) to (4), we obtain

$$ikxe^{ikx} = \int_{-c}^{c} \frac{ikR}{R^3}(1 - ikR)(a^2 - x\xi)f(\xi)d\xi\,, \qquad (x, \rho) \in S\,. \tag{29}$$

Next we use the same expansions as in the case of Dirichlet problem except that now

$$f(x) = \sum_{n=0}^{\infty} f_n(x)(ik)^{n+1}\,. \tag{30}$$

Proceeding as in the previous case, we obtain the following system of simple integral equations:

$$\int_{-c}^{c} \frac{f_0(\xi)(a^2 - x\xi)}{R^3}\,d\xi = x\,, \tag{31}$$

$$\int_{-c}^{c} \frac{f_1(\xi)(a^2 - x\xi)}{R^3}\,d\xi = x^2\,, \tag{32}$$

$$\int_{-c}^{c} \frac{f_2(\xi)(a^2 - x\xi)}{R^3}\,d\xi = \frac{x^3}{2} + \frac{1}{2}\int_{-c}^{c} \frac{f_0(\xi)(a^2 - x\xi)}{R^3}\,d\xi\,, \tag{33}$$

$$\begin{aligned}\int_{-c}^{c} \frac{f_3(\xi)(a^2 - x\xi)}{R^3}\,d\xi =& \frac{1}{6}x^4 + \frac{1}{2}\int_{-c}^{c} \frac{f_0(\xi)(a^2 - x\xi)}{R^3}\,d\xi \\ &+ \frac{1}{3}\int_{-c}^{c} f_0(\xi)(a^2 - x\xi)d\xi\,,\end{aligned} \tag{34}$$

where $(x, \rho) \in S$.

The values of $f_i(x)$, $i = 0, 1, 2, 3$, are derived by a method similar to that of earlier sections. The result is

$$f_0(x) = \left[\frac{2e}{1 - e^2} - \alpha\right]^{-1} x\,, \tag{35}$$

$$\begin{aligned}f_1(x) =& -\frac{a^2}{2e}(\alpha - 2e)[2e + 2e(3 - 2e^2) - 3(1 - e^2)\alpha]^{-1}(1 - e^2)^2 \\ &+ \left[\frac{2e}{1 - e^2} + 4e - 3\alpha\right]^{-1} x^2\,,\end{aligned} \tag{36}$$

$$\begin{aligned}f_2(x) =& \frac{a^2[-6e + (3 - e^2)\alpha][22e + 9e^2\alpha - 22e/(1 - e^2)]}{2[4e(1 - e^2) + 26e - 3(5 - e^2)\alpha][2e/(1 - e^2) - \alpha]^2}\,x \\ &- \frac{(5 - e^2)\alpha - 6e - 4e/(1 - e^2)}{2(2e/(1 - e^2) - \alpha)[4e/(1 - e^2) + 26e - 3(5 - e^2)\alpha]}\,x^3\,,\end{aligned} \tag{37}$$

and

$$f_3(x) = G_0 + G_1 x + G_2 x^2 + G_4 x^4\,, \tag{38}$$

where

$$
\begin{aligned}
G_0 =& -\frac{1-e^2}{2e}a^2\left\{\frac{a^2}{2}\left(\frac{1-e^2}{2e}\alpha^2-\frac{1-e^2}{2}\alpha-e\right)\left(\frac{2e}{1-e^2}+4e-3\alpha\right)^{-1}\right.\\
&\left.+(\alpha-2e)G_2+\frac{1}{2}a^2[6e-4e^3-3(1-e^2)\alpha^2]G_4\right\}\\
G_1 =& -\frac{2a^3e^3}{9}\left(\frac{2e}{1-e^2}-\alpha\right)^{-2},\\
G_2 =& \left(\frac{2e}{1-e^2}+4e-3\alpha\right)^{-1}a^2\left\{\frac{1}{2}\left[\frac{\alpha^2(1-e^2)}{2e}+\alpha-4e\right]\right.\\
&\left.\times\left[\frac{2e}{1-e^2}+4e-3\alpha\right]^{-1}-[8e^3-30e+3(5-e^2)\alpha]G_4\right\},\\
G_4 =& -2\left[5(7-3e^2)\alpha-66e+\frac{32}{3}e^3-\frac{4e}{1-e^2}\right]^{-1}\\
&\times\left\{\frac{1}{6}-\frac{1}{4}[-10e+(5-3e^2)\alpha+\frac{8}{3}e^3]\left\{\frac{2e}{1-e^2}+4e-3\alpha\right\}^{-1}\right\},
\end{aligned}
$$

and α is defined by (20).

Putting this value of $f(x)$ in (4) and using (22)–(24) with j_1 replaced by j_2, we obtain the far-field behavior

$$
\begin{aligned}
u^s(x) \simeq& -\frac{e^{ikr}}{r}k^2\left[\left[\left\{\frac{a^3(1-e^2)}{3}-\frac{2a^3e^3}{3}\left(\frac{2e}{1-e^2}-\alpha\right)^{-1}\cos\vartheta\right\}\right.\right.\\
&-k^2\left\{2aeG_0+\frac{2}{3}a^3e^3G_2+\frac{2}{5}a^5e^5G_4\right.\\
&-\frac{a^5e^5}{30}\left(\frac{2e}{1-e^2}-\alpha\right)^{-2}\left[\frac{4e}{1-e^2}+26e-3(5-e^2)\alpha\right]^{-1}\\
&\times\left[\frac{1392e^2}{1-e^2}+\frac{48e^2}{(1-e^2)^2}-e\alpha\left\{808+\frac{512}{1-e^2}\right\}\right.\\
&\left.+6(50-16e^2)\alpha^2\right]\cos\vartheta-\frac{a^5e^2}{30}\left[\frac{2e}{1-e^2}+4e-3\alpha\right]^{-1}\\
&\times[5(1-e^2)\alpha-10e+4e^3]\cos^2\vartheta\\
&\left.\left.\left.-\frac{1}{15}a^5e^5\left(\frac{2e}{1-e^2}-\alpha\right)^{-1}\cos^3\vartheta\right\}\right]\right]. \qquad (39)
\end{aligned}
$$

The scattering cross-section σ_2 is given by

$$\sigma_2 \simeq \frac{4\pi}{3}(ka)^4 a^2 \left\{ \frac{(1-e^2)^2}{3} + \frac{4}{9}e^6 \left(\frac{2e}{1-e^2} - \alpha\right)^{-2} \right.$$
$$- k^2 a^2 \left[\frac{2e(1-e^2)}{a^4} \left(G_0 + \frac{a^2 e^2}{3} G_2 + \frac{a^4 e^4}{5} G_4 \right) \right.$$
$$- \frac{e^2(1-e^2)}{90} [5(1-e^2)\alpha - 10e + 4e^3] \left[\frac{2e}{1-e^2} + 4e - 3\alpha \right]^{-1}$$
$$- \frac{e^8}{45} \left(\frac{2e}{1-e^2} - \alpha \right)^{-3} \left[\frac{4e}{1-e^2} + 26e - 3(5-e^2)\alpha \right]^{-1}$$
$$\times \left[\frac{1392e^2}{1-e^2} + \frac{48e^2}{(1-e^2)^2} - e\alpha \left(808 + \frac{512}{1-e^2} \right) \right.$$
$$\left.\left. \left. + \, 6(50 - 16e^2)\alpha^2 \right] + \frac{2}{75} e^8 \left(\frac{2e}{1-e^2} - \alpha \right)^{-2} \right] \right\} . \tag{40}$$

11.11. Stokes Flow

Let us now study the flow of a viscous fluid in which a solitary body or a large number of bodies of microscopic scale are moving. They may be carried about passively by the flow, like solid particles in sedimentation, or moving actively, as in the locomotion of microorganisms. The linear differential equations (the inertia forces have a negligible effect on the flow) that describe their motion are the Stokes equations

$$\nabla \cdot u(x) = 0 , \tag{1}$$

$$\nabla p(x) = \mu \nabla^2 u(x) + F(x) , \tag{2}$$

where u is the velocity vector, p is the pressure, μ is the constant viscosity, and F is the external forcing function. We shall study only rotary motion. For this purpose we assume F to be solenoidal, that is, $\nabla \cdot F = 0$, so that

$$F(x) = 4\pi\mu \nabla \times \Omega(x) , \tag{3}$$

where Ω is the vector potential. The far-field behavior of u and p is

$$u \to 0 , \qquad p \to p_\infty , \text{ a constant,} \qquad \text{as} \quad r = |x| \to \infty . \tag{4}$$

Taking the divergence of both sides of (2), we find that

$$\nabla^2 p = \mu \nabla^2 (\nabla \cdot u) + \nabla \cdot F = 0 ,$$

which means that p is a harmonic function having a constant value p_∞ at infinity. Accordingly, $p = p_\infty$ in the entire space, and the differential equation (2) reduces to

$$\nabla^2 u = -(1/\mu)F(x) = -4\pi \nabla \times \Omega(x) . \tag{5}$$

The fundamental solution of (5) is obtained by taking $\Omega(x) = \gamma\delta(x)$ where γ is a constant vector. It is readily verified that

$$u = \nabla \times \frac{\gamma}{r} = \frac{1}{r}\nabla \times \gamma - \frac{1}{r^3}(x \times \gamma) = \frac{\gamma \times x}{r^3} \tag{6}$$

is then the solution of (5). This solution is called a *rotlet* or *couplet*. Physically, this solution provides the velocity generated by a sphere of radius a rotating about the γ axis with angular velocity $\omega_0 = |\gamma|/a^3$.

For a volume distribution $\Omega(x)$ that is a superposition of delta functions the solution is

$$u = \nabla \times A , \qquad A = \int \frac{\Omega(x')}{|x - x'|}\, d^3x' . \tag{7}$$

Following the method of this chapter, let us take an axial distribution of rotlets of the form

$$\Omega = e_x f(x)\delta(y)\delta(z), \qquad (-c_1 \le x \le c - 2) , \tag{8}$$

where $f(x)$ is the line distribution and e_x is the unit vector along the x axis. Thus

$$\begin{aligned} A &= \int_V \frac{e_x f(x')\delta(y')\delta(z')}{[|x - x'|^2 + |y - y'|^2 + |z - z'|^2]^{1/2}}\, dx'dy'dz' \\ &= \int_{-c_1}^{c_2} \frac{e_x f(x')\, dx'}{[|x - x'|^2 + y^2 + z^2]^{1/2}} . \end{aligned} \tag{9}$$

For a body with fore-and-aft symmetry, this equation yields

$$A = e_x \int_{-c}^{c} \frac{f(\xi)d\xi}{[(x - \xi)^2 + \rho^2]^{1/2}} , \tag{10}$$

where $\rho^2 = y^2 + z^2$. Thus

$$u = \nabla \times A = (e_x y - e_y z) \int_{-c}^{c} \frac{f(\xi)d\xi}{[(x - \xi)^2 + \rho^2]^{3/2}} . \tag{11}$$

In terms of cylindrical polar coordinates (ρ, ϑ, z), (11) is equivalent to $u = (0, 0, u_\vartheta)$, and

$$u_\vartheta = \rho \int_{-c}^{c} \frac{f(\xi)d\xi}{[(x - \xi)^2 + \rho^2]^{3/2}} . \tag{12}$$

The boundary condition is

$$u_\vartheta(x, \rho) = \omega_0 \rho , \qquad \text{on} \quad \rho = \rho_0(x) , \tag{13}$$

where ω_0 is the angular velocity of the body. From (12) and (13) we get the Fredholm integral equation.

$$\omega_0 = \int_{-c}^{c} \frac{f(\xi)d\xi}{[(x-\xi)^2+\rho^2]^{3/2}}\,, \qquad \rho = \rho_0(x)\,. \tag{14}$$

This equation is a special case of (11.1.7). Accordingly, we can obtain the solution of the present problem by taking $f(\xi)$ given in Sections 11.2–11.6. For various examples in fluid mechanics solved by this method, the reader is referred to the work of Chwang andWu [51].

The analysis for the rectilinear motion of bodies in Stokes flow can be obtained as an appropriate limit of their displacement field in elastostatics as given in the next section.

11.12. Displacement-Type Boundary Value Problems in Elastostatistics

For a homogeneous, linearly elastic, and isotropic medium, the displacement equations of equilibrium are [48]

$$\mu\left[\frac{1}{1-2\nu}\nabla\,\nabla\cdot u + \nabla^2 u\right] + f = 0\,, \tag{1}$$

where $u = (u_1, u_2, u_3)$ denotes the displacement field, μ the shear modulus, ν the Poisson ratio of the material, and f the body force per unit of volume in a three-dimensional Euclidean space. In Exercise 13 of Chapter 10 we found that by taking f to be the Kelvin force

$$f^k = 16\pi\mu(1-\nu)\delta(x)\alpha\,, \tag{2}$$

where x is the position vector of the field point and α is a constant vector that characterizes the direction and magnitude of the force, we can obtain the fundamental solution

$$U^k(x;\alpha) = \frac{(3-4\nu)\alpha}{r} + \frac{(\alpha\cdot x)x}{r}\,, \qquad r = |x|\,. \tag{3}$$

This solution expresses the displacement field corresponding to a concentrated force located at the origin of the coordinates. Let us note in passing that

$$U^k(x;\alpha) = O(1/r) \qquad \text{as} \quad r\to\infty \text{ or } r\to 0\,, \tag{4a}$$

and, for any constant A,

$$U^k(x; A\alpha) = AU^k(x;\alpha)\,. \tag{4b}$$

The net force F experienced by the medium bounded by a control surface S enclosing the singular point is given by

$$F = \int_S \widehat{n}\cdot T\, dS = \int_V \nabla\cdot T\, dV = \int_V -f^k DV = -16\pi\mu(1-\nu)\alpha\,, \tag{5}$$

where $\widehat{n}$ is the unit outward normal at S, T is the stress tensor, and V is the control volume enclosed by S. The net moment M is obviously zero.

In the absence of boundaries, a derivative of (3) of any order in an arbitrary fixed direction is also a solution of (1), the corresponding forcing function being the derivative of f^k of the same order and in the same arbitrary direction. These derivatives are readily obtained by considering the displacement field corresponding to a concentrated force located at a point $\beta \neq 0$. The formal multipole expansion (similar to that in classical potential theory) in a Taylor series about x yields

$$U^k(x-\beta;\alpha) = U^k(x;\alpha) - (\beta\cdot\nabla)U^k(x;\alpha) + \frac{1}{2}(\beta\cdot\nabla)^2U^k(x;\alpha) + \cdots . \tag{6}$$

The interpretation of various terms on the right of (6) is obvious: The first term has already been explained in (3), the second term represents a Kelvin dipole characterized by the vectors α and β, the third term represents a Kelvin quadrupole, etc. We write a Kelvin dipole

$$U^{kd}(x;\alpha,\beta) = -(\beta\cdot\nabla)U^k(x;\alpha) = \frac{(3-4v)(\beta\cdot x)\alpha - (\alpha\cdot x)\beta - (\alpha\cdot\beta)x}{r^3} + \frac{3(\alpha\cdot x)(\beta\cdot x)x}{r^5}, \tag{7a}$$

the corresponding forcing function being

$$f^{kd} = -16\pi\mu(1-\nu)(\beta\cdot\nabla)[\delta(x)\alpha] = -16\pi\mu(1-\nu)[\beta\cdot\nabla\delta(x)]\alpha , \tag{7b}$$

where the superscript kd stands for Kelvin dipole.

For fixed x, a Kelvin dipole (7a) is, in general, not symmetric with respect to an interchange of α and β. It is more convenient to handle a Kelvin dipole in terms of its symmetric and antisymmetric parts. The antisymmetric part is itself a physical entity of special significance;

$$\frac{1}{2}[U^{kd}(x;\alpha,\beta) - U^{kd}(x;\beta,\alpha)] = -2(1-\nu)(\alpha\times\beta)\times x/r^3 . \tag{8a}$$

This singularity is called a *center of rotation*, and we denote it U^r. Next, set

$$\gamma = -2(1-\nu)(\alpha\times\beta) . \tag{8b}$$

Then (8a), in view of (8b), can be rewritten as

$$U^r(x;\gamma) = \gamma\times x/r^3 = \nabla\times(\gamma/r) = \frac{1}{2}\nabla\times U^k(x;\beta\times\alpha) , \tag{8c}$$

where the superscript r indicates the center of rotation. The corresponding forcing function is

$$f^r = \frac{1}{2}[f^{kd}(x;\alpha,\beta) - f^{kd}(x;\beta,\alpha)] = 4\pi\mu\nabla\times[\delta(x)\gamma] . \tag{8d}$$

The displacement field (8c) has a vector potential γ/r and no scalar potential. The net force F and the torque M experienced by a control volume V containing the singular point and bounded by a surface S are given by

$$F = \int_S \widehat{n} \cdot T \, dS = \int_V \nabla \cdot T \, dV = \int_V -f^r \, dV = 0 , \tag{9a}$$

and

$$\begin{aligned} M &= \int_S x \times (\widehat{n} \cdot T) dS = -\int_V (\nabla \cdot T) \times x \, dV \\ &= \int_V (f^r \times x) dV = -8\pi\gamma , \end{aligned} \tag{9b}$$

respectively.

The symmetric part of a Kelvin dipole yields another fundamental singularity, called a *stresslet*. Its displacement field and the forcing function are

$$\begin{aligned} U^{ks}(x; \alpha, \beta) &= \frac{1}{2}[U^{kd}(x; \alpha, \beta) + U^{kd}(x; \beta, \alpha)] \\ &= \frac{(1-2\nu)[(\beta \cdot x)\alpha + (\alpha \cdot x)\beta] - (\alpha \cdot \beta)x}{r^3} \\ &\quad + \frac{3(\alpha \cdot x)(\beta \cdot x)x}{r^5} , \end{aligned} \tag{10a}$$

and

$$f^{ks} = -8\pi\mu(1-\nu)[\beta \cdot \delta(x)\alpha + \alpha \cdot \nabla\delta(x)\beta] , \tag{10b}$$

respectively. Physically, a stresslet represents a self-equilibrating system and hence contributes no net force or moment to the medium. It is of the nature of two double forces with moments in the plane determined by the vectors α and β, the moment of one double force being equal in magnitude but opposite in direction to that of the other double force.

Yet another type of singularity is useful in the present investigation. It is called a *center of dilation*, and it is characterized by a scalar. In this case the displacement field and the forcing function are

$$U^d(x) = -\nabla(1/r) = x/r^3 , \tag{11}$$

and

$$f^d = -\frac{8\pi\mu(1-\nu)}{1-2v} \nabla\delta(x) , \tag{12}$$

respectively. Thus, the displacement field is derivable from a scalar potential $1/r$ and, like a stresslet, is a self-equilibrating singularity.

A dipole formed by two centers of dilation yields

$$U^{dd}(x; \alpha) = -\nabla(\alpha \cdot x/r^3) = -(\alpha \cdot \nabla)(-1/r) = -\alpha/r^3 + 3(\alpha \cdot x)x/r^5 , \tag{13}$$

whereas a quadrupole formed by two dipoles or by four centers of dilation leads to

$$\begin{aligned} U^{d4}(x; \alpha, \beta) &= -(\beta \cdot \nabla)U^{dd}(x; \alpha) = (\beta \cdot \nabla)(\alpha \cdot \nabla)U^d(x) = (\alpha \cdot \nabla)(\beta \cdot \nabla)U^d(x) \\ &= U^{d4}(x; \beta, \alpha) = -\nabla(\beta \cdot \nabla)(\alpha \cdot \nabla)(1/r) . \end{aligned} \tag{14a}$$

This shows that the quadrupole U^{d4} is symmetric with respect to an interchange of α and β. In particular,

$$U^{d4}(x; e_x, e_y) = -\nabla \frac{\partial^2}{\partial x \partial y}\left(\frac{1}{r}\right) = -\nabla\left(\frac{3xy}{r^3}\right) . \tag{14b}$$

These singular solutions are useful in constructing solutions to several boundary value problems involving different modes of displacement of rigid prolate and oblate spheroids, including their limiting configurations — the sphere, slender body, and thin circular disk. We now proceed to demonstrate this by an example. Various other examples are given in reference [48].

Translation of a prolate spheroid

Consider a rigid prolate spheroid, as defined by (11.3.3), embedded in an unbounded elastic medium. It is given the displacement

$$U = U_x e_x + U_y e_y . \tag{15}$$

There is no loss of generality in assuming the displacement to be in the form of (15) because the y and z axes can be so chosen that the z component of the specified displacement is zero.

The displacement field in the elastic medium must
(a) satisfy the homogeneous equations of equilibrium (1) (with $f = 0$),
(b) vanish at infinity, and
(c) meet the boundary condition (15) on the surface S of the spheroid. We attempt to construct the appropriate solution by employing a line distribution of Kelvin solutions (3) and dipoles formed by centers of dilation (13) between the foci $x = -c$ and $x = c$; i.e.,

$$\begin{aligned} u(x) &= \int_{-c}^{c} [\alpha_1 U^k(x-\xi; e_x) + \alpha_2 U^k(x-\xi; e_y)]\, d\xi \\ &\quad + \int_{-c}^{c} (c^2 - \xi^2)[\beta_1 U^{dd}(x-\xi; e_x) + \beta_2 U^{dd}(x-\xi; e_y)]\, d\xi . \end{aligned} \tag{16}$$

where $\xi = \xi e_x$. The first integral in (16) represents a line distribution of Kelvin solutions of constant strength α_1 and α_2 oriented in the x and y directions, respectively. The second integral represents a line distribution of dipoles formed by centers of dilation, each of parabolic density and pointing in the x and y directions.

It is thus obvious that (16) satisfies the homogeneous (with $f = 0$) equations (1) of equilibrium in the elastic medium and vanishes as $|x| \to \infty$. To satisfy boundary condition

(2) on the surface S of the spheroid, we use the integrated form of (16), namely,

$$u(x) = \left[\left\{-\frac{x+c}{R_1} + \frac{x-c}{R_2} + 4(1-\nu)B_{1,0}\right\} e_x - \left(\frac{1}{R_1} - \frac{1}{R_2}\right)\rho e_\rho\right]\alpha_1$$
$$- \left[y\left(\frac{1}{R_1} - \frac{1}{R_2}\right) e_x - (3-4\nu)B_{1,0}e_y - yB_{3,0}\rho e_\rho\right]\alpha_2$$
$$+ \nabla\left[-2B_{1,1} + \beta_2 y\left\{B_{1,0} + \frac{x-c}{\rho^2}R_1 - \frac{x-c}{\rho^2}R_2\right\}\right], \tag{17}$$

where $e_\rho = (ye_y + ze_z)/\rho$ is the unit radial vector in the y, z plane, R_1 and R_2 are defined in (11.2.2), and the quantities $B_{m,n}$ are given by (11.6.3). The particular values $B_{1,0}$, $B_{3,0}$, $B_{3,1}$, and $B_{5,0}$ have already been evaluated in (11.6.7). The remaining particular values needed are

$$B_{1,1} = R_2 - R_1 + xB_{1,0}, \qquad B_{5,1} = \frac{1}{3}(1/R_1^3 - 1/R_2^3) + xB_{5,0}. \tag{18}$$

Next we evaluate these quantities on the surface of the spheroid S. We find from (17) that the displacement on S is

$$u|_S = 2\left[\left\{-\frac{1}{e} + 2(1-\nu)L\right\}\alpha_1 - L\beta_1\right]e_x + 2\left(\frac{2e^2\beta_1}{1-e^2} + \alpha_1\right)\frac{b^2e_x + e^2x\rho e_\rho}{e(a^2 - e^2x^2)}$$
$$+ \left[\left(L - \frac{2e}{1-e^2}\right)\beta_2 + (3-4\nu)L\alpha_2\right]e_y$$
$$+ 2e\left(\frac{2e^2\beta_2}{1-e^2} + \alpha_2\right)\frac{b^2xye_\rho + a^2y\rho e_\rho}{b^2(a^2 - e^2x^2)}, \tag{19}$$

where

$$L = \ln\frac{1+e}{1-e} = 2\tanh^{-1}e.$$

Accordingly, boundary condition (15) on the surface S is satisfied if we choose

$$\alpha_1 = -2e^2\beta_1/(1-e^2) = U_xe^2[-2e + 1 + 3e^2 - 4\nu e^2)L]^{-1}, \tag{20a}$$
$$\alpha_2 = -2e^2\beta_2/(1-e^2) = 2U_ye^2[2e - (1 - 7e^2 + 8\nu e^2)L]^{-1}. \tag{20b}$$

This choice of α_1, β_1, α_2, and β_2 in (17) furnishes the required solution.

Next, we compute the net force F experienced by the spheroid by superposition of (5),

$$F = -16\pi\mu(1-\nu)\int_{-c}^{c}(\alpha_1 e_x + \alpha_2 e_y)d\xi = F_xe_x + F_ye_y, \tag{21a}$$

where the longitudinal and transverse components of the force are

$$F_x = -32\pi\mu(1-\nu)U_xae^3[-2e + (1 + 3e^2 - 4\nu e^2)L]^{-1}, \tag{21b}$$
$$F_y = -64\pi\mu(1-\nu)U_yae^3[2e - (1 - +7e^2 + 8\nu e^2)L]^{-1}, \tag{21c}$$

respectively.

By the limiting processes $\nu \to \frac{1}{2}$ and $2\mu\nu\nabla \cdot u(1-2\nu) \to -p$ (where p is the pressure), the elastostatic equations (1) reduce to the equations of motion for Stokes flow as given in Section 11.11. In this case u denotes the velocity field. By setting $\nu = \frac{1}{2}$ in (21b) and (21c), we obtain the corresponding drag formulas when a rigid spheroid moves in a viscous fluid with velocity U.

In the limiting case of a sphere ($b \to a$. or $e \to 0$), the displacement field (17) and the net force (21) reduce, respectively, to

$$u(x) = \frac{3a}{10-12\nu}\left[\frac{(3-4\nu)U}{r} + \frac{(U \cdot x)x}{r^3}\right] + \frac{a^3}{10-12\nu}\nabla\left(\frac{U \cdot x}{r^3}\right). \tag{22}$$

and

$$F_x = -24\pi\mu(1-\nu)aU_x(5-6\nu), \tag{23a}$$

$$F_y = -24\pi\mu(1-\nu)aU_y(5-6\nu). \tag{23b}$$

These results can be summarized as follows: An embedded rigid sphere of radius a is given a displacement U. The field in the elastic medium is the same as would be produced by
(i) a concentrated force $-24\pi\mu(1-\nu)aU\ (5-6\nu)$ and
(ii) a dipole of strength $-a^3U(10-12\nu)$ formed by centers of dilation, both located at the center of the sphere, namely,

$$u(x) = U^k\left(x; \frac{-3aU}{10-12\nu}\right) + U^{dd}\left(x; \frac{-a^3U}{10-12\nu}\right). \tag{24}$$

For the other extreme case of a very slender spheroid,

$$F_x = -\frac{16\pi\mu(1-\nu)aU_x}{4(1-\nu)\ln(2/\varepsilon)-1}[1+O(\varepsilon)] \quad \text{and}, \tag{25}$$

$$F_y = -\frac{32\pi\mu(1-\nu)aU_y}{(6-8\nu)\ln(2/\varepsilon)+1}[1+O(\varepsilon)]. \tag{26}$$

When the elastic medium is under torsion, the displacement field is quadratic even without any inclusion. The problem is therefore rather complicated when a rigid inclusion is embedded in it. However, by this method this problem can be solved effectively, as explained in reference [52].

11.13. The Extension to Elastodynamics

The concepts in the previous section can be extended to cases of steady state dynamic singularities [53–55]. The singularities can then be applied to obtain the solutions for the rectilinear oscillations of rigid inclusions in an infinite homogeneous isotropic medium.

The dynamical equations of elasticity are

$$(\lambda+\mu)\ \text{grad}\ (\text{div}\ u) + \mu\nabla^2 u - \rho\frac{\partial^2 u}{\partial t^2} + f = 0, \tag{1}$$

where u is the displacement vector, λ and μ are Lamé's constants of the elastic medium, ρ is the density of the medium, and f is the body force per unit volume. Let us assume that $f = f_0 e^{i\omega t}$, $u = u_0 e^{i\omega t}$. Substituting these values in (1) and dropping the zero subscript, we have

$$(\lambda + \mu) \text{ grad } (\text{div } u) + \mu \nabla^2 u + \rho \omega^2 u + f = 0\,, \tag{2}$$

or

$$\mu \left[\frac{1}{1-2\nu} \text{ grad } (\text{div } u) + \nabla^2 u + m^2 u \right] + f = 0\,, \tag{3}$$

where $m^2 = \rho\omega^2/\mu$, and ν is the Poisson ratio.

The fundamental solution U^{dk} corresponds to the force

$$f^{dk}(x) = 4\pi\mu\delta(x)\alpha\,, \tag{4}$$

where α is a constant vector, and the superscript dk stands for dynamical Kelvin, located at the original of the coordinate system as explained in the previous section. Also

$$U^{dk}(x;\alpha) = \left[\frac{e^{-imr}}{r}\,\alpha + \frac{1}{m^2}\,\nabla(\alpha\cdot\nabla)\left(\frac{e^{-imr}}{r} - \frac{e^{-im\tau r}}{r} \right) \right], \tag{5}$$

where $r = |x|$ and $\tau^2 = (1-2\nu)/2(1-\nu)$ and is called the *dynamical Kelvin solution*.

The net dynamical force F experienced by a control surface S enclosing the singular point is given by

$$\begin{aligned} F &= \int_S \widehat{n}\cdot T\,ds = -\int_V \nabla\cdot T\,dV = \int_V f^{dk}dV + m^2\mu \int_V \overline{U}^{dk}(x;\alpha)dV \\ &= 4\pi\mu\alpha + m^2\mu \int_V \overline{U}^{dk}(x,\alpha)\,dV\,, \end{aligned} \tag{6}$$

where $\widehat{n}$ is the unit outward normal to S, T is the stress tensor, and V is the control volume enclosed by S. The net moment, M is clearly zero.

In free space a derivative of (5) of any order in arbitrary direction is also a solution of (3). The corresponding forcing function is the derivative of f^{dk} of the same order in the same direction. These derivatives can be obtained easily by expanding $\overline{U}^{dk}(x,\beta;\alpha)$, where β is a given vector, in a Taylor series about x. This expansion is

$$U^{dk}(x,\beta;\alpha) \simeq U^{dk}(x,\alpha) - (\beta\cdot\nabla)U^{dk}(x;\alpha) + \frac{1}{2}(\beta\cdot\nabla)^2 U^{dk}(x;\alpha) + \cdots\,. \tag{7}$$

The interpretation of various terms on the right-hand side of (7) is as follows: The first term is the dynamical Kelvin solution just discussed, the second term represents the dynamical Kelvin dipole characterized by the vectors α and β, the third term gives the corresponding quadruple, and so on. The dynamical Kelvin dipole can be written

$$\begin{aligned} U^{dkd}(x;\alpha,\beta) &= -(\beta\cdot\nabla)U^{dk}(x;\alpha) \\ &= -\alpha(\beta\cdot\nabla)\frac{e^{imr}}{r} - \frac{1}{m^2}\nabla(\beta\cdot\nabla)(\alpha\cdot\nabla)\left[\frac{e^{-imr}}{r} - \frac{e^{im\tau r}}{r} \right], \end{aligned} \tag{8}$$

while the corresponding forcing function is

$$f^{dkd} = -4\pi\mu(\beta\cdot\nabla)\delta(x)\alpha = -4\pi\mu[\beta\cdot\nabla\delta(x)]\alpha\,, \tag{9}$$

where the superscript *dkd* stands for dynamical Kelvin dipole.

From (8) it follows that a dynamical Kelvin dipole is not symmetric with respect to α and β. Its symmetric and antisymmetric parts yield other important fundamental solutions. The antisymmetric part is

$$\begin{aligned}\frac{1}{2}[U^{dkd}(x;\alpha,\beta) - U^{dkd}(x;\beta,\alpha)] &= \frac{1}{2}[(\beta\cdot\nabla)e^{-imr}/r - \beta(\alpha\cdot\nabla)e^{-imr}/r]\\ &= -\nabla\times\frac{1}{2}(\alpha\times\beta)e^{-imr}/r\,.\end{aligned}$$

We shall call it the *dynamical center of rotation* and denote it U^{dr}. If we set $\gamma = -\frac{1}{2}(\alpha\times\beta)$, the foregoing relation becomes

$$U^{dr}(x,\gamma) = (\nabla\times\gamma)e^{-imr}/r\,. \tag{10}$$

The corresponding forcing function is

$$f^{dr} = \frac{1}{2}[f^{dkd}(x;\alpha,\beta) - f^{dkd}(x,\beta,\alpha)] = 4\pi\mu\nabla\times[\delta(x)\gamma]\,. \tag{11}$$

Since U^{dr} has only a vector potential $\gamma e^{-imr}/r$ and no scalar potential, the net force F vanishes, while the torque, M, experienced by the control volume, V, containing the singular point and bounded by S is

$$\begin{aligned}M &= \int_S x\times(\widehat{n}\cdot t)dS = -\int_V(\nabla\cdot T)\times x\,dV\\ &= \int_V f^{dr}\times x\,dV + m^2\mu\int_V U^{dr}(x;\alpha,\beta)\times x\,dV\\ &= -8\pi\mu\gamma + m^2\mu\int_V U^{dr}(x;\alpha,\beta)\times x\,dV\,.\end{aligned} \tag{12}$$

The symmetric part of the dynamical Kelvin dipole gives another fundamental solution, which will be called the *dynamical stresslet* (dks). The corresponding displacement field and the forcing function respectively are

$$\begin{aligned}U^{dks}(x;\alpha,\beta) &= \frac{1}{2}[U^{dkd}(x;\alpha,\beta) + U^{dkd}(x,\beta,\alpha)]\\ &= -\frac{1}{2}[\alpha(\beta\cdot\nabla) + \beta(\alpha\cdot\nabla)]e^{-imr}/r\\ &\quad - (1/m^2)\nabla(\beta\cdot\nabla)(\alpha\cdot\nabla)(e^{-imr}/r - e^{-im\tau r}/r)\end{aligned} \tag{13}$$

and

$$f^{dks} = -2\pi\mu[\{\beta\cdot\nabla\delta(x)\}\alpha + \{\alpha\cdot\nabla\delta(x)\}\beta]\,. \tag{14}$$

This fundamental solution represents a self-equilibrating system and accordingly contributes neither a net force nor a moment to the medium.

Another useful fundamental solution in the present investigation is the potential dipole, which is characterized by a scalar function and is also useful in discussing vibrations in Stokes flow. The displacement field due to this potential is

$$U^{dd}(x;\alpha) = \nabla\left(\nabla\cdot\frac{e^{-imr}}{r}\alpha\right) + m^2\frac{e^{-imr}}{r}\alpha\,. \tag{15}$$

We now demonstrate that these fundamental solutions are very effective in solving boundary value problems in elastodynamics. To fix the ideas we consider the case of a spheroid.

Translation of a prolate spheroid

Let the rigid spheroid S,

$$x^2/a^2 + \rho^2/b^2 = 1\,, \qquad \rho^2 = y^2 + x^2\,, \quad c^2 = (a^2 - b^2) = a^2e^2\,, \tag{16}$$

where $e(0 \le e \le 1)$ is the eccentricity and $2c$, the focal length, be embedded in an isotropic and homogeneous elastic medium. It is excited by a periodic force with period $2\pi/\omega$ acting in the direction of its axis of symmetry, so the displacement of the points on S is

$$U = Ue_x\,, \tag{17}$$

where e_x is the unit vector along the x axis and U is a constant.

Our aim is to find the displacement field in the elastic medium so that (3) and the boundary condition (17) as well as the far-field radiation condition are satisfied. For this purpose we apply the technique of distributing the appropriate singularities. In the present situation we have the following line distributions between the foci $x = -c$ and $x = c$:

$$u(x) = \int_{-c}^{c} F_1(\xi)U^{dk}(x-\xi, e_x)dx + \int_{-c}^{c} F_2(\xi)(c^2-\xi^2)U^{dd}(x-\xi, e_x)d\xi\,, \qquad x \in S\,, \tag{18}$$

where ξ stands for $\xi\, e_x$. The first integral in (18) represents the line distribution of dynamical Kelvin solutions (5) of variable strength $F_1(\xi)$, while the second integral is the line distribution of potential dipoles (15) with strength $(c^2 = \xi^2)\, F_2(\xi)$. Clearly (18) satisfies the differential equation and vanishes as $|x| \to \infty$.

To find $F_1(\xi)$ and $F_2(\xi)$ we apply boundary condition (17) of the surface, S, of the spheroid and obtain

$$Ue_x = \int_{-c}^{c} F_1(\xi)U^{dk}(x-\xi, e_x)d\xi + \int_{-c}^{c} F_2(\xi)(c^2-\xi^2)U^{dd}(x-\xi, e_x)d\xi\,, \qquad x \in S\,. \tag{19}$$

It is obvious that we cannot easily find a closed form solution for equation (19). Accordingly, we use a perturbation technique for small values of the parameter m. For this purpose, we expand all the functions in (19) in powers of m:

$$F_1(\xi) = f_{11}(\xi) + mf_{12}(\xi) + m^2 f_{13}(\xi) + O(m^3), \tag{20a}$$

$$F_2(\xi) = f_{21}(\xi) + mf_{22}(\xi) + m^2 f_{23}(\xi) + O(m^3), \tag{20b}$$

$$\begin{aligned} U^{dk}(x - \xi e_x) =& \frac{1}{4(1-\upsilon)} U^k(x - \xi, e_x) - i\left(\frac{2+\tau^3}{3}\right) me_x \\ &+ \left[-\frac{(3+\tau^4)}{8} |x-\xi| e_x - \frac{(1-\tau^4)(x-\xi)(x-\xi)}{8|x-\xi|}\right] m^2 \\ &+ O(m^3), \end{aligned} \tag{20c}$$

$$\begin{aligned} U^{dd}(x - \xi, e_x) =& U_0^{dk}(x-\xi, e_x) + \frac{1}{2}\left[\frac{e_x}{|x-\xi|} + \frac{(x-\xi)(x-\xi}{|x-\xi|^3}\right] m^2 \\ &+ O(m^3), \end{aligned} \tag{20d}$$

where

$$U^k(x, \alpha) = \frac{(3-4\upsilon)\alpha}{r} + \frac{(\alpha\cdot x)x}{r^3}, \quad \text{and} \quad U_0^{dd}(x,\alpha) = -\frac{\alpha}{r^3} + \frac{3(\alpha\cdot x)x}{r^5}$$

are the Kelvin solution and potential dipole, respectively, for the elastostatic field, as derived in the previous section.

Substituting the expansions (20) in (19) and equating equal powers of m, we get the following system of equations:

$$\begin{aligned} &\frac{1}{4(1-\nu)} \int_{-c}^{c} f_{11} U^k(x-\xi, e_x) d\xi \\ &\qquad + \int_{-c}^{c} (c^2 - \xi^2) f_{21} U_0^{dd}(x-\xi, e_x) d\xi = U e_x, \end{aligned} \tag{21}$$

$$\begin{aligned} &\frac{1}{4(1-\nu)} \int_{-c}^{c} f_{12} U^k(x-\xi, e_x) d\xi + \int_{-c}^{c} (c^2-\xi^2) f_{22} U_0^{dd}(x-\xi, e_x) d\xi \\ &\quad = e_x \int_{-c}^{c} \frac{i(2+\tau^3)}{3} f_{11} d\xi, \end{aligned} \tag{22}$$

$$\begin{aligned} &\frac{1}{4(1-\nu)} \int_{-c}^{c} f_{13} U^k(x-\xi, e_x) d\xi + \int_{-c}^{c} (c^2-\xi^2) f_{23} U_0^{dd}(x-\xi, e_x) d\xi \\ &\quad = \int_{-c}^{c} f_{11} \left[\frac{3+\tau^4}{8} |x-\xi| e_x - \frac{(1-\tau^4)(x-\xi)(x-\xi)}{8|x-\xi|}\right] d\xi \\ &\quad -\frac{1}{2} \int_{-c}^{c} (c^2-\xi^2) f_{21} \left[\frac{e_x}{|x-\xi|} + \frac{(x-\xi)(x-\xi)}{|x-\xi|^3}\right] d\xi \\ &\quad + \frac{i(2+\tau^3)}{3} e_x \int_{-c}^{c} f_{12} d\xi. \end{aligned} \tag{23}$$

The solutions of (21) and (22) are available in Section 11.12, and thus we have

$$\frac{1}{4(1-\nu)}f_{11} = -\frac{2e^2}{1-e^2}f_{21} = Ue^2[-2e + (1+3e^2-4\nu e^2)L]^{-1}, \tag{24}$$

$$\begin{aligned}\frac{1}{4(1-\nu)}f_{12} = -\frac{2e^2}{1-e^2}f_{22} = \frac{2i(2+\tau^3)}{3}\, f_{11}ae^3 \\ \times[-25e + (1+3e^2-4\nu e^2]^{-1},\end{aligned} \tag{25}$$

where $L = \ln[(1+e)/(1-e)]$.

Next we substitute (24) and (25) in (23) and obtain

$$\begin{aligned}&\frac{1}{4(1-\nu)}\int_{-c}^{c} f_{13}U^k(x-\xi, e_x)d\xi + \int_{-c}^{c}(c^2-\xi^2)f_{23}U_0^{dd}(x-\xi, e_x)d\xi \\ &= (\psi_0 + \psi_1 x^2) + \psi_2 x\rho e_\rho,\end{aligned} \tag{26}$$

where

$$\begin{aligned}\psi_0 &= \frac{1}{4}a^2 f_{11}[e(1+\tau^4) + (1-e^2)L] - a^2 f_{21}(-2e+L) \\ &\quad + \frac{2}{3}iae(2+\tau^3)f_{12},\end{aligned} \tag{27a}$$

$$\psi_1 = \frac{1}{4}f_{11}[e(1+\tau^4) - (1-e^2)L] - f_{21}[4e - (2-e^2)L], \tag{27b}$$

$$\psi_2 = -\left[\frac{1}{4}(1-\tau^4)f_{11}e + f_{21}(-2e+L)\right], \tag{27c}$$

and $e_\rho = (ye_y + ze_z)/\rho$ is the unit radial vector in the y, z plane. To solve (26) we set

$$f_{13}(x) = A_0 + A_2x^2, \qquad |x| \le c, \tag{28a}$$

$$f_{23}(x) = C_0 + C_2x^2, \qquad |x| \le c. \tag{28b}$$

Substitution of (28) in (26) and simplification yields

$$\begin{aligned}&\frac{2}{e}\left[\frac{1}{4(1-\nu)}(A_0 + c^2A_2) + \frac{2e^2}{1-e^2}(C_0 + c^2C_2)\right]\frac{b^2e_x + e^2x\rho e_\rho}{a^2 - e^2x^2} \\ &\quad + [\phi_0A_0 + \phi_1A_2 - 2C_0L + \phi_2C_2]e_x \\ &\quad - x^2(A_2\lambda_1 + C_2\lambda_2)e_x - (A_2\vartheta_1 + C_2\vartheta_2)x\rho e_\rho \\ &= (\psi_0 + \psi_1x^2)e_x + \psi_2 x\rho e_\rho,\end{aligned} \tag{29}$$

where

$$\phi_0 = -1/2(1-\nu)e + L\,, \tag{30a}$$
$$\phi_1 = a^2[e - \{(1-e^2)(3-2\nu)/4(1-\nu)\}L]\,, \tag{30b}$$
$$\phi_2 = -2a^3[6e - (3-2e^2)L]\,, \tag{30c}$$
$$\lambda_1 = [(14-12\nu)e - (7-6\nu-3e^2+2\nu e^2)L]/4(1-\nu)\,, \tag{30d}$$
$$\lambda_2 = -36e + 2(9-3e^2)L\,, \tag{30e}$$
$$\vartheta_1 = -(2e+L)/2(1-\nu)\,, \tag{30f}$$
$$\vartheta_2 = 16e + 8e/(1-e^2) - 12L\,. \tag{30g}$$

Equation (29) is satisfied if we choose

$$\frac{1}{4(1-\nu)}(A_0 + c^2 A_2) = -\frac{2e^2}{1-e^2}(C_0 + c^2 C_2)\,, \tag{31}$$
$$\phi_0 A_0 + \phi_1 A_2 - 2LC_0 + \phi_2 C_2 = \psi_0\,, \tag{32}$$
$$\vartheta_1 A_2 + \vartheta_2 C_2 = -\psi_2\,, \tag{33}$$
$$\lambda_1 A_2 + \lambda_2 C_2 = -\psi_1\,. \tag{34}$$

This is a linear system of equations with the following solution:

$$A_2 = \frac{\lambda_2\psi_2 - \vartheta_2\psi_1}{\lambda_1\vartheta_2 - \lambda_2\vartheta_1}\,, \tag{35}$$
$$C_2 = \frac{\vartheta_1\psi_1 - \lambda_1\psi_2}{\lambda_1\vartheta_1 - \lambda_2\vartheta_1}\,, \tag{36}$$

$$C_0 = \left[\frac{8e^2(1-\nu)\phi_0}{1-e^2} + 2L\right]^{-1}\left[\left\{\phi_2 - \frac{8a^2(1-\nu)e^2}{1-e^2}\right\} C_2 - \psi_0 + (\phi_1 - c^2\phi_0)A_2\right]\,, \tag{37}$$
$$A_0 = -\frac{8e^2(1-\nu)}{1-e^2}\,(C_0 + c^2 C_2) - c^2 A_2\,. \tag{38}$$

The net force F experienced by the spheroid can be computed by superposition of (6);

$$F = 4\pi\mu e_x \int_{-c}^{c} [f_{11}(\xi) + m f_{12}(\xi) + m^2 f_{13}(\xi) + O(m^3)]d\xi + \mu m^2 \int_V \bar{u}\, dV\,,$$

where the volume integral is taken over the spheroid. Substituting the values of f_{11}, f_{12}, f_{13}, and u in this formula, we have

$$F = P_0\left[\left\{1 + \frac{P_0 i(\tau^3+2)}{12\pi\mu a U} am\right\} + \frac{8\pi\mu a e}{3P_0}\left\{\frac{3A_0}{a^2} + A_2e^2 + f_{21}[2-4(1-\nu)e^2 - (e^{-1}+4e-5\nu e)L]\right\}(am)^2\right]e_x + O(am)^3, \tag{39}$$

where

$$P_0 = 32\pi\mu a(1-\nu)Ue^3[-2e+(1+3e^2-4\nu e^2)L]^{-1}.$$

Various other rotary and rectilinear oscillation problems can be solved by this method and are presented in references [54,55].

11.14. Distributions on Arbitrary Lines

The concepts of the distribution of the delta function and its derivatives can be extended to include any straight line in the plane. Indeed, in the theory of bending and buckling of elastic plates one has to apply concentrated loads on lines inside the plate, the so-called *line loads*. The classical approach is to take a strip of finite width, so that the problem is first solved for the case of loads on an area; then the width is made to approach zero. The problem can also be solved easily with the help of the theory of distributions. As an illustration, let us find the bending of a rectangular plate $0 \le x \le a$ $0 \le y \le b$ of small thickness h when a periodic concentrated load of uniform strength P is applied on its diagonal line $x_2 = (b/a)x_1$ (see Figure 11.5). This load $p(x_1, x_2, t)$ can therefore be written as [50]:

$$p(x_1, x_2, t) = Pe^{i\omega t}\delta(x_2 - bx_1/a)(1+b^2/a^2)^{1/2}, \tag{1}$$

where the factor $(1+b^2/a^2)^{1/2}$ has been added for dimensional reasons.

The equation of motion for the plate in the foregoing situation is

$$\mathcal{D}\nabla^4 v + h\, d^2v/dt^2 = Pe^{i\omega t}\delta(x_2 - bx_1/a)(1+b^2/a^2)^{1/2}, \tag{2}$$

where v denotes the deflection of the plate in the x_3 direction, $\mathcal{D} = Eh^3/12(1-\nu^2)$ is the flexural rigidity (also called bending stiffness), E is Young's modulus, ρ is the density, and the differential operator ∇^4 is

$$\nabla^4 = \partial^4/\partial x_1^4 + 2\partial^4/\partial x_1^2\partial x_2^2 + \partial^4/\partial x_2^4.$$

Let us assume that the plate has simply supported edges so that the boundary conditions are

$$\begin{aligned} v &= \partial^2 v/\partial x_1^2 = 0, \qquad x_1 = 0 \quad\text{and}\quad x_1 = a, \\ v &= \partial^2 v/\partial x_2^2 = 0, \qquad x_2 = 0 \quad\text{and}\quad x_2 = b. \end{aligned} \tag{3}$$

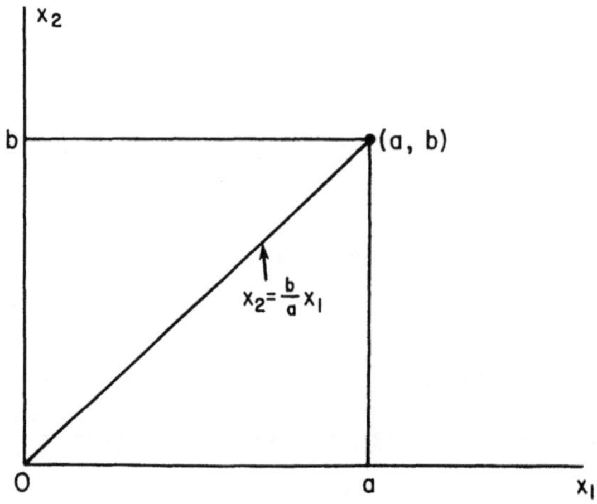

Figure 11.5. $x_2 = (b/a)x_1$.

The solution of boundary value problem (2) and (3) is given by the double series

$$v(x_1, x_2, t) = \sum_{m=1}^{\infty}\sum_{n=1}^{\infty} \Lambda_{mn} e^{i(\omega t - \phi_{mn})} V_{mn}(x_1, x_2), \tag{4}$$

where Λ_{mn} is the amplitude, ϕ_{mn} is the phase, and

$$V_{mn}(x_1, x_2) = \sin(m\pi x_1/a)\sin(n\pi x_2/b)$$

are eigenfunctions for this boundary. The corresponding eigenvalues λ_{mn} are

$$\lambda_{mn} = \pi^2[m^2a^2 + n^2b^2](\mathcal{D}\rho h)^{1/2} .$$

Substituting (4) in (2), we obtain the following values for the amplitude and phase, respectively.

$$\Lambda_{mn} = (F_{mn}/\lambda_{mn}^2)\{[1 - (\omega/\lambda_{mn})^2]^2 + 4\zeta_{mn}^2(\omega/\lambda_{mn})^2\}^{-1/2},$$
$$\phi_{mn} = \tan^{-1}[2\zeta_{mn}\,\omega\,\lambda_{mn}(\lambda_{mn}^2 - \omega^2)^{-1}],$$

where

$$F_{mn} = 4P\left[\rho hab\left(1 + \frac{b^2}{a^2}\right)^{1/2}\right]^{-1} \int_0^a \int_0^b W_{mn}(x_1, x_2)\delta\left(x_2 - \frac{bx_1}{a}\right) dx_2 dx_2 ,$$

ξ_{mn} is the modal damping factor. Using the sifting property of the delta function, we have

$$F_{mn} = 2P(a^2 + b^2)^{1/2}\rho hab(1 + b^2/a^2)^{1/2}\delta_{mn} ,$$

where δ_{mn} is the Kronecker delta. We have thereby obtained all the values of the required solution (4) to the deflection v of the plate.

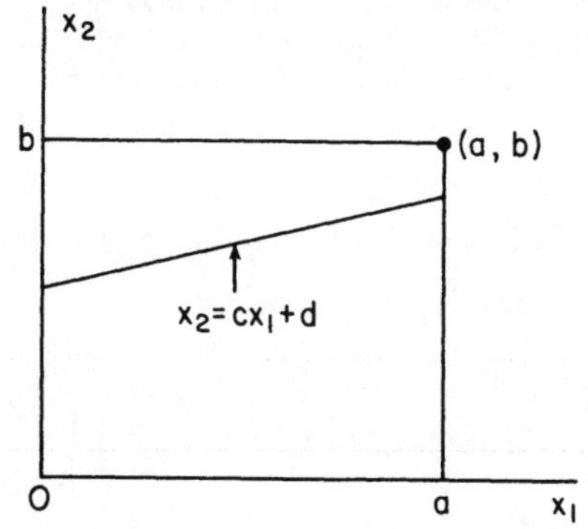

Figure 11.6. $x_2 = cx_1 + d$.

By the same token the load can be distributed on any straight line $x_2 = cx_1 + d$ that intersects the plate (see Figure 11.6). In this case the corresponding load is $Pe^{i\omega t}\delta[x_2 - (cx_1 + d)](1 + c^2)^{1/2}$.

11.15. Distributions on Plane Curves

Let us now distribute the generalized functions on a plane curve $x_2 = f(x_1)$ that intersects the elastic plate as in Figure 11.7. Accordingly, the value of the load is

$$p(x_1, x_2, t) = Pe^{i\omega t}\delta[x_2 - f(x_1)][1 + (f'(x_1))^2]^{1/2}.$$

We have already discussed the delta function $\delta[x_2 - f(x_1)]$ in Section 5.8.

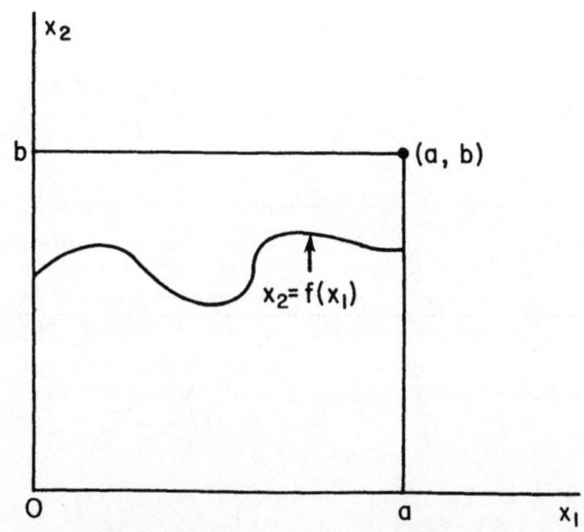

Figure 11.7. $x_2 = f(x_1)$.

As an example, let us take the curve to be a parabola, $x_2 = b(x_1/a)^2$. In this case the differential equation for the deflection v of the plate is

$$\mathcal{D}\nabla^4 v + h\frac{d^2v}{dt^2} = Pe^{i\omega t}\left(\frac{4b^2x_1^2}{a^4}+1\right)^{1/2}\delta\left(x_2 - \frac{bx_1^2}{a^2}\right).$$

Proceeding as in Section 11.14 [see (11.14.4)], we again obtain the value of the deflection $v(x_1, x_2, t)$, where the quantities F_{mn} are now

$$\begin{aligned} F_{mn} &= 4P\int_0^a\int_0^b V_{mn}(x_1, x_2)\delta\left(x_2 - \frac{bx_1^2}{a^2}\right)\left(\frac{4b^2x_1^2}{a^4}+1\right)^{1/2}\frac{dx_2dx_1}{\rho hab} \\ &= 4P\int_0^a\left(\frac{4b^2x_1^2}{a^4}+1\right)^{1/2}\sin\frac{m\pi x_1}{a}\sin\frac{n\pi x_1^2}{a^2}\frac{dx_1}{\rho hab}. \end{aligned}$$

The next natural step is to distribute the generalized functions on laminae and surfaces and to solve the corresponding boundary value problems for configurations such as shells. This is easily accomplished with the help of the surface distributions explained in Chapter 5. In the next section, we content ourselves with discussing the distribution on a circular disk.

11.16. Distributions on a Circular Disk

The ideas of the previous sections can be extended by distributing the singularities on a circular disk. We take

$$F(x) = f(\rho)\delta(x), \qquad 0 \le \vartheta \le 2\pi, \quad 0 \le \rho \le b, \tag{1}$$

where (x, ρ, ϑ) are cylindrical polar coordinates and $f(\rho)$ is defined over the disk $0 \le \rho \le b, 0 \le \vartheta \le 2\pi$. Then (11.1.3) reduces to

$$u(x, \rho) = \int_0^b\int_0^{2\pi}\frac{f(\xi)\xi\, d\xi\, d\vartheta_1}{[x^2+\rho^2+\xi^2-2\xi\rho\cos(\vartheta-\vartheta_1)]^{1/2}}. \tag{2}$$

If we set $\alpha = \vartheta_1 - \vartheta$, it is easy to see that

$$\int_0^{2\pi}\frac{d\vartheta_1}{[x^2+\rho^2+\xi^2-2\xi\rho\cos(\vartheta-\vartheta_1)]^{1/2}} = \int_0^{2\pi}\frac{d\alpha}{[x^2+\rho^2+\xi^2-2\xi\rho\cos\alpha]^{1/2}}.$$

Thus (2) takes the form

$$u(x, \rho) = \int_0^b\int_0^{2\pi}\frac{f(\xi)\xi\, d\xi\, d\alpha}{[x^2+\rho^2+\xi^2-2\xi\rho\cos\alpha]^{1/2}}. \tag{3}$$

Next, we use the expansion

$$[x^2 + \rho^2 + \xi^2 - 2\xi\rho\cos\alpha]^{1/2}$$
$$= \int_0^\infty \left[\sum_{m=0}^{\infty} (2 - \delta_{0m}) \cos m\alpha \, J_m(p\zeta) J_m(p\xi) e^{-|x|p} dp \right],$$

in (3), where δ_{0m} is the Kronecker delta and we use the orthogonality of the cosine functions. The result is

$$u(x, \rho) = 2\pi \int_0^b \int_0^\infty f(\xi)\xi J_0(p\rho) J_0(p\xi) e^{-|x|p} d\xi \, dp \,. \tag{4}$$

Suppose that $u(x, \rho) = \lambda(x, \rho)$ on the surface S, $\rho = \rho_0(x)$, so that (4) becomes

$$\lambda(x, \rho) = \int_0^b \xi f(\xi) K_0(x, \xi, \rho) d\xi \,, \qquad (\rho, x) \in S \,, \tag{5}$$

where the kernel $K_0(x, \xi, \rho)$ is

$$K_0(x, \xi, \rho) = 2\pi \int_0^\infty e^{-|x|p} J_0(p\rho) J_0(p\xi) d\xi \,. \tag{6}$$

Equation (5) is a Fredholm integral equation of the first kind for $f(\xi)$, provided the equation $\rho = \rho_0(x)$ of the surface S is known. The inverse problem pertains to finding the equation of the surface when $\lambda(x, \rho)$, $f(\xi)$, and b are prescribed.

For the special case $f(\xi) = \delta(\xi - b)$, (4) reduces to

$$u(x, \rho) = 2\pi b \int_0^b J_0(p\rho) J_0(pb) e^{-|x|p} dp \,, \tag{7}$$

which is the electrostatic potential due to point sources on a ring.

CHAPTER 12

Applications to Wave Propagation

12.1. Introduction

In Chapter 10 we have discussed various properties of the homogeneous and inhomogeneous wave equations in two and three dimensions. We derived their fundamental solutions and studied moving point, line, and surface sources. In Chapter 5 we considered various kinematic and geometrical aspects of the wave propagation in the context of surface distributions. In this chapter we consider some applications of these results and study partial differential equations whose solutions can be defined as regular singular functions. In problems of these kinds, singular surfaces can play two essentially different roles: one relates to the propagation of wave fronts, and the other to regular singular functions that arise in the study of boundary value problems. In the latter case, the surface Σ is given along with some data on it, and the solution is to be found in only some region of the space. A powerful method of attacking these problems is to embed them in the whole space. This is achieved by extending the solution to the other side of the surface in some suitable fashion, as we did in deriving the Poisson integral formula in Chapter 10. We then obtain a regular singular function that satisfies the equation in the complement of Σ.

If a generalized solution of an equation is a vector of regular singular functions with respect to Σ, then we say that Σ is a singular wave of that equation. More precisely, if $F \in E(\Sigma)$ is such that LF is continuous across Σ for all differential operators L of order less than k, but there exists an operator of order k such that LF is discontinuous, then it is said that Σ is a singular wave of order k. For $k = 0$ it is called a shock wave, and for $k = 1$ it is an acceleration wave.

In general, F describes the behavior of a mechanical system with different states on the two sides of Σ. The problem is to examine how F behaves at the wavefront. In this situation, F satisfies the equation in the whole space, even across Σ, and the discontinuities arise naturally [8].

12.2. The wave Equation

One of the simplest cases in which singular waves exist is provided by the wave equation

$$\overline{\Box}^2 F = \overline{\nabla}^2 F - \frac{1}{c^2}\frac{\overline{\partial}^2 F}{\partial t^2} = 0\,. \tag{1}$$

We assume that $\Sigma(t)$ is a sonic discontinuity; that is, A, the jump in F, cannot vanish.

From the results of Chapter 5 we obtain

$$\overline{\Box}^2 F = \Box^2 F + \left[B - 2\Omega A - \frac{1}{c^2}\left(BG^2 - 2G\frac{\delta A}{\delta t} - A\frac{\delta G}{\delta t}\right)\right]\delta(\Sigma)$$
$$+ \left(A - \frac{AG^2}{c^2}\right)\delta'(\Sigma) = 0. \tag{2}$$

For this equation to hold, both the regular part and the coefficients of $\delta(\Sigma)$, and $\delta'(\Sigma)$ must vanish. Since $A \neq 0$, the coefficient of $\delta'(\Sigma)$ shows that $G = \pm c$. We choose the value $G = c$. This amounts to choosing a particular orientation. The fact that G is a constant implies that the propagation must be parallel, that is, that the wave fronts form a family of parallel surfaces. Indeed, if we let s be the arc length as measured along the orthogonal trajectories of $\Sigma(t)$, then

$$\delta s/\delta t = c\,, \qquad dx_i/ds = n_i\,,$$

and

$$\frac{d^2 x_i}{ds^2} = \frac{dn_i}{ds} = \frac{1}{c}\frac{\delta n_i}{\delta t} = \frac{1}{c}\frac{\delta(-G)}{\delta x_i} = 0\,.$$

This proves that the orthogonal trajectories of the wave fronts are straight lines, which is another way of saying that the propagation is parallel. Accordingly, if the wave front is known at a time $t = t_0$, then we can construct it for all subsequent times.

Next, we equate the coefficients of $\delta(\Sigma)$ to zero and obtain

$$-2\Omega A + \frac{2}{c}\frac{\delta A}{\delta t} = 0\,, \qquad \text{or} \qquad \frac{dA}{ds} = \Omega A\,. \tag{3}$$

We can integrate this equation to obtain

$$A = \alpha \exp\left(\int_{s_0}^{s} \Omega(\sigma)d\sigma\right)\,, \tag{4}$$

where α is the distribution on the surface that corresponds to $s = s_0$. In the three-dimensional case.

$$\Omega(s) = (\Omega_0 - sK_0)/(1 - 2s\Omega_0 + s^2K_0)\,, \tag{5}$$

where Ω_0 and K_0 are the mean and the Gaussian curvatures, respectively, corresponding to $s = s_0$. Thus (4) becomes

$$A = \alpha(1 - 2s\Omega_0 + s^2K_0)^{-1/2}. \tag{6}$$

To derive the jump in the normal derivative of F we appeal to (5.6.9) and obtain

$$0 = [\Box^2 F] = C - 2\Omega B + \nabla^2 A - \frac{1}{c^2}\left(CG^2 - 2G\frac{\delta B}{\delta t} + \frac{\delta^2 A}{\delta t^2}\right)\,,$$

or

$$\frac{dB}{ds} - \Omega B = \frac{1}{2}\left[\left(\frac{d\Omega}{ds} + \Omega^2\right)A - \nabla^2 A\right]\,. \tag{7}$$

Let T be a tube of rays that cuts the surface $\Sigma(t_0)$ and $\Sigma(t_1)$ at elements of area S_1 and S_2 and let R be the volume enclosed thereby. Then the divergence theorem applied to the vector field $A^2\widehat{n}$ yields

$$\left[\int_T + \int_T + \int_T\right] A^2\widehat{n} \cdot dS = \int_R \operatorname{div}\,(A^2\widehat{n})dV\,. \tag{8}$$

Because

$$\operatorname{div}\,(A^2\widehat{n}) = \frac{\partial}{\partial x_i}(A^2 n_i) = 2A\frac{\partial A}{\partial x_i}n_i + A^2\frac{\delta n_i}{\delta x_i} = 2A\left(\frac{1}{c}\frac{\partial A}{\partial t} - \Omega A\right) = 0\,,$$

where we have used (3), from (8) we derive

$$\int_{S_1} A^2 dS = \int_{S_2} A^2 dS\,. \tag{9}$$

Finally, let us consider the case of higher-order waves. Suppose that F and all its derivatives of order less than k are continuous but LF is discontinuous, where L is the partial derivative of the form

$$L = \frac{\partial^k}{\partial x_1^{k_1}\cdots\partial x_p^{k_p}\partial t^{k_t}}\,, \qquad k = k_1 + \cdots + k_p + k_t\,. \tag{10}$$

Since $LF = \overline{L}F$, we have

$$\overline{\Box}^2 LF = \overline{\Box}^2\overline{L}F = \overline{L}\,\overline{\Box}^2 F = 0\,, \tag{11}$$

and, accordingly, the foregoing analysis can be applied to the functions LF. The conclusions about the surface $\Sigma(t)$ remain the same, as do the results for the quantity A, except that we now designate $A = [LF]$.

The value of the expansion coefficient of wave fronts can also be derived by the analysis of this section [56].

12.3. First-Order Hyperbolic Systems

Let us now consider the existence of the regular singular solutions of a system of quasilinear partial differential equations of the form

$$\frac{\overline{\partial} F_i}{\partial t} + a_{ij\beta}\frac{\overline{\partial} F_j}{\partial x_\beta} + b_i = 0\,, \qquad i = 1, \ldots, q\,, \tag{1}$$

where $a_{ij\beta}$ and b_i are given infinitely differentiable functions of $x_1, \ldots, x_p, t$, and $F_1, \ldots, F_q$. Here Latin indices run from 1 to q and Greek indices from 1 to p. Furthermore, we shall employ the notation

$$A_{ijk\beta} = \partial a_{ij\beta}/\partial F_k\,, \qquad B_{ik} = \partial b_i/\partial F_k\,. \tag{2}$$

Our aim is to examine the existence of first-order singular waves. If we denote the jump of dF_i/dn by Λ_i and recall the analysis of Section 5.5, we have

$$[\partial F_i/\partial x_\beta] = \Lambda_i n_\beta\,, \tag{3}$$

$$[\partial F_i/\partial t] = -\Lambda_i G\,, \tag{4}$$

since F is supposed to be continuous.

Next we take the derivative of (1) with respect to x_γ and obtain

$$\begin{aligned} 0 &= \frac{\overline{\partial}^2 F_i}{\partial x_\gamma \partial t} + \frac{\overline{\partial}}{\partial x_\gamma}\left[a_{ij\beta}\frac{\overline{\partial} F_j}{\partial x_\beta}\right] + \frac{\overline{\partial} b_i}{\partial x_\gamma} \\ &= \frac{\partial^2 F}{\partial x_\gamma \partial t} + \frac{\partial}{\partial x_\gamma}\left[a_{ij\beta}\frac{\partial F_j}{\partial x_\beta}\right] + \frac{\partial b_i}{\partial x_\gamma} + [-\Lambda_i G n_\gamma + a_{ij\beta}\Lambda_j n_\gamma n_\beta]\delta(\Sigma)\,. \end{aligned} \tag{5}$$

Both the regular and the singular part of this equation must vanish. Multiplication of the singular part by n_γ yields

$$-\Lambda_i G + a_{ij\beta} n_\beta \Lambda_j = 0\,, \tag{6}$$

which says that G is an eigenvalue and Λ_i is an eigenvector of the matrix $a_{ij\beta}n_\beta$. For a singular wave to exist, this matrix should have at least one real eigenvalue. In this case the system is called *hyperbolic*.

If G is an eigenvalue with multiplicity $m \leq q$, then (6) enables us to write the strength Λ_i as a linear combination of m of them, say, $\Lambda_1, \ldots, \Lambda_m$:

$$\Lambda_i = \Lambda_i(\Lambda_1, \Lambda_2, \ldots, \Lambda_m)\,, \qquad i = 1, \ldots, q\,. \tag{7}$$

It follows that the decay and growth of these discontinuities can be studied merely by determining the propagation of $\Lambda_1, \ldots, \Lambda_m$. For this purpose we write the jump conditions for the regular part of (5),

$$\begin{aligned} 0 &= \left[\frac{\partial^2 F}{\partial x_\gamma \partial t} + \frac{\partial}{\partial x_\gamma}\left(a_{ij\beta}\frac{\partial F_j}{\partial x_\beta}\right) + \frac{\partial b_i}{\partial x_\gamma}\right] \\ &= -\Gamma_i G n_\gamma + \frac{\delta \Lambda_i}{\delta t} n_\gamma - \frac{\delta \Lambda_i}{\delta x_\gamma} G - \Lambda_i \frac{\delta G}{\delta x_\gamma} + A_{ijk\beta}\left[\frac{\partial F_k}{\partial x_\gamma}\frac{\partial F_j}{\partial x_\beta}\right] \\ &\quad + a_{ij\beta}\left(\Gamma_j n_\gamma n_\beta + \frac{\delta \Lambda_j}{\delta x_\beta} n_\gamma + \frac{\delta \Lambda_j}{\delta x_\gamma} n_j + \Lambda_j \frac{\delta n_\beta}{\delta x_\gamma}\right) + B_{ij}\Lambda_j n_\gamma\,, \end{aligned} \tag{8}$$

where $\Gamma_j = [d^2 F_j/dn^2]$. Multiplying (8) by n_γ, we obtain

$$0 = -\Gamma_i G + a_{ij\beta}\Gamma_j n_\beta + \frac{\delta \Lambda_i}{\delta t} + A_{ijk\beta}\left[\frac{\partial F_k}{\partial x_\gamma}\frac{\partial F_j}{\partial x_\beta}\right] n_\gamma + a_{ij\beta}\frac{\delta \Lambda_j}{\delta x_\beta} + B_{ij}\Lambda_j\,. \tag{9}$$

In order to eliminate the $\Gamma_i s$ from (9), we take m linearly independent right eigenvalues of the matrix $a_{ij\beta}n_\beta$, which we designate $\widetilde{\Lambda}_j^{(r)}$, $r = 1, \ldots, m$, and multiply (9) by each of them. This leads to

$$0 = \widetilde{\Lambda}_i^{(r)} \left\{ \frac{\delta \Lambda_i}{\delta t} + A_{ijk\beta} \left[\frac{\partial F_k}{\partial x_\gamma} \frac{\partial F_j}{\partial x_\beta} \right] n_\gamma + a_{ij\beta} \frac{\delta \Lambda_j}{\delta x_\beta} + B_{ij} \Lambda_j \right\}, \qquad r = 1, \ldots, m. \tag{10}$$

These m equations determine the propagation of $\Lambda_1 \ldots, \Lambda_m$, and (7) will provide us with the rest.

Let us also observe that we have

$$\left[\frac{\partial F_k}{\partial x_\gamma} \frac{\partial F_j}{\partial x_\beta} \right] n_\gamma = \left(\frac{dF_k}{dn} \right) \Bigg|_+ n_\beta \Lambda_j + \left(\frac{\partial F_j}{\partial x_\beta} \right) \Bigg|_+ \Lambda_k - n_\beta \Lambda_j \Lambda_k, \tag{11}$$

and, accordingly, if F is known in the positive part, so will be the coefficients of this expression. A particular, but quite important case, is that in which the positive part is at rest, that is, for which $F = 0$ in the positive part; then $-n_\beta \Lambda_j \Lambda_k$ is the only nonzero term left in (11).

Let us illustrate these results with the help of the equations of inviscid fluid dynamics, namely,

$$\frac{\partial F_i}{\partial t} + F_\beta \frac{\partial F_i}{\partial x_\beta} + a^2 \frac{\partial R}{\partial x_i} = 0, \tag{12}$$

$$\frac{\partial R}{\partial t} + F_\beta \frac{\partial R}{\partial x_\beta} + \frac{\partial F_\beta}{\partial x_\beta} = 0, \tag{13}$$

where $i, \beta = 1, 2, 3$, F_i are the components of velocity, $R = \ln \rho$, ρ is the density, $a^2(R) = p'(\rho)$; and p is the pressure. We assume that a sonic discontinuity is propagating into a region at rest and at constant pressure. In this case (6) leads to

$$-G\Lambda_i + F_\beta n_\beta \Lambda_i + a^2 r n_i = 0, \tag{14}$$

$$-Gr + F_\beta n_\beta r + \Lambda_\beta n_\beta = 0, \tag{15}$$

where $\Lambda_i = [dF_i/dn]$, $r = [dR/dn]$. From these equations we obtain

$$G = F_\beta n_\beta \pm a = \pm a, \tag{16}$$

since $F_\beta = 0$ on the surface. We take the positive sign to fix the ideas. Another relation that follows is

$$\Lambda_i = arn_i, \tag{17}$$

and so only r needs to be determined. Furthermore, since the velocity G is a constant, it follows from the analysis of Section 12.2 that propagation is parallel.

Next, we attend to (9), which, in the present case, reduces to

$$-\Gamma_i G + a^2 \widetilde{r} n_i + \frac{\delta \Lambda_i}{\delta t} + \Lambda_i \Lambda_\beta n_\beta + a^2 \frac{\delta r}{\delta x_i} + c r^2 n_i = 0 \text{ and}, \tag{18}$$

$$-\widetilde{r} G + \Gamma_\beta n_\beta + \frac{\delta r}{\delta t} + a \Lambda_\beta n_\beta r + a \frac{\delta \Lambda_\beta}{\delta x_\beta} = 0, \tag{19}$$

where $\widetilde{r} = [d^2R/dn^2]$ and $c = \partial a^2/\partial R$ in the positive part. Multiplying (18) by n_i and (19) by a, adding the resulting equations, and using (17), we derive the transport equation

$$dr/ds - r\,\Omega + (1+c)r^2 = 0\,, \tag{20}$$

which determines the propagation of r.

The solution of (20) is

$$r(s) = (1 - 2s\,\Omega_0 + s^2K_0)^{-1/2}[(1+c)w(s) + \text{const}\,]^{-1}\,,$$

where

$$w(s) = \begin{cases} \left(1/\sqrt{K_0}\right)\ln|K_0s - 2\Omega_0K_0 \\ \qquad + [K_0(1-2s\,\Omega_0 + s^2K_0)]^{1/2}|\,, & \text{for} \quad K_0 > 0\,, \\ \dfrac{1}{\sqrt{-K_0}}\sin^{-1}\dfrac{\Omega_0 - K_0s}{(\Omega_0^2 - K_0)^{1/2}}\,, & \text{for} \quad K_0 < 0\,, \\ -(1/\Omega_0)(1-2\Omega_0 s)^{1/2} & \text{for} \quad K_0 = 0\,. \end{cases} \tag{21}$$

12.4. Aerodynamic Sound Generation

We can effectively extend the results of Section 12.3 by including the dissipative terms in (12.3.12). Such a system has been studied by Ffowcs-Williams and Hawkings [57]. To facilitate the comparison, we write the mass and momentum equations in their usual notation, namely,

$$\frac{\overline{\partial}\rho}{\partial t} + \frac{\overline{\partial}(\rho v_i)}{\partial x_i} = 0\,, \tag{1}$$

$$\frac{\overline{\partial}(\rho v_i)}{\partial t} + \frac{\overline{\partial}(\rho v_i v_j - \sigma_{ij})}{\partial x_j} = 0\,, \tag{2}$$

where ρ is the density, the v_i are the components of the velocity vector, and σ_{ij} is the symmetric stress tensor.

Let $\Sigma(t)$ be a singular wave of order zero. Then equating the singular part of (1) with zero gives

$$[\rho(v_n - G)] = 0\,. \tag{3}$$

Let Q denote the common value of $\rho(v_n - G)$ on Σ. Then for any function F, discontinuous across Σ, we have

$$[F\rho(v_n - G)] = Q[F]\,. \tag{4}$$

Similarly, equating the singular part of (2) with zero yields

$$Q[v_i] = [\sigma_{ij}]n_j\,. \tag{5}$$

The particular values of the stress tensor lead to interesting results.

The following type of problem associated with (1) and (2) is also of great importance. Let us suppose that a body of arbitrary shape and surface $\Sigma(t)$ is surrounded by a continuum medium. Then the normal velocity v_n of a point on the surface of the body is equal to G, the velocity of the fluid. This means that we have to solve the partial differential equations (1) and (2) subject to the boundary condition

$$v_n = v_i n_i = G \qquad \text{on} \quad \Sigma(t) .$$

This boundary value problem can be transformed to an integral equation by means of the following embedding procedure.

Let us extend the functions appearing in this boundary value problem to the inside of the body in such a way that the equations are satisfied there. A simple way to do this is to assume that all the flow parameters vanish inside the body. In that case the jumps across $\Sigma(t)$ reduce to the boundary values from the fluid side. Because $[F_1 F_2] = [F_1][F_2]$, we obtain the following system of equations for the extended functions:

$$\frac{\overline{\partial}\rho}{\partial t} + \frac{\overline{\partial}(\rho v_i)}{\partial x_i} = (-\rho G + \rho v_n)\delta(\Sigma) = 0\,, \tag{6}$$

$$\frac{\overline{\partial}(\rho v_i)}{\partial t} + \frac{\overline{\partial}(\rho v_i v_j - \sigma_{ij})}{\partial x_j} = (-\rho v_i G + \rho v_i v_n - \sigma_{ij} n_j)\delta(\Sigma) = -\sigma_{ij} n_j \delta(\Sigma)\,. \tag{7}$$

Taking the time derivative of (6) and the space derivative of (7) and subtracting the resulting relations, we obtain

$$\frac{\overline{\partial}^2 \rho}{\partial t^2} = \frac{\overline{\partial}^2(\rho v_i v_j - \sigma_{ij})}{\partial x_i \partial x_j} + \frac{\overline{\partial}}{\partial x_i}[\sigma_{ij} n_j \delta(\Sigma)]\,. \tag{8}$$

It is convenient to subtract ρ_0 from ρ so that $\widetilde{\rho} = \rho - \rho_0$ vanishes at infinity. Clearly, (8) holds if ρ is changed to $\widetilde{\rho}$ on the left side. Subtracting the term $c^2 \overline{\nabla}^2 \widetilde{\rho}$ from both sides of (8), we obtain

$$-c^2 \overline{\Box}^2 \widetilde{\rho} = \frac{\overline{\partial}^2 T_{ij}}{\partial x_i \partial x_j} + \frac{\overline{\partial}}{\partial x_i}[\sigma_{ij} n_j \delta(\Sigma)]\,, \tag{9}$$

where $T_{ij} = \widetilde{\rho} v_j - \sigma_{ij} - c^2 \widetilde{\rho} \delta_{ij}$ is the Lighthill stress tensor.

It is interesting to observe that we can have extra source terms in (9) if we extend the parameters in a different fashion. For instance, if the density is assumed to be a constant, say, Λ, inside the body, then (6) is replaced by

$$\overline{\partial}\rho/\partial t + \overline{\partial}(\rho v_i)/\partial x_i = \Lambda G \delta(\Sigma)\,, \tag{10}$$

while (7) remains unchanged. Then (9) becomes

$$-c^2 \overline{\Box}^2 \widetilde{\rho} = \frac{\overline{\partial}^2 T_{ij}}{\partial x_i \partial x_j} + \frac{\overline{\partial}}{\partial x_i}(\sigma_{ij} n_j \delta(\Sigma)) + \frac{\overline{\partial}}{\partial t}(\Lambda G \delta(\Sigma))\,. \tag{11}$$

The extra term on the right side of this equation will, however, not change the solution in the fluid because $\widetilde{\rho}$ depends on Λ only inside the body. By giving other values to the parameters inside the body, we can get various other equations, with source terms that affect the solution only inside the body.

Finally, the convolution with the fundamental solution

$$E(x,t) = (1/4\pi|x|)\delta(t - |x|/c)\,, \tag{12}$$

of the wave equation gives the integral equation

$$\begin{aligned} c^2\widetilde{\rho}(x,t) = \int_{-\infty}^{\infty}\int \left[\frac{\overline{\partial}^2 T_{ij}}{\partial x_i \partial x_j} + \frac{\overline{\partial}}{\partial x_i}(\sigma_{ij} n_i \delta(\Sigma)) + \frac{\overline{\partial}}{\partial t}(\Lambda G \delta(\Sigma))\right] \\ \times E(\tau - t, y - x) dy\, d\tau\,. \end{aligned} \tag{13}$$

Assigning the value ρ_0 to Λ and taking derivatives outside of the integral sign, we obtain the Ffowcs-Williams-Hawkings integral equation [57].

12.5. The Rankine-Hugoniot Conditions

When the hyperbolic system (12.3.1) is of the form

$$\overline{\partial} F_i/\partial t + \overline{\partial} M_{ij}/\partial x_j = 0\,, \tag{1}$$

where M_{ij} is a suitable tensor, we can obtain the jump conditions for a shock front by applying (5.5.4) and (5.5.5). Then (1) becomes

$$\partial F_i/\partial t - [F_i]G\delta(\Sigma) + \partial M_{ij}/\partial x_j + [M_{ij}]n_j\delta(\Sigma) = 0\,,$$

or

$$\partial F_i/\partial t + \partial M_{ij}/\partial x_j + \{[M_{ij}]n_j - G[F_i]\}\delta(\Sigma) = 0\,.$$

The singular part immediately yields the jump conditions

$$[M_{ij}]n_j = G[F_i]\,. \tag{2}$$

Now we show that the system of equations (12.3.12) and 12.3.13) can be readily put in the form of (1), which leads to the well-known Rankine-Hugoniot conditions across a shock wave in fluid mechanics. For this purpose we use the conventional notation of Section 12.4, so that the equations for the conservation of mass, momentum, and energy are

$$\frac{\partial \rho}{\partial t} + \frac{\partial}{\partial x_j}(\rho v_j) = 0\,, \tag{3}$$

$$\frac{\partial}{\partial t}(\rho v_j) + \frac{\partial}{\partial x_j}(\rho v_i v_j + p\delta_{ij}) = 0\,, \tag{4}$$

$$\frac{\partial}{\partial t}\left(\frac{1}{2}\rho v_i v_i + \rho e\right) + \frac{\partial}{\partial x_j}\left(\rho v_j\left(\frac{1}{2}v_i v_i + e + \frac{p}{\rho}\right)\right) = 0\,, \tag{5}$$

respectively, where all the terms have already been defined except for e, which is the internal energy. This system can be put in the form of (1) if we set

$$F_i = \begin{pmatrix} \rho \\ \rho v_1 \\ \rho v_2 \\ \rho v_3 \\ \frac{1}{2}\rho v_i v_i + \rho e \end{pmatrix} \tag{6}$$

$$M_{ij} = \begin{pmatrix} \rho v_1 & \rho v_2 & \rho v_3 \\ \rho v_1 v_1 + p & \rho v_1 v_2 & \rho v_1 v_3 \\ \rho v_2 v_1 & \rho v_2 v_2 + p & \rho v_2 v_3 \\ \rho v_3 v_1 & \rho v_3 v_2 & \rho v_3 v_3 + p \\ \rho v_1 Z & \rho v_2 Z & \rho v_3 Z \end{pmatrix}, \tag{7}$$

where $Z = \frac{1}{2}v_i v_i + e + p/\rho$. Substituting (6) and (7) in (2), setting $\nu = [\rho]$, which gives the strength of the shock front, and simplifying, we obtain

$$[p] = \frac{\nu}{1+\nu}\rho_1(G - v_{1n})^2\,, \qquad [v_k] = \frac{\nu}{1+\nu}(G - v_{1n})n_k\,, \tag{8}$$

$$[h] = \frac{1+\frac{1}{2}\nu}{(1+\nu)^2}(G - v_{1n})^2\,,$$

where $h = e + p/\rho$. These are the required jumps.

If we apply the concepts of Theorem 3 as given in Section (5.5) we can evaluate the jumps the gradients of p, v_k and h, the jump in vorticity and various other results [58–60]. Furthermore, by taking the $\delta/\delta t$ of the jumps of these flow quantities we get relations of the form (12.3.20).

12.6. Wave Fronts that Carry Infinite Singularities

In Chapter 5 we discussed the theory of distributional derivatives with jump discontinuities across points, curves, and surfaces. In the process we obtained the values of the jumps of various derivatives across these interfaces. These results do not, in general, apply if the fields have infinite singularities at a moving and a deforming surface. This happens in the theories of electromagnetism and magnetohydrodynamics, where surface densities can be generated by the sudden creation of electromagnetic multipoles in free space. Take, for instance, the Maxwell equations:

$$\overline{\text{curl}}\, E = -\overline{\partial} B/\partial t\,, \tag{1}$$

$$\overline{\text{div}}\, B = 0\,, \tag{2}$$

$$\overline{\text{curl}}\, H - \bar{\partial} D/\partial t = J\,, \tag{3}$$

$$\overline{\text{div}}\, D = \rho\,, \tag{4}$$

where E is the electric field, D is the displacement current, B is the density of magnetic flux, H is the magnetic field, J is the current density, and ρ is the charge density.

The corresponding equations for the electromagnetic potentials ϕ and A are

$$E = \overline{\text{grad}}\, \phi = -\frac{\bar{\partial} A}{\partial t}\,, \tag{5}$$

$$B = \overline{\text{curl}}\, A\,, \tag{6}$$

$$-\frac{1}{c^2}\frac{\bar{\partial}\phi}{\partial t} = \overline{\text{div}}\, A\,, \tag{7}$$

where the speed, c, is a constant. Now, if the fields E, B, D, and H have only the jump discontinuities across an interface Σ, then from (3) and (4) it follows that the current density, J, and the change density, ρ, are of the order of the Dirac delta function ($O(\delta)$). Similarly, from (5) and (6) it follows that if the potentials ϕ and A are $O(\delta)$, then E and B are $O(\delta')$. Then from (1)–(4) we find that the densities J and ρ are $O(\delta'')$. The same relationship is valid for higher-order infinite discontinuities. For example, if E, B, D, H are $O(\delta^{(n)})$ then J and ρ are $O(\delta^{n+1})$. Similar situations arise in the study of vortex sheets, atmospheric disturbances, and magnetohydrodynamic flows.

Our aim in this chapter is to use and extend the results of Chapter 5 to the present context. We follow the notation of that chapter and confine ourselves to R_3 so that $x = (x_1, x_2, x_3)$ denotes the field point and t denotes time. Some of the results of this chapter have been derived by Costen [61] by the classical technique of expressing the partial differential equations in their integral forms and then taking the usual limiting processes across the surface of discontinuity. We follow the method of generalized functions [62, 63].

For the present study we need a few results from Chapter 5. Let $f(x, t)$ be a quantity defined only on the surface $\Sigma(x, t)$. Then the first order derivatives of f with respect to the space and time variables are defined by the formulas (5.1.6) and (5.1.7) which we write as follows

$$\frac{\delta f}{\delta x_i} = \left(\frac{\partial \tilde{f}}{\partial x_i} - n_i \frac{d\tilde{f}}{dx_i}\right)\bigg|_{\Sigma}\,, \tag{8}$$

and

$$\frac{\delta f}{\delta t} = \left(\frac{\partial \tilde{f}}{\partial t} + G\frac{d\tilde{f}}{dx_i}\right)\bigg|_{\Sigma}\,, \tag{9}$$

where $\tilde{f}$ is any smooth extension of f to a neighborhood of $\Sigma(t)$ in $R_n \times R$. In the sequel we shall drop tilda from f. The distributional derivatives of a generalized function $F(x, t)$ are given by (5.5.4) and (5.5.5);

$$\frac{\bar{\partial} F}{\partial x_i} = \frac{\partial F}{\partial x_i} + [F] n_i \delta(\Sigma)\,, \tag{10}$$

$$\frac{\overline{\partial} F}{\partial t} = \frac{\partial F}{\partial t} - [F] G\delta(\Sigma) . \tag{11}$$

Formulas (5.5.9) and (5.5.8) are

$$\frac{\overline{\partial}}{\partial x_i}\delta(\Sigma) = n_i \delta'(\Sigma) = n_i (2\Omega\delta(\Sigma) + d_n \delta(\Sigma)) , \tag{12}$$

$$\frac{\overline{\partial}}{\partial t}\delta(\Sigma) = -G\delta'(\Sigma) = -G(2\Omega\delta(\Sigma) + d_n \delta(\Sigma)) , \tag{13}$$

wherein we have used formula (5.3.9), namely

$$\delta'(\Sigma) = 2\Omega\delta(\Sigma) + d_n \delta(\Sigma) . \tag{14}$$

With the help of relations (8)–(14) it follows that for a differentiable function $f(x, t)$ defined on $\Sigma(t)$ we have

$$\begin{aligned} \frac{\overline{\partial}}{\partial x_i}(f\delta(\Sigma)) &= \frac{\delta f}{\delta x_i}\delta(\Sigma) + n_i \, f \, \delta'(\Sigma) \\ &= \left(\frac{\partial f}{\partial x_i} - n_i \frac{df}{dn} + 2\Omega n_i f \right) \delta(\Sigma) + n_i \, f \, d_n \delta(\Sigma) , \end{aligned} \tag{15}$$

and

$$\begin{aligned} \frac{\overline{\partial}}{\partial t}(f\delta(\Sigma)) &= \frac{\delta f}{\delta t}\delta(\Sigma) - G \, f \, \delta'(\Sigma) \\ &= \left(\frac{\partial f}{\partial t} + G\frac{df}{dn} - 2\Omega G f \right) \delta(\Sigma) + G \, f \, d_n \delta(\Sigma) . \end{aligned} \tag{16}$$

The values of the quantities $\overline{\mathrm{curl}}(f\delta(\Sigma))$ and $\overline{\mathrm{div}}(f\delta(\Sigma))$ can be read off from (15) as

$$\overline{\mathrm{curl}}(f\delta(\Sigma)) = \left[\left(\mathrm{curl} \; - \widehat{n} \times \frac{d}{dn} \right) f + 2\Omega \widehat{n} x f \right] \delta(\Sigma) + \widehat{n} \times \, f d_n \delta(\Sigma) , \tag{17}$$

and

$$\begin{aligned} \mathrm{div}\, [f\delta(\Sigma)] = &\left[\left(\mathrm{div} \; - \widehat{n} \cdot \frac{d}{dn} \right) f + 2\Omega (\widehat{n} \cdot f) \right] \delta(\Sigma) \\ &+ \widehat{n} \cdot \, f d_n \delta(\Sigma) . \end{aligned} \tag{18}$$

From equations (1–7) we observe that we have to study the system of distributional equations of the first order such as

$$\lambda = -\frac{\overline{\partial}\vartheta}{\partial t} , \tag{19a}$$

$$m = \overline{\mathrm{grad}}\; \psi , \tag{19b}$$

$$b = -\frac{\overline{\partial} w}{\partial t} , \tag{19c}$$

$$j = \overline{\mathrm{curl}}\; p , \tag{19d}$$

$$\xi = \overline{\mathrm{div}} q , \tag{19e}$$

where the functions ϑ, ψ, w, p and q have jump discontinuities across a moving and deforming surface $\Sigma(t)$ which also carries a single layer potential. Accordingly, for any of these functions we can set [62, 63]

$$A(x,t) = A_0(x,t) + A_1(x,t)\delta(\Sigma)\,, \tag{20}$$

where the function ϕ_0 has the representation $\phi_0 = \mathcal{F} + H\,A$ (see (5.5.14)), $A = [F]$ and H is the Heaviside function. Then, the functions λ, m, b, j, ξ have a decomposition of the form

$$B(x,t) = B_0(x,t) + B_1(x,t)\delta(\Sigma) + B_2(x,t)d_n\delta(\Sigma)\,. \tag{21}$$

In view of relation (14), we can write (21) as

$$B(x,t) = B_0(x,t) + \overline{B}_1(x,t)\delta(\Sigma) + \overline{B}_2(x,t)\delta'(\Sigma)\,, \tag{22}$$

where

$$\overline{B}_1 = B_1 - 2\Omega B_2\,, \qquad \overline{B}_2 = \psi_2\,. \tag{23}$$

Let us now consider equation (19a) and use relations (16) and (20). Thereby, we get

$$\lambda = -\left\{\frac{\partial\vartheta_0}{\partial t} + \left(-G[\vartheta_0] + \left(\frac{\partial\vartheta_1}{\partial t} + G\frac{d\vartheta_1}{dn} - 2\Omega G\vartheta_1\right)\right)\delta(\Sigma) - G\vartheta_1 d_n\delta(\Sigma)\right\}\,. \tag{24}$$

Comparing this with (21) we find the following interfacial conditions

$$\lambda_0 = -\partial\vartheta_0/\partial t \tag{25a}$$

$$\lambda_1 = G[\vartheta_0] - \frac{\partial f}{\partial t} - G\frac{df}{dn} + 2\Omega G\vartheta_1\,, \tag{25b}$$

$$\lambda_2 = \vartheta_1 G\,. \tag{25c}$$

The distributional time derivative of the vector w, as given by (19d), is processed in the same fashion. Next we process equation (19b) and obtain

$$m_i = \frac{\partial\psi_0}{\partial x_i} + \left(n_i[\psi_0] + \frac{\partial\psi_1}{\partial x_i} - n_i\frac{d\psi_1}{dn} + 2\Omega\,n_i\,\psi_1\right)\delta(\Sigma) + \psi_1 n_i\,d_n\,\delta(\Sigma)\,. \tag{26a}$$

In order to deduce the corresponding relations for (19d) and (19e) we write, (26a) in the vectorial notation,

$$m = \text{grad}\,\psi_0 + \left(\widehat{n}[\psi_0] + 2\Omega\widehat{n}\psi_1 + \left(\text{grad}\, - \widehat{n}\frac{d}{dn}\right)\psi_1\right)\delta(\Sigma) + \psi_1\widehat{n}\,d_n\,\delta(\Sigma)\,. \tag{26b}$$

This yields the corresponding interfacial relations

$$m_{i_0} = \partial\psi_0/\partial x_i\,, \tag{27a}$$

$$m_{i_1} = [\psi_0]n_i + \left(\frac{\partial\psi_1}{\partial x_i} - n_i\frac{d\psi_1}{dn}\right) + 2\Omega_i\psi_1\,, \tag{27b}$$

$$m_{i_2} = n_i\,\psi_1\,. \tag{27c}$$

The value of b follows from relation (19c) and by putting w for λ in (24). Similarly, the formulas for j and ξ as given by equations (19d) and (19e) can be read off from (26b). These formulas are

$$j = \overline{\text{curl}}\, p = \text{curl}\ p_0 + \Bigg\{ (\widehat{n} \times [p_0]) + 2\Omega(\widehat{n} \times p_1) \\ \left(\text{curl}\ - \widehat{n} \times \frac{d}{dn}\right) p_1 \Bigg\} \delta(\Sigma) + (\widehat{n} \times \psi_1) d_n \delta(\Sigma)\,, \tag{28}$$

and

$$\xi = \overline{\text{div}} q = \text{div}\ q_0 + \left\{ \widehat{n} \cdot [q_0] + 2\Omega(\widehat{n} \cdot q_1) + \left(\text{div}\ - \widehat{n} \cdot \frac{d}{dn}\right) q_1 \right\} \delta(\Sigma) + \widehat{n} \cdot \psi_1 d_n \delta(\Sigma)\,. \tag{29}$$

The coefficients of $\delta(\Sigma)$ in (24), (26), (28) and (29) give the surface densities. Take for example (24) which yields

$$\lambda_\delta = G[\vartheta_0] - 2\,\Omega\, G\, \vartheta_1 - (\partial/\partial t + G\, \partial/\partial n)\vartheta_1\,. \tag{30}$$

The terms in this formula have interesting physical interpretations. The term $G[\vartheta_0]$ in (30) represents the rate of increase of density due to the snow plow action. The term $2\Omega\, G\, \vartheta_1$ accounts for an expanding (receding) bulge on Σ that tends to cause local dilution (concentration) of ϑ_1 as it appears to the observer moving with the surface Σ. Similarly, the last term on the right side of (30) represents the time rate of change (namely $\delta\vartheta_1/\delta t$) as it appears to the observer moving with the surface Σ.

The term $g\, \vartheta_1\, d_n \delta(\Sigma)$ in (24) also has an interesting interpretation. It gives the dipole distribution on Σ.

The terms on the right sides of (26), (28) and (29) can be interpreted in a similar manner.

Example 1. Let us apply the above analysis to the expanding sphere $r - ct = 0$. In spherical polar coordinates (r, ω_1, ω_2), relation (20) for ϑ is

$$\vartheta = f(r, \omega_1, \omega_2, t) + g(r, \omega_1, \omega_2, t) H(r - t) + h(r, \omega_1, \omega_2, t) \delta(r - ct)\,.$$

Thus

$$[\vartheta] = g\,,\ \ \vartheta_1 = h\,,\ \ G = c\,,\ \ \widehat{n} = r\,,\ \ \text{and}\ \ \Omega = -1/r\,,$$

and (19a) becomes

$$\begin{aligned} \lambda &= -\frac{\partial f}{\partial t} - \frac{\partial g}{\partial t} H(r - ct) + \left(cg - \frac{\delta h}{\delta t}\right) - h\delta'(r - t) \\ &= -\frac{\partial f}{\partial t} - \frac{\partial g}{\partial t} H(r - ct) + \left[cg - \frac{2ch}{r} - \left(\frac{\partial}{\partial t} + c\frac{\partial}{\partial r}\right) h\right] \delta(r - ct) \\ &\quad + ch\, d_r\, \delta(r - ct)\,. \end{aligned}$$

Example 2. With the analysis of this section we can study equations (1)–(7) of electromagnetism. For instance, by applying formula (29), equation (2) yields

$$[\widehat{n}.B] + 2\Omega\left(\widehat{n}.B_1 + \text{div } - \widehat{n}.\frac{\partial}{\partial n}\right) B_1 + \widehat{n}B_1 d_n \delta(\Sigma) = 0\,. \tag{31}$$

The above discussion is also applicable to the case when the function $f(x,t)$, appearing in relations (15) and (16), is discontinuous across a submanifold $\Delta(t)$ embedded in the surface $\Sigma(t)$. In this case we interpret the terms $\delta f/\delta x_i = \partial f/\partial x_i - n_i df/dn$ and $\delta f/\delta t = \partial f/\partial t + G\, df/dn$ as the distributional derivatives $\overline{\delta} f/\delta x_i$ and $\overline{\delta} f/\delta t$. Let the submanifold $\Delta(t)$ divide the surface $\Sigma(t)$ in two parts, called positive and negative so that $[f] = (f_+ - f_-)$ denotes the jump of ψ_1 across $\Delta(t)$. Just as we defined the manifold $\Sigma(t)$ by equation (5.2.1) we define the submanifold $\Delta(t)$ as by

$$\widetilde{u}(x_1, \ldots, x_m, t) = 0\,. \tag{32}$$

Then we have

$$\begin{aligned}\frac{\overline{\delta} f}{\delta x_i} &= g^{\alpha\beta}\frac{\partial x_i}{\partial v_\beta}\frac{\overline{\partial} f}{\partial v_\beta} \\ &= g^{\alpha\beta}\frac{\partial x_i}{\partial v_\beta}\frac{\partial f}{\partial v_\beta} + g^{\alpha\beta}\frac{\partial x_i}{\partial v_\beta}[f]\hat{n}_\alpha\delta(\Delta)\,,\end{aligned} \tag{33}$$

where the v's are the Gaussian coordinates as defined by (5.2.2) and $\hat{n}_\alpha$ are the components of the unit normal vector to the submanifold Δ. Accordingly, we can write (33) as

$$\frac{\overline{\delta} f}{\delta x_i} = \frac{\delta f}{\delta x_i} + w_i[f]\delta(\Delta)\,, \tag{34}$$

where the w_i are the components of the unit normal to $\Delta(t)$ which points in the positive direction and is tangent to $\Sigma(t)$. From (32), their values are

$$w_i = \frac{1}{u}\frac{\delta\overline{u}}{\delta x_i}\,, \qquad U = \frac{\delta\overline{u}}{\delta x_j}\frac{\delta\overline{u}}{\delta x_j} \tag{35}$$

When we substitute (34) for $\delta f/\delta x_i$ in (15) and observe that $\delta(\Delta)\delta(\Sigma) = \delta(\Delta)$, we obtain

$$\begin{aligned}\frac{\overline{\partial}}{\partial x_i}(f\delta(\Sigma)) &= \left(\frac{\partial f}{\partial x_i} - n_i\,\frac{df}{dn} + 2\Omega\, n_i\, f\right)\delta(\Sigma) \\ &\quad + w_i[f] + f\, n_i\, d_n \delta\,(\Sigma)\,.\end{aligned} \tag{36}$$

Similarly (16) becomes

$$\frac{\overline{\partial}}{\partial t}(f\,\delta(\Sigma)) = \left(\frac{\partial f}{\partial t} + G\frac{df}{dn} - 2\Omega G\right)\delta(\Sigma) - w_t[f]\delta(\Delta) + f\, G\, d_n\,\delta(\Sigma)\,, \tag{37}$$

where

$$w_t = -\frac{1}{U}\frac{\delta\overline{u}}{\delta t} \tag{38}$$

From relations (36) and (37) it follows that we must add the terms $w_t[\vartheta_1]\delta(\Sigma)$, $w_i[\psi_1]\delta(\Delta)$, $w_\times[\psi_1]\delta(\Delta)$ and $w_i[\psi_1]\delta(\Delta)$, respectively, in relations (24), (26), (28) and (29).

12.7. Kinematics of Wavefronts

Our aim in this section is to study the evolution of the surface measures, such as the fundamental forms, as the wave front propagates [64]. Let us restrict ourselves to the three-dimensional case and write some of the formulas discussed in Section 5.2 for the case $n = 3$.

The wave front $\Sigma(x, t)$ is represented as

$$x_i = x_i(v^1, v^2, t), \qquad i = 1, 2, 3, \tag{1}$$

where v^α, $\alpha = 1, 2$ are the curvilinear coordinates on the surface. We assume that the Jacobian $(\partial x^i/\partial v^\alpha)$ has the rank two. The quantities $x^i_\alpha = \partial x^i/\partial v^\alpha$ compute the tangent vectors on $\Sigma(t)$. Accordingly, the normal vectors n_i and x^i_α satisfy the relations

$$n_i n_i = 1\,, \quad \text{and } x^i_\alpha n_i = 0\,. \tag{2}$$

The first fundamental form $g_{\alpha\beta}$ and its inverse matrix $g^{\alpha\beta}$ are defined by relations (5.2.19):

$$g_{\alpha\beta} = x^i_\alpha\, x^i_\beta\,, \quad \text{and } g_{\alpha\beta}\, g^{\beta\gamma} = \delta^\gamma_\alpha\,. \tag{3}$$

Thereafter we define the quantities μ_{ij} in relation (5.2.20), which are the components of the second fundamental form. Here we write them as $b_{\alpha\beta}$ to conform to the usual notation in R_3. They are defined by the formulas

$$x^i_{\alpha,\beta} = b_{\alpha\beta} n_i\,, \quad \text{and } n^i = b^\beta_\alpha x^i_\beta\,. \tag{4}$$

Indeed, formula (5.2.19) and (4) are equivalent.

In the present notation, the mean curvature Ω and the Gaussian curvature K are given as

$$\Omega = -\frac{1}{2} b^\alpha_\alpha = -\frac{1}{2} g^{\alpha\beta} b_{\alpha\beta}\,, \tag{5}$$

and

$$K = \frac{\det(b_{\alpha\beta})}{\det(g_{\alpha\beta})} = \frac{b}{g}\,, \tag{6}$$

respectively. In terms of the principal curvature κ_1 and κ_2 of the surface $\Sigma(t)$, the values of Ω and K are

$$\Omega = \frac{1}{2}(\kappa_1 + \kappa_2)\,, \qquad K = \kappa_1\kappa_2\,. \tag{7}$$

The Christoffel symbols are defined as

$$\Gamma^\mu_{\alpha\beta} = \frac{1}{2} g^{\mu\gamma}\left[\frac{\partial g_{\beta\gamma}}{\partial_v \alpha} + \frac{\partial g_{\alpha\gamma}}{\partial_v \beta} - \frac{\partial g_{\alpha\beta}}{\partial_v \gamma}\right]. \tag{8}$$

Similarly, the curvature tensor $R_{\alpha\beta\gamma\delta}$ is given as

$$R_{\alpha\beta\gamma\delta} = b_{\alpha\delta} b_{\beta\gamma} - b_{\gamma\delta} b_{\beta\gamma} \,. \tag{9}$$

Let us now study the evolution of the above-mentioned surface quantities. For this prupose we appeal to the transport operator $\delta/\delta t$ as discussed in Section (5.2) and formula (5.2.8), namely, $\delta x^i/\delta t = G\,n_i$. Then we get

$$\frac{\delta x^i_\alpha}{\delta t} = \left(\frac{\delta x^i}{\delta t}\right)_{,\alpha} = (G\,n^i)_{,\alpha} = G_{,\alpha} n_i - G\,b^\beta_\alpha x^i_\beta \,, \tag{10}$$

where we have used formula (4). Next, we apply the $\delta/\delta t$ operator to the formula (6), and use (10). The result is

$$\frac{\delta g_{\alpha\beta}}{\delta t} = \left(\frac{\delta x^i_\alpha}{\delta t}\right) x^i_\beta + x^i_\alpha \left(\frac{\delta x^i_\beta}{\delta t}\right) = -2\,G\,b_{\alpha\beta} \,. \tag{11}$$

Because the determinant g of $g_{\alpha\beta}$ is $g = g_{11}g_{22} - g^2_{12}$, we find from (11) that

$$\frac{\delta g}{\delta t} = -2\,G\,(g_{22}b_{11} + g_{11}b_{22} - 2\,g_{12}b_{12}) \,. \tag{12}$$

This relation can be put in an elegant form if we use the contravariant tensor $g^{\alpha\beta}$ as defined in (3) so that $g_{11} = g\,g^{22}$, $g_{22} = g\,g^{11}$, $g_{12} = -g\,g^{12}$. Substitution of these values in (12) yields

$$\frac{\delta g}{\delta t} = -2\,Gg(g_{11}b_{11} + g^{22}b_{22} + 2\,g^{12}b_{12}) = -4\,G\,\Omega g \,. \tag{13}$$

Next we differentiate formula (3) and use formula (11) so that we get

$$\frac{\delta g^{\alpha\beta}}{\delta t} = 2\,G\,b^{\alpha\beta} \,. \tag{14}$$

The evolution of the second fundamental form $b_{\alpha\beta}$ is derived when we take the following two steps. First we differentiate (4) so that

$$\frac{\delta x^i_{\alpha,\beta}}{\delta t} = \left(\frac{\delta x^i_\alpha}{\delta t}\right),\beta = \frac{\delta b_{\alpha\beta}}{\delta t} n^i - (g^{\mu\gamma} b_{\alpha\beta} G_{,\gamma})\,x^i_\mu \,. \tag{15}$$

Next we substitute for $\delta x^i_\alpha/\delta t$, from (10), in the above relation, multiply both sides by n_i and sum over i. Thereby, we get the required transport equation

$$\begin{aligned} \frac{\delta b_{\alpha\beta}}{\delta t} &= G_{,\alpha\beta} - G b^\gamma_\alpha b_{\gamma\beta} \\ &= G_{,\alpha\beta} - 2G\,\Omega\,b_{\alpha\beta} + G\,K\,g_{\alpha\beta}, \end{aligned} \tag{16}$$

where we have used the Hamilton-Cayley formula

$$b^\alpha_\gamma\, b^\gamma_\beta - 2\,\Omega\,b^\alpha_\beta + K\,\delta^\alpha_\beta = 0 \,. \tag{17}$$

Since the determinant b of $B_{\alpha\beta}$ is $b_{11}b_{22} - b_{12}^2$ and $b_{11} = b\, b^{22}$, $b_{22} = b\, b^{11}$, $b_{12} = -b\, b^{12}$, we find from the above relations that

$$\frac{\delta b}{\delta t} = -2G\, g\, K\, \Omega + K\, g\, b^{\alpha\beta} G,_{\alpha\beta}\ . \tag{18}$$

Similarly, the value of $\delta b^{\alpha\beta}/\delta t$ follows by appealing to the relation $b^{\alpha\beta} b_{\beta\gamma} = \delta^\alpha_\gamma$ and formulas (16) and (18) so that

$$\frac{\delta b^{\alpha\beta}}{\delta t} = -G,_{\mu\gamma}\, b^{\alpha\mu} b^{\beta\gamma} + 2\,G\,\Omega\, b^{\alpha\beta}(1+K) - G\,K^2 g^{\alpha\beta}. \tag{19}$$

The evolution of the Gaussian curvature K follows by differentiating the formula $K = b/g$. Thus

$$\frac{\delta K}{\delta t} = \frac{\delta}{\delta t}\left(\frac{b}{g}\right) = \frac{1}{g}\frac{\delta b}{\delta t} - \frac{b}{g^2}\frac{\delta g}{\delta t}\ .$$

When we substitute the values of $\delta g/\delta t$ and $\delta b/\delta t$ from (12) and (18) in the above relation we obtain

$$\frac{\delta K}{\delta t} = 2\,G\,K\Omega + K\, b^{\alpha\beta} G,_{\alpha\beta}\ . \tag{20}$$

Similarly, if we differentiate the formula

$$\Omega = -\frac{1}{2} g^{\alpha\beta} b_{\alpha\beta} = -\frac{1}{2g}(g_{11}b_{22} + g_{22}b_{11} - 2g_{12}b_{12})\,,$$

we derive the relation

$$\frac{\delta\Omega}{\delta t} = -\left[(2\Omega^2 - K)G + \frac{1}{2} g^{\alpha\beta} G,_{\alpha\beta}\right]. \tag{21}$$

Proceeding in a similar manner we can derive the transport equations for the principal curvatures, the Christoffel symbols and the curvature tensor by differentiating formulas (7), (8) and (9).

Let us now demonstrate how generalized functions enter this picture. For this purpose we consider the two-phase flows. Let the interface $\Sigma(t)$ separate the region $V(t)$ occupied by two-phase continuum so that V_- and V_+ are the subregions for these two phases. The picture is essentially the same as we encountered in Figure 5.2.

Next, we define the phase function $\chi(x,t)$

$$\chi(x,t) = \begin{cases} 0\,, & x \in V_-(t)\,, \\ 1\,, & x \in V_+(t)\,, \end{cases} \tag{22}$$

which is similar to the Heaviside function. By appealing to the distributional derivatives (5.24) and (5.25) we derive the formulas

$$\frac{\overline{\partial}\chi}{\partial x_i} = n^i \delta(\Sigma)\,, \quad \text{and} \quad \frac{\overline{\partial}\chi}{\partial t} = -G\,\delta(\Sigma)\,. \tag{23}$$

Thus, it follows that

$$\frac{\overline{\partial}\chi}{\partial t} + G n^i \frac{\overline{\partial}\chi}{\partial n_i} = 0\,, \tag{24}$$

which is called the equation of motion for $\chi(x, t)$. Similarly, an interesting equation for the function $\overline{\partial}\chi/\partial n$ arises if we observe from (23) that

$$\frac{\overline{\partial}\chi}{\partial n} = \frac{\overline{\partial}\chi}{\partial x_i} n_i = \delta(\Sigma)\,. \tag{25}$$

Substituting this value of $\delta(\Sigma)$ in relation (5.2.9) we get

$$\frac{\partial}{\partial t}\left(\frac{\partial\chi}{\partial n}\right) + G\frac{\partial}{\partial n}\left(\frac{\partial\chi}{\partial n}\right) = 0\,. \tag{26}$$

Now we apply the relation $n_i n_{i,\alpha} = 0$ and formula (5.2.21), which can be written as

$$\frac{\delta n^i}{\delta t} = \frac{\delta}{\delta x_i}(-G) = -g^{\alpha\beta} x^i_\alpha G_{,\beta}\,, \tag{27}$$

so that formula (26) takes the form

$$\frac{\overline{\partial}}{\partial t}\left(\frac{\partial\chi}{\partial n}\right) + \overline{\nabla}\cdot\left(G n^i \frac{\partial\chi}{\partial n}\right) + G\Omega\frac{\overline{\partial}\chi}{\partial n} = 0\,. \tag{28}$$

This formula has been derived before by the classical techniques [65]. The generalized functions have made the derivation easy.

With the help of the analysis in the previous section we can extend these results. Given a function $f(x, t)$, the quantity $f\chi$ has jump f at the interface between V_- and V_+. By setting $F = f\chi$, we have the distributional derivatives

$$\frac{\overline{\partial}}{\partial x_i}(f\chi) = \frac{\partial f}{\partial x_i}\chi + n_i\, f\,\delta(\Sigma)\,, \tag{29}$$

and

$$\frac{\overline{\partial}}{\partial t}(f\chi) = \frac{\partial f}{\partial t}\chi - G\, f\,\delta(\Sigma)\,. \tag{30}$$

To get another interesting result we rewrite (21) as

$$\frac{\delta\Omega}{\partial t} + G n^i \frac{\partial\Omega}{\partial x_i} = -(2\Omega^2 - K)G - \frac{1}{2} g^{\alpha\beta} G_{,\alpha\beta}\,. \tag{31}$$

Next, we return to relations (5.2.8) and (5.2.9). When we multiply (5.2.8) by $G\,n_i$ and sum over the index i and then add the resulting relation to (5.2.9) we obtain

$$\frac{\partial}{\partial t}(\delta(\Sigma)) + G\,n_i\frac{\partial}{\partial x_i}(\delta(\Sigma)) = 0\,. \tag{32}$$

Finally, with the help of relations (31) and (32), we readily obtain

$$\frac{\partial}{\partial t}(\Omega\delta(\Sigma)) + \text{div}\,(G\widehat{n}\Omega\delta(\Sigma)) = \left(-(\Omega^2 - K)G - \frac{1}{2} g^{\alpha\beta} G_{,\alpha\beta}\right)\delta(\Sigma)\,. \tag{33}$$

This formula yields the flow equation for the mean curvature density spread over the wavefront.

Evolution of surfaces by prescribed vector fields

The evolution of hypersurfaces by normal vector fields based on their various curvature functions is being studied extensively. This concept rests on finding a family of maps $F(x,t)$ which deform and evolve the initial surface $F_0(x) = F(x,0)$ by the equation

$$\frac{\partial F(x,t)}{\partial t} = \text{a function of surface measures.}$$

For instance, for the mean curvature flow, we have

$$\frac{\partial F(x,t)}{\partial t} = 2\Omega\, n^i\,. \tag{34}$$

If the surface is defined as $F = F(x,y)$, then the mean curvature Ω is

$$\Omega = \operatorname{div}\left(\frac{\operatorname{grad} F}{(1+|\operatorname{grad} F|^2)^{1/2}}\right), \tag{35}$$

so that (34) becomes the non-linear partial differential equation

$$\frac{\partial F(x,t)}{\partial t} - \left[2\operatorname{div}\left(\frac{\operatorname{grad} F}{(1+|\operatorname{grad} F|^2)^{1/2}}\right)\right] n^i = 0 \tag{36}$$

with the initial condition $F(x,0) = F_0(x)$.

More recent studies are devoted to the Ricci flow where

$$\frac{\partial g_{ij}(x,t)}{\partial t} = (r - R(x,t))\; g_{ij}(x,t)\,, \tag{37}$$

where R is the scalar Riemann curvature and r is the average

$$r = \frac{\int R\, d\Sigma}{\int d\,\Sigma}\,. \tag{38}$$

There is an extensive literature on the subject. These studies are based on closed surfaces which contain no singularities on them. However, if the singularities exist on the surfaces or if they are open and have edges, then the derivatives in these studies become distributional derivatives and require the results developed above.

12.8. Derivation of the Transport Theorems for Wave Fronts

1. Volume Integral. In Example 3 of Section 5.1 we used the distributional derivatives and obtained the Leibnitz formula for the differentiation of integrals with varying limits of integration. We now extend that formulation and derive the transport theorem for wavefronts [66].

Let $f(x,t)$ be a sufficiently smooth function defined in a moving solid $V(t)$ and assume that its boundary is a moving hypersurface $\Sigma(t)$. Then we consider the function $F(t)$

$$F(t) = \int_{V(t)} f(x,t)dx = \langle f(x,t), 1\rangle_x\,, \tag{1}$$

and set $V(t) = V_+$, and $R_n \backslash \overline{V}(t) = V_-$ so that n is the outward unit normal. Next, we extend $f(x,t)$ to all R_n by setting $f(x,t) = 0$ in V_- and differentiate both sides of (1) so that

$$\begin{aligned} F'(t) &= \left\langle \frac{\overline{\partial}}{\partial t} f(x,t), 1 \right\rangle \\ &= \left\langle \frac{\partial}{\partial t} f(x,t) - G[f]\delta(\Sigma(t)), 1 \right\rangle . \end{aligned} \tag{2}$$

Then relations (1), (2) and the fact that $[f] = -f_- = -f$, yield the transport theorem.

$$\frac{d}{dt}\int_{V(t)} f(x,t)dx = \int_{V(t)} \frac{\partial f}{\partial t}\, dx + \int_{\Sigma(t)} G\, f\, d\,\Sigma\,. \tag{3}$$

2. Surface integral. Let $f(x,t)$ be a quantity defined only on the moving and deforming surface $\Sigma(t)$ and recall formulas (5.2.16) and (5.2.17) which we write as we did in Section 12.6, namely,

$$\frac{\delta f}{\delta x_i} = \left(\frac{\partial \tilde{f}}{\partial x_i} - n_i \frac{\partial \tilde{f}}{\partial n}\right)\Bigg|_{\Sigma}, \tag{4}$$

and

$$\frac{\delta f}{\delta t} = \left(\frac{\partial \tilde{f}}{\partial t} + G \frac{\partial \tilde{f}}{\partial n}\right)\Bigg|_{\Sigma}, \tag{5}$$

where $\tilde{f}$ is an extension of f to $R_n \times R$. Next, we appeal to formula (12.6.15),

$$\frac{\overline{\partial}}{\partial t}(f\delta(\Sigma)) = \left(\frac{\delta f}{\delta t} - 2\Omega\, G\, f\right)\delta(\Sigma) - f\, G\, d_n\, \delta(\Sigma)\,. \tag{6}$$

The next step is to set

$$F(t) = \int_{\Sigma(t)} f(x,t)d\,\Sigma_x = \langle f\delta(\Sigma), 1\rangle_x\,, \tag{7}$$

where Σ_x signifies that integration is with respect to x, and use (6) so that

$$\begin{aligned} F'(t) &= \left\langle \frac{\overline{\partial}}{\partial t}(f\,\delta(\Sigma))\,, 1 \right\rangle_x \\ &= \left\langle \left(\frac{\delta f}{\delta t} - 2\Omega\, G\, f\right)\delta(\Sigma) - f\, G\, d_n\, \delta(\Sigma)\,, 1 \right\rangle . \end{aligned} \tag{8}$$

Thus, we get the required transport theorem,

$$\frac{d}{dt}\int_{\Sigma(t)} f(x,t)d\Sigma_x = \int_{\Sigma(t)}\left(\frac{\delta f}{\delta t} - 2\Omega G f\right)d\Sigma_x . \tag{9}$$

Discontinuous integrands on the wave front

Suppose that $f(x,t)$ is of class C^2 on $\Sigma(t)$ except for a jump discontinuity across the $(n-2)$-dimensional moving submanifold $\Delta(t)$ of $\Sigma(t)$. Then we must use formula (12.6.37), namely

$$\begin{aligned}\frac{\bar{\partial}}{\partial t}(f\,\delta(\Sigma)) = \left(\frac{\delta f}{\delta t} - 2\Omega G f\right)\delta(\Sigma) - w_t[f]\delta(\Delta)\\ - f\,G\,d_n\delta(\Sigma)\end{aligned} \tag{10}$$

instead of (6) so that (9) is replaced by the formula.

$$\begin{aligned}\frac{d}{dt}\int_{\Sigma(t)} f(x,t)d\,\Sigma_x = \int_{\Sigma(t)}\left(\frac{\delta f}{\delta t} - 2\Omega G f\,\delta\,(\Sigma)\right)\\ - \int_{\Delta(t)} w_t[f]\,d\Delta_x .\end{aligned} \tag{11}$$

12.9. Propagation of Wave Fronts Carrying Multilayer Densities

Recall the definition of the single and dipole layer densities as defined by relations (5.3.1) and (5.3.7). We extend these concepts and define the multilayer distribution $d_n^P\delta(\Sigma)$ as

$$\langle d_n^P\delta(\Sigma),\phi\rangle = (-1)^P\int_{-\infty}^{\infty}\int_{\Sigma(t)}\frac{d^P\phi(x,t)}{dn^P}\,d\Sigma\,dt\,, \tag{1}$$

$P = 1,2,3\ldots.$ It reduces to the single layer for $P = 0$ and the dipole layer for $P = 1$. When we introduce the density $f(x,t)$ spread over the layer, we obtain the multilayer $f\,d_n^P\delta(\Sigma)$,

$$\langle f\,d_n^P\delta(\Sigma),\phi\rangle = (-1)^P\left\langle f,\frac{d^P\phi}{dn^P}\right\rangle . \tag{2}$$

Observe that we can write this multilayer also as $d_n^P(f\,\delta(\Sigma))$. We have already come across these quantities for $P = 1$, in the previous sections.

Higher order fundamental forms. Let us recall the surface quantities μ_{ij}, μ_{it} and μ_{tt} defined by relations (5.3.20) to (5.3.24). Our aim is to build up the higher order fundamental forms with the help of these quantities. Accordingly, we mention them for a ready reference. They are

$$\mu_{ij}^{(0)} = \delta_{ij} - n_i n_j\,, \qquad \mu_{ij} = \delta n_i/\delta x_i\,, \tag{3a}$$

$$\mu_{it} = \frac{\delta n_i}{\delta t} = \frac{\delta}{\delta x_i}(-G) = -g^{\alpha\beta}\, x^i_\alpha\, G_{,\beta}, \quad \mu_{tt} = -\frac{\delta G}{\delta t}\,. \tag{3b}$$

Next, we define the matrices μ_{ij}^P as

$$\mu_{ij}^{(P+1)} = \mu_{ik}^{(P)}\mu_{kj} \quad , \quad P = 0, 1, 2, \ldots, \tag{4}$$

where $\mu_{ij}^{(0)}$ and $\mu_{ij}^{(1)} = \mu_{ij}$ are defined by (3a). Similarly, $\mu_{it}^{(P+1)}$ and $\mu_{tt}^{(P)}$ are defined by replacing j by t and (i, j) by (t, t) so that $\mu_{it}^{(1)} = \mu_{it}$ as given by (3b).

The higher order analogs of the second form are $\lambda^{(N,P)}$ defined as follows:

$$\lambda_i^{(1,1)} = n_i\,, \quad \lambda_i^{(1,P)} = 0\,, \qquad P \neq 1\,, \tag{5}$$

and for $n \geq 1$ we have

$$\lambda_j^{(N+1,P)} = \frac{\delta}{\delta x_j}(\lambda^{(N,P)}) + \mu_{jk}\lambda_k^{(N+1,P-1)} = 0\,. \tag{6}$$

From these relations we obtain $\lambda^{(N,0)} = 0$, for all N. For $N = 2$,

$$\lambda_{ij}^{(2,P)} = \mu_{ij}^{(P)}, \quad P = 1, 2, 3, \ldots, \tag{7}$$

while for $N = 3$, we have

$$\lambda_{ijk}^{(3,P)} = \sum_{Q=1}^{P} \mu_{is}^{(P-Q)} \frac{\delta \mu_{jk}^{(Q)}}{\delta x_s} - P n_i\, \mu_{jk}^{P+1}\,. \tag{8}$$

Most of the quantities that appear in this analysis are symmetric tensors. Accordingly, we use the notation ab to denote the symmetric product of a and b. For instance, if $a = (a_i)$, $b = (b_i)$ are vectors then ab is the symmetric matrix

$$(ab)_{ij} = \frac{1}{2}(a_i b_j + a_j b_i)\,. \tag{9}$$

Similarly, if $c = (c_{ij})$ is a symmetric matrix, then

$$(ac)_{ijk} = \frac{1}{3}(a_i c_{jk} + a_j c_{ik} + a_k c_{ij})\,. \tag{10}$$

The product of two symmetric matrices $c = (c_{ij})$ and $d = (d_{ij})$ is

$$\begin{aligned}(cd)_{ijk\ell} = \frac{1}{6}(&c_{ij}d_{k\ell} + c_{ik}d_{j\ell} + c_{i\ell}d_{kj} \\ &+ c_{jk}d_{i\ell} + c_{j\ell}d_{ik} + c_{k\ell}d_{ij})\end{aligned} \tag{11}$$

and so on. Recall that in Section 5.10 we defined the quantity a^R, R a positive integer, as the symmetric product $a \ldots$, R times.

Let us now use the concept of symmetric product to generalize the operator D^2_{ij} as defined by the relation (5.2.30). Indeed, an N-th order symmetric operator D^N can be defined by means of the recursive formula

$$D_j^{N+1} = D_j(D^N) - n_j \sum_{P=1}^{N} \binom{N}{P-1} \lambda_k^{(N-P+2,1)} D_k^P \,. \tag{12}$$

Observe that the coefficient $\binom{N}{P-1}$ is the number of different terms in the symmetric product of the tensors $\lambda_k^{(N-P+2,1)}$ and D_k^P (here k is a fixed index). In particular

$$D^3_{ijk} = \frac{\delta}{\delta x_i}(D^2_{jk}) - n_i \left(\mu_{js} D^2_{sk} + \mu_{ks} D^2_{sj} + \lambda^{(3,1)}_{jks} \frac{\delta}{\delta x_s} \right). \tag{13}$$

Interestingly, an analog of the Leibnitz rule is valid for the operators D^N. Indeed,

$$D^n(fg) = \sum_{P=0}^{N} \binom{N}{P} D^{N-P} f \, D^P g \,, \tag{14}$$

where the symmetric product is used between the tensors $D^{N-P} f$ and $D^P g$. Since it can be proved that

$$\lambda_j^{(N+1,1)} = D^N(n_j) \,, \tag{15}$$

it follows from (15) that

$$D^N(n_i f) = \sum_{P=0}^{N} \binom{N}{P} \lambda_j^{(P+1,1)} D^{N-P} f \,, \tag{16}$$

so that in particular

$$\sum_{P=0}^{N} \binom{N}{P} \lambda_j^{(P+1,1)} \lambda_j^{(N-P+1,1)} = 0 \,. \tag{17}$$

Similarly, from (13) and (17) we obtain

$$\lambda^{(N+1,2)} = -n_j \, \lambda_j^{(N+2,1)} = -n_j \, D_{n_j}^{N+1}(n_j) \,. \tag{18}$$

For many more interesting relations of this nature and their detailed proofs readers should consult reference [13]. We mention a couple of examples below.

(i) For a plane, $\lambda^{(N,P)} = 0$ for $N \geq 2$.

(ii) When Σ is a sphere of radius r, the surface derivative takes the form

$$\frac{\delta f}{\delta x_i} = \left(\delta_{ij} - \frac{x_i x_j}{r} \right) \frac{\delta \tilde{f}}{\delta x_j} \bigg|_{\Sigma} \,, \tag{19}$$

where $\widetilde{f}$ is any extension of f to R_n. Since, $n_i = x_i/r$, we find, from (19), that

$$\mu_{ij} = \frac{\delta n_i}{\delta x_j} = \frac{1}{r}(\delta_{ij} - n_i n_j)\,, \tag{20}$$

so that (4) yields

$$\mu^{(P)} = \frac{1}{r^P}\,(\delta - n^2)\,, \qquad P = 0, 1, 2, \ldots, \tag{21}$$

where $\delta = \delta_{ij}$ is the Kronecker delta and we have used the notation as defined by (9). By applying these results in (6) we can derive the values of the higher order fundamental forms λ. For instance, for $P > 1$ we have

$$\lambda^{(3,P)} = -3\,P\,n\,\mu^{(P+1)}\,, \tag{22}$$

and

$$\lambda^{(4,P)} = \binom{P+1}{2}(12n^2 - 3\,\mu^{(0)})\mu^{P+2}\,, \tag{23}$$

and so on.

Inter-facial conditions across $\Sigma(t)$. Let F be a function which is defined and has derivatives of all orders in R_n/Σ such that all these derivatives have boundary values on $\Sigma(t)$ from both sides. In the study of such functions in Section 5.7 we defined the quantities A^Q:

$$A^{(Q)} = \left[\frac{d^Q F}{dn^Q}\right] = \left(\frac{d^Q F}{dn^Q}\right)_+ - \left(\frac{d^Q F}{dn^Q}\right)_-\,, \tag{24}$$

$Q = 0, 1, 2, \ldots,$ to be the jumps of the normal derivatives of order Q of F across Σ. We found that we can evaluate the interfacial relations of the first and second order of F across Σ in terms of the unit normal, n, and the second fundamental form μ_{ij}. They are given by relations (5.5.23) and (5.6.11). In the notation of (24), they can be written as

$$[DF] = \left[\frac{\partial F}{\partial x_i}\right] = A^{(1)}\,n_i + \frac{\delta A^{(0)}}{\delta x_i}\,, \tag{25}$$

and

$$[D^2 F] = \left[\frac{\partial^2 F}{\partial x_i \partial x_j}\right] = A^{(2)} n_i n_j + n_i \frac{\delta A^{(1)}}{\delta x_j} + n_j \frac{\delta A^{(1)}}{\delta x_i} + \mu_{ij} A^{(1)} + D^2_{ij} A^{(0)}\,. \tag{26}$$

If we multiply both sides of (26) by n_j and sum over j we obtain

$$\left[\frac{d\,DF}{dn}\right] = A^{(2)} n_i + \frac{\delta A^{(1)}}{\delta x_i} - \mu_{ik}\frac{\delta A^{(0)}}{\delta x_k}\,. \tag{27}$$

When $A^{(Q)}$ are constants, the above formulas reduce to

$$[DF] = A^{(1)} n_i\,,\ [D^2 F] = \mu_{ij} A^{(1)} + n_i n_j A^{(2)},\ \text{and}\ \left[\frac{d\,DF}{dn}\right] = A^{(2)} n^i\,. \tag{28}$$

Observe that they are linear relations in A^Q.

In order to find the higher order jump conditions we need the values of the higher order fundamental forms, otherwise the analysis is the same as in Chapter 5. The details and proofs of these derivations are given by Estrada and Kanwal [13,67]. Here we merely state them. The general formula is

$$\left[\frac{d^P D^N F}{dn^P}\right] = \sum_{Q=0}^{N+P} S_Q^{(N,P+q)}(A^{(Q)}), \tag{29}$$

where P and Q are integers. The quantities $S^{(N,P)}$ are certain operators that act on quantities $A^{(Q)}$ and whose values are tensors of order N. For instance, the values of $S^{(0,1)}$, $S^{(0,2)}$ and $S^{(1,1)}$ are given by the right hand sides of (25) to (27).

When $A^{(Q)}$ are constants, relations (29) reduce to

$$\left[\frac{d^P D^N F}{dn^P}\right] = A^{(Q)} C_Q^{(N,P+Q)}. \tag{30}$$

For the special cases of $(P, Q) = (0, 1), (0, 2), (1, 1)$, the values of $C_Q^{(N,P+Q)}$ can be read off from (28). Clearly, these operators are linear in the quantities $A^{(Q)}$.

Observe that $S_Q^{(N,P+Q)}$ is a differential operator of order N and acts on $A^{(Q)}$. The constant term is $C_Q^{(N,P+Q)}$. For $N \geq 3$, both these operators contain the fundamental forms of order greater then two in addition to the surface quantities n_i and μ_{ij}. Let us now give several special cases of $S_Q^{(N,P+Q)}$.

(i) $Q = 0, \quad P = 0$:

$$S_Q^{(N,0)}(A^{(0)}) = D^N(A^{(0)}), \tag{31a}$$

(ii) $N = 1, \quad P \geq 0$,

$$S_Q^{(1,P+Q)}(A^Q) = \frac{(-1)^P\ (P+Q)!}{Q!}\ \mu_k^{(P)} D^k(A^Q), \tag{31b}$$

while

$$S_Q^{(1,Q-1)} = n\,I. \tag{31c}$$

Then for, $P \geq 0$ and $N = 1$, we have

$$\left[\frac{d^P F, i}{dn\ P}\right] = \sum_{Q=\infty}^{P} \frac{P!}{Q!}\,(-1)^{P-Q}\mu_{ij}^{P-Q}\frac{\delta A^Q}{\delta x_j} + A^{(P+1)}n_i. \tag{32}$$

For $P = 0$ and $P = 1$, we recover (25) and (27) respectively.

Similarly, for $P > 0$, $N = 2$. Formula (29) is

$$\left[\frac{d^P F_{,ij}}{dn^P}\right] = \sum_{Q=0}^{P} \frac{(-1)^{(P-Q)} P!}{Q!} \left\{ \sum_{B=0}^{P-Q} \overline{\mu}_{ir}^{(P-Q-B)} \overline{\mu}_{js}^{(B)} D_{rs}^2(A^{(Q)}) + \lambda_{ijr}^{(3,P-Q)} \frac{\delta A^{(Q)}}{\delta x_r} \right.$$
$$\left. - Q(n_i \mu_{jr}^{(P-Q+1)} \frac{\delta A^{(Q)}}{\delta x_r} + n_j \mu_{ir}^{(P-Q_1)} \frac{\delta A^{(Q)}}{\delta x_r} + \mu_{ij}^{(P-Q+2)} A^{(Q)}) \right\}$$
$$+ n_i \frac{\delta A^{(P+1)}}{\delta x_r} + \mu_{ij} A^{(P+1)} + n_j \frac{\delta A^{(P+1)}}{\delta x_i} n_i n_j A^{(P+2)}, \tag{33}$$

where $\overline{\mu}_{ij}^{(P)} = \mu_{ij}^{(P)}$ for $P \geq 1$ and $\overline{\mu}_{ij}^{0} = \delta_{ij}$, and for $P = 0$, $N = 3$ we have

$$[F_{,ijk}] = D_{ijk}^3 A^{(0)} + n_i D_{jk}^2 A^{(1)} + n_j D_{jk}^2 A^{(1)} + n_k D_{ij}^2 A^{(1)} + \mu_{ij} \frac{\delta A^{(1)}}{\delta x_k} + \mu_{ik} \frac{\delta A^{(1)}}{\delta x_j}$$
$$+ \mu_{jk} \frac{\delta A^{(1)}}{\delta x_i} + \lambda_{ijk}^{(3,1)} A^{(1)} + n_i n_j \frac{\delta A^{(2)}}{\delta x_k} + n_i n_k \frac{\delta A^{(2)}}{\delta x_j} + n_j n_k \frac{\delta A^{(2)}}{\delta x_i}$$
$$+ (n_i \mu_{jk} + n_j \mu_{ik} + n_k \mu_{ij}) A^{(2)} + n_i n_j n_k A^{(3)}, \tag{34}$$

and so on.

Observe that the corresponding relations for the time derivatives can be read off from the above analysis as explained previously.

Distributional derivatives of multilayers. With the help of the results of this section we can extend formulas (12.6.15) and (12.6.16), which give the first order distributional derivatives of the single layer $f(\Sigma)$. Indeed, the present analysis enables us to obtain the formula

$$\overline{D}^N (f\, d_n^P \delta(\Sigma)) = \sum_{Q=0}^{N+P} T_Q^{N,P}(f) d_n^Q \delta(\Sigma), \tag{35}$$

where the tensorial quantities are functions of $f(x, t)$, the quantities $S_Q^{N,P}$ and the higher fundamental forms of $\Sigma(t)$ as derived previously. For instance, when $N = 1$, we have

$$T_Q^{(1,P)}(f) = \frac{P!}{Q!} \frac{\delta}{\delta x_k} (\mu_k^{P-Q} f), \quad Q \leq P, \tag{36}$$

while

$$T_{P+1}^{(1,P)}(f) = fn. \tag{37}$$

Let us present a few examples.

Example 1. $N = 1$. We have

$$\frac{\overline{\partial}}{\partial x_i}(f\, d_n^P \delta(\Sigma)) = \sum_{Q=0}^{P} \frac{P!}{Q!} \frac{\delta}{\delta x_i} (\mu_{ik}^{P-Q} f) d_n^Q \delta(\Sigma)$$
$$+ f\, n_i\, d_n^{P+1} \delta(\Sigma), \tag{38}$$

and

$$\frac{\bar{\partial}}{\partial t}(f\, d_n^P \delta(\Sigma)) = \sum_{Q=0}^{P} \frac{P!}{Q!} \frac{\delta}{\delta t}(\mu_{ik}^{P-Q} f) d_n^Q \delta(\Sigma)$$
$$- G\, f\, d_n^{P+1} \delta(\Sigma)\,. \tag{39}$$

For the special case of $P = 0$ these formulas reduce to (12.6.15) and (12.6.16) respectively.

2. $N = 2$, $P = 0$. This yields

$$\frac{\partial^2}{\partial x_i \partial x_j}(f\, \delta(\Sigma)) = D_{ij}^{*^2} f\, \delta(\Sigma)$$
$$+ \left\{ \frac{\delta^*}{\delta x_i}(n_j f) + \frac{\delta^*}{\delta x_j}(n_i f) - \mu_{ij} f \right\} d_n \delta(\Sigma)$$
$$+ n_i n_j\, f\, d_n^2 \delta(\Sigma)\,. \tag{40}$$

3. $N = 3$, $P = 0$.

$$\frac{\partial^3}{\partial x_i \partial x_j \partial x_k}(f\delta(\Sigma)) = D_{ijk}^{*3}(f)\delta(\Sigma) +$$
$$+ \left\{ D_{jk}^{*2}(n_i f) + D_{ij}^{*2}(n_k f) + D_{ik}^{*}(n_j f) - \frac{\delta^*}{\delta x_k}(\mu_{ij} f) \right.$$
$$\left. - \frac{\delta^*}{\delta x_j}(\mu_{ik} f) - \frac{\delta^*}{\delta x_i}(\mu_{jk} f) + \lambda_{ijk}^{(3,1)} f \right\} d_n \delta(\Sigma)$$
$$+ \left\{ \frac{\delta^*}{\delta x_k}(n_i n_j f) + \frac{\delta^*}{\delta x_j}(n_i n_k f) + \frac{\delta^*}{\delta x_i}(n_j n_k f) \right.$$
$$\left. - (n_i \mu_{jk} + n_j \mu_{ik} + n_k \mu_{ij}) f \right\} d_n^2 \delta(\Sigma)$$
$$+ n_i n_j n_k f\, d_n^3 \delta(\Sigma)\,, \tag{41}$$

and so on.

12.10. Generalized Functions with Support on the Light Cone

In this section we construct the generalized functions of type $f(t, y) d_n^P \delta(t - r)$ supported by the light cone $t \pm r = 0$. The formulas, as given in the previous sections, cannot be applied for this surface because it is singular at the point $(0, 0)$. As such, the previous formulas should be modified by the addition of suitable distributions concentrated at $(0, 0)$, i.e., the distributions of the form

$$\delta^{(k)}(t) \delta^{(m_1)}(x_1) \delta^{m_2}(x_2) \delta^{m_3}(x_3)\,.$$

The present discussion is based on the analysis as given by Estrada and Kanwal [68]. Instead of considering functions $g(t, x)$ defined on the cone, they consider functions of the form of $f(t, y) = g(t, ty)$ defined in the cylinder $t \geq 0$, $|y| = 1$. Then the multilayers supported in the light cone take the form $f(t, y)d_n^P \delta(t - r)$, where f is a generalized function in the cylinders.

Let $\xi = (t, x)$ so that

$$\langle \xi, \xi \rangle = t^2 - |x|^2 = t^2 - r^2 = (t - r)(t + r)$$

is the square of the length in the Minkowski space from the origin $(0, 0)$ to the point ξ. Accordingly, the Dirac delta function $\delta(\xi, \xi)$ has support on the light cone $t \pm r = 0$. Then formula (3.1.8) yields

$$\delta(\xi, \xi) = \delta(t - r)\delta(t + r) = \frac{\delta(t - r)}{2r} + \frac{\delta(t + r)}{2r}, \tag{1}$$

if either t is fixed (and $t \neq 0$) or r is fixed (and $r \neq 0$).

Next, we adapt the analysis of the previous sections when the surface $\Sigma(t)$ is the light cone $t - r = 0$. Thus, the basic surface distribution is

$$\langle \delta(t - r), \phi(t, x) \rangle = \int_0^\infty \int_{t=r} \phi(t, x) dS(x) dt$$
$$\int_0^\infty t^2 \int_{S_1} \phi(t, ty) dS(y) dt, \tag{2}$$

where $x = ty$ and S_1 is the sphere of radius unity x in R_3. Similarly, the distribution $d_n^P \delta(t - r)$, $P \geq 1$, is

$$\langle d_n^P \delta(t - r),\ \phi(t, x) \rangle = (-1)^P \langle \delta(t - r), d^P \phi / dn^P \rangle. \tag{3}$$

Although the normal derivatives are continuous at $(0, 0)$ but this discontinuity causes no problem because $d^P \phi / dn^P$ is integrable near $t \pm r = 0$.

As in relation (3) it is convenient to use the coordinates (t, y) with $|y| = 1$, $t \geq 0$ to describe the points of the forward cone $t = r$ by putting $x = ty$. This defines a map from the half-cylinder $S_1 \times [0, \infty)$ onto the cone. If ϕ is a member of the class $D(\mathbb{R}_1 \times \mathbb{R}_3)$ of test functions then its restriction to the cone gives rise to an element $\widetilde{\phi}(t, y) = \phi(t, ty)$ of $D(S_1 \times [0, \infty))$. Actually, the same is true of each normal derivative $d^P \phi / dn^P$, because even though the function is discontinuous at $(0, 0)$ the associated function in the cylinder is smooth. These considerations suggest that the appropriate multilayers $f d_n^P \delta(t - r)$ are those where $f \in D'(S_1 \times [0, \infty])$, and where D' is the space dual to D. They are defined as

$$\langle f(t, y) d_n^P \delta(t - r), \phi(t, x) \rangle = (-1)^P \langle t^2 f(t, y), d^P \phi(t, ty) / dn^P \rangle, \tag{4}$$

where the last operation takes place in $D'(S_1 \times [0, \infty]) \times D(S_1 \times [0, \infty])$.

Let us now observe that relation (4) immediately gives

$$\delta(t) d_n^P \delta(t - r) = \delta'(t) d_n^P \delta(t - r) = 0. \tag{5}$$

Next, we attempt to find an expression for $\delta^{(k)}(t)d_n^P\delta(t-r)$ for $k \geq 2$. For this purpose, we start with $\delta^{(k)}(t)\delta(t-r)$ by introducing the function $\Phi(t)$,

$$\Phi(t) = \int_{S_1} \phi(t, ty)\, dS(y)\,, \tag{6}$$

so that

$$\langle \delta^{(k)}(t)\delta(t-r), \phi(t,x)\rangle = \langle t^2\delta^{(k)}(t)\,, \Phi(t)\rangle = \frac{(-1)^k k!}{(k-2)!}\Phi^{(k-2)}(0)\,. \tag{7}$$

But using the value [16]

$$c_m = c_{m,3} = \int_{S_1} y_i^{2m}\, dS(y) = \frac{4\pi}{2m+1}\,, \tag{8}$$

we obtain

$$\Phi^{(q)}(0) = \sum_{j=0}^{[q/2]} \binom{q}{2j} c_j \frac{\partial^{q-2j}}{\partial t^{q-2j}} \nabla^{2j}\phi(0,0)\,, \tag{9}$$

where $[q/2]$ is the greatest integer less than or equal to $q/2$. Thus,

$$\delta''(t)\delta(t-r) = 9\pi\delta(t)\delta(x)\,, \tag{10}$$

and more generally

$$\delta^{(k)}(t)\delta(t-r) = \frac{4\pi k!}{(k-2)!}\sum_{j=0}^{[(k-2)/2]}\binom{k-2}{2j}\frac{1}{(2j+1)}\delta^{(k-2-2j)}(t)\nabla^{2j}\delta(x)\,. \tag{11}$$

A similar analysis yields

$$\delta_k(t)d_n^P\delta(t-r) = \frac{4\pi k!}{(k-2)!}\sum_{j=[(P+1)/2]}^{[(k-2+P)/2]}\binom{k-2}{2j-P}\left(\frac{1}{2j+1}\right)\delta^{(k+P-2-2j)}(t)\nabla^{2j}\delta(x)\,. \tag{12}$$

For the cone $t - r = 0$, the required geometrical quantities take the following simple form:

$$\begin{aligned} &n_i = x_i/t \quad \mu_{ij}^{(P)} = t^{-P}(\delta_{ij} - n_i n_j)\,, \\ &\Omega = -1/t\,, \quad G = 1\,, \quad \text{and} \quad \mu_{it} = \mu_{tt} = 0\,. \end{aligned} \tag{13}$$

Next, we use formulae (12.9.29) and (13) and get

$$\frac{\bar{\partial}}{\partial t}\left(f d_n^P\delta(t-r)\right) = -f d_n^{P+1}\delta(t-r) + \left(\frac{\delta f}{\delta t} + \frac{2f}{t}\right)d_n^P\delta(t-r)\,. \tag{14}$$

Observe that the division $f(t,y)/t$ does not give a uniquely determined distribution since the general solution of the division problem contains an arbitrary multiple of $\delta(t)$. But this causes no problem because of relation (5). Similarly, from relations (12.9.39) and (13) we find that

$$\frac{\overline{\partial}}{\partial t}\left(f d_n^P \delta(t-r)\right) = f n_i d_n^{P+1}\delta(t-r) + \sum_{M=0}^{P} \frac{P!}{M!}\frac{1}{t^{P-M}}\left(\frac{\delta f}{\delta x_i} - \frac{2n_i f}{t}\right) d_n^M \delta(t-r). \quad (15)$$

Here also, we have to be careful because we again have a division problem of the form

$$\frac{1}{t^{P-M}}\left(\frac{\delta f}{\delta x_i} - \frac{2n_i f}{t}\right)$$

and such a problem gives rise to $P - M$ arbitrary constants. However, a moment's reflection will convince the reader that it really does not matter which solutions of the division problem are taken, as long as the solutions are consistent in the sense that if g_k is the solution of the division f/t^k then tg_k is the solution that should be taken for the division f/t^{k-1}.

Let us denote by $(1/t^k)d_n^P\delta(t-r)$ the distribution

$$Pf\left(\frac{H(t)}{t^k}\right) d_n^P \delta(t-r).$$

The time derivatives of $(1/t^k)d_n^P\delta(t-r)$ can be obtained from (14) as follows:

$$\begin{aligned}\frac{\overline{\partial}}{\partial t}\left[\left(\frac{1}{t^k}\right) d_n^P \delta(t-r)\right] &= -\frac{1}{t^k} d_n^{P+1}\delta(t-r) + \left[\frac{\overline{\delta}}{\delta t}\left(\frac{1}{t^k}\right) + \frac{2}{t}\left(\frac{1}{t^k}\right)\right] d_n^P \delta(t-r) \\ &= -\frac{1}{t^k} d_n^{P+1}\delta(t-r) + \left(\frac{2-k}{t^{k+1}} + \frac{(-1)^k \delta^{(k)}(t)}{h!}\right) d_n^P \delta(t-r),\end{aligned}$$

or

$$\begin{aligned}\frac{\overline{\partial}}{\partial t}\left(\frac{1}{t^k} d_n^P \delta(t-r)\right) &= -\frac{1}{t^k} d_n^{P+1}\delta(t-r) + \frac{2-k}{t^{k+1}} d_n^P \delta(t-r) \\ &\quad + \frac{(-1)^k 4\pi}{(k-2)!} \sum_{j=[(P+1)/2]}^{[(k-2+P)/2]} \binom{k-2}{2j-P} \frac{\delta^{k+P-2-2j)}(t)\nabla^{2j}(x)}{2j+1},\end{aligned} \quad (16)$$

where we have used (12) and relation (4.2.5)

$$\frac{\overline{d}}{dt}\left[Pf\left(\frac{H(t)}{t^k}\right)\right] = -kPf\left(\frac{H(t)}{t^{k+1}}\right) + \frac{(-1)^k \delta^{(k)}(t)}{k!}.$$

As a particular case we obtain

$$\frac{\overline{\partial}}{\partial t}\left(\frac{\delta(t-r)}{t}\right) = -\frac{1}{t} d_n \delta(t-r) + \frac{1}{t^2}\delta(t-r). \quad (17)$$

Higher-order time derivatives can be obtained by repeated application of formula (16). As a special case we get

$$\frac{\overline{\partial}^2}{\partial t^2}\left(\frac{\delta(t-r)}{t}\right) = \frac{1}{t} d_n^2 \delta(t-r) - \frac{2}{t^2} d_n \delta(t-r) + 4\pi\delta(t)\delta(x) \quad (18)$$

and more generally,

$$\left(\frac{\overline{\partial}}{\partial t}\right)^{2Q}\left(\frac{\delta(t-r)}{t}\right)=\frac{1}{t}\,d_n^{2Q}\delta(t-r)-\frac{2}{t^2}\,d_n^{2Q-1}\delta(t-r)+4\pi\sum_{j=0}^{Q-1}\frac{\delta^{(2Q-2j-2)}(t)\nabla^{2j}\delta(x)}{2j+1}\,, \tag{19}$$

$$\left(\frac{\overline{\partial}}{\partial t}\right)^{2Q+1}\left(\frac{\delta(t-r)}{t}\right)$$
$$=-\frac{1}{t}\,d_n^{2Q+1}\delta(t-r)+\frac{1}{t^2}d_n^{2Q}\delta(t-r)+4\pi\sum_{j=0}^{Q-1}\frac{\delta^{(2Q-2j-1)}(t)\nabla^{2j}\delta(x)}{2j+1}\,. \tag{20}$$

A similar analysis yields

$$\left(\frac{\overline{\partial}}{\partial t}\right)^{P}\left(\frac{\delta(t-r)}{t}\right)=\frac{(-1)}{t^2}\,d_n^{P}\delta(t-r)+4\pi\sum_{j=0}^{[(P-1)/2]}\frac{\delta^{(P-2j-2)}(t)\nabla^{2j}\delta(x)}{2j+1}\,. \tag{21}$$

Let us now consider the space derivatives. Using (15) we obtain

$$\left(\frac{\overline{\partial}}{\partial t}\right)\left(\frac{1}{t^k}\,d_n^{P}\delta(t-r)\right)=\frac{n_i}{t^k}\,d_n^{P+1}\delta(t-r)-2\sum_{M=0}^{P}\frac{P!}{M!}\frac{n_i}{t^{k+P-M+1}}\,d_n^{M}\delta(t-r)\,. \tag{22}$$

In particular,

$$\left(\frac{\overline{\partial}}{\partial t}\right)\left(\frac{\delta(t-r)}{t^k}\right)=\frac{n_i}{t^k}d_n\delta(t-r)-\frac{2n_i}{t^{k+1}}\,\delta(t-r)\,. \tag{23}$$

The mixed space and time derivatives can then be obtained by combining formulas (16) and (22). In order to do this it will be to our advantage to compute the value of the generalized function

$$n_i\delta^{(k)}(t)d_n^{P}\delta(t-r)\,.$$

In the case $P=0$ and $\phi(t,x)$ is a test function of $D(\mathbb{R}_1\times\mathbb{R}_3)$ we obtain

$$\langle n_i\delta^{(k)}(t)\delta(t-r),\phi(t,x)\rangle=\frac{(-1)^k k!}{(k-2)!}\Psi^{(k-2)}(0)\,, \tag{24}$$

where

$$\Psi(t)=\int_{S_1}\phi(t,ty)y_i dS(y)\,. \tag{25}$$

But since

$$\Psi^{(q)}(0)=4\pi\sum_{j=0}^{[(q-1)/2]}\binom{q}{2j+1}\frac{\partial^{q-2j-1}}{\partial t^{q-2j-1}}D_i\nabla^{2j}\phi(0,0)\frac{1}{2j+3}$$

we obtain

$$n_i\delta^{(k)}(t)\delta(t-r)=\frac{4\pi k!}{(h-2)!}\sum_{j=0}^{[(k-3)/2]}\binom{k-2}{2j+1}\frac{\delta^{(k-3-2j)}(t)D_i\nabla^{2j}\delta(x)}{2j+3}\,. \tag{26}$$

We mention the special cases

$$n_i \delta^{(k)}(t)\delta(t-r) = 0 \text{ and } k \leqslant 2,$$
$$n_i \delta'''(t)\delta(t-r) = 8\pi\delta(t)D_i\delta(x).$$

A similar study gives

$$n_i \delta^{(k)}(t) d_n^P \delta(t-r) = \frac{4\pi k!}{(k-2)!} \sum_{j=[P/2]}^{[(k-3+P)/2]} \binom{k-2}{2j-P+1} \frac{\delta^{(k-3+P-2j)}(t) D_i \nabla^{2j} \delta(x)}{2j+3}. \tag{27}$$

Using (27) we readily obtain the mixed derivatives as

$$\begin{aligned}
&\frac{\overline{\partial}^2}{\partial t\, \partial x_i}\left(\frac{1}{t^k} d_n^P \delta(t-r)\right) \\
&= -\frac{n_i}{t^k} d_n^{P+2}\delta(t-r) + \frac{(4-k)n_i}{t^{k+1}} d_n^{P+1}\delta(t-r) \\
&\quad + 2\sum_{M=0}^{P} \frac{(P+k-1)P!}{M!} \frac{n_i}{t^{k+2+P-M}} d_n^M \delta(t-r) \\
&\quad + \frac{(-1)^k 4\pi}{(k-2)!} \sum_{j=[(P+1)/2]}^{[(k-2+P)/2]} \binom{k-2}{2j-P} \frac{\delta^{(k+P-2-2j)}(t) D_i \nabla^{2j}\delta(x)}{2j+1}.
\end{aligned} \tag{28}$$

In particular,

$$\frac{\overline{\partial}^2}{\partial t \partial x_i}\left(\frac{\delta(t-r)}{t}\right) = -\frac{n_i}{t} d_n^2 \delta(t-r) + \frac{3n_i}{t^2} d_n \delta(t-r). \tag{29}$$

The second-order space derivatives take the form

$$\begin{aligned}
&\frac{\overline{\partial}^2}{\partial x_i \partial x_j}\left(\frac{1}{t^k} d_n^P \delta(t-r)\right) \\
&= \frac{n_i n_j}{t^k} d_n^{P+2}\delta(t-r) + \frac{\delta_{ij} - 5n_i n_j}{t^{k+1}} d_n^{P+1}\delta(t-r) \\
&+ \sum_{M=0}^{P} \left(\frac{(2M-P-1)}{N!}(\delta_{ij} - 3n_i n_j) - \frac{2P!}{(M-1)!} n_i n_j\right) \frac{1}{t^{k+P-M+2}} d_n^M \delta(t-r).
\end{aligned} \tag{30}$$

Putting $i = j$ and summing we get

$$\overline{\nabla}^2\left(\frac{1}{t^k} d_n^P \delta(t-r)\right) = \frac{1}{t^k} d_n^{P+2}\delta(t-r) - \frac{2}{t^{k+1}} d_n^{P+1}\delta(t-r) - \sum_{M=1}^{P} \frac{2P! d_n^M \delta(t-r)}{(M-1)! t^{k+P-M+2}}. \tag{31}$$

Special cases include:

$$\frac{\overline{\partial}^2}{\partial x_i \partial x_j}\left(\frac{\delta(t-r)}{t}\right) = \frac{n_i n_j}{t} d_n^2 \delta(t-r) + \frac{(\delta_{ij} - 5n_i n_j)}{t^2} d_n \delta(t-r) - \frac{(\delta_{ij} - 3n_i n_j)}{t^3} \delta(t-r), \tag{32}$$

and

$$\overline{\nabla}^2 \left(\frac{\delta(t-r)}{t}\right) = \frac{1}{t} d_n^2 \delta(t-r) - \frac{2}{t^2} d_n \delta(t-r). \tag{33}$$

As a check on our formulas we subtract (33) from (18) and get

$$-\overline{\Box}^2 \left(\frac{\delta(t-r)}{t}\right) = \left(\frac{\overline{\partial}^2}{\partial t^2} - \overline{\nabla}^2\right)\left(\frac{\delta(t-r)}{t}\right) = 4\pi\delta(t)\delta(x) \tag{34}$$

which agrees with formula (10.10.15).

The analysis for the second factor in (1) is simlar. Thus, we have completed a distributional analysis for delta functions concentrated on the light cone.

12.11. Examples

Example 1. In Section 10.6 we studied the double layer potentials in the theory of harmonic functions. The results in this chapter enable us to extend that analysis. Let us start with the single layer potential due to the distribution $\sigma(x)$ on Σ, as defined by the integral (10.6.19) and write it as

$$G(x) = \int_\Sigma \sigma(y) E(x-y) dy, \qquad x \in R_p/\Sigma, \tag{1}$$

where

$$E(x) = \begin{cases} \frac{1}{2\pi} \ln|x|, & p = 2, \\ \frac{1}{(n-2)S_p|x|^{p-2}} \ln|x|, & p \geqslant= 3, \end{cases} \tag{2}$$

and S_p is the area of the unit sphere in R_p.

Similarly, the double layer potential due to the distribution $\rho(x)$ on Σ, is defined by the integral (10.6.28) which we now write as

$$H(x) = -\int_\Sigma \rho(y) \frac{dE(x-y)}{dn} dy, \qquad x \in R_p/\Sigma. \tag{3}$$

Both functions, G and H, are harmonic in the complement of the surface Σ and their jumps across Σ can be expressed in terms of σ and ρ. Indeed, we presented the interfacial formulas (10.6.21) to (10.6.24) for the single layer potential and the formulas (10.6.29) to (10.6.33) for the double layer potential. With the help of the general interfacial formulas we can extend that analysis. For instance, in relation (12.9.32) we set $i = j$ and sum on j and use the fact that $\nabla^2 F = 0$. This yields

$$0 = \left[\frac{d^P}{dnP}(\nabla^2 F)\right] = \sum_{Q=0}^{P} \frac{(-1)^{P-Q} P!}{Q!}$$
$$\times \left\{(P - Q + 1)\mu_{rs}^{(P-Q)} D_{rs}^2 A^{(Q)} + \frac{\delta\omega_{P-Q}}{\delta x_r}\frac{\delta A^0}{\delta x_r}\right.$$
$$\left. - Q\omega_{P-Q+2}A^0\right\} + \omega_1 A^{(P+1)} + A^{(P+2)}, \tag{4}$$

which is the recurrence relation that helps us to compute $A^{(P+2)}$ once the previous $A^{(Q)}$ have been found. Thereby we obtain

$$A^{(0)} = \rho, \quad A^{(1)} = \sigma, \quad A^{(2)} = -\nabla^2\rho - \omega_1\sigma, \tag{5}$$

$$A^{(3)} = \omega_1\nabla^2\rho + 2\mu_{rs}D_{rs}^2\rho + \frac{\delta\omega_1}{\delta x_r}\frac{\delta\rho}{\delta x_r}$$
$$- \nabla^2\sigma + (\omega_2 + \omega_1^2)\sigma, \tag{6}$$

and so on.

In case F is a single layer we have $\rho = 0$. Then we obtain not only the jumps (10.6.20) to (10.6.24) but also the jumps $\left[\frac{d^P}{dn^P} D^N F\right]$ for an arbitrary P and N. For instance, for $P = 0$, $N = 3$, we have

$$\left[\frac{\partial^3 F}{\partial x_i \partial x_j \partial x_k}\right] = - n_i n_j n_k \nabla^2\sigma + n_i D_{jk}^2\sigma + n_j D_{ik}^2\sigma + n_k D_{ij}^2\sigma$$
$$+ (\mu_{ij} - \omega_1 n_i n_j)\frac{\delta\sigma}{\delta x_k} + (\mu_{ik} - \omega_1 n_i n_K)\frac{\delta\sigma}{\delta x_i}$$
$$+ (\mu_{jk} - \omega_1 n_j n_k)\frac{\delta\sigma}{\delta x_i}$$
$$+ \left[\lambda_{ijk}^{(3,1)} + (\omega_2 + \omega_1^2)n_i n_j n_k - n_i n_j \frac{\delta\omega_1}{\delta x_k}\right.$$
$$\left. - n_i n_k \frac{\delta\omega_1}{\delta x_i} - (n_i\mu_{jk} + n_j\mu_{ik} + n_k\mu_{ij})\omega_1\right]\sigma. \tag{7}$$

The same is true for the double layer density, that is $\rho \neq 0$, $\sigma = 0$. The formula corresponding to (7) is

$$\left[\frac{\partial^3 F}{\partial x_i \partial x_j \partial x_k}\right] = D^3_{ijk}\rho - n_i n_j \frac{\delta}{\delta x_k}(\nabla^2 \rho) - n_j n_k \frac{\delta}{\delta x_i}(\nabla^2 \rho) - n_i n_k \frac{\delta}{\delta x_j}(\nabla^2 \rho)] + [\omega_1 n_i n_j n_k - (n_i \mu_{jk} + n_j \mu_{ik} + n_k \mu_{ij})]\nabla^2 \rho + 2 n_i n_j n_k \mu_{rs} D^2_{rs}\rho + n_i n_j n_k \frac{\delta \omega_1}{\delta x_r}\frac{\delta \rho}{\delta x_r} . \tag{8}$$

Example 2. Deformation of the wave fronts. In Section 12.2 we studied how the jumps A, B, C across the sonic discontinuity deformed as the wave front propagated. Now that we have studied the jumps $A^{(Q)}$ of arbitrary order we can extend that study. For this purpose we follow the analysis of Estrada and Kanwal [67] and suppose that $F(x, t)$ is a solution of the equation

$$L(F) = 0 , \tag{9}$$

where L is the partial derivative operator given by

$$L(F) = \sum_{Q=0}^{N} \langle a^{(Q)} , D^Q F \rangle . \tag{10}$$

Here $a^{(Q)}(x)$ is a smooth symmetric tensor of order Q and the scalar product between symmetric tensors is given by

$$\langle a^{(Q)}, b^{(Q)} \rangle = a_{i_1 \ldots i_Q} b_{i_1 \ldots i_Q} . \tag{11}$$

If F is smooth except for a discontinuity across the surface Σ then the strength of the discontinuity can be measured by the quantities $A^{(Q)} = [d^Q F / dn^Q]$. Since for every $P \geqslant 1$ we have

$$\left[\frac{d^P}{dn^P}(L(F))\right] = 0 , \tag{12}$$

use of the interfacial relations (12.9.30) yields the transport equations for the quantities $A^{(Q)}$.

We illustrate these ideas with a simple example. Let L be the operator

$$L(F) = \nabla^2 F - \frac{1}{c^2}\frac{\partial^2 F}{\partial t^2} + a_i \frac{\partial F}{\partial x_i} + a \frac{\partial F}{\partial t} + bF , \tag{13}$$

and let F be a discontinuous solution, whose discontinuity is located on the moving surface $\Sigma(t)$. Since

$$\overline{L}(F) = L(F) + \left[\left[1 - \frac{G^2}{c^2}\right] A^{(1)} + \frac{2G}{c^2}\frac{\delta A^{(0)}}{\delta t} + \left\{\omega_1 - \frac{\mu_{tt}}{c^2} + a_i n_i - aG\right\} A^{(0)}\right]\delta(\Sigma) + \left(1 - \frac{G^2}{c^2}\right)\delta'(\Sigma) , \tag{14}$$

where $\delta'(\Sigma) = -\omega_1\delta(\Sigma) + d_n\delta(\Sigma)$, $\omega_1 = \mu_{ii} = -2\Omega$, and Ω, is the mean curvature of Σ. It follows that $G^2 = c^2$; let us say $G = c$. The transport equation for $A^{(0)}$ becomes

$$\frac{2}{c}\frac{\delta A^{(0)}}{\delta t} + \left(\omega_1 - \frac{\mu_{tt}}{c^2} + a_i n_i - ac\right) A^{(0)} = 0\,. \tag{15}$$

The jump $[L(F)] = 0$ gives us the transport equation for $A^{(1)}$, namely

$$\begin{aligned}\frac{2}{c}\frac{\delta A^{(1)}}{\delta t} &+ \left(\omega_1 - \frac{\mu_{tt}}{c^2} + a_i n_i - ac\right) A^{(1)} + \nabla^2 A^{(0)} \\ &- \frac{1}{c^2}\frac{\delta^2 A^{(0)}}{\partial t^2} + a_i\frac{\delta A^{(0)}}{\delta x_i} - c\frac{\delta A^{(0)}}{\delta t} + bA^{(0)} = 0\,,\end{aligned} \tag{16}$$

where we have used the fact that the surface moves according to $G = c$. The transport equation for $A^{(P+1)}$ follows by using formulas (12.8.32) and (12.8.33) to compute the jump $[(d^P/dn^P)(L(F))]$, which gives

$$\begin{aligned}0 &= \left[\frac{d^P}{dn^P}(L(F))\right] \\ &= \frac{2}{c}\frac{\delta A^{(P+1)}}{\delta t} + \left(\omega_1 - \frac{\mu_{tt}}{c^2} + n_i a_i - ac\right) A^{(P+1)} + bA^{(P)} + \sum_{Q=0}^{P}\frac{(-1)^{P-Q}P!}{Q!} \\ &\times \left\{\left[(P-Q+1)\mu_{rs}^{(P-Q)} - \frac{1}{c^2}\sum_{B=0}^{P-Q}\overline{\mu}_{tr}^{(P-Q-B)}\overline{\mu}_{ts}^{(B)}\right] D^2_{rs}(A^{(Q)})\right. \\ &+ \left[\frac{\delta\omega_{P-Q}}{\delta x_r} - \frac{\lambda_{ttr}^{(3,P-1)}}{c^2} - \frac{Q}{c}\mu_{tr}^{(P-Q+1)} + a_i\overline{\mu}_{tr}^{(P-Q)}\right]\frac{\delta A^{(0)}}{\delta x_r} \\ &+ \left. Q\left(-\omega_{P-Q+2} + \frac{\mu_{tt}^{(P-Q+2)}}{c^2}\right) A^{(Q)}\right\}\,,\end{aligned} \tag{17}$$

where we used the convention that $\overline{\mu}_{rt}^{(0)}$ means that r has to be replaced by t in the given expression.

Example 3. Two-dimensional analysis. In Section 5.8 we worked out the two-dimensional analog for the analysis that we presented for the n-dimensional ($n > 2$) case in the previous sections of Chapter 5. We can do precisely the same now. The values of the quantities $\mu^{(P)}$ and ω^P are given by (5.8.14) and (5.8.15), namely

$$\mu_{ij}^{(P} = -\kappa^P \dot{x}_i \dot{x}_j\,; \qquad \omega^P = (-\kappa)^P\,. \tag{18}$$

To find the formulas for $\lambda^{N,P}$ we follow the same steps as in Section (12.8). For instance, the value of the quantity $\lambda^{(3,P)}$ in two dimensions is

$$\begin{aligned}\lambda^{3,P} =&(-1)^P \left\{ \frac{P(P+1)}{2} \kappa^{P-1} \dot{\kappa} \dot{x}^3 \right. \\ &\left. - 3\, P\, \kappa^{P+1} \dot{x}^2 n \right\} . \end{aligned} \tag{19}$$

Similarly, the formulas for the quantities for $[d^P D^N F/dn^P]$ can also be read off from the Section (12.9). For instance, the formula corresponding to (12.9.32) is

$$\left[\frac{d^P DF}{dn^P} \right] = \sum_{Q=0}^{P} \frac{P!}{Q!} \kappa^{P-Q} \dot{A}^Q \dot{x}_i + A^{P+1} n_i \ . \tag{20}$$

Continuing in the same fashion we can find the two-dimensional results corresponding to Example 1 above.

CHAPTER 13

Interplay Between Generalized Functions and the Theory of Moments

13.1. The Theory of Moments

Estrada and Kanwal have recently [6, 69–71] developed a distributional approach to asymptotic analysis. Their study is based on the interplay between the generalized functions and the theory of moments. Thus, they have not only succeeded in presenting a simplified approach to various known aspects of asymptotics but have also found many new results. They have applied their technique to many different branches of asymptotic expansions, such as asymptotic evaluation of divergent integrals, boundary layer theory and singular perturbations. Our aim in this chapter is to present the basic concepts of their methods and illustrate them with representative examples.

The moments μ_n of a function $f(x)$ are defined as

$$\mu_n = \langle f(x), x^n \rangle = \int_{-\infty}^{\infty} f(x) x^n dx \,, \tag{1}$$

$n = 1, 2, \ldots$. Now consider a test function $\phi(x)$ whose Taylor series is

$$\phi(x) = \sum_{n=0}^{\infty} \phi^{(n)}(0) \frac{x^n}{n!} \,. \tag{2}$$

Then it follows from (1) that

$$\begin{aligned} \langle f(x), \phi(x) \rangle &= \left\langle f(x), \sum_{n=0}^{\infty} \phi^{(n)}(0) \frac{x^n}{n!} \right\rangle \\ &= \sum_{n=0}^{\infty} \frac{\phi^{(n)}(0) \mu_n}{n!} \,. \end{aligned} \tag{3}$$

Since $\phi^{(n)}(0) = (-1)^n \langle \delta^{(n)}(x), \phi(x) \rangle$, the above relation can be written as

$$\langle f(x), \phi(x) \rangle = \left\langle \sum_{n=0}^{\infty} (-1)^n \frac{\mu_n}{n!} \delta^{(n)}(x), \phi(x) \right\rangle \,. \tag{4}$$

Thus,

$$f(x) = \sum_{n=0}^{\infty} \frac{(-1)^n \mu_n \delta^{(n)}(x)}{n!} \,. \tag{5}$$

Let us recall that the collections $\{(-1)^m \delta^{(m)}(x)\}_{m=0}^{\infty}$ and $\{x^n/n!\}_{n=0}^{\infty}$ form a biorthogonal series [see Exercise 14, Chapter 2], that is

$$\langle (-1)^m \delta^m(x), (x^n/n!) \rangle = \begin{cases} 0, & \text{if } n \neq m, \\ 1, & \text{if } n = m. \end{cases} \tag{6}$$

For this reason expansion (5) is called the dual Taylor series.

Next, we use the formula

$$\delta^{(n)}(\lambda x) = \frac{1}{\lambda^{n+1}} \delta^{(n)}(x), \tag{7}$$

and get from (5) the required asymptotic formula

$$f(\lambda x) \sim \sum_{n=0}^{\infty} \frac{(-1)^n \mu_n \delta^{(n)}(x)}{n!\, \lambda^{n+1}}, \qquad \lambda \to \infty. \tag{8}$$

The Moment Problem. Given the sequence $\{\mu_n\}$ of real or complex numbers, we are required to find a function $f(x)$ that satisfies relation (1).

Our contention is that the function $f(x)$ defined by the infinite series (5) of delta functions is a formal solution of the moment problem. The proof follows by multiplying both sides of (5) by x^n and using the orthogonality relation (6).

Occasionally the Fourier transform facilitates the evaluation of the moments. This is accomplished by recalling the formula.

$$[\delta^{(n)}(u)]^\wedge(x) = (-i\, x)^n.$$

Then we can write (1) as

$$\begin{aligned} \mu_n &= \int_{-\infty}^{\infty} (-i)^n [\delta^n(u)]^\wedge(x) f(x) \\ &= (-i)^n \int_{-\infty}^{\infty} \delta^{(n)}(u) \widehat{f}(u) du \\ &= (-i)^n (-1)^n \frac{d^n}{du^n} \widehat{f}(u)\Big|_{u=0}, \end{aligned}$$

so that

$$\mu_n = i^n \frac{d^n}{du^n} \widehat{f}(u)\Big|_{u=0}.$$

Accordingly, we can write the moment expansion formula (5) as

$$f(x) = \sum_{n=0}^{\infty} \frac{(-i)^n \{(d^n/du^n) \widehat{f}(u)\}\big|_{u=0} \delta^n(x)}{n!}. \tag{9}$$

However, we must point out that the moment asymptotic expansion neither holds in the space D' nor in S' but it holds in E' and also in the following spaces [6]:

(i) The space P', that is dual of the space P of test functions $\phi(x)$ which satisfy the condition $\phi^{(n)}(x) = 0(e^{\gamma |x|})$ as $|x| \to \infty$ for each $n = 1, 2, \ldots$, and $\gamma > 0$.

(ii) The space $\mathcal{O}'_M$ that is the dual of the space $\mathcal{O}_M$ of test functions $\phi(x)$ which are such that if there exist constants k_n with $\phi^{(n)}(x) = 0(|x|^{k_n})$, as $|x| \to \infty$ for all $n \in \mathbb{N}$.

(iii) The space $\mathcal{O}'_C$ that is dual of the space $\mathcal{O}_C$ which is the subspace of $\mathcal{O}_M$ as given in (ii) above such that all k_n can be taken equal.

(iv) The space K' that is dual of the space K which is the subspace of $\mathcal{O}_M$ where $k_n = \gamma - n$ for some $n \in \mathbb{N}$.

Example 1. $f(x) = e^{-x^2}$.
The moments are

$$\mu_n = \int_{-\infty}^{\infty} e^{-x^2} x^n dx = \begin{cases} \Gamma\left(\frac{n+1}{2}\right), & n \text{ even}, \\ 0, & n \text{ odd}. \end{cases} \tag{10}$$

The moment expansion is

$$\begin{aligned} e^{-x^2} &= \sum_{n=0}^{\infty} \frac{\Gamma((2n+1)/2)\,\delta^{(2n)}(x)}{(2n)!} \\ &= \sqrt{\pi} \sum_{n=0}^{\infty} \frac{1}{4^n n!}\,\delta^{(2n)}(x). \end{aligned} \tag{11}$$

The asymptotic expansion is

$$e^{-\lambda x^2} \sim \sum_{n=0}^{\infty} \frac{\Gamma((2n+1)/2)\,\delta^{(2n)}(x)}{(2n)!\lambda^{(2n+1)/2}} = \sqrt{\pi} \sum_{n=0}^{\infty} \frac{1}{4^n n!} \frac{1}{\lambda^{2n+1}} \delta^{(2n)}(x), \quad (\lambda \to \infty). \tag{12}$$

For $\lambda = 1/\epsilon$, this becomes

$$e^{x^2/\epsilon} \sim \sqrt{\pi} \sum_{n=0}^{\infty} \frac{1}{4^n n!} \epsilon^{2n+1} \delta^{2n}(x), \quad \epsilon \to 0. \tag{13}$$

Example 2. $f(x) = H(x)e^{-x}$.
The moments are

$$\mu_n = \int_0^{\infty} e^{-x} x^n dx = n!. \tag{14}$$

The moment expansion is

$$H(x)e^{-x} = \sum_{n=0}^{\infty} (-1)^n \delta^{(n)}(x). \tag{15}$$

The asymptotic expansion is

$$H(x)e^{-\lambda x} \sim \sum_{n=0}^{\infty} (-1)^n \frac{\delta^{(n)}(x)}{\lambda^{n+1}} \qquad \lambda \to \infty. \tag{16}$$

When we set $\lambda = 1/\epsilon$ in (16), we have

$$H(x)e^{-x/\epsilon} = \sum_{n=0}^{\infty} (-1)^n \epsilon^{n+1} \delta^{(n)}(x), \qquad \epsilon \to 0. \tag{17}$$

From this formula we observe that

$$H(x)e^{-x/\epsilon} = \epsilon\delta(x) + O(\epsilon^2)$$

or

$$\epsilon\,\delta(x) = H(x)e^{-x/\epsilon} + O(\epsilon^2)\,. \tag{18}$$

We shall find this relation very useful in the discussion of the boundary layer theory.

Example 3. $f(x) = e^{-|x|}$
The moments are

$$\begin{aligned}\mu_n &= \int_{-\infty}^{\infty} e^{-|x|}x^n dx \\ &= \int_{-\infty}^{0} e^{-|x|}x^n dx + \int_{0}^{\infty} e^{-|x|}x^n dx \\ &= \int_{0}^{\infty} e^{-|x|}(-x)^n dx + n! \\ &= 2\begin{cases}(2n)!\,, & n \text{ even}\,, \\ 0 & n \text{ odd}\,.\end{cases}\end{aligned} \tag{19}$$

The moment expansion is

$$e^{-|x|} = 2\sum_{n=0}^{\infty}(2n)!\;\delta^{(2n)}(x)\,. \tag{20}$$

The asymptotic expansion is

$$e^{-\lambda|x|} \sim 2\sum_{n=0}^{\infty}\frac{\delta^{(2n)}(x)}{\lambda^{2n+1}}\,, \quad \lambda \to \infty\,. \tag{21}$$

For $\lambda = 1/\epsilon$, we have

$$e^{-|x|/\epsilon} \sim 2\sum_{n=0}^{\infty}\epsilon^{(2n+1)}\delta^{(2n)}(x)\,, \quad \epsilon \to 0\,. \tag{22}$$

Example 4. $f(x) = H(x)e^{ix}$.
The moments are

$$\mu_n = \int_0^{\infty} e^{ix}x^n dx = n!\; e^{\pi i(n+1)/2}\,. \tag{23}$$

The moment expansion is

$$H(x)e^{ix} = \sum_{n=0}^{\infty}(-1)^n e^{\pi i(n+1)/2}\delta^{(n)}(x)\,. \tag{24}$$

Similarly,

$$H(x)e^{-ix} = \sum_{n=0}^{\infty}(-1)^n e^{-\pi i(n+1)/2}\delta^{(n)}(x)\,. \tag{25}$$

From relations (24) and (25) we can derive the moment expansions for $H(x)\sin x$ and $H(x)\cos x$. For instance,

$$H(x)\sin x = \sum_{n=1}^{\infty}(-1)^{n+1}\delta^{(2n-2)}(x) . \tag{26}$$

These relations, in turn, yield the asymptotic expansions

$$H(x)e^{\pm i\lambda x} \sim \sum_{n=0}^{\infty}\frac{(-1)^n e^{\pm i\pi(n+1)/2}}{\lambda^{n+1}}\delta^{(n)}(x) , \quad \lambda \to \infty , \tag{27}$$

and

$$H(x)\sin(\lambda x) \sim \sum_{n=1}^{\infty}\frac{(-1)^{n+1}\delta^{(2n-2)}(x)}{\lambda^{2n-1}} , \quad \lambda \to \infty . \tag{28}$$

For $\lambda = 1/\epsilon$, these formulas become

$$H(x)e^{\pm ix/\epsilon} \sim \sum_{n=0}^{\infty}(-1)^n\epsilon^{n+1}e^{\pm i\pi(n+1)/2}\delta^{(n)}(x) , \quad \epsilon \to 0 , \tag{29}$$

and

$$H(x)\sin(x/\epsilon) \sim \sum_{n=1}^{\infty}(-1)^{n+1}\epsilon^{2n-1}\delta^{(2n-2)}(x) , \quad \epsilon \to 0 . \tag{30}$$

Changing ϵ to $\sqrt{\epsilon}$ in the above formula and multiplying the resulting expression by $\sqrt{\epsilon}$, we derive the useful result

$$H(x)(\sqrt{\epsilon}\sin(x/\sqrt{\epsilon})) \sim \sum_{n=1}^{\infty}(-1)^n\epsilon^{2n-1}\delta^{(2n-2)}(x), \quad \epsilon \to 0 . \tag{31}$$

13.2. Asymptotic Approximation of Integrals

Let $\phi(x)$ and $h(x)$ be sufficiently smooth real valued functions on $[a, b]$ then the principle of stationary phase asserts that the main contribution to the integral

$$I(\lambda) = \int_a^b \phi(x)e^{-\lambda h(x)}dx , \quad \lambda \to \infty , \tag{1}$$

arises from the points where the function $h(x)$ is stationary. If x_0 is the only global minimum of $h(x)$ then we have the Laplace formula

$$I(\lambda) \sim \left(\frac{2\pi}{\lambda h''(x_0)}\right)^{1/2}\phi(x_0)e^{\lambda h(x_0)} . \tag{2}$$

A heuristic proof is as follows. For $\epsilon > 0$, we have

$$\begin{aligned} I(\lambda) &\sim \int_{x_0-\epsilon}^{x_0+\epsilon} \phi(x) e^{-\lambda h(x)} dx \\ &\sim \phi(x_0) \int_{x_0-\epsilon}^{x_0+\epsilon} e^{-\lambda[h(x_0)+h''(x_0)(x-x_0)^2/2]} dx \\ &\sim \phi(x_0) e^{-\lambda h(x_0)} \int_{-\infty}^{\infty} e^{-\lambda[h''(x_0)(x-x_0)^2/2]} dx \\ &= \phi(x_0) \left(\frac{2\pi}{\lambda h''(x_0)} \right)^{1/2} e^{-\lambda h(x_0)}, \quad \lambda \to \infty . \end{aligned}$$

Suppose that we can prove that $e^{-\lambda h(x)}$ has an asymptotic series of delta functions so that

$$e^{-\lambda h(x)} = \left(\frac{2\pi}{\lambda h''(x_0)} \right)^{1/2} e^{-\lambda h(x_0)} \delta(x - x_0) + O\left(\frac{1}{\lambda^2}\right) . \tag{3}$$

Then we can substitute (3) in (1) and use the sifting property of the delta function to obtain formula (2). This is accomplished with the help of the moment asymptotic expansion as explained in the previous section. Indeed, we start with the function $f(x) = e^{-\lambda x^2}$ whose asymptotic expansion is given by (13.1.12);

$$e^{-\lambda x^2} \sim \sum_{n=0}^{\infty} \frac{\Gamma\left(\frac{2n+1}{2}\right) \delta^{(2n)}(x)}{(2n)! \lambda^{\left(\frac{2n+1}{2}\right)}}, \quad \text{as } \lambda \to \infty . \tag{4}$$

Suppose now that a smooth function $h(x)$ has a minimum at $x = x_0$ so that $h'(x_0) = 0$, $h''(x_0) > 0$ and consider the Laplace integral (1)

$$I(\lambda) = \int_{-\infty}^{\infty} e^{-\lambda h(x)} \phi(x) dx = \langle e^{-\lambda h(x)}, \phi(x) \rangle , \tag{5}$$

where $\phi(x)$ is chosen so that the support of $\phi(x)$ is a small enough neighborhood of x_0 so that it contains no other critical point of $h(x)$. Accordingly, we can find an increasing smooth function $\psi(x)$ with $\psi(x_0) = 0$, $\psi'(x_0) > 0$ for all $x \in \mathbb{R}_1$ such that

$$h(x) = h(x_0) + (\psi(x))^2 ,$$

in the support of $h(x)$. Then $h'(x) = 0 + 2\psi(x)\psi'(x)$ and $h'(x_0) = 2\psi(x_0)\psi'(x_0) = 0$. Therefore we set

$$u = \psi(x) , \qquad du = \psi'(x) dx \qquad \text{or} \qquad dx = du/\psi'(x) .$$

Then (5) becomes

$$I(\lambda) = e^{-\lambda h(x_0)} \int_{-\infty}^{\infty} e^{-\lambda(\psi(x))^2} \phi(x) \frac{1}{\psi'(x)} \psi'(x) dx$$

or

$$I(\lambda) = e^{-\lambda h(x_0)} \int_{-\infty}^{\infty} e^{-\lambda u^2} \phi_1(u) du \,, \tag{6}$$

where

$$\phi_1(u) = \frac{\phi(u)}{\psi'(u)} \,.$$

For $e^{-\lambda u^2}$ in the integral (6) we substitute the moment formula (4) and we have

$$I(\lambda) \sim e^{-\lambda h(x_0)} \int_{-\infty}^{\infty} \sum_{n=0}^{\infty} \frac{\Gamma\left(\frac{2n+1}{2}\right)}{(2n)1} \frac{\delta^{2n}(u)}{\lambda^{\frac{2n+1}{2}}} \phi_1(u) du \,, \tag{7}$$

as $\lambda \to \infty$.

The first term of formula (7) is

$$\begin{aligned} I(\lambda) &\sim e^{\lambda h(x_0)} \int_{-\infty}^{\infty} \Gamma\left(\frac{1}{2}\right) \frac{\delta(u)}{\sqrt{\lambda}} \phi_1(u) du \\ &= e^{-\lambda h(x_0)} \sqrt{\pi} \, \frac{\phi_1(0)}{\sqrt{\lambda}} \\ &= e^{-\lambda h(x_0)} \sqrt{\pi} \, \frac{1}{\sqrt{\lambda}} \frac{\phi(x_0)}{\psi'(x_0)} \,. \end{aligned} \tag{8}$$

To put this formula in the form of formula (2), we observe that

$$h''(x) = 2(\psi'(x))^2 + 2\psi(x)\psi''(x) \,,$$

so that

$$h''(x_0) = 2(\psi'(x_0))^2 \,.$$

When we substitute this value of $\psi'(x_0)$ in (8) we recover formula (2), namely

$$I(\lambda) \sim e^{-\lambda h(x_0)} \left(\frac{2\pi}{h''(x_0)}\right)^{\frac{1}{2}} \frac{\phi(x_0)}{\sqrt{\lambda}} \,.$$

Oscillatory integrals. In this case we have to evaluate the integral

$$I(\lambda) = \int_{-\infty}^{\infty} e^{i\lambda h(x)} \phi(x) dx \,, \tag{9}$$

as $\lambda \to \infty$. This integral is processed in the same fashion as the Laplace integral. Indeed, the steps leading (9) to the relation

$$I(\lambda) = \int_{-\infty}^{\infty} e^{i\lambda u^2} \phi_1(u) du \tag{10}$$

are the same. A slight difference arises in the values of the moments μ_n which now are

$$\mu_n = \int_{-\infty}^{\infty} e^{i\lambda u^2} u^n du = \begin{cases} \Gamma\left(\frac{n+1}{2}\right) e^{\pi i \left(\frac{2n+1}{4}\right)} \,, & n \text{ even} \,, \\ 0 \,, & n \text{ odd} \,. \end{cases} \tag{11}$$

This yields the moment expansion of $e^{i\lambda u^2}$ as

$$e^{i\lambda u^2} \sim \sum_{n=0}^{\infty} \frac{\Gamma\left(\frac{2n+1}{2}\right) e^{\pi i\left(\frac{2n+1}{4}\right)}}{(2n)!} \frac{\delta^{(2n)}(u)}{\lambda^{\frac{2n+1}{2}}} . \tag{12}$$

When we substitute this value for $e^{i\lambda u^2}$ in integral (10) we obtain

$$\begin{aligned} I(\lambda) \sim e^{i\,h(x_0)} \int_{-\infty}^{\infty} \sum_{n=0}^{\infty} \frac{\Gamma\left(\frac{2n+1}{2}\right) e^{-\pi i\left(\frac{2n+1}{4}\right)}}{(2n)!} \\ \times \frac{\delta^{(2n)}(u)}{\lambda^{\frac{2n+1}{2}}} \phi_1(u) du . \end{aligned} \tag{13}$$

The first term of this formula yields

$$I(\lambda) \sim e^{i\left(\lambda h(x_0)+\frac{\pi}{4}\right)} \left(\frac{2\pi}{h''(x_0)}\right)^{\frac{1}{2}} \frac{\phi(x_0)}{\sqrt{\lambda}} . \tag{14}$$

Watson's Lemma.. To derive this famous lemma we appeal to the moment asymptotic expansion (13.1.16)

$$H(x)e^{-\lambda x} \sim \sum_{n=0}^{\infty} (-1)^n \frac{\delta^{(n)}(x)}{\lambda^{n+1}} , \qquad \lambda \to \infty . \tag{15}$$

For a smooth function $\phi(x)$, the above formula yields

$$\begin{aligned} \int_0^{\infty} e^{-\lambda x} \phi(x) dx &\sim \sum_{n=0}^{\infty} (-1)^n \left\langle \frac{\delta^{(n)}(x)}{\lambda^{n+1}} , \quad \phi(x) \right\rangle \\ &= \frac{\phi(0)}{\lambda} + \frac{\phi'(0)}{\lambda^2} + \frac{\phi''(0)}{\lambda^2} + \cdots + \frac{\phi^{(N)}(0)}{\lambda^{N+1}} , \ldots , \end{aligned} \tag{16}$$

as $\lambda \to \infty$. This is precisely Watson's lemma.

13.3. Applications to the Singular Perturbation Theory

We shall illustrate the concepts of this section with the help of certain initial and boundary value problems. For the general theory of distributional approach to the boundary layer theory the reader is referred to Estrada and Kanwal [71].

Initial Value Problem. Let us start with the initial value problem

$$\epsilon y'' + y = 0 , \qquad x > 0 , \tag{1}$$

$$y(0) = 0 , \qquad y'(0) = 1 , \tag{2}$$

where $\epsilon \ll 1$ is a real parameter. We cannot write the solution of this problem as a regular perturbation series $\sum_{n=0}^{\infty} \epsilon^n y_n(x)$ since when $\epsilon = 0$, the differential equation (1) becomes of first order and both the initial values $y(0)$ and $y'(0)$ cannot be satisfied. Our aim is to demonstrate that many times a distributional regular series exists.

In the time domain the variable x in equation (1) is the time and the initial time is $x = 0$. Accordingly, as a first step we set

$$z(x) = y(x)\, H(x)\,, \tag{3}$$

then we proceed as we did in Chapter 9. Indeed, when we differentiate (3) twice and use the initial values (2) we obtain

$$z'(x) = y'(x)\, H(x)\,, \qquad z''(x) = y''(x)\, H(x) + \delta(x)\,. \tag{4}$$

Next we multiply both sides of equation (1) by $H(x)$ and substitute for $y''(x)\, H(x)$ and $y(x)\, H(x)$ from (3) and (4) in it. Thus, the initial value problem (1) and (2) reduce to the single inhomogeneous differential equation

$$\epsilon\, z''(x) + z(x) = \epsilon\, \delta(x)\,. \tag{5}$$

Let us now solve the differential equation (5) by setting

$$z(x) = \sum_{n=0}^{\infty} z_n(x)\, \epsilon^n\,, \tag{6}$$

so that we have

$$\sum_{n=0}^{\infty} z_n''(x)\, \epsilon^{n+1} + \sum_{n=0}^{\infty} z_n(x)\epsilon^n = \epsilon\, \delta(x)\,. \tag{7}$$

When we equate the equal powers of ϵ in (7) we have the following system of equations:

$$0(1) \;:\; z_0(x) = 0\,; \tag{8}$$

$$0(\epsilon) \;:\; z_1(x) = \epsilon\, \delta(x) \tag{9}$$

and

$$0(\epsilon^{n-1}) \;:\; z_n(x) = -z_{n-1}''(x) \qquad n \geq 2\,. \tag{10}$$

Substituting these values of z_n in (6) we have the delta series solution as

$$z(x) \sim \sum_{n=1}^{\infty} (-1)^{n+1} \epsilon^{2n} \delta^{(2n-2)}(x)\,, \qquad \epsilon \to 0\,. \tag{11}$$

How can we interpret this series? In this problem the question can be easily answered because the explicit solution is known to be

$$y(x) = \sqrt{\epsilon}\, \sin\left(\frac{x}{\sqrt{\epsilon}}\right), \tag{12}$$

so that

$$z(x) = \sqrt{\epsilon}\, \sin\left(\frac{x}{\sqrt{\epsilon}}\right) H(x)\,. \tag{13}$$

Now recall formula (13.1.31);

$$H(x)\left(\sqrt{\epsilon}\sin\frac{x}{\sqrt{\epsilon}}\right) \sim \sum_{n=1}^{\infty}(-1)^{n+1}\epsilon^{2n-1}\delta^{(2n-2)}(x)\,, \qquad \epsilon\to 0\,,$$

which when substituted in (13) agrees with relation (11).

With this analysis we have proved two points by introducing the delta series. First we have used the regular perturbation series and secondly the delta series yields the asymptotic series.

Boundary Value Problem. Let us now solve the boundary value problem:

$$\epsilon\, y''(x) + y'(x) + y(x) = 0\,, \qquad 0 \le x \le 1\,, \tag{14}$$

$$y(0) = 0\,, \qquad y'(1) = 1\,. \tag{15}$$

We again set $z(x) = H(x)y(x)$ as in the previous example and find that system (14) and (15) reduce to the inhomogeneous differential equation

$$\epsilon\, z''(x) + z'(x) + z(x) = \epsilon\, C\,\delta(x)\,, \tag{16}$$

such that

$$z(1) = 1\,, \qquad y'(0) = C\,, \tag{17}$$

where C is an unknown constant.

Next, we substitute

$$z(x) = \sum_{n=0}^{\infty} z_n(x)\,\epsilon^n\,, \tag{18}$$

and

$$C = \frac{C_{-1}}{\epsilon} + C_0 + C_1\epsilon + \cdots\,, \tag{19}$$

in (16) and compare the equal powers of ϵ on both sides of the resulting equation. The terms of $O(1)$ yield the equation

$$z_0'(x) + z_0(x) = C_{-1}\delta(x)\,. \tag{20}$$

For assigning the boundary value at $x = 1$, we appeal to (17) and set

$$z_0(1) = 1\,. \tag{21}$$

Equation (20) has solution

$$z_0(x) = C_{-1}H(x)e^{-x}\,. \tag{22}$$

Applying boundary condition (21) in (22) we obtain $C_{-1} = e$. Thus,

$$z_0(x) = H(x)\,e^{1-x}\,. \tag{23}$$

We return to equations (16) and (19) and proceed in the same fashion and find that the $O(\epsilon)$ term is

$$z_0''(x) + z_1'(x) + z_1(x) = C_0\delta(x)\,. \tag{24}$$

To this we adjoin the boundary value $z_1(1) = 0$, and substitute the value of $z_0''(x)$ from (23). This yields the system

$$z_1'(x) + z_1(x) = C_0\,\delta(x) - e^{1-x} H(x) + e\,\delta(x) - e\,\delta'(x)\,, \tag{25a}$$

$$z_1(0) = 1\,. \tag{25b}$$

Its solution is $C_0 = -e$ and

$$z_1(x) = H(x)\,e^{1-x}(1-x) - e\,\delta(x)\,. \tag{26}$$

We can continue this process and can evaluate as many terms as desired. Thus,

$$z(x) = H(x)\,e^{1-x} + \epsilon(H(x)\,e^{1-x}(1-x) - e\,\delta(x)) + O(\epsilon^2)\,. \tag{27}$$

Inner and Outer Decomposition. The singular perturbation problems are usually treated by giving one asymptotic expansion in the inner region and the second in the outer region. These two expansions are then matched in the overlapping region. From the present technique it emerges that the distributional asymptotic expansion of the solution contains the information in the entire region. We demonstrate this with the initial value problem

$$\epsilon\, y'(x) + y(x) = f(x)\,, \quad x > 0\,, \quad 0 < \epsilon \ll 1\,, \tag{28}$$

$$y(0) = y_0\,, \tag{29}$$

where $f(x)$ is a generalized function. As in the previous two cases we set $z(x,\epsilon) = y(x,\epsilon)H(x)$. This transforms the above initial value problem to the single inhomogeneous differential equation

$$\epsilon\, z'(x) + z(x) = f(x) + \epsilon\, y_0\,\delta(x)\,. \tag{30}$$

Next, we substitute

$$z(x) = \sum_{n=0}^{\infty} z_n(x)\epsilon^n\,, \tag{31}$$

and

$$z'(x) = \sum_{n=0}^{\infty} z_n'(x)\epsilon^n\,, \tag{32}$$

in (30) and get

$$\sum_{n=0}^{\infty} z_n'(x)\,\epsilon^{n+1} + \sum_{n=0}^{\infty} z_n(x)\epsilon^n = f(x) + \epsilon y_0\delta(x)\,.$$

Or

$$\sum_{n=1}^{\infty}(z_{n-1}'(x) + z_n(x))\epsilon^n + z_0(x) = f(x) + \epsilon\, y_0\delta(x)\,. \tag{33}$$

Now we compare the equal powers of ϵ on both sides of (33) and obtain

$$z_0(x) = f(x)\,, \tag{34}$$

$$z_0'(x) + z_1(x) = y_0\delta(x)\,, \tag{35}$$

$$z_{n-1}'(x) + z_n(x) = 0\,, \qquad n \geq 2\,. \tag{36}$$

When we substitute the value of $z_0(x)$ from (34) in (35) we obtain the value

$$z_1(x) = -\frac{\overline{d} f}{dx} + y_0\delta(x)\,. \tag{37}$$

Proceeding in this fashion we find that the values of $z_2(x)$ and $z_3(x)$ are

$$z_2(x) = \frac{\overline{d}^2 f}{dx^2} - y_0\delta'(x)\,, \tag{38}$$

and

$$z_3(x) = -\frac{\overline{d}^2 f}{dx^3} + y_0\delta''(x)\,. \tag{39}$$

By continuing this process we find the values of all the z_n. Substituting them in (31) we finally have

$$z(x) = \sum_{n=0}^{\infty}(-1)^n \frac{\overline{d}^n}{dx^n} f(x)\epsilon^n + y_0 \sum_{n=0}^{\infty}(-1)^{n+1}\delta^{(n)}(x)\,. \tag{40}$$

To derive the classical inner and outer expansions from (40) we suppose that $f(x)$ is smooth in $[0, \infty)$. Accordingly, $f(x)$ can be considered as a distribution supported in $[0, \infty)$ so that it is smooth everywhere except at the origin, where $f(x)$ and all the derivatives of $f(x)$ have jump discontinuities. This leads to the distributional derivatives

$$\frac{\overline{d}^n f(x)}{dx^n} = \frac{d^n f(x)}{dx^n} + f(0)\delta^{n-1}(x) + f'(0)\delta^{n-2}(x) \\ + \cdots + f^{(n-1)}(0)\,\delta(x)\,. \tag{41}$$

Substituting the value (41) in (40) yields

$$z(x;\epsilon) = z^{(out)}(x;\epsilon) + z^{(in)}(x;\epsilon)\,, \tag{42}$$

where

$$z^{(out)}(x;\epsilon) \sim \sum_{N=0}^{\infty}(-1)^n f^{(n)}(x)\epsilon^n\,, \tag{43}$$

and

$$z^{(in)}(x;\epsilon) \sim \sum_{n=1}^{\infty}(-1)^n \left(\sum_{n=0}^{\infty} f^{(k)}(0)\delta^{(n-1-k)}\right)\epsilon^n \\ + y_0 \sum_{n=1}^{\infty}(-1)^{n+1}\delta^{(n)}(x) \tag{44}$$

Thus $z^{(in)}$ to $O(\epsilon^2)$ is

$$z^{(in)}(x;\epsilon) = (y_0 - f(0))\epsilon\,\delta(x) + O(\epsilon^2)\,. \tag{45}$$

13.4. Applications to Number Theory

Series of the form

$$\sum_{n=1}^{\infty} a_n \phi(n\epsilon) \tag{1}$$

were studied by Ramanujan [72] for functions of the type

$$\phi(x) \sim b_1 x^{\alpha_1} + b_2 x^{\alpha_2} + \cdots, \quad \text{as} \quad x \to 0,$$

where $-1 < \alpha_1 < \alpha_2 < \cdots$. But he gave no proofs. The proofs were subsequently presented by classical analysis. We present the proofs with the help of generalized functions [6, 69, 73].

Let

$$f(x) = \sum_{n=1}^{\infty} a_n \delta(x - b_n), \tag{2}$$

so that

$$\begin{aligned} \langle f(\lambda x), \phi(x) \rangle &= \frac{1}{\lambda} \left\langle f(x), \phi\left(\frac{x}{\lambda}\right) \right\rangle \\ &= \frac{1}{\lambda} \left\langle \sum_{n=1}^{\infty} a_n \delta(x - b_n), \phi\left(\frac{x}{\lambda}\right) \right\rangle. \end{aligned} \tag{3}$$

For $\lambda = 1/\epsilon$, the above relation becomes

$$\left\langle f\left(\frac{x}{\epsilon}\right), \phi(x) \right\rangle = \epsilon \sum_{n=1}^{\infty} a_n \, \phi(\epsilon \, b_n). \tag{4}$$

Thus, we have formulated the Ramanujan problem with the help of generalized functions.

Let us now provide the solution of the above problem with the help of generalized functions. For this purpose we appeal to the moment expansion for $f(x)$, namely

$$f(x) \sim \sum_{n=0}^{\infty} (-1)^n \mu_n \delta^{(n)}(x) \, \frac{1}{n!}. \tag{5}$$

Thus,

$$\begin{aligned} \langle f(\lambda x), \phi(x) \rangle &= \frac{1}{\lambda} \left\langle f(x), \phi\left(\frac{x}{\lambda}\right) \right\rangle \\ &\sim \frac{1}{\lambda} \left\langle \sum_{n=0}^{\infty} (-1)^n \mu_n \delta^{(n)}(x) \frac{1}{n!}, \phi\left(\frac{x}{\lambda}\right) \right\rangle \\ &= \frac{1}{\lambda} \sum_{n=0}^{\infty} \mu_n \phi^{(n)}(0) \frac{1}{n!}. \end{aligned}$$

With $\lambda = 1/\epsilon$, this relation becomes

$$\left\langle f\left(\frac{x}{\epsilon}\right), \phi(x) \right\rangle = \epsilon \sum_{n=0}^{\infty} \mu_n \phi^{(n)}(0) \, \frac{1}{n!}. \tag{6}$$

Comparing (4) and (6) we obtain the result

$$\sum_{n=1}^{\infty} a_n \, \phi\,(\epsilon b_n) \sim \sum_{n=0}^{\infty} \mu_n \, \phi^{(n)}(0) \frac{1}{n!} \,. \tag{7}$$

Let us use this analysis to prove the Ramanujan formula

$$\sum_{n=1}^{\infty} (-1)^n \, n^q \, e^{-\epsilon n^p} \sim \sum_{n=0}^{\infty} \frac{(2^{np+q-1}-1)\zeta(-np-q)}{n!} \,, \tag{8}$$

where ζ is the Riemann zeta function

$$\zeta(s) = \sum_{n=1}^{\infty} \frac{1}{n^s} \,, \qquad \operatorname{Re} s > 1 \,. \tag{9}$$

To prove relation (8) we take

$$f(x) = \sum_{n=1}^{\infty} (-1)^n \delta(x-n) \,, \tag{10}$$

so that we have

$$a_n = (-1)^n \quad \text{and} \quad b_n = n \,,$$

and the left side of relation (7) becomes

$$\sum_{n=1}^{\infty} (-1)^n \phi(\epsilon n) \,. \tag{11}$$

Similarly, comparing it with left side of (8) we find that

$$\phi(x) = x^q e^{-xp} \tag{12}$$

Next, the right side of (8) is obtained from the right side of (7). Indeed, we evaluate the moments of the function given by (10);

$$\begin{aligned} \mu_k = \langle f(x), x^k \rangle &= \left\langle \sum_{n=1}^{\infty} (-1)^n \delta(x-n), x^k \right\rangle \\ &= (-1)^n n^k = 1^k + 2^k - 3^k + 4^k - 5^k + 6^k + \cdots \\ &= -(1^k + 3^k + 5^k + \cdots) + 2^k(1^k + 2^k + 3^k + \cdots) \\ &= -(1^k + 2^k + 3^k + 4^k + 5^k + \cdots) + 2^{k+1}(1^k + 2^k + 3^k + \cdots) \,, \end{aligned}$$

so that

$$\mu_k = (2^{k+1} - 1)\, \zeta(-k) \,. \tag{13}$$

Finally, we substitute the above values of a_n, b_n, $\phi(\epsilon b_n)$, μ_n and $\phi^{(n)}(0)$ in (7) and obtain Ramanujan's formula (8).

13.5. Distributional Weight Functions for Orthogonal Polynomials

Motivated by the failure of classical methods in the study of the weights for the Bessel polynomials [74,75], Morton and Krall [76] introduced the series of delta functions into the analysis. These concepts were subsequently extended and improved so much so that this analysis has become very elegant and simple. We present here the salient features of this subject.

A sequence of orthogonal polynomials $P_n(x)$ may be defined as follows. Let $\{\mu_n\}_{n=0}^{\infty}$ be a collection of real numbers with the property that $\Delta_0 = \mu_0 \neq 0$ and the determinant

$$\Delta_n = \begin{vmatrix} \mu_0 & \cdots & \mu_n \\ \vdots & & \vdots \\ \mu_n & \cdots & \mu_{2n} \end{vmatrix} \neq 0 , \qquad n = 1, \ldots . \tag{1}$$

Then we set $P_0 = 1$, and

$$P_n(x) = \frac{1}{\Delta_{n-1}} \begin{vmatrix} \mu_0 & \cdots & \mu_n \\ \vdots & & \vdots \\ \mu_{n-1} & \cdots & \mu_{2n-1} \\ 1 & \cdots & x^n \end{vmatrix} , \qquad n \geq 1 . \tag{2}$$

The polynomials $\{P_n(x)\}$ can be made orthogonal by a weight function $w(x)$ that satisfies the property

$$\int_a^b P_m(x) P_n(x) w(x) dx = \begin{cases} 0 , & m \neq n , \\ \neq 0, & m = n . \end{cases} \tag{3}$$

The function $w(x)$ satisfies certain conditions. For instance

$$\int_a^b w(x) dx > 0 .$$

Because $P_n(x)$ is a polynomial, orthogonality relation (3) leads us to the moments

$$\mu_n = \int_a^b x^n w(x) dx . \tag{4}$$

We stipulate that $\mu_n < \infty$ for all n. The moments play an important role in the theory of orthogonal polynomials. Indeed, every polynomial can be expressed in terms of its moments.

From relation (4) it is clear that if we know the weight function of an orthogonal polynomial sequence, then we can calculate the moments. The theory of generalized functions helps us in solving the inverse problem. That is, given the moments, we can determine the corresponding weight functions. For this purpose, we consider the concept of a linear functional w such that (4) can be written as

$$\mu_n = \langle w, x^n \rangle \qquad \text{for all} \quad n = 0, 1, 2, \ldots . \tag{5}$$

This is clearly a linear functional.

By setting $f(x) = w(x)$ in relations (13.1.1) to (13.1.5) we can write $w(x)$ as a dual Taylor series;

$$w(x) = \sum_{n=0}^{\infty} (-1)^n \frac{\mu_n \delta^{(n)}(x)}{n!} . \tag{6}$$

When the moments $\{\mu_i\}_{i=0}^{\infty}$ are those associated with the various classical orthogonal polynomials — the Legendre polynomials, the Laguerre polynomials, or the Hermite polynomials — the weight functions w yield the same results as the classical weight functions concerning orthogonality and norms. However, when the moments $\{\mu_i\}_{i=0}^{\infty}$ are those associated with the Jacobi polynomials or the generalized Laguerre polynomials, then w remains a suitable distributional weight function.

The weight $w(x)$ can be obtained by solving an ordinary differential equation [37]. The proof is as follows. If the polynomials satisfy a differential equation of the second order, it should be of the form

$$p(x)y''(x) + q(x)y' = \lambda y(x) , \tag{7}$$

$$p(x) = ax^2 + bx + c , \qquad q(x) = dx + e , \tag{8}$$

if $p(x)$ and $q(x)$ are to be independent of n. Now let $y(x)$ and $z(x)$ represent two different polynomials satisfying the differential equations

$$p(x)y''(x) + q(x)y(x) = \lambda y(x) ,$$

and

$$p(x)z''(x) + q(x)z(x) = \mu z(x) ,$$

respectively. Then the Wronskian, $W = yz' - y'z$, satisfies the differential equation

$$p(x)W'(x) + q(x)W(x) = (\mu - \lambda)y(x)z(x) . \tag{9}$$

Suppose $w(x)$ satisfies the equation

$$-(p(x)w(x))' + q(x)w(x) = 0 , \tag{10}$$

then eliminating the terms involving q between (9) and (10) we obtain

$$(p(x)w(x)W(x))' = (\mu - \lambda)y(x)z(x)w(x) . \tag{11}$$

Integrating both sides of this equation we have

$$(\mu - \lambda) \int_{-\infty}^{\infty} y(x)z(x)w(x)dx = [p(x)w(x)W(x)]_{-\infty}^{\infty} . \tag{12}$$

From this relation we find that the polynomials $y(x)$ and $z(x)$ will be orthogonal with respect to the weight function $w(x)$ if the right side of equation (12) vanishes. Accordingly, we have proved that the weight function $w(x)$ satisfies equation (10) subject to the condition that it vanishes at $\pm\infty$.

Let us now study the distributional weight function for some well known polynomials.

Legendre Polynomials. The Legendre differential equation is

$$(1-x^2)y''(x) - 2xy'(x) + n(n+1)y(x) = 0\,. \tag{13}$$

Then from equation (10) we find that $w(x)$ satisfies the differential equation

$$(1-x^2)w'(x) = 0\,. \tag{14}$$

To solve this distributionally, we observe that dividing by $(1-x)$ and $(1+x)$ yields

$$w'(x) = A\delta(x+1) + B\delta(x-1)\,, \tag{15}$$

where A and B are arbitrary constants.

The solution of this first order equation is

$$w(x) = AH(x+1) + BH(x-1) + C\,, \tag{16}$$

where C is another arbitrary constant. The only way $w(x)$ can vanish at $\pm\infty$ is to set $A = -B = 1$ (say) and $C = 0$. Thus

$$w(x) = H(x+1) - H(x-1) = \begin{cases} 1\,, & |x| \le 1\,, \\ 0\,, & \text{otherwise}\,. \end{cases} \tag{17}$$

The moments μ_n associated with the Legendre polynomials are $\mu_{2n} = 2/(2n+1)$, and $\mu_{2n+1} = 0$. Accordingly, the dual Taylor series (6) becomes

$$w(x) = \sum_{n=0}^{\infty} \frac{2\,\delta^{(2n)}(x)}{2n+1}\,. \tag{18}$$

Its inverse Fourier transform is obtained from (6.4.7) as

$$F^{-1}w(x) = -\frac{1}{\pi}\sum_{n=0}^{\infty} \frac{(it)^{2n}}{(2n+1)}\,, \tag{19}$$

which is a Fourier series representation for

$$f(t) = \frac{(e^{it} - e^{-it})}{2\pi i t} = \frac{\sin t}{\pi t}\,. \tag{20}$$

This function has the classical Fourier transform

$$w(x) = \begin{cases} 1\,, & -1 \le x \le 1\,, \\ 0\,, & |x| > 1\,, \end{cases} \tag{21}$$

which agrees with the classical weight function for the Legendre polynomials.

Hermite Polynomials. These polynomials satisfy the equation

$$y''(x) - 2xy'(x) + 2ny = 0\,, \tag{22}$$

so that the differential equation (10) for the weight function becomes

$$w'(x) + 2xw(x) = 0\,. \tag{23}$$

Its solution is $w(x) = e^{-x^2}$, $-\infty < x < \infty$. This is the well-known classical weight function.

The dual Taylor series (6) in this case becomes

$$w(x) = \sum_{n=0}^{\infty} \frac{\sqrt{\pi}\,\delta^{(2n)}(x)}{4^n\, n!}\,, \tag{24}$$

which agrees with the moment expansion of e^{-x^2} as given in formula (13.1.11).

The Laguerre Polynomials. The generalized Laguerre polynomials satisfy the equation

$$xy''(x) + (\alpha + 1 - x)y' + ny = 0\,. \tag{25}$$

Then the differential equation (10) for $w(x)$ is

$$xw'(x) + (x - \alpha)w = 0\,. \tag{26}$$

As long as $\alpha > -1$, the above equation has the solution

$$w(x) = x^\alpha e^{-x}[AH(x) + B]\,, \tag{27}$$

which vanishes at $x = \infty$. For $w(x)$ to vanish at $-\infty$, we should set $B = 0$. Thus, setting $A = 1$, relation (27) becomes

$$w(x) = x^\alpha e^{-x} H(x) = \begin{cases} x^\alpha e^{-x}\,, & 0 \le x < \infty\,, \\ 0\,, & -\infty < x < 0\,. \end{cases} \tag{28}$$

Since, the moments associated with the Laguerre polynomials are $\mu_n = \Gamma(\alpha + n + 1)/\Gamma(\alpha + 1)$, the dual Taylor series (6) is

$$w(x) = \sum_{n=0}^{\infty} \frac{(-1)^n \Gamma(\alpha + n + 1)}{\Gamma(\alpha + 1) n!}\, \delta^{(n)}(x)\,. \tag{29}$$

It is easily proved that the right side of the above relation is the moment expansion of the function (28).

The Bessel Polynomials. As mentioned in the introduction, it was this problem which motivated the distributional approach to the weight function $w(x)$.

For the Bessel polynomials, the differential equation is

$$x^2y''(x) + (2x + 2)y'(x) + n(n + 1)y(x) = 0\,. \tag{30}$$

Equation (10) for $w(x)$ becomes

$$x^2 w'(x) - 2w(x) = 0\,, \tag{31}$$

whose solution is $w(x) = e^{-2/x}$ which does not vanish at $\pm\infty$. Accordingly, it does not satisfy an essential requirement and, as such, is not a genuine weight function. Observe that $e^{-2/x}$ has an essential singularity at $x = 0$ and cannot be regularized at this point.

Now the moments in this case are

$$\mu_n = \frac{(-2)^{n+1}}{(n+1)!}\,, \qquad n = 0, 1, \ldots, \tag{32}$$

so that the dual Taylor series (6) becomes

$$w(x) = \sum_{n=0}^{\infty} \frac{2^{n+1}\delta^{(n)}(x)}{(n+1)!\, n!}\,. \tag{33}$$

Although (33) does not define a distribution it has been established that the series defines a hyperfunction concentrated at $\{0\}$. For studying the mathematical structure of the delta series solution such as (33) consult Hernández-Urená and Estrada [77].

13.6. Convolution Type Integral Equation Revisited

In Section 7.9 we employed the Fourier transform methods to solve certain integral equations of the convolution type. In this section we use the moment asymptotic expansion methods to accomplish that and much more.

In their book [78], Hirchman and Widder describe the differential inversion method for solving convolution type integral equations. They mainly use the bilateral Laplace transform as their tool. We explain their basic idea with the help of the Fourier transform. Let us study the integral equation

$$\int_{-\infty}^{\infty} K(x-y)\psi(y)dy = \phi(x)\,, \tag{1}$$

which can be written as

$$K(x) * \psi(x) = \phi(x)\,. \tag{2}$$

We take the Fourier Transform of this equation and get

$$\widehat{K}(u)\widehat{\psi}(u) = \widehat{\phi}(u)\,. \tag{3}$$

Let $\widehat{K}$ have the form $E(u)$;

$$E(u) = \left(1 - \frac{u}{a_1}\right)\left(1 - \frac{u}{a_2}\right)\cdots\left(1 - \frac{u}{a_n}\right), \tag{4}$$

so that (3) becomes

$$\widehat{\psi}(u) = \frac{1}{E(u)}\,\widehat{\phi}(u)\,. \tag{5}$$

Now we use the correspondence relation

$$\frac{df(x)}{dx} \longleftrightarrow -i\,u\,\widehat{f}(u) \tag{6}$$

between the differential of a function $f(x)$ and its Fourier transform, $\widehat{f}(u)$. Then we can invert (5) and get

$$\psi(x) = E\left(i\frac{d}{dx}\right)\phi(x)\,, \tag{7}$$

which is precisely the basic principle of differential inversion.

Apparently unaware of this analysis, Hohlfeld et al [79] have presented the solution of the convolution type integral equation (1) but they combine it with the method of moment expansions. Subsequently, Murthy [80] pointed out the similarity in their approach and that of Hirchman and Widder and added his own insight also. Our aim is to integrate the foregoing concepts into the theme of this chapter.

The main step in solving equation (1) is to introduce the concept of Green's function $g(x)$ as is done in the study of differential equations. Accordingly, we first solve the integral equation

$$\int_{-\infty}^{\infty} K(x-y)g(x)dx = K(x) * g(x) = \delta(x) \tag{8}$$

If we can solve equation (3) for Green's function $g(x)$, then by convolution we obtain the solution of equation (1) as

$$\psi(x) = g(x) * \phi(x)\,. \tag{9}$$

The advantage in solving equation (8) is that when we take the Fourier transform of both sides we obtain

$$\widehat{K}(u)\widehat{g}(u) = 1\,, \tag{10}$$

which, if invertible, yields the value of $\widehat{g}(u)$. Then we take the inverse of $\widehat{g}(u)$ and substitute it in (9) to obtain the solution $\psi(x)$ of the integral equation (1).

Let us now demonstrate how the moment expansions enter the picture. Indeed, the moment expansions of Green's function $g(x)$ and the kernel $K(x)$ are

$$g(x) = \sum_{n=0}^{\infty} \frac{(-1)^n \mu_n}{n!}\,\delta^{(n)}(x)\,, \tag{11}$$

and

$$K(x) = \sum_{n=0}^{\infty} \frac{(-1)^n \nu_n}{n!}\,\delta^{(n)}(x)\,, \tag{12}$$

where

$$\mu_n = \int_{-\infty}^{\infty} x^n g(x)dx \quad \text{and} \quad \nu_n = \int_{-\infty}^{\infty} x^n K(x)dx\,. \tag{13}$$

Observe that the values of ν_n are known because $K(x)$ is known while the moments μ_n have to be evaluated. For this purpose we need the formula

$$[\delta^k(x)]^\wedge(u) = (-i\,u)^k\,. \tag{14}$$

When we take the Fourier transform of relations (11) and (12) and use (14) we have

$$\widehat{g}(u) = \sum_{n=0}^{\infty} \frac{\mu_n}{n!}\,(i\,u)^n\,, \tag{15}$$

$$\widehat{K}(u) = \sum_{n=0}^{\infty} \frac{\nu_n}{n!}\,(i\,u)^n\,. \tag{16}$$

We now substitute these values of $\widehat{g}(u)$ and $\widehat{K}(u)$ in equation (10) and get

$$\left(\sum_{n=0}^{\infty} \frac{\mu_n}{n!}\,(i\,u)^n\right)\left(\sum_{m=0}^{\infty} \frac{\nu_m}{m!}\,(i\,u)^m\right) = 1\,. \tag{17}$$

Comparing the equal powers of $(i\,u)$ on both sides of (17) we get

$$\mu_0\nu_0 = 1\,,\quad n = 0\,,\qquad \sum_{k=0}^{n} \mu_k\,\nu_{n-k} = 0\,,\qquad n > 0\,. \tag{18}$$

Since the values of ν_n are known, we can evaluate the moments $\mu_0, \mu_1, \ldots, \mu_n$ of $\widehat{g}(u)$. Subsequently, we invert $\widehat{g}(u)$ and substitute this value of $g(x)$ in (9). The result is

$$\begin{aligned}\psi(x) = g * \phi &= \sum_{n=0}^{\infty}\left(\frac{(-1)^n\mu_n}{n!}\,\frac{d^n\delta(x)}{dx^n}\right) * \phi \\ &= \sum_{n=1}^{\infty} \frac{(-1)^n\mu_n}{n!}\,\frac{d^n\phi}{dx^n}\,,\end{aligned} \tag{19}$$

where we have used the relation $(d^n\delta(x)/dx^n) * \phi = d^n\phi/dx^n$.

This method appears to be tedious because we have to solve an infinite number of algebraic equations. Let us see how we can improve on it. We started this section with the concept of differential inversion. Let us now demonstrate how the method of moment expansions helps us in achieving that goal. For this purpose we return to equation (1) and put the moment expansion (12) of the kernel $K(x-y)$ in it so that we have

$$\phi(x) = \int_{-\infty}^{\infty} \sum_{n=0}^{\infty} \frac{(-1)^n\mu_n}{n!}\,\frac{d^n\delta(x-y)}{dx^n}\,\psi(y)dy\,. \tag{20}$$

Now, if we could express the moment expansion of $K(x-y)$ as $L(x)\delta(x-y)$, where L is a differential operator, then equation (20) can be written as

$$\phi(x) = \int_{-\infty}^{\infty} \{L\delta(x-y)\}\psi(y)dy\,. \tag{21}$$

We can solve this equation if the operator L is invertible. In that case we have

$$L^{-1}\phi(x) = \int_{-\infty}^{\infty} \delta(x-y)\psi(y)dy\,. \tag{22}$$

The sifting property of the delta function then yields the solution

$$\psi(x) = L^{-1}\phi(x)\,. \tag{23}$$

Let us illustrate this method with a few examples.

Example 1. Gaussian kernel $K(x) = e^{-x^2}/\sqrt{\pi}$. The moment expansion of this kernel is provided by (13.1.11);

$$\frac{e^{-x^2}}{\sqrt{\pi}} = \sum_{n=0}^{\infty} \frac{1}{4^n n!}\,\delta^{2n}(x)\,, \tag{24}$$

which can be written as

$$\frac{e^{-x^2}}{\sqrt{\pi}} = e^{(\frac{1}{4}D^2)}\delta(x)\,. \tag{25}$$

Thus the required operator L is $e^{(-\frac{1}{4}D^2)}$ whose inverse is $e^{\frac{1}{4}D^2}$. Then from (23) we have the solution

$$\psi(x) = \sum_{n=0}^{\infty}\left(-\frac{1}{4}\right)^n \frac{1}{n!}\,\frac{d^{2n}\phi(x)}{dx^{2n}}\,. \tag{26}$$

Example 2. $K(x) = e^{-|x|}$. Its moment expansion is given by the formula (13.1.20);

$$e^{-|x|} = 2\sum_{n=0}^{\infty}(2n)!\delta^{2n}(x)\,, \tag{27}$$

which can be written as

$$K(x) = \left(\frac{2}{1-D^2}\right)\delta(x)\,. \tag{28}$$

In this case $L^{-1} = \frac{1}{2}(1-D^2)$ so that from formula (23) we obtain the solution

$$\psi(x) = \frac{1}{2}\left(\phi(x) - \frac{d^2\phi}{dx^2}\right)\,. \tag{29}$$

Example 3. Let us now consider the three-dimensional kernel

$$K(x-y) = -\frac{1}{4\pi r} = -\frac{1}{4\pi|x-y|} \tag{30}$$

which is the solution of the equation

$$\nabla^2 K(x-y) = \delta(x)\,. \tag{31}$$

To find the moments of the kernel (30) we appeal to the relation

$$\mu_n = -\int_{-\infty}^{\infty} x^{\otimes n} \frac{1}{4\pi |x-y|} dx\,, \tag{32}$$

where $x^{\otimes n}$ stands for the n-times tensor product of the vector x with itself, and find that the required moment expansion is

$$\begin{aligned} -\frac{1}{4\pi |x-y|} &= \left(\sum_{n=0}^{\infty} \frac{1}{n!} (-y.D)^n \right) \delta(x-y) \\ &= e^{(-y.D)} \delta(x-y)\,. \end{aligned} \tag{33}$$

Accordingly, the solution of the integral equation

$$\phi(x) = \int_{-\infty}^{\infty} -\frac{1}{4\pi |x-y|} \psi(y)\, dy \tag{34}$$

is

$$\psi(x) = e^{(y.D)} \phi(x)\,. \tag{35}$$

13.7. Further Applications

The methods explained in this chapter can be applied to various other problems. We shall demonstrate this by presenting the delta series solution for the inhomogeneous heat equation.

Recall from Section 10.7, that the fundamental solution of the Heat operator for $n = 3$ is

$$E(x, t) = \frac{1}{(4\pi t)^{3/2}} \exp(-r^2/4t)\,, \tag{1}$$

where $r = |x|$. Accordingly, the solution of the inhomogeneous Heat equation

$$\frac{\partial u(xt)}{\partial t} - \nabla^2 u(x, t) = f(x) \tag{2}$$

follows from the convolution principle, and is

$$u(x, t) = \frac{1}{(4\pi t)^{3/2}} \exp\left(-\frac{r^2}{4t}\right) * f(x)\,. \tag{3}$$

Next, we use the moment asymptotic expansion [6]

$$\exp\left(-\left(\frac{r}{2\sqrt{t}}\right)^2\right) = \sum_{N=0}^{\infty} \frac{\pi^{3/2}}{2^{2N}} \sum_{|k|=N} \frac{D^{2k}\delta(x)}{k!\, t^{N+3/2}}\,. \tag{4}$$

Substituting (4) in (3) and using the relation

$$\delta^{2k}(x) * f(x) = D^{2k} f(x)\,, \tag{5}$$

we have

$$u(x,t) = \frac{1}{8}\sum_{N=0}^{\infty}\frac{1}{2^{2N}}\sum_{|k|=N}\frac{D^{2k}f(x)}{k!\,t^{N+3}}\,. \tag{6}$$

Thus we have obtained the behavior of $u(x,t)$ as $t \to \infty$.

For simple expository notes on the subject matter of this chapter the reader may consult references [81,82].

CHAPTER 14

Linear Systems

14.1. Operators

Recall that we define a function as a rule that maps (transforms) numbers into numbers, whereas a functional maps functions into numbers. An *operator* maps functions into functions. Let a class of functions be given, all defined for a variable, say, time t, $-\infty < t < \infty$. Then an operator (transformation) L assigns a member of this class (inputs, excitations, or signals) to members of a second class of functions (outputs or responses). We shall use the symbol $x(t)$ for an input and $y(t)$ for the corresponding output. A system is a mathematical model of a physical device and is represented by an operator L. Then $y(t)$ is called the response of $x(t)$ due to the given system. A system is often represented by a box as shown in Figure 14.1.

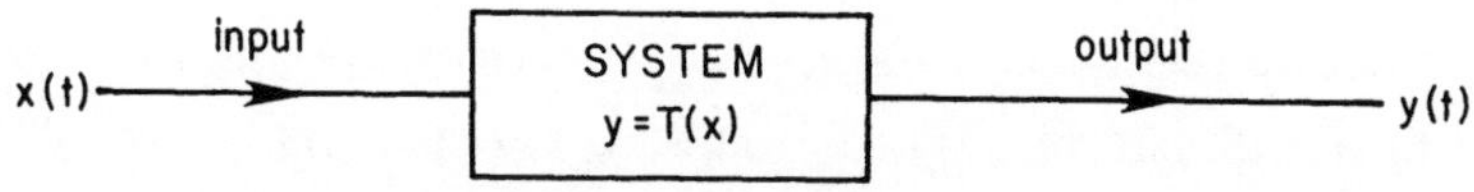

Figure 14.1. A system.

Inputs, such as a force applied to a mass, a voltage applied to an electrical circuit, or a heat source applied to a vessel filled with a liquid, are given; outputs are to be calculated or measured. The outputs corresponding to the inputs we have mentioned might be the velocity of the mass, the voltage across a resistor, and the rate at which the liquid flows through a pipe.

If the operator L satisfies the identity

$$L[c_1x_1 + c_2x_2] = c_1L[x_1] + c_2L[x_2] = c_1y_1 + c_2y_2,$$

for all inputs x_1 and x_2 and for all constants c_1 and c_2, then L is a *linear operator*. A system is called linear if and only if the operator which represents it is a linear operator. This means that a linear system obeys the *superposition* principle, which requires that the following two tests be satisfied:

(1) Multiplying the input by any constant c must multiply the output by c.

(2) The response to several inputs applied simultaneously must be the sum of the individual responses to each input applied separately.

It follows that if $x(t) = 0$ for all t, then so is $y(t)$. Thus, there can be no output without an input. Accordingly, there is no stored energy. A linear system is therefore also called a *relaxed* system.

Let us write down a few examples of linear systems:

(1) reflection, $x(t) \to y(t) = x(-t)$,

(2) a constant gain amplifier, $x(t) \to y(t) = cx(t)$, where c is constant,

(3) the difference, $x(t) \to y(t) = x(t+1) - x(t)$,

(4) an ideal time delay of duration a, $x(t) \to y(t) = x(t-a) = x_a(t)$, where a is a positive number,

(5) the multiplier, $x(t) \to y(t) = f(t)x(t)$, where $f(t)$ is a given fixed function,

(6) the differential, $x(t) \to y(t) = f(D)x(t)$, where $f(D)$ is a linear differential operator,

$$f(D) = a_0 \frac{d^n}{dt^n} + a_1 \frac{d^{n-1}}{dt^{n-1}} + \cdots + a_n\,,$$

with $a_0, a_1, \ldots, a_n$ functions of t only;

(7) the integrator $x(t) \to y(t) = \int_{-\infty}^{t} x(\tau)d\tau$;

(8) the integrodifferential system

$$x(t) \to y(t) = \frac{dx}{dt} + ax + b\int_{-\infty}^{t} x(\tau)d\tau\,,$$

where a and b are constants;

(9) the Fourier transform, $x(t) \to y(t) = \int_{-\infty}^{\infty} x(\tau)e^{it\tau}d\tau$;

(10) the Laplace transform, $x(t) \to y(t) = \int_{0}^{\infty} x(\tau)e^{-t\tau}d\tau$.

So far we have implicitly assumed that all the functions and constants involved in the discussion are real. However, the analysis holds for complex quantities as well. In that case we write $x(t) = x_1(t) + ix_2(t)$ and $y(t) = y_1(t) + iy_2(t)$, so that $x_1(t) \to y_1(t)$ and $x_2(t) \to y_2(t)$. Of course, in a physical system we expect all the input and output signals to be real valued.

An operator is called *stationary* if for each fixed real number a we have

$$L[x_a] = (L[x])_a = y(t-a)\,.$$

The corresponding system is called *fixed* or *time-invariant*. Accordingly, for a time-invariant system, the only effect of delaying (or advancing) and input singal is to produce a corresponding delay (advance) of the output signals.

We shall be mainly concerned with time-invariant linear systems. A system will be associated with a specific class of functions that are admissible inputs in the sense that the corresponding output functions are well defined. For instance, for a system defined by a differential operator, we expect $x(t)$ to be differentiable.

An operator is called *continuous* if, whenever a sequence of input signals converges to zero, then the corresponding sequence of outputs signals also converges to zero.

14.2. The Step Response

Let $U(t)$ be the response to the Heaviside function $H(t)$, as shown in Figure 14.2. The

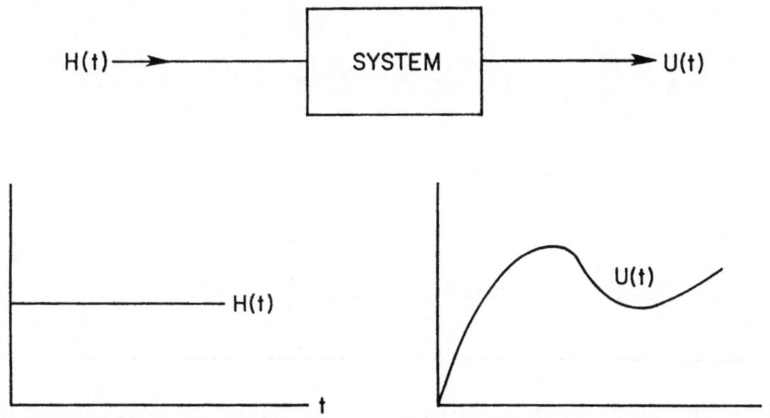

Figure 14.2. The step response function $U(t)$.

function $U(t)$ is called the *step response*.

Example 1. Let us find the step response for the system

$$(a_0 D + a_1)y(t) = x(t), \qquad a_1 \neq 0, \qquad x(t) = 0 \text{ for } t < 0. \tag{1}$$

This means that we have to solve the differential equation

$$(a_0 D + a_1)U(t) = H(t), \qquad a_1 \neq 0, \qquad U(t) = 0, \ t < 0. \tag{2}$$

Its solution is

$$U(t) = (1/a_1)(1 - e^{-a_1 t/a_0})H(t). \tag{3}$$

Example 2. Consider the system

$$[D^2 + (a+b)D + ab]y(t) = x(t), \qquad a \neq 0, \quad ab \neq 0, \tag{4}$$

where $x(t) = 0$ for $t < 0$.

The step response for this system is given by

$$[D^2 + (a+b)D + ab]U(t) = H(t). \tag{5}$$

Its solution is

$$U(t) = \frac{1}{ab}\left[1 + \frac{be^{-at} - ae^{-bt}}{a-b}\right] H(t). \tag{6}$$

14.3. The Impulse Response

From the analysis of Section 14.2 we find that the response to rectangular pulse of height A and width a, as shown in Figure 14.3, is

$$y(t) = A\left[U\left(t + \frac{1}{2}a\right) - U\left(t - \frac{1}{2}a\right)\right]. \tag{1}$$

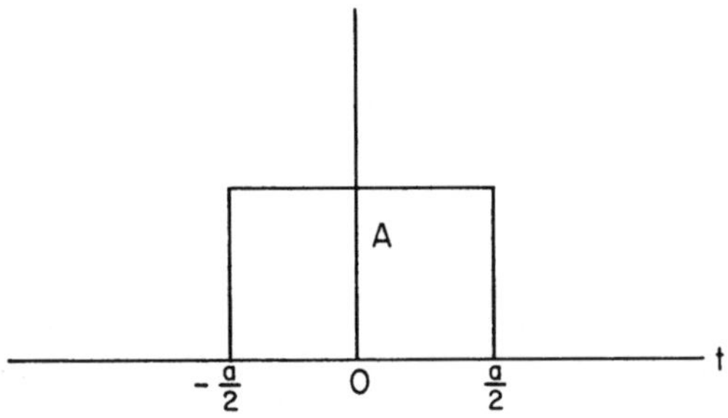

Figure 14.3. Response to a rectangular pulse of height A and width a.

If $U(t)$ is a continuously differential function we write

$$y(t) = Aa\left[\frac{U\left(t+\frac{1}{2}a\right) - U\left(t-\frac{1}{2}a\right)}{a}\right]$$

and let $a \to 0$, $A \to \infty$ such that $Aa = 1$. We then have

$$y(t) = dU(t)/dt\,.$$

This is called a unit *impulse response* and is denoted by $h(t)$. We know from Chapter 1 that the rectangular pulse (1) yields $\delta(t)$ in the limit, the response to the delta function is $h(t)$. We have already encountered these functions in the previous chapters where we called them the fundamental solutions. Just as in the case of differential equations we can obtain the output for the equation $Ly = f$ by appealing to the impulse response h of L and using the principle of convolution. Indeed, $y = f * h$. When we take the Fourier transform of this equation we get

$$[y]^\wedge = \widehat{f}\widehat{h}\,.$$

The function $\widehat{h}$ is called the *system function.*

Example 1. In Example 1 of Section 14.2 we found that the step response to the system (14.2.1) is (14.2.3). Hence, the impulse response of the present system is

$$h(t) = (e^{-a_1 t/a_0}/a_0)H(t)\,. \tag{3}$$

This is precisely the solution of the system

$$(a_0 D + a_1)y = \delta(t)\,, \tag{4}$$

as is easily verified by setting $y(t) = G(t)H(t)$ and following the method explained in Section 9.8.

Example 2. Let us now find the impulse response of the system of Example 2 of Section 14.2, namely,

$$[D^2 + (a+b)D + ab]y(t) = x(t)\,. \tag{5}$$

Differentiating the step response (14.2.6), we obtain

$$h(t) = \frac{e^{-bt} - e^{-at}}{a - b} H(t), \tag{6}$$

which is precisely the solution of the system

$$[D^2 + (a+b)D + ab]h(t) = \delta(t). \tag{7}$$

14.4. The Response to an Arbitrary Input

Having found the response function corresponding to the delta function, we can convolute to get the response to an arbitrary input. Thus

$$y(t) = x(t) * h(t) = \int_{-\infty}^{\infty} x(\tau)h(t-\tau)d\tau = \int_{-\infty}^{\infty} x(t-\tau)h(\tau)d\tau. \tag{1}$$

Example 1. In Example 1 of Section 14.3 we found that the impulse response to the system (14.2.1) is given by (14.3.3). Thus the output $y(t)$ for an arbitrary input $x(t)$ is

$$y(t) = \frac{1}{a_0} e^{-a_1 t/a_0} \int_0^t x(\tau) e^{a_1 t/a_0} d\tau. \tag{2}$$

Example 2. Proceeding as in Example 1, we find that the output $y(t)$ for system (14.3.5) is

$$y(t) = \frac{1}{a-b} \int_0^t x(\tau)[e^{b(\tau-t)} - e^{a(\tau-t)}]d\tau. \tag{3}$$

By using the associative property of the convolution, we derive the impulse response h_3 due to the impulse responses h_1 and h_2 connected in cascade, as in Figure 14.4;

$$h_3(t) = h_1 * h_2 = \int_{-\infty}^{\infty} h_1(t-\tau)h_2(\tau)d\tau. \tag{4}$$

Stability: A system is called stable if a bounded input produces a bounded output. Clearly, this is desirable, because otherwise the output would increase without limit. Suppose that

$$|x(t)| < K_1,$$

then from the relation (14.4.1)

$$y(t) = \int_{-\infty}^{\infty} h(u)x(t-u)du,$$

where h is the impulse response, we have

$$\begin{aligned}|y(t)| &= \left|\int_{-\infty}^{\infty} h(u)x(t-u)du\right| \\ &\leq \int_{-\infty}^{\infty} |h(u)|\,|x(t-u)|du \\ &\leq K_1 \int_{-\infty}^{\infty} |h(u)|du\,.\end{aligned}$$

Accordingly, a sufficient condition for the system to be stable is that

$$\int_{-\infty}^{\infty} |h(u)|du < K_2\,,$$

when K_2 is a finite constant.

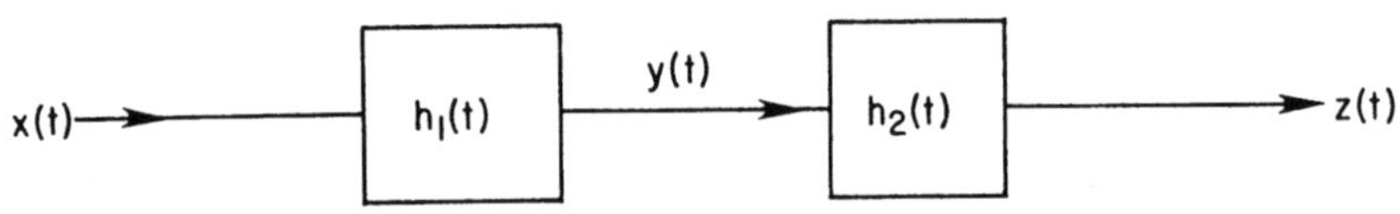

Figure 14.4. The output $z(t)$ due to the impulse responses h_1 and h_2 connected in cascade.

14.5. Generalized Functions as Impulse Response Functions

From the sifting property, $x(t) = \int_{-\infty}^{\infty} x(t-\tau)\delta(\tau)dt$, we infer that $\delta(t)$ is itself the impulse response function of the time-invariant linear system that leaves every input signal $x(t)$ unchanged. Similarly, from

$$x(t-a) = \int_{-\infty}^{\infty} x(t-\tau)\delta(\tau-a)dt$$

we find that the impulse response function associated with an ideal time delay system of duration a is $\delta(t-a)$.

Next we appeal to the differential property of the delta function;

$$\frac{d^n x}{dt^n} = \int_{-\infty}^{\infty} x(t-\tau)\delta^{(n)}(\tau)d\tau\,,$$

and infer that the generalized function $\delta^{(n)}(t)$ is the impulse response function characterising the time-invariant linear system transforming every n-times continuously differentiable input $x(t)$ into the nth derivative $x^{(n)}(t)$.

Since the generalized functions are impulse responses for various systems, they can be convoluted with other impulse responses, leading to very interesting cascade connections, such as

(1) $h * \delta_a = \delta_a * h$,
(2) $\delta_a(t) * \delta_b(t) = \int_{-\infty}^{\infty} \delta_a(t-\tau)\delta_b(\tau)d\tau = \delta_{a+b}(t)$, and
(3) $\delta^{(m)}(t) * \delta^{(n)}(t) = \int_{-\infty}^{\infty} \delta^{(n)}(t-\tau)\delta^{(m)}(\tau)d\tau = \delta^{(m+n)}(t)$,

where $m \geq 0$, $n \geq 0$.

14.6. The Transfer Function

The response $y(t)$ for an exponential input $x(t)$,

$$x(t) = e^{st}, \qquad s = \sigma + i\omega,$$

plays a key role in the analysis of linear systems. It is usually denoted $H_s(t)$. For a time-invariant linear system it can be deduced from the following two considerations:

(1) From the time invariance we observe that for any given fixed value of τ the response to the input $x(t+\tau)$ is $H_s(t+\tau)$ and for $x(t) = e^{st}$ we have

$$x(t+\tau) = e^{s(t+\tau)} = e^{st}e^{s\tau} = e^{s\tau}x(t).$$

(2) In view of the linearity of the system, we know that if $x(t) \to H_s(t)$, then

$$x(t+\tau) \to e^{s\tau} H_s(t). \tag{1}$$

Specifically, for $t = 0$, (1) becomes

$$x(\tau) \to e^{s\tau} H_s(0) = e^{s\tau} H(s), \tag{2}$$

where we have set $H_s(0) = H(s)$. Thus, the response of a time-invariant linear system to the input $x(t) = e^{st}$ is

$$y(t) = H(s)e^{st}. \tag{3}$$

The quantity $H(s)$ depends only on the parameter $s = \sigma + i\omega$ and on the particular system concerned. It is called the *transfer function* of the system and is defined in some suitable region of the complex plane (σ, ω).

The response to the input $x(t) = e^{st}$, obtained from the convolution integral, is

$$\int_{-\infty}^{\infty} e^{st}h(t-\tau)d\tau = \int_{-\infty}^{\infty} e^{(t-\tau)}h(\tau)d\tau, \tag{4}$$

where $h(t)$ is the impulse response of the system. Comparing (3) and (4), we obtain

$$H(s)e^{st} = e^{s\tau}\int_{-\infty}^{\infty} e^{-st}h(\tau)d\tau,$$

or

$$H(s) = \int_{-\infty}^{\infty} e^{-s\tau} h(\tau) d\tau \,. \tag{5}$$

This relation yields the following two results:

(1) If $\sigma = 0$, so that $s = i\omega$, we have

$$H(s) = \int_{-\infty}^{\infty} e^{i\omega\tau} h(\tau) d\tau = \widehat{H}(\omega) \,, \tag{6}$$

where $\widehat{H}(\omega)$ is the Fourier transform of $h(t)$. In this case the transfer function is called the *frequency response function.*

(2) For a system whose impulse response function $h(t)$ vanishes for all negative values of t, we obtain from (5)

$$H(s) = \int_{0}^{\infty} e^{-st} h(\tau) d\tau = \mathcal{L}\{h(t)\} \,, \tag{7}$$

where $\mathcal{L}$ stands for the Laplace transform. This means that if the input $x(t)$ vanishes for all $t < 0$, then so does the output $y(t)$. Accordingly, no output signal can ever precede its input signal. That is, the signal is *causal* (or *non-anticipative*).

Equation (7), which gives the transform function $H(s)$ as the Laplace transform of the impulse response, facilitates its evaluation. Taking the Laplace transform of

$$y(t) = \int_{-\infty}^{\infty} h(t-\tau) x(\tau) d\tau \,,$$

we have

$$\mathcal{L}\{y(t)\} = H(s)\mathcal{L}\{x(t)\} \,. \tag{8}$$

We illustrate these concepts by the help of the following example.

Example. The zero-state response. Let the system be in the form of a general nth order input-output ordinary differential equation:

$$a_n \frac{d^n y}{dt^n} + a_{n-1} \frac{d^{n-1} y}{dt^{n-1}} + \cdots + a_1 \frac{dy}{dt} + a_0 y$$

$$= b_m \frac{d^m x}{dt^m} + b_{m-1} \frac{d^{m-1} x}{dt^{m-1}} + \cdots + b_0 x \,, \tag{9a}$$

$$y(0) = 0 \,, \qquad y'(0) = 0 \,, \qquad \ldots, \qquad y^{n-1}(0) = 0 \,, \tag{9b}$$

$$x(0) = 0 \,, \qquad x'(0) = 0 \,, \qquad \ldots, \qquad x^{m-1}(0) = 0 \,. \tag{9c}$$

Since, the initial conditions are zero, the stored energy is zero. This is called a *zero-state response.*

Taking the Laplace transform of both sides of (9), we find

$$(a_n s^n + a_{n-1}s^{n-1} + \cdots + a_1 s + a_0)\widetilde{y}(s)$$
$$= (b_m s^m + b_{m-1}s^{m-1} + \cdots + b_0)\widetilde{x}(s)\,,$$

where we have used the notation of Chapter 8, so from (8) we find that

$$\widetilde{y}(s) = \frac{b_m s^m + b_{m-1}s^{m-1} + \cdots + b_0}{(a_n s^n + a_{n-1}s^{n-1} + \cdots + a_1 s + a_0)}\widetilde{x}(s) = H(s)\widetilde{x}(s)\,.$$

Thus

$$H(s) = \frac{b_m s^m + b_{m-1}s^{m-1} + \cdots + b_0}{a_n s^n + a_{n-1}s^{n-1} + \cdots a_1 s + a_0}\,. \tag{10}$$

By factoring the two polynomials in (10) we can write

$$H(s) = K\frac{(s-z_1)(s-z_2)\cdots(s-z_m)}{(s-p_1)(s-p_2)\cdots(s-p_n)}\,, \tag{11}$$

where $K = b_m/a_n$. The quantities $z_1, z_2, \ldots, z_m$ are the zeros of $H(s)$, and $p_1, p_2, \ldots, p_n$ are the poles. From this relation it is apparent that $H(s)$ can be completely specified by its zeros, poles, and the multiplying factor K, which is assumed to be nonzero. The zeros and poles may be complex numbers and can be represented graphically in a complex plane, which is called the *complex frequency plane*.

Let us assume that the Laplace transform of the input $x(t)$ can be written as a rational function of s,

$$\widetilde{x}(s) = N(s)/D(s)\,, \tag{12}$$

where $N(s)$ and $D(s)$ are polynomials. Then from (8), (11), and (12) it follows that

$$\widetilde{y}(s) = K\frac{(s-z_1)\cdots(s-z_m)}{(s-p_1)(s-p_2)\cdots(s-p_n)}\frac{N(s)}{D(s)}\,.$$

By factoring $N(s)$ and $D(s)$ we can obtain all the zeros and poles of $\widetilde{y}(s)$ from this expression. This facilitates the inversion of $\widetilde{y}(s)$, and we thereby obtain the required solution $y(t)$.

This analysis can be readily extended to the case of the nonzero initial conditions.

14.7. Discrete-Time Systems

A *discrete-time system* contains a sampler (a switch) that closes every T second (Figure 14.5). For this reason it is also called *sampler data system*. The output of an *ideal sampler* is a string of impulses starting at $t = 0$ spaced at intervals of T seconds of amplitude (power) $x(nT)$, so that

$$y(t) = \sum_{n=0}^{\infty} x(nT)\delta(t - nT)\,. \tag{1}$$

We take the Laplace transform of both sides of this relation, obtaining

$$\mathcal{L}\{y(t)\} = \sum_{n=0}^{\infty} x(nT)e^{-nTs}\,. \tag{2}$$

This series is different from the one we encountered in Example 4 of Section 8.4, where we discussed the periodic functions. We handle this series by introducing the Z transform, which is defined as follows. Let the function $f(t)$ be known to us only through its sample values at $t = 0, 1, 2, \ldots, n$, namely

$$f(0)\,, \quad f(1)\,, \quad f(2)\,, \quad \ldots, \quad f(n)\,.$$

Then the polynomial

$$F(z) = Z[f] = f(0) + f(1)z^{-1} + f(2)z^{-2} + \cdots + f(n)z^{-n}$$

defines the Z transform. This definition can be extended to infinite series. To make use of this transform we set

$$z = e^{st}\,, \tag{3}$$

so that right side of (2) becomes

$$Z[x(t)] = X(z) = \sum_{n=0}^{\infty} x(nT)z^{-n}\,, \tag{4}$$

which is the Z transform of $x(t)$.

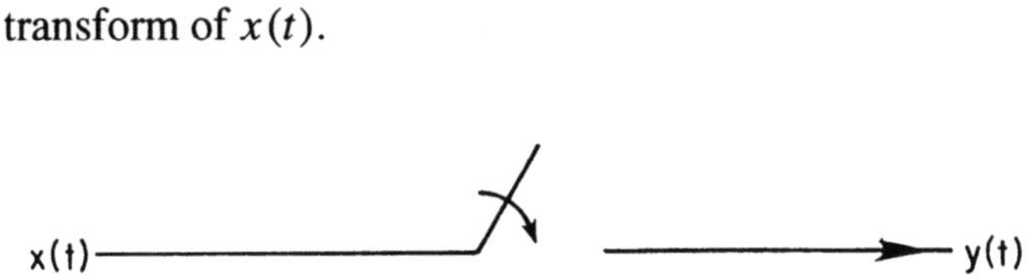

Figure 14.5. A discrete-time system.

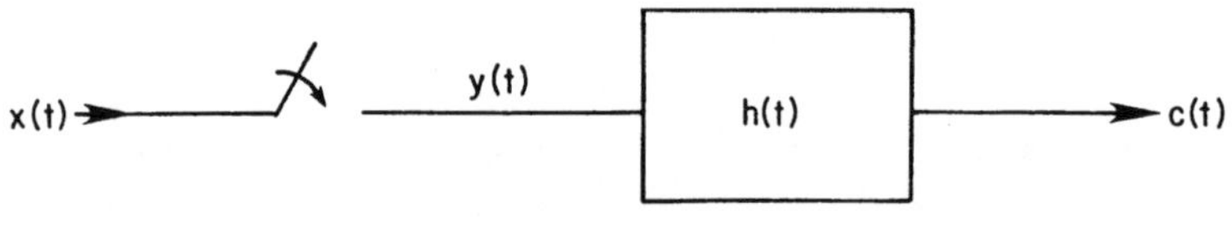

Figure 14.6. A linear discrete-time system with impulse response $h(t)$.

Let us consider the system of Figure 14.6, a linear system with impulse response $h(t)$. Then

$$\begin{aligned} c(t) &= \int_0^{\infty} y(\tau)h(t-\tau)d\tau = \int_0^{\infty} \sum_{n=0}^{\infty} x(nT)\delta(\tau - nT)h(t-\tau)d\tau \\ &= \sum_{n=0}^{\infty} x(nT)h(t-nT)\,. \end{aligned} \tag{5}$$

Because the Z transform of $c(t)$ is $C(z) = Z[c(t)] = \sum_{n=0}^{\infty} c(nT)z^{-n}$, we find from (5) that

$$C(z) = \sum_{n=0}^{\infty}\left[\sum_{m=0}^{\infty} x(mT)h(tn-m)T)\right]z^{-n} = \sum_{m=0}^{\infty} x(mT)\left[\sum_{n=0}^{\infty} h(n-m)T)z^{-n}\right]$$

$$= \sum_{m=0}^{\infty}\left[x(mT)\sum_{m=0}^{\infty} h(kT)z^{-k-m}\right] = \sum_{m=0}^{\infty} x(mT)z^{-m}\sum_{k=0}^{\infty} h(kT)z^{-k},$$

where $k = n - m$, and we have used the fact that $h(kT) = 0$ for $k < 0$. Then,

$$C(z) = X(z)H(z). \tag{6}$$

This relation is very useful in determining the Z transform function for various systems connected in cascade.

14.8. The Sampling Theorem

A basic theorem of the signal theory is the Shannon sampling theorem. It permits us to recover a signal $f(x)$ at $x \in \mathbb{R}$ from a sample of its values at the integers. We need to state a few concepts before we discuss this theorem. The first relates to the band-limited function $f(x)$. It means that its Fourier transform $\widehat{f}(u)$ has a compact support. A simple example is the characteristic function

$$\widehat{f}(u) = \begin{cases} 1, & |u| < \omega, \\ 0, & \text{otherwise}. \end{cases} \tag{1}$$

Its inverse Fourier transform is

$$f(x) = \frac{1}{2\pi}\int_{-\infty}^{\infty} e^{-iux}\,du = \frac{\sin \omega x}{2\pi\omega x}. \tag{2}$$

The second concept in the signal theory is the power of the signal $f(x)$. It is proportional to $|f(x)|^2$, the square of the amplitude, so that the total energy of the signal is proportional to $\int |f(x)|^2 dx$. Accordingly, the condition that the total energy be finite is that $f(x) \in \mathcal{L}^2$, the class of square-integrable functions.

We are now ready to discuss the celebrated Shannon sample theorem. Let the function $f(x) \in \mathcal{L}^2$ be such that $\widehat{f}(u)$ has compact support in $[-\pi, \pi]$. Then $f(x)$ is completely determined by the values it takes at the integers. The mathematical formula is

$$f(x) = \sum_{k=-\infty}^{\infty} f(k)\frac{\sin \pi(x-k)}{\pi(x-k)}, \qquad -\infty < x < \infty. \tag{3}$$

The proof follows by an interplay between the Fourier transforms and the Fourier series [83]. Indeed when we expand $\widehat{f}(u)$ in the Fourier series in the interval $[-\pi, \pi]$, we have

$$\widehat{f}(u) = \sum_{n=-\infty}^{\infty} c_k\, e^{iku}, \qquad |u| < \pi, \tag{4}$$

where the coefficients c_k are

$$c_k = \frac{1}{2\pi}\int_{-\pi}^{\pi} \widehat{f}(u)\, e^{-iku} du = f(k)\,, \tag{5}$$

and we have used the fact that $\widehat{f}(u) = 0$ for $|u| > \pi$. Thus, (4) becomes

$$\widehat{f}(u) = \sum_{k=-\infty}^{\infty} f(k)\, e^{iku}\,. \tag{6}$$

Taking the inverse transform of this relation yields

$$\begin{aligned} f(x) &= \frac{1}{2\pi}\int_{-\infty}^{\infty} \widehat{f}(u)\, e^{-iux} du \\ &= \frac{1}{2\pi}\int_{-\pi}^{\pi} \sum_{k=-\infty}^{\infty} f(k)\, e^{iku}\, e^{-iux} dx \\ &= \frac{1}{2\pi}\sum_{k=-\infty}^{\infty} \left[f(k) \frac{e^{-i\pi(x-k)u}}{i\pi(x-k)} \right]_{-\pi}^{\pi} \\ &= \sum_{k=-\infty}^{\infty} f(k) \frac{\sin \pi(x-k)}{\pi(x-k)} \end{aligned}$$

as desired. Observe that the termwise integration in the above result is possible because the Fourier series of $\widehat{f}(u)$ converge in $\mathcal{L}^2$.

The Shannon sampling theorem has many extensions. For instance, when the Fourier transform $\widehat{f}(u)$ has support contained in $[-A, A]$, where $A > \pi$, sampling formula (3) is not valid but the variant

$$f(x) = \sum_{k=-\infty}^{\infty} f(k\epsilon) \frac{\sin \pi(x/\epsilon - k)}{\pi(x/\epsilon - k)} \tag{7}$$

with $\epsilon = \pi/A$ holds.

In case $\widehat{f}(u)$ does not hold for any $\epsilon > 0$, Estrada [84] has shown that the series on the right side of (7) provides an asymptotic approximation of $f(x)$ as $\epsilon \to 0$. Indeed, he has shown that under mild conditions on functions $f(x)$ and $g(x)$ we have the more general approximation formula

$$f(x) \sim \sum_{k=-\infty}^{\infty} f(k\epsilon) g\left(\frac{x}{\epsilon} - k\right) \qquad \text{as } \epsilon \to 0\,. \tag{8}$$

To derive this formula we start with the Dirac comb as given by (2.4.3): $h(x) = \sum_{k=-\infty}^{\infty} \delta(x - k)$. This generalized function does not vanish at ∞ but we can write it as $h(x) = 1 + h_0(x)$. Fortunately, the moment asymptotic expansion holds for $h_0(x)$. Moreover, the moments of any periodic generalized function vanish. Thus, the moment expansion for $h_0(x)$ yields

$h_0(x) = o(\lambda^{-\infty})$ so that

$$\sum_{k=-\infty}^{\infty} \delta(\lambda x - k) = 1 + o(\lambda^{-\infty}), \qquad \text{as } \lambda \to \infty. \tag{9}$$

This concept is discussed in more detail in Section 15.3. Now we set $\lambda = 1/\epsilon$ in (9) and evaluate the resulting relation for a test function $\phi(x)$ and get

$$\epsilon \sum_{k=-\infty}^{\infty} \phi(\epsilon k) = \int_{-\infty}^{\infty} \phi(x)dx + o(\epsilon^{\infty}), \qquad \text{as } \epsilon \to 0. \tag{10}$$

Recall that if ϕ is a test function then so is $f\phi$. Accordingly, we can replace ϕ by $f\phi$ in (10) so that

$$\epsilon \sum_{k=-\infty}^{\infty} f(k\epsilon)\phi(k\epsilon) = \int_{-\infty}^{\infty} f(x)\phi(x)dx + o(\epsilon^{\infty}), \qquad \text{as } \epsilon \to 0, \tag{11}$$

which can be written as

$$\begin{aligned} f_\epsilon(x) &= \sum_{k=-\infty}^{\infty} f(k\epsilon)\delta\left(\frac{x}{\epsilon} - k\right) = \sum_{k=-\infty}^{\infty} f(x)h\left(\frac{x}{\epsilon}\right) \\ &= f(x) + o(\epsilon^{\infty}), \qquad \text{as } \epsilon \to 0. \end{aligned} \tag{12}$$

This comb contains the information of $f(x)$ at the sample points $\{k\epsilon\}_{k\in\mathbb{Z}}$ and relation (12) yields $f_\epsilon(x) \sim f(x)$ as $\epsilon \to 0$.

Let us now follow the analysis of Chapter 13 and take $g(x)$ that decays so fast at infinity that it has the moment asymptotic expansion

$$g\left(\frac{x}{\epsilon}\right) \sim \mu_0\delta(x) - \mu_1\delta'(x)\epsilon^2 + \frac{\mu_2\delta''(x)}{2!}\epsilon^2 + \cdots, \qquad \epsilon \to 0, \tag{13}$$

where μ_k are the moments of $g(x)$. Since

$$g\left(\frac{x}{\epsilon}\right) * \delta\left(\frac{x}{\epsilon} - k\right) = g(x/\epsilon - k),$$

the convolution of $g(x/\epsilon)$ with $f_\epsilon(x)$ in (12) yields

$$\sum_{k=-\infty}^{\infty} f(k\epsilon)g\left(\frac{x}{\epsilon} - k\right) \sim \mu_0 f(x) - \mu_1\, f'(x)\epsilon + \frac{\mu_2 f''(x)\epsilon^2}{2!} + \cdots, \qquad \epsilon \to 0. \tag{14}$$

Let us now take $g(x)$ such that $\mu_0 = 1$. Then (14) is the formula for the approximation of $f(x)$

$$\sum_{k=-\infty}^{\infty} f(k\epsilon)g\left(\frac{x}{\epsilon} - k\right) = f(x) + O(\epsilon). \tag{15}$$

Actually, if $g(x)$ is chosen appropriately, relation (15) can be improved. Indeed, if $\mu_k = 0$, $k > 1$ (which holds if $g = \widehat{G}$ with $G(u) = 1$ for $|u| \le a$) then

$$\sum_{k=-\infty}^{\infty} f(k\epsilon) g\left(\frac{x}{\epsilon} - k\right) = f(x) + o(\epsilon) . \tag{16}$$

In particular, if $g(x) = (\sin \pi x)/\pi x$, we recover formula (7).

Relation (16) is not valid when the moments μ_k of $g(x)$ do not vanish for $k \ge 1$. That is the case if $g(x)$ itself has a compact support as is the case when it is a spline function. In that case we replace $f(x)$ by $f'(x)$ in (14) and obtain

$$\sum_{k=-\infty}^{\infty} f'(k\epsilon) g\left(\frac{x}{\epsilon} - k\right) \sim f'(x) - \mu_1 f''(x)\epsilon + \frac{\mu_1 f'''(x)}{2!}\epsilon^2 + \cdots ,$$

so that

$$\sum_{k=-\infty}^{\infty} [f(k\epsilon) + \mu_1 \epsilon f'(k\epsilon)] g\left(\frac{x}{\epsilon} - k\right) = f(x) + O(\epsilon^2) . \tag{17}$$

Note that relation (15) is an interpolation formula and the error is of order ϵ. However, in (17) we have achieved the answer to the order ϵ^2 by using the quasi-interpolate

$$\sum_{k=-\infty}^{\infty} [f(k\epsilon) + \mu_1 \epsilon f'(x)] g\left(\frac{x}{\epsilon} - k\right) .$$

We can continue this process to obtain

$$\sum_{k=-\infty}^{\infty} \left[f(k\epsilon) + \mu_1 \epsilon f'(k\epsilon) + \left(\mu_1^2 - \frac{\mu_2}{2}\right)\right] g\left(\frac{x}{\epsilon} - k\right) = f(x) + O(\epsilon^3) . \tag{18}$$

and so on.

CHAPTER 15

Miscellaneous Topics

15.1. Applications to Probability and Random Processes

In order to present the applications of generalized functions to the theories of probability and random processes let us start with some basic concepts.

Let Ω be the set of elementary events and let U be a class of the subsets of Ω such that

(1) the family U contains Ω,

(2) if $A \in U$ than $A^c \in U$, where A^c is the complement of A,

(3) if the sets $A_1, A_2, \ldots, A_n$ belong to U, then their union $\bigcup_{n=1}^{\infty} A_n$ belongs to U.

The probability measure P on Ω is the mapping from U to a set of real numbers $\mathbb{R}$ with the following properties:

(1) $P(\Omega) = 1$,

(2) $P(A) \geq 0$, for every $A \in U$, and

(3) If the sets $A_1, A_2, \ldots, A_n, \ldots$, belong to U and are mutually disjoint, that is $A_i \cap A_j = \emptyset$, the empty set, $i \neq j$; then

$$P\left(\bigcup_{n=1}^{\infty} A_n\right) = \sum_{n=1}^{\infty} P(A_n) .$$

The triplet (Ω, U, P) is called the probability space.

For a random variable X (a measurable function from Ω to $\mathbb{R}$) we define its probability distribution function $F(x)$ as

$$F(x) = P\{X < x\} = P\{X^{-1}(-\infty, x)\} , \qquad x \in \mathbb{R} . \tag{1}$$

The function $F(x)$ has the following properties:

(1) F is monotone,

(2) F is continuous from the left, and

(3) $\lim_{x\to\infty} F(x) = 1, \quad \lim_{x\to-\infty} F(x) = 0.$

In the sequel we shall call the probability distribution function only as probability distribution. We shall keep in mind that the terms probability distribution and distribution refer to different entities. However, the probability distribution, function $F(x)$, being a locally integrable function, defines the distribution

$$\langle F(x), \phi(x)\rangle = \int_{-\infty}^{\infty} F(x)\phi(x)dx , \qquad \phi \in D . \tag{2}$$

Accordingly, the distributional derivative $F'(x)$ is

$$\begin{aligned}\langle F'(x), \phi(x)\rangle &= -\langle F(x), \phi'(x)\rangle = -\int_{-\infty}^{\infty} \phi'(x)F(x)dx \\ &= -[\phi(x)F(x)]_{-\infty}^{\infty} + \int_{-\infty}^{\infty} \phi(x)dF(x) = \langle f(x), \phi(x)\rangle\,, \end{aligned} \tag{3}$$

where $f(x) = dF(x)/dx$ is called the probability density function (density function for short). The density function $f(x)$ has the following properties:

1. $f(x) \geq 0,\ x \in \mathbb{R}$,
2. By setting $f(-\infty) = 0,\ f(\infty) = 1$ we find that

$$\int_{-\infty}^{\infty} f(t)dt = 1\,, \quad \text{and} \tag{4}$$

X is a continuous random variable if $f(x)$ is a function in the classical sense, otherwise it is called a random variable.

Example 1. Let us examine the tossing of a coin. We assign the value $x = 0$ if we obtain heads and the value $x = 1$ if we get tails. To evaluate the probability distribution $F(x)$ we make the following observations:

(a) Since we have only two possibilities, namely $\{x = 0\}$ and $\{x = 1\}$, the probability that $x < 0$ yields $F(x) = 0,\ x \leq 0$,

(b) For a fair coin, on a single toss there is a probability of $1/2$ of obtaining heads, with an equal probability for tails. Hence, for $0 < x \leq 1$, $F(x) = 1/2$, and

(c) For the case $x > 1$ we appeal to the definition of $F(x)$ which states that the outcome is less than x. Since both the possibilities ($x = 0$ and $x = 1$) are less than x in this case $F(x) = 1$.

Summing up these observations we find that

$$\begin{aligned} F(x) &= \begin{cases} 0\,, & x \leq 0 \\ 1/2\,, & 0 < x \leq 1 \\ 1\,, & x > 1 \end{cases} \\ &= \frac{1}{2}[H(x) + H(x-1)]\,, \end{aligned} \tag{5}$$

where $H(x)$ is the Heaviside function. Then from (3) and (5) it follows that in the present case the probability density function is

$$f(x) = \frac{1}{2}[\delta(x) + \delta(x-1)]\,. \tag{6}$$

The equation (4) is clearly satisfied by this density function.

In case the random variable X takes the values $a_1, a_2, \ldots, a_n$ with the probabilities

$p_1, p_2, \ldots, p_n$ respectively, such that $\sum_{k=1}^{n} p_k = 1$, then the generalized function

$$f(x) = \sum_{k=1}^{n} p_k \delta(x - a_x) \tag{7}$$

is the probability density of X.

Example 2. If X has the binomial distribution then its probability density is

$$f(x) = \sum_{k=0}^{n} \binom{n}{k} p^k q^{n-k} \delta(x - k), \quad 0 \leq p \leq 1, \quad q = 1 - p. \tag{8}$$

It follows that the distribution function is

$$F(x) = \sum_{k=0}^{n} \binom{n}{k} p^k q^{n-k} H(x - k).$$

Example 3. X has the Poisson distribution if it has the probability density

$$f(x) = e^{-\lambda} \sum_{k=0}^{n} \frac{\lambda^k}{k!} \delta(x - k), \qquad \lambda > 0. \tag{9}$$

Then the distribution function has the value

$$F(x) = e^{-\lambda} \sum_{k=0}^{\infty} \frac{\lambda^k}{k!} H(x - k). \tag{10}$$

Example 4. For the case when X has the normal (Gaussian) distribution then the probability density and the distribution function are

$$f(x) = \frac{1}{\sqrt{2\pi\sigma^2}} e^{\frac{-(x-\mu)^2}{2\sigma^2}} \tag{11}$$

and

$$F(x) = \frac{1}{\sqrt{2\pi\sigma^2}} \int_{-\infty}^{x} e^{\frac{-(t-\mu)^2}{2\sigma^2}} dt, \tag{12}$$

respectively. Observe that in this case, X is a continuous random variable.

Example 5. The random variable X has exponential distribution if the probability density is

$$f(x) = \begin{cases} 0, & x \leq 0 \\ \lambda e^{-\lambda x}, & x > 0, \quad \lambda > 0. \end{cases} \tag{13}$$

Then the distribution function is

$$F(x) = \begin{cases} 0, & x \leq 0, \\ (1 - e^{\lambda x}), & x > 0. \end{cases} \tag{14}$$

The Cauchy representation of the probability density. Let us recall from Section 7.6 that the Cauchy representation of a distribution t in E' is $t_c(z) = 1/2\pi i\langle t, 1/t-z\rangle$. Because $\delta(x) \in E'$, the Cauchy representation of the probability density

$$f(x) = \sum_{k=1}^{n} p_k\delta(x - x_k)$$

is

$$\widetilde{f}_c(z) = \frac{1}{2\pi i}\left\langle \sum_{k=1}^{n} p_k\delta(x - x_k),\ \frac{1}{x - z}\right\rangle = \frac{1}{2\pi i}\sum_{k=1}^{n} p_k\frac{1}{x_k - z}. \tag{15}$$

For instance, for the binomial distribution studied in Example 2 we have

$$f(x) = \sum_{k=0}^{n}\binom{n}{k}p^k(1 - p)^{n-k}\delta(x - k).$$

Thus in this case the Cauchy representation is

$$f_c(z) = \frac{1}{2\pi}\sum_{k=0}^{n}\binom{n}{k}p^k(1 - p)^{n-k}\frac{1}{x_k - z}. \tag{16}$$

The Cauchy representation exists for any probability density; the proof follows on showing that if $f(x)$ is monotonic, then it is measurable. Since the number of discontinuities of a monotonic function is at most countable, $f(x)$ is continuous almost everywhere and hence measurable almost everywhere. Furthermore, if $f(x)$ is measurable on disjoint intervals A and B, then it is measurable on their union. Accordingly, $f(x)$ is measurable on R_1.

Next, we show that since $0 \le F(x) \le 1$ and $1/(x-z)^2$ is in L^1, then $F(x)/(x-z)^2 \in L^1$. This follows on observing that

$$\left|\int_{-\infty}^{\infty}\frac{F(x)}{(x - z)^2}dt\right| \le \left|\max F(x)\int_{-\infty}^{\infty}\frac{dt}{(x - z)^2}\right| = 1\left[-\frac{1}{x - z}\right]_{-\infty}^{\infty} = 0.$$

Accordingly, the functional $\langle F(x),\ 1/(x-z)^2\rangle$ is defined. From this relation we infer that $1/2\pi i\langle f(x), 1/x - z\rangle$ is defined, which proves the assertion.

The characteristics of a random variable. They are defined as follows:

1. The *expectation value*, also called *mean*, of X is

$$E(X) = \int_{\Omega} X(\omega)d\,P(\omega) = \int_{-\infty}^{\infty} x\,dF(x) = \langle x, f\rangle. \tag{17}$$

2. The *variance* of X is

$$D(X) = \int_{\Omega}[X(\omega) - E(X)]^2 d\,P(\omega) = \int_{-\infty}^{\infty}(x - E(X))^2 dF(x) = \langle (x - E(X))^2, f(x)\rangle. \tag{18}$$

3. The m-th moment of X is

$$E(X^m) = \int_\Omega (X(\omega))^m dP(\omega) = \int_{-\infty}^{\infty} x^m dF(x) = \langle x^m, f(x)\rangle . \tag{19}$$

For $m = 1$ we recover (17).

4. The central m-th moment of X is

$$\begin{aligned} E(x - E(X))^m &= \int_\Omega (X(\omega) - E(X))^m dP(\omega) \\ &= \int_{-\infty}^{\infty} (X - E(X))^m dF(x) = \langle X - E(X)^m, f(x)\rangle . \end{aligned} \tag{20}$$

For $m = 2$, we recover (18) while by setting $m = 1$ in the above relation we find that the central moment of first order is zero.

Let us now apply these characteristics to the probability distributions as studied in Examples 2–5. The results are as follows:

1. The *binomial distribution*: $E(x) = np$; $D(X) = np^2$.
2. The *Poisson distribution*: $E(x) = \lambda$; $D(X) = \lambda$.
3. The *Gaussian distribution*: $E(x) = \mu$; $D(X) = \sigma^2$.
4. The *exponential distribution*: $E(x) = \frac{1}{\lambda}$; $D(X) = 1/\lambda^2$.

The characteristic function of a random variable. Since we have found that the probability distribution $f(x)$ is a generalized function we can find its Fourier transform. Accordingly, we define its characteristic function $\chi(t)$ as

$$\chi(u) = E(e^{iux}) = \int_{-\infty}^{\infty} e^{iux} f(x)dx = \widehat{f}(u) . \tag{21}$$

Conversely, given a characteristic function $\chi(u)$, we derive

$$f(x) = F^{-1}(\chi(u)) = \frac{1}{2\pi}\int_{-\infty}^{\infty} \chi(u)e^{-iux}du . \tag{22}$$

Thus the random variable X is completely determined by its characteristic function $\chi(u)$. Let us illustrate this concept with the help of a few examples.

Example 6. Let us take $\chi(u) = e^{i\lambda u}$. Then from formula (22) we have

$$f(x) = \frac{1}{2\pi}\int_{-\infty}^{\infty} e^{i\lambda u}e^{-iux}du = \frac{1}{2\pi}\int_{-\infty}^{\infty} e^{-iu(x-\lambda)}du = \delta(x - \lambda) ,$$

where we have used relation (3.5.3).

Example 7. From the binomial distribution, as given in Example 2, we have

$$\chi(u) = \sum_{k=0}^{n} \binom{n}{k} p^k q^{n-k} e^{iuk} , \quad 0 \le p \le 1 , \quad q = 1 - p . \tag{23}$$

Then from relation (22) we obtain

$$\begin{aligned} f(x) &= \frac{1}{2\pi}\int_{-\infty}^{\infty} e^{-iux}\left(\sum_{k=0}^{n}\binom{n}{k}p^k 2^{n-k}e^{iuk}\right)du \\ &= \left(\sum_{k=0}^{n}\binom{n}{k}p^k q^{n-k}\right)\frac{1}{2\pi}\int_{-\infty}^{\infty} e^{-iu(x-k)}du \\ &= \sum_{k=0}^{n}\binom{n}{k}p^k q^{n-k}\delta(x-k)\,, \end{aligned}$$

which agrees with equation (8).

Example 8. For the Gaussian distribution, as given in Example 4 we find that the characteristic function is

$$\begin{aligned} \chi(u) &= \frac{1}{\sqrt{2\pi\sigma^2}}\int_{-\infty}^{\infty} e^{iux}e^{-(x-\mu)^2/2\sigma^2}dx \\ &= \exp\left\{e^{i\mu u}-\frac{1}{2}\sigma^2u^2\right\}. \end{aligned} \tag{24}$$

For the special case $\mu = 0$, $\sigma^2 = 1$, we obtain

$$\chi(u) = e^{-u^2/2}\,. \tag{25}$$

***n*-dimensional random variable.** We can readily extend the above discussion to the system of random variables $X_1, X_2, \cdots, X_n$. This system may be considered as a mapping from the set Ω to the n-dimensional space R_n. Such a mapping is called an n-dimensional random variable. In the present case the probability distribution function is

$$F(X_1, X_2, \ldots, X_n) = P(X_1(\omega) < x_1, \ldots, X_n(\omega) < x_n)\,. \tag{26}$$

The moments are given by the formula

$$\int_{R_n}(x_1^{r_1}\cdots x_n^{r_n})d\,F(x) = \int_{\Omega}(X_1(\omega))^{r_1}\cdots(X_n(\omega))^{r_n}d\,P(\omega)\,. \tag{27}$$

The characteristic function of an n-dimensional random variable $X = (X_1, \ldots, X_n)$ is

$$\chi(u) = E(e^{iu.x}) = \int_{-\infty}^{\infty} e^{iu.x}\,dF(x)\,, \tag{28}$$

where $u = (u_1 \ldots u_n)$ and $x = (x_1 \ldots x_n)$. This relation introduces the Fourier transform, and hence the generalized functions are a useful tool in the study of probability theory. Another connection is given by the sum $S = X_1 + X_2$ of two random variables. Let F_1 and F_2 denote the probability functions of X_1 and X_2 and let f_1 and f_2 denote the corresponding densities. Then the density of S is given by the Fourier convolution

$$(f_1 * f_2)(s) = \int_{-\infty}^{\infty} f_2(s-x)f_1(x)dx\,,$$

whereas the probability distribution function is

$$P\{X+Y \le s\} = \int_{-\infty}^{\infty} F_2(s-x)F_1(x)dx\,.$$

Example 9. A random variable $X = (X_1, \ldots, X_n)$ has an n-dimensional Gaussian distribution if its characteristic function is

$$\chi(u) = \exp\left\{i\; m_k\, u_k - \frac{1}{2}V_{jk}u_j u_k\right\}\,, \quad j = 1, \ldots, n\,, \quad k = 1, \ldots, n\,, \tag{29}$$

where $m = (m_1, \ldots, m_n)$, $m_k = E(X_k)$,

$$V_{jk} = E(X_j - m - j)(X_k - m_k)\,,$$

and we have used the summation convention in relation (29).

The probability density is

$$f(x) = (2\pi)^{-n/2}|V|^{-1/2}\exp\left\{-\frac{1}{2}\sum_{j,k=1}^{n} B_{jk}(x_j - m_j)(x_k - m_k)\right\}\,, \tag{30}$$

where the matrix B_{jk} is the inverse of the matrix V_{jk}.

Random processes. A random process is a family $\{X(t), t \in T\}$, $T \in \mathbb{R}$, of random variables. In view of the relation (26) it amounts to the following definition in n-dimensions.

Definition. A random process $X(t)$ is defined by the family of n-dimensional distribution functions

$$F_{t_1,\ldots,t_n}(X_1, \ldots, X_n) = P\{X(t_1) < x_1, \ldots, X(t_n) < x_n\}\,, \tag{31}$$

where $n \in \mathbb{N}$, $t_1, \ldots t_n \in T$. In our notation this relation is, $F_t(x) = P\{X(t) < x\}$.

For a fixed $\omega \in \Omega$, $X(t, \omega)$ is a real function of $t \in T$ called a sample path of $X(t)$.

Example 10. A Gaussian random process is a process such that for every $n \in \mathbb{N}$, its n-dimensional distribution is given by the density (30).

At this stage we need to set up a Hilbert space. Recall that a Hilbert space is an inner product space that is complete in its natural metric, namely, $d(x, y) = \|x - y\|$. We shall be dealing with the Hilbert space $\mathcal{L}^2(\Omega)$, which has the inner product

$$\langle X, Y\rangle = \int_{\sim} X(\omega)Y(\omega)d\,P(\omega) = E(x, y)\,,$$

so that the norm is

$$\|X\|^2 = \langle X, X\rangle\,.$$

The variable $X(t)$ is called a second order process if for every $t \in T$, $X(t) \in \mathcal{L}^2(\Omega)$. When the mapping $T \to \mathcal{L}^2(\Omega)$ is continuous, we say that $X(t)$ is continuous in the mean.

A quantity of great interest for the second order process is the correlation function which is defined as

$$B(t,s) = E(X(t)X(s)) = \int_\Omega X(t,\omega)X(s,\omega)dP(\omega) = \langle X(t), X(s)\rangle . \tag{32}$$

The correlation function of the Wiener process is

$$B_W(t,s) = \min\{t,s\} . \tag{33}$$

The sample paths of the Wiener processes are continuous but are nowhere differentiable functions. The reason is that the random variable $(W(t+h) - W(t))/h$ has Gaussian distribution with mean zero and variance $\sigma^2/|h|$ which tends to ∞ as $h \to 0$, an impossibility.

The generalized random process. A natural way to generalize the notion of a random process is to allow the sample paths of a process to belong to some set of distributions. Although the concept of a random process is based upon the assumption that it is possible to measure the value of the process at every moment of time t without calculating the value of the process at other units of time, every measurement is accomplished by means of an apparatus which has some inertia. Accordingly, the information that the apparatus yields is not the value of the random variable $X(t)$ but rather a certain average value (a functional) $\Phi(\phi) = \langle X(t), \phi(t)\rangle = \int_{-\infty}^{\infty} X(t)\phi(t)dt$, where the function $\phi(t)$ characterises the apparatus. Moreover, this functional is linear and continuous. Thus, we have obtained a linear continuous random functional, namely, a generalized linear random process — a distribution.

Definition. Let M be a test functions space $\phi(t)$ and $\mathbb{B}$ a space of random variables. A *generalized random process* $\Phi(\phi)$ is a linear and continuous mapping from M to $\mathbb{B}$.

Example 12. Let $X(t)$ be a random process such that it is a continuous mapping of T to $L^2(\Omega)$ so that it defines a generalized random process through the formula

$$\Phi(\phi) = \langle X(t), \phi(t)\rangle = \int_{-\infty}^{\infty} X(t)\phi(t)dt , \qquad \phi(t) \in D . \tag{34}$$

This functional is well defined because the test function $\phi(t)$ has compact support and therefore the integral (34) is finite. It is clearly linear. The continuity is ensured because

$$\begin{aligned}\|\Phi(\phi)\| = \|\langle X(t), \phi(t)\rangle\| &= \left\| \int_a^b X(t)\phi(t)dt \right\| \\ &\le (b-a) \max_{t\in[a,b]} |\phi(t)| \max_{t\in[a,b]} \|X(t)\|\end{aligned}$$

when $[a,b]$ is the support of $\phi(t)$ and we have used the fact that the functions $X(t)$ and $\phi(t)$ are continuous. Now, when $\phi_n(t) \to 0$ in D, then so does $\max|\phi_n(t)|$ and we conclude that $\|X(t), \phi_n(t)\| \to 0$, as $n \to \infty$.

Definition. The mean $m(\phi)$ of a generalized random process Φ is defined as

$$m(\phi) = E(\Phi(\phi)), \quad \phi \in M. \tag{35}$$

The correlation functional is

$$B(\phi, \psi) = E(\Phi(\phi), \Phi(\psi)) \quad \phi, \psi \in M. \tag{36}$$

Example 13. If a generalized random process Φ is given by formula (34). Then its correlation functional is

$$\begin{aligned} B(\phi, \psi) &= E \int_{-\infty}^{\infty} \int_{-\infty}^{\infty} X(t)X(s)\phi(t)\psi(s)\, dt \\ &= \int_{-\infty}^{\infty} \int_{-\infty}^{\infty} B(t, s)\phi(t)\psi(s) dt\, ds, \end{aligned} \tag{37}$$

where $B(t, s) = E[x(t)\, x(s)]$. Specifically, if W is a Wiener random process then by formula (34) it defines a generalized Wiener random process $\overline{W}$. Its correlation functional is

$$\begin{aligned} B_{\overline{W}}(\phi, \psi) &= \int_0^{\infty} \int_0^{\infty} \phi(t)\psi(s) \min\{t, s\} dt ds \\ &= \int_0^{\infty} (\phi_1(t) - \phi_1(\infty))(\psi_1(t) - \psi_1(\infty)) dt, \end{aligned} \tag{38}$$

where

$$\phi_1(t) = \int_0^t \phi(t) dt, \quad \psi_1(t) = \int_0^t \psi_1(t) dt. \tag{39}$$

The characteristic functional of a generalized random process. Recall definition (21) for the characteristic function $\chi(u)$ for a probability density $f(x)$ of a one-dimensional random variable. Conversely, given the characteristic function $\chi(u)$, we can recover the density $f(x)$. These concepts can be extended to a generalized random process $\Phi(\phi)$, defined by its n-dimensional probability distribution, $n \in \mathbb{N}$, i.e. by its characteristic function:

$$\chi(u) = \int_{\Omega} \exp\left\{ i \sum_{k=1}^{n} u_k \Phi(\phi_k, w) \right\} d P(w), \quad u = (u_1, \ldots, u_n). \tag{40}$$

Observe that the functional Φ is linear so that $\chi(u)$ as given by (40) is the characteristic function of the random variable $\Phi\left(\sum_{k=1}^n u_k \phi_k\right)$. Since $\sum_{k=1}^n u_k \phi_k \in M$, we find that the distribution density of a generalized random process is completely determined by the functional

$$C(\phi) = \int_{\Omega} \exp\{i \Phi(\phi, w)\} d P(w) = E(e^{i\langle \Phi, \phi \rangle}). \tag{41}$$

Accordingly, $C(\phi)$ is called the characteristic functional and has the following properties:

(1) it is continuous in ϕ,
(2) it is positive definite, i.e., for every $(\alpha_1, \dots \alpha_n) \in \mathbb{C}$ and $(\phi_1, \dots, \phi_n) \in M$, $\sum_{j,k} \alpha_j \overline{\alpha}_k C(\phi_j, \phi_k) \geq 0$, and
(3) $C(0) = 1$.

Example 14. A generalized random process ϕ with mean zero and correlation functional $B(\phi, \psi)$ is a Gaussian generalized random process if its characteristic functional $C(\phi)$ is

$$C(\phi) = \exp\left\{-\frac{1}{2}E\Phi^2(\phi)\right\} = \exp\left\{-\frac{1}{2}B(\phi, \phi)\right\}. \tag{42}$$

Definition. The derivative of a generalized random process Φ is defined the same way as is done in the case of a generalized function, namely,

$$\Phi'(\phi) = -\Phi(\phi'). \tag{43}$$

If a generalized random process Φ has correlation functional $B_\Phi(\phi, \psi)$, then the correlation functional of its derivative is

$$B_{\Phi'}(\phi, \psi) = E(\Phi'(\phi), \Phi'(\psi)) = E(\Phi(\phi'), \Phi(\psi')) = B_\Phi(\phi', \psi'). \tag{44}$$

Specifically, it follows that the derivative of a Gaussian random process is again a Gaussian generalized process.

Example 15. The derivative of the Wiener generalized random process $\overline{W}$ is the generalized random process $\dot{W}$ with the correlation functional

$$\begin{aligned} B_{\dot{W}}(\phi, \psi) = B_{\overline{W}}(\phi', \psi') &= \int_0^\infty \phi(t)\psi(t)dt \\ &= \int_0^\infty \int_1^\infty \delta(t-s)\phi(t)\psi(s)dt\,ds. \end{aligned} \tag{45}$$

Thus, the correlation functional, $B(t, s)$, of the derivative of the Wiener process is the delta function, $\delta(t-s)$. The derivative of the Wiener process is the simplest generalized process of a Gaussian type and is called the unit generalized random process. From relations (45) and (42) we obtain

$$B_{\dot{W}}(\phi, \phi) = \int_0^\infty \phi^2(t)dt, \tag{46}$$

and

$$C_{\dot{W}}(\phi) = \exp\left\{-\frac{1}{2}\int_0^\infty \phi^2(t)dt\right\}, \tag{47}$$

respectively.

Stationary generalized random processes. A generalized random process is called stationary if for any functions $\phi_1(t), \ldots, \phi_n(t)$ in M and any $h \in \mathbb{R}$, the functionals $(\Phi(\phi_1(t+h)), \ldots, \Phi(\phi_n(t+h)))$ and $\Phi(\phi_1(t), \ldots, \phi_n(t))$ are identically distributed.

If Φ is stationary then its mean is invariant under translation so that

$$E(\Phi(\phi)) = a \int_{\mathbb{R}} \phi(t)dt \,. \tag{48}$$

There is an important theorem regarding the correlation functional of stationary generalized random process. We merely state it.

Theorem. *The correlation functional $B(\phi, \psi)$ of a stationary generalized random process Φ has the form*

$$B(\phi, \psi) = \int_R \widehat{\phi}(u)\widehat{\psi}(u)d\sigma(u) \,, \tag{49}$$

where $\widehat{\phi}(u)$ is the Fourier transform of $\phi(t)$ and σ is a positive measure such that

$$\int_R (1 + |u|^2)^{-p} d\sigma(u) < \infty \,, \qquad \textit{for some } p \geq 0 \,. \tag{50}$$

The measure σ is called the spectral measure of the process ϕ. For example, it follows from relation (45) that the spectral measure of the unit generalized random process is Lebesgue measure, i.e., $d\sigma(u) = du$ because

$$B_{\dot{W}}(\phi, \psi) = \int_{-\infty}^{\infty} \phi(t)\psi(t)dt = \int_{-\infty}^{\infty} \delta(t) \left[\int_{-\infty}^{\infty} \phi(s)\psi(s-t)d \right] dt \,. \tag{51}$$

The reader may consult references [85–89] for a detailed study.

The Fokker-Plank operator. $\partial/\partial t = \partial/\partial x(\partial/\partial x + x)$. This operator describes the evolution of probability distribution function in statisical physics [90]. The fundamental solution $E(x, \xi)$ of this operator satisfies the differential equation

$$[\partial/\partial t - \partial/\partial x(\partial/\partial x + x)]\, E(x, \xi; t, \tau) = \delta(x - \xi)\delta(t - \tau) \,. \tag{52}$$

Its solution is derived by the transformation $x = ye^{-t}$, $E = ve^{t}$. Then the homogeneous part of equation (52) becomes

$$\frac{\partial v}{\partial t} = e^{2t} \frac{\partial^2 v}{\partial y^2} \,. \tag{53}$$

Next, we set $e^{2t} = 2s$ in (53) and get

$$\frac{\partial v}{\partial s} = \frac{\partial^2 v}{\partial y^2} \,,$$

which is the heart operator. Accordingly, we appeal to the analysis of Section 10.7 and then return to the x and t variables. It is left as an exercise for the reader to write down the value of $E(x, \xi; t, \tau)$. We merely present the limit of the fundamental solution as $t \to \infty$. This limit is

$$\lim_{t\to\infty} E(x, \xi; t, \tau) = \frac{1}{\sqrt{2\pi}} e^{-(x-\xi)^2/2}, \tag{54}$$

the Gaussian probability density.

15.2. Applications to Economics

Several studies of the optimal path of capital accumulation of a firm facing costs of adjustment have been made using the techniques of control theory. The mathematical analysis in these models is interesting in itself since economic reasons suggest that the investment schedule is not necessarily a function of time but rather a general Radon measure. The cost of adjustment, however, is a nonlinear functional of the investment schedule, and extension to the generalized schedules is not straightforward.

There has been some controversy as to which is the more adequate form of the cost of adjustment function. Arguments for both a convex and a concave cost function and the corresponding analyses have been made.

To learn the basic concepts and definitions of mathematical economic analysis, the reader is referred to the bibliography given in reference [91] from which this material is taken.

In this study we obtain the weak lower semicontinuous extension of the cost of adjustment functional. Weak lower semicontinuity accords well with the minimal properties that economics suggests that a cost should have. Weak continuity does not hold unless the cost of adjustment function is linear, since the extension should be linear with respect to the Dirac delta function. In deriving this extension functional, it will be shown that the cost of adjustment function, whatever its form, must be replaced by a convex one.

We first give the rationale for using generalized functions in economic models. We then devote discussion to the basic model, where the discontinuous character of the cost of adjustment functional is portrayed. The next two topics of interest will be the derivation of the extension functional and some of its properties. We conclude this section with some comments on the uses of this model, in particular, when we should expect jumps in the capital stock and why.

The use of generalized functions. Most models of the dynamical behavior of an economic system usually assume (implicitly) that the variables of the system are functions of time. This is a very reasonable assumption for certain types of variables, namely, those that can be observed at almost each instant of time, such as the price of a certain commodity or the prevailing interest rate.

For other variables, however, such an interpretation is problematic. Consider, for instance, the output produced at a certain factory (measured in physical units). Given a time interval $I = [t_1, t_2]$, we can evaluate the output $Y(I)$ produced during that period, but the output $Y(t)$ at time $t = t_1$ is not clearly defined. Continuing with this example, if the price of each unit of output at any time is given, the dollar value of output during any interval can be evaluated, or, alternatively, the discounted or present value of such an output can be evaluated.

This example shows that certain types of economic variables are not ordinary functions of time but rather distributions: They give a definite number only when multiplied with a function (as prices or discounted prices) and integrated. Distributions also arise when certain analytic operations, such as derivative, are applied to observable functions.

In our particular problem the investment schedule gives the additions to the capital stock. The size of the capital stock can be observed at almost all times, but it can suffer jumps when additions are made in a very short perod. A Dirac delta function placed at the instant of the jump is the best description of the investment.

The basic model. The basic dynamic model for the investment decisions of a firm postulates that the investment schedule $I(t)$ for $t \geq t_0$ is chosen at time $t = t_0$ in such a way as to maximize the present value of the future stream of profits.

$$\pi = \pi(I) = \int_{t_0}^{\infty} V(t - t_0)\{R(t, K(t)) - r(I(t))\}dt\,, \tag{1}$$

where $K(t)$ is the capital stock at time t, which, depreciation disregarded, is formed according to

$$\dot{K}(t) = \frac{dK}{dt} = I(t)\,, \tag{2}$$

$R(t, K)$ is the expected quasi-rent to be obtained from a capital stock of size K at time t, $r(I)$ is the cost of adjustment; and $V(t)$ is the discount factor.

We assume that $R(t, K)$ is smooth enough. We also assume that for each value of t it is a strictly concave function of K in the interval $[0, \infty)$ with a unique maximum at $K^*(t)$ and with two zeros, at $K = 0$ and $K = K_1(t)$. The discount factor $V(t)$ is chosen by the firm and is assumed to be continuous, decreasing, and having compact support. The last assumption is made to explain the way the firm handles the uncertainty of $R(t, K)$ for large t. The cost of the adjustment function $r(I)$ is positive if $I \neq 0$ and zero if $I = 0$. It increases for I positive and decreases for I negative. Smoothness for $I \neq 0$ is assumed, but discontinuities of r or any of its derivatives at $I = 0$ are allowed. Finally, we assume the asymptotic relations

$$\lim_{I \to \infty} r'(I) = a\,, \qquad \lim_{I \to -\infty} r'(I) = b\,, \tag{3}$$

where

$$0 < b < a < \infty\,. \tag{4}$$

Let us observe that, for a given discount factor $V(t)$, the cost of adjustment is a nonlinear functional of the investment schedule $I(t)$ given by

$$C(V, I) = \int_{t_0}^{\infty} V(t - t_0) r(I(t)) dt\,. \tag{5}$$

This integral functional is well defined if $I(t)$ is a locally integrable function of time. When $I(t)$ is a delta function $I = \delta(t - t_1)$, direct substitution in (5) is not possible, and thus a limit process is needed to evaluate this cost. Let

$$f_n(t) = \begin{cases} nI\,, & t_1 \le t \le t_1 + 1/n\,, \\ 0\,, & \text{otherwise.} \end{cases} \tag{6}$$

Then it is natural to set

$$C(V, \delta(t - t_1)) = \lim_{n \to \infty} C(V, f_n) = \begin{cases} aIV(t_1 - t_0)\,, & I \ge 0\,, \\ -bIV(t_1 - t_0)\,, & I \le 0\,. \end{cases} \tag{7}$$

Thus the cost of a discrete change in the capital stock can be computed as a limit of the cost of continuous changes. One would expect the reverse also to be possible — to recover the cost of a continuous change as a limit of the costs of a sequence of discrete approximations. This is not possible, however, since the cost functional cannot be continuous under the weak topology of signed measures unless $r(I)$ is linear for $I > 0$ and $I < 0$. (The weak topology is the one for which $\lim_{n\to\infty} f_n(t) = \delta(t - t_1)$.) For example, let $g(t)$ be a positive continuous function on $[t_0, \infty)$, and let the support of V be contained in the interval $[0, \tau]$. Let us then consider the discrete approximations

$$g_n(t) = \sum_{k=0}^{n} f\left(t_0 + \frac{k}{n}\tau\right) \delta\left(t - t_0 - \frac{k}{n}\tau\right)\,. \tag{8}$$

We then have $\lim_{n\to\infty} g_n = g$ weakly in $[t_0, t_0 + \tau]$, but

$$\lim_{n \to \infty} C(V, g_n) = a \int_{t_0}^{\infty} V(t - t_0) g(t) dt\,, \tag{9}$$

which, in general, differs from (5) as long as r is nonlinear.

The minimum cost function. In order to derive a cost functional with some interesting analytic properties, it is necessary to introduce a new cost function. Let us assume that no discounting is made ($V = 1$).

The function $r(I)$, as implied by (5), gives the cost of augmenting the capital stock I units in one unit of time using a constant rate of investment. The cost of augmenting the capital stock I units in one unit of time, however, will, in general, be smaller than $r(I)$. Let us define the cost function $r^*(I, \tau)$ giving the cost of investing I units of capital in τ units of time, by

$$r^*(I, \tau) = \inf\left\{ \int_0^t r(f(t)) dt : f \in L^1[0, \tau]\,, \quad \int_0^t f(t) dt = I \right\}. \tag{10}$$

A change of variables in definition (10) shows that $r^*(I, \tau)$ is homogeneous of the first degree. If we denote by $r^*(I)$ the unit cost function $r^*(I, 1)$, we then have

Lemma 1. *The function $r^*(I, \tau)$ satisfies*

$$r^*(I, \tau) = \tau r^*(I/\tau). \tag{11}$$

We can therefore limit our attention to the function $r^*(I)$. This function can be constructed from $r(I)$ by a very simple method, as our next lemma shows.

Lemma 2. *The function $r^*(I)$ is the lower convex envelope of the function $r(I)$; that is, it satisfies the following three properties:*

(a) *$r^*(I)$ is convex,*

(b) *$r^*(I) \le r(I)$, and*

(c) *If $h(I)$ is a convex function that satisfies $h(I) \le r(I)$. Then $h(I) \le r^*(I)$.*

Proof. Direct use of the definition of $r^*(I, \tau)$ shows that if $\tau_1 + \tau_2 = 1$, then

$$\begin{aligned} r^*(\tau_1 I_1 + \tau_2 I_2) &= r^*(\tau_1 I_1 + \tau_2 I_2, 1) \le r^*(\tau_1 I_1, \tau_2) + r^*(\tau_2 I_2, \tau_2) \\ &= \tau_1 r^*(I_1) + \tau_2 r^*(I_2). \end{aligned}$$

Condition (b) is clear. For (c), let h be a convex function satisfying the stated conditions. Let $I \in R$. Then for any constants $\tau_1, \ldots, \tau_n, a_1, \ldots, a_n$ such that

$$\sum_{i=1}^{n} \tau_i = 1, \qquad \tau_i \ge 0, \tag{12}$$

and

$$\sum_{i=1}^{n} \tau_i a_i = I, \tag{13}$$

we have

$$h(I) \le \sum_{i=1}^{n} \tau_i h(a_i) \le \sum_{i-1}^{n} \tau_i r(a_i) = \int_0^1 r\left(\sum_{i=1}^{n} a_i g_i(t)\right) dt, \tag{14}$$

where the function $g_i(t)$ is given by

$$g_i(t) = \begin{cases} 1, & \text{if } \displaystyle\sum_{j=1}^{i-1} \tau_j \le t \le \sum_{j=1}^{i} \tau_j, \\ 0, & \text{otherwise.} \end{cases} \tag{15}$$

Now we can take the infimum of (14) over all constants a_i, τ_i that satisfy (12) and (13). This infimum is equivalent to the infimum of $\int_0^1 r(f(t))dt$ over all piecewise constant functions whose integral from 0 to 1 equals I, which (according to (10)) reduces to $r^*(I)$.

Example. A particularly appealing type of basic cost function, $r(I)$, is one that is convex for $I > 0$ but whose right limit at $I = 0$ is a strictly positive constant (indicative of fixed costs). Let

$$c_0 = \inf\{r(I)/I : I > 0\}. \tag{16}$$

Then if $c_0 < a$, there exists a largest number $I^* > 0$ such that $r(I^*) = c_0 I^*$, and the cost function is given by

$$r^*(I) = \begin{cases} c_0 I\,, & I \le I^*\,, \\ r(I)\,, & I \ge I^*\,. \end{cases} \tag{17}$$

If $c_0 = a$, then $r^*(I) = aI$ for all I.

In either case, $r^*(I, \tau)$ has the interesting property that if the investment is small relative to the time interval, then the cost will not depend on the time interval.

Extension of the cost functional. We shall use the notation

$$\lim_{n\to\infty} f_n = f(w)\,, \tag{18}$$

to indicate that f_n converges weakly to f, that is, for every continuous function ϕ defined on $[\tau_0, \tau_1]$ we have

$$\lim_{n\to\infty} \int_{t_0}^{t_1} \phi(t) f_n(t)dt = \int_{t_0}^{t_1} \phi(t) f(t)dt\,. \tag{19}$$

Economic considerations suggest that the cost of adjustment functional should be a minimal property of the following type: If $\lim_{n\to\infty} f_n = \mu(w)$, then

$$\lim_{n\to\infty} \inf \overline{C}(V, f_n) \ge \overline{C}(V, \mu)\,, \tag{20}$$

where $\overline{C}$ is the extension of the cost functional. It is thus natural to define the cost of adjustment of a generalized investment schedule represented by a Radon measure μ as

$$\overline{C}(V, \mu) = \inf\left\{ \lim_{n\to\infty} \inf C(V, f_n) : \lim_{n\to\infty} f_n = \mu(w) \right\}\,. \tag{21}$$

When definition (21) is applied to an ordinary function, the original value $C(V, \mu)$ is not necessarily recovered. Before we study generalized investments we shall obtain a formula for $\overline{C}(V, \mu)$ when μ is a piecewise continuous function. We start with the convex case.

Lemma 3. *Let $r(I)$ be convex. Let ϕ be a continuous positive function with compact support on $[0, \infty)$. Then if f is piecewise continuous on $[0, \infty)$, we have*

$$\overline{C}(\phi, f) = C(\phi, f) = \int_0^\infty \phi(t) r(f(t))dt\,. \tag{22}$$

Proof. The inequality $\overline{C} \le C$ is clear.

Since ϕ has compact support, we can assume that its support is contained in $[0, 1]$. Let f_n be a sequence of functions that converge weakly to f. Let $J_j = [(j-1)/k, j/k]$ for $j = 1, \ldots, k$ and let $\phi_1, \ldots, \phi_k$ be a set of positive continuous functions on $[0, 1]$ that satisfy the following conditions:

$$\sum_{i=1}^{n} \phi_i(t) = \phi(t)\,, \tag{23}$$

$$\int_0^1 \phi_j(t)dt = \int_{J_j} \phi(t)dt\,, \tag{24}$$

and

$$\operatorname{supp} \phi_j \subseteq [(j-1)/k - \varepsilon_k, j/k + \varepsilon_k]\,, \tag{25}$$

where ε_k is a positive constant.

Let $\varepsilon > 0$. There exists $N = N(\phi_1, \ldots, \phi_k)$ such that if $n \geq N$, then

$$\left|\int_0^1 \phi_j(t)dt\right| \left|\int_0^1 \phi_j(t)(f_n(t) - f(t))dt\right| \leq \varepsilon\,. \tag{26}$$

Since r is convex, one has

$$\int_0^1 \psi(t)r(g(t))dt \geq \left(\int_0^1 \psi(t)dt\right) r\left(\int_0^1 \psi(t)f(t)dt \Big/ \int_0^1 \psi(t)dt\right), \tag{27}$$

for any pair of piecewise continuous functions ψ, g with $\psi \geq 0$. Hence

$$\begin{aligned}
\int_0^1 r(f_n(t))\phi(t)dt &= \sum_{j=1}^k \int_0^1 r(f_n(t))\phi_j(t)dt \\
&\geq \sum_{j=1}^k \left(\int_0^1 \phi_j(t)dt\right) r\left(\int_0^1 f_n(t)\phi_j(t)dt \Big/ \int_0^1 \phi_j(t)dt\right) \\
&\geq \sum_{j=1}^k \left(\int_0^1 \phi_j(t)dt\right) r\left(\int_0^1 f(t)\phi_j(t)dt \Big/ \int_0^1 \phi_j(t)dt - \varepsilon\right) \\
&\geq \sum_{j=1}^k \left(\frac{1}{k}\right) \phi(\tau_j)[r(f(\tau_j')) - a\varepsilon]\,,
\end{aligned}$$

for some constants $\tau_j \in J_j$, $\tau_j' \in \operatorname{supp} \phi_j$ $(j = 1, \ldots, n)$. Therefore, for any k and ε we obtain

$$\liminf_{n\to\infty} \int_0^1 r(f_n(t))\phi(t)dt \geq \frac{1}{k}\sum_{j=1}^k \phi(\tau_j)[r(f(\tau_j')) - a\varepsilon]\,. \tag{28}$$

As $k \to \infty$, if the right-hand side of (28) approaches

$$\int_0^1 \phi(t)[r(f(t)) - \varepsilon]dt\,,$$

and if we then let $\varepsilon \to 0$, the result follows.

We can now derive a formula for $\overline{C}(\phi, f)$ for a general function $r(I)$.

Lemma 4. *Let ϕ be a continuous positive function with compact support on* $[0, \infty)$. *Then if f is piecewise continuous on* $[0, \infty)$, *we have*

$$\overline{C}(\phi, f) = \int_0^\infty \phi(t)r^*(f(t))dt\,. \tag{29}$$

Proof. Let us suppose, as in the previous lemma, that supp $\phi \subseteq [0, 1]$ and let $J_j = [(j-1)/k, j/k]$ as before. We shall suppose that f is positive. Let $g_k(t)$ be a piecewise constant function that satisfies the following conditions for $j = 1, \ldots, k$:

$$\int_{J_j} g_k(t)dt = \int_{J_j} f(t)dt\,, \tag{30}$$

and

$$\int_{J_j} r(g_k(t))dt < r^*\left(\int_{J_j} f(t)dt, \frac{1}{k}\right) + \frac{1}{k^2}\,. \tag{31}$$

Let g_k be given by constants $a_{m,j}$ on intervals $J_{m,j}(m = 1, \ldots, m(j))$ whose union over all m is J_j. Thus

$$\sum_{m=1}^{m(j)} a_{m,j}|J_{m,j}| = \int_{J_j} f(t)dt\,, \tag{32}$$

and

$$\sum_{m=1}^{m(j)} r(a_{m,j})|J_{m,j}| < r^* = \left(\int_{J_j} f(t)dt, \frac{1}{k}\right) + \frac{1}{k^2}\,. \tag{33}$$

We first show that $\lim_{k\to\infty} g_k = f(w)$. Let $\psi \in C[0, 1]$ be positive. Let $\varepsilon > 0$ and let $\delta > 0$ be such that $|\psi(x) - \psi(y)| \leq \varepsilon$ for $|x - y| \leq \delta$. Then if $k \geq 1/\delta$, we have

$$\begin{aligned}\left|\int_0^1 \psi(t)(g_k(t) - f(t))dt\right| &= \left|\sum_{j=1}^{k}\sum_{m=1}^{m(j)}\int_{J_{m,j}} \psi(t)(a_{m,j} - f(t))dt\right| \\ &= \left|\sum_{j=1}^{k}\sum_{m=1}^{m(j)} a_{m,j}\right| J_{m,j}\left|[\psi(\tau_{m,j}) - \psi(\tau_j)]\right|\,,\end{aligned}$$

for some constants $\tau_{m,j\in J_{m,j}}$ and $\tau_j \in J_j$. Thus

$$\left|\int_0^1 \psi(t)(g_k(t) - f(t))dt\right| \leq \varepsilon\int_0^1 f(t)dt\,.$$

A similar argument shows that $C(\phi, g_k)$ converges to $\int_0^1 \phi(t)r^*(f(t))dt$ since, because of (33), $C(\phi, g_k)$ is almost a Riemann sum of this integral. Therefore,

$$\overline{C}(\phi, f) \leq \int_0^\infty \phi(t)r^*(f(t))dt\,. \tag{34}$$

For the reverse inequality, let $C(r^*, \phi, f)$ be the integral functional formed with the function r^* and let $\overline{C}(r^*, \phi, f)$ be the corresponding extension. Then $C(r^*, \phi, f) = \overline{C}(r^*, \phi, f)$ since r^* is convex. On the other hand, since $r \leq r^*$ we get

$$C(r^*k, \phi, f) = \overline{C}(r^*, \phi, f) \leq \overline{C}(\phi, f)\,,$$

which, combined with (34), gives the desired result.

We now turn to generalized investments. We study a single jump first.

Lemma 5. *Let ϕ be a continuous, positive function with compact support on* $[0, \infty)$. *Let f be piecewise continuous on* $[0, \infty)$ *and let $\tau \geq 0$, $c \geq 0$. Then*

$$\overline{C}(\phi, f(t) + c\delta(t - \tau)) = \overline{C}(\phi, f) + ac\phi(\tau). \tag{35}$$

Proof. Let f_n be a sequence of functions converging weakly to $f(t) + c\delta(t - \tau)$. Let $\phi_1, \ldots, \phi_k$ be chosen as in Lemma 3, except that $\tau \in \operatorname{supp} \phi_i$ for only one i and $\phi_i(\tau) = \phi(\tau)$. Then proceeding in a similar fashion we obtain

$$\begin{aligned}\liminf_{n\to\infty} \int_0^1 f^*(f_n(t))\phi(t)dt \geq & \frac{1}{k}\sum_{j=1}^{n} \phi(\tau_j)[r^*(f(\tau_j')) - a\varepsilon] \\ & - \frac{1}{k}(\phi(\tau_i)[r^*(t(\tau_i')) - a\varepsilon]) \\ & + \left(\int_0^1 \phi_i(t)dt\right) r^*\left(c\phi(\tau) \middle/ \int_0^1 \phi_i(t)dt \right. \\ & \left. + f(\tau_i) - a\varepsilon\right).\end{aligned} \tag{36}$$

As $k \to \infty$ and $\varepsilon \to 0$, the first term approaches $\overline{C}(\phi, f)$, the second approaches zero, and the third, by L'Hospital's rule, approaches $ac\phi(\tau)$. Thus

$$\overline{C}(\phi, f + c\delta(t - \tau) \geq \overline{C}(\phi, f) + ac\phi(\tau).$$

The reverse inequality is obtained by taking a sequence $g_n \to f$ with $C(\phi, g_n) \to C(\phi, f)$ and then adding a term g_n' that is n on the interval $[\tau, \tau + 1/n]$ and zero everywhere else, since $g_n + cg_n' \to f + c\delta(t - \tau)$ and $C(\phi, g_n + cg_n') \to \overline{C}(\phi, f) + ac\phi(\tau)$.

We thus obtain

Theorem 1. *Let ϕ be a continuous, positive function with compact support on* $[0, \infty)$. *Let*

$$f(t) = f_0(t) + \sum_{j=1}^{\infty}(\alpha_j\delta(t - \tau_j) - \beta_j\delta(t - \tau_j')). \tag{37}$$

where $\alpha_j, \beta_j \geq 0$, where $\tau_i \neq \tau_j'$ for all i, j, and where $f_0(t)$ is locally integrable on $[0, \infty)$. *Let*

$$\overline{C}(\phi, f) = \int_0^{\infty} \phi(t)r^*(f(t))dt + \sum_{j=1}^{\infty}(a\alpha_j\phi(\tau_j) + b\beta_j\phi(\tau_j')). \tag{38}$$

The rationale for using the functional $\overline{C}$ instead of C is that, given an investment schedule $f(t)$, one can obtain an arbitrarily close schedule $g(t)$ whose cost is as close to $\overline{C}(V, f)$ as desired. In considering the maximization of (1), the use of $\overline{C}$ would be justified only if the revenue part of the profits is weakly continuous with respect to investment. We thus provide the following theorem:

Theorem 2. *Let $\phi(t)$ be a continuous, positive function with compact support in $[0, \infty)$. Suppose $\partial R/\partial K$ is bounded on each finite time interval. Then the revenue functional,*

$$Y(\phi, f, K_0) = \int_0^\infty \phi(t) R\left(t, K_0 + \int_0^t f(s)ds\right) dt\,, \tag{39}$$

is weakly continuous on f for every constant K_0.

Proof. If $f_n \to f(w)$, then

$$F_n(t) = K_0 + \int_0^t f_n(s)ds \to K_0 + \int_0^t f(s)ds \tag{40}$$

strongly as measures. Let A be a bound for $\partial R/\partial K$ for t in the support of ϕ. We then have

$$\begin{aligned}|Y(\phi, f, K_0) - Y(\phi, f_n, K_0)| &= \left|\int_0^t \phi(t) R(t, F(t)) - R(t, F_n(t)) dt\right| \\ &\leq A \sup\{|\phi(t)| : t \geq 0\} \cdot \int_0^\infty |d(F_n - F)| \to 0\,.\end{aligned}$$

Further comments. The foregoing analysis applies to a very broad class of functions R, r, and V. It is clear that unless more information is available only very general aspects of the solution can be obtained. The behavior of $r(I)$ for large I and the relative size of costs with respect to revenues are particularly important. Different types of investments, however, seem to behave quite differently, and thus a wide range of functions R, r, and V are found in practice. The model we have presented is then the basic framework upon which more detailed case studies should be made.

It is very difficult to prescribe conditions that guarantee the existence or nonexistence of generalized solutions given this level of generality. A few remarks on the likelihood of such solutions can be made. If the function $r(I)$ is linear for $I > 0$, then upward adjustments in the capital stock will probably be made in jumps, not continuously. The reason is, clearly, that $r^*(I\tau) = aI(I \geq 0)$, and thus the benefits of making a given adjustment quickly can be derived without having to pay extra costs.

A cost function like $r(I) - aI - \ln(I + 1)$, on the other hand, would preclude jumps in almost all cases, since adjusting very fast is very expensive.

15.3. Periodic Distributions

Recall that we encountered the distribution $t(x)$

$$t(x) = \sum_{m=-\infty}^{\infty} \delta(x - 2m\pi) = \frac{1}{2\pi} \sum_{m=-\infty}^{\infty} e^{imx} \tag{1}$$

in Sections 2.4, 3.4 and 5.2 in different contexts. This is a periodic distribution according to the following definition:

Definition. A distribution $t(x)$ is called periodic with period p if $t(x+p) = t(x)$ in the sense that

$$\langle t(x+p), \phi(x)\rangle = (t(x), \phi(x))\,, \tag{2}$$

where $\phi(x)$ is a test function.

Let us recall, from section 3.4, the condition for the coefficients c_m of the Fourier series (1), is that $|c_m| \leq M/m_k$, $k \geq 2$. Then this series converges uniformly. It therefore converges distributionally according to definition (2). We found in that section that by appealing to certain higher order permitives of series (1) we can circumvent the conditions on $|c_m|$. These ideas were extended by Estrada and Kanwal [6] by using the theory of moments. We present here a few salient features and use the notation of Chapter 13.

Let $t(x)$ be a periodic distribution with zero mean, that is, the integral of $t(x)$ over any interval p is zero (equivalently $\langle t(x), 1\rangle = 0$). Then all the moments of $t(x)$ vanish because for each n there exists another periodic distribution $s(x)$ with zero mean such that $s^{(n)}(x) = t(x)$. Indeed, if n is large enough then $s(x)$ is continuous. Accordingly, the formula

$$\langle t, \phi\rangle = (-1)^n \langle s, \phi^{(n)}\rangle \tag{3}$$

defines the value of t at a test function ϕ in the space K because if n is large enough then $\phi^{(n)}(x) = O(|x|^{-2})$ as $|x| \to \infty$ so that $\langle s(x), \phi^{(n)}(x)\rangle = \int_{-\infty}^{\infty} s(x)\phi^{(n)}(x)dx$ is a convergent integral and $t(x) \in K'$. This means that for $\phi(x) = x^k$, $n > k$,

$$\langle t(x), x^k\rangle = 0\,, \quad k = 0, 1, 2, 3, \ldots, \tag{4}$$

that is, all the moments of $t(x)$ vanish. This amounts to the asymptotic formula $t(\lambda x) = o\,(\lambda^{-\infty})$ where by $o(\lambda^{-\infty})$ we mean that $t(\lambda x)$ is $o(\lambda^{-n})$ for each $n \in \mathbb{N}$. The simplest example is

$$\widehat{\phi}(\lambda) = \int_{-\infty}^{\infty} e^{i\lambda x}\phi(x)dx = o(\lambda^{-\infty}) \quad \text{as} \quad \lambda \to \infty\,, \tag{5}$$

which is the Riemann Lebesgue lemma mentioned in Example 18 of Section 6.4.

In the case when the mean of the periodic distribution t is a nonzero constant c, $t(x)$ does not belong to the space K'. However, we can write $t(x) = c + t_0(x)$, where $t_0 \in K'$ and we have

$$t(\lambda x) = c + o(\lambda^{-\infty})\,, \quad \text{as} \quad \lambda \to \infty\,. \tag{6}$$

An important example of such a periodic distribution is given by (1) where $c = 1/2\pi$.

For interplay between these concepts and the analysis of summability and the distributionally small sequences the reader is referred to Estrada and Kanwal [6].

We illsutrate these concepts with three examples.

Example 1. Time-domain asymptotics. Chapman [92] considered the asymptotic approximation of the integrals of the form

$$I(\lambda) = \int_a^b p(\lambda h(x))\phi(x)dx\,, \tag{7}$$

where p is a periodic function with zero mean, $h(x)$ and $\phi(x)$ are smooth functions, λ is a large and positive parameter, and $a < b$. Accordingly, the integrand oscillates rapidly and almost cancels out in I except near the endpoints a, b and the stationary points within the range of integration. If $x = x_0$ is a stationary point such that $h'(x_0) = 0$, $h''(x_0) \neq 0$, then Chapman gave the approximation

$$I(\lambda) \sim \mu^{\sigma}(\lambda h(x_0)) \left(\frac{2}{|h''(x_0)|\lambda} \right)^{1/2} \phi(x_0), \tag{8}$$

where $\sigma = \text{sign}\,(h''(x_0))$ and

$$\mu^{\sigma}(c) = \int_{-\infty}^{\infty} p(c + \sigma x^2)dx, \tag{9}$$

for fixed c. Formula (8) generalizes the celebrated principle of stationary phase when $p(x) = e^{ix}$ and has been extended by Prentice [93] to two-dimensional case.

The results of Chapman and Prentice have been derived with the distributional asymptotic theory by Estrada and Kanwal [94] and then extended to the multidimensional case. We give here a brief account of their analysis and restrict our presentation to R_1. We start with the formula (6) which in the present case is

$$p(\lambda x) = o(\lambda^{-\infty}), \quad \text{as} \quad \lambda \to \infty, \tag{10}$$

because $p(x)$ has zero mean. Estrada and Kanwal show that if $f(x)$ is a measurable function in the neighborhood of $(-\varepsilon, \varepsilon)$ of $x = 0$ and that $f(x^{2k})$ is integrable in this neighborhood so that it is a distribution then it satisfies the moment asymptotic expansion (13.1.8)

$$f(\lambda x^{2k}) \sim \sum_{j=0}^{\infty} \frac{(-1)^j \mu_j \delta^{(j)}(x)}{j!\, \lambda^{(j+1)/2k}}, \quad \text{as} \quad \lambda \to \infty, \tag{11}$$

where $\mu_j = \langle f(x^{2k}), x^j \rangle$ are the moments of $f(x^{2k})$.

We can apply formula (11) to $f(x) = p(c + \sigma x^{2k})$ where p is a periodic measurable function of zero mean, c is fixed, $\sigma = 1$ or $\sigma = -1$, and where $p(c + \sigma x^{2k})$ is locally integrable so that

$$p(\lambda(c + \sigma x^{2k})) \sim \sum_{j=0}^{\infty} \frac{(-1)^j \mu_j^{\sigma,2k}(c) \delta^{(j)}(x)}{j!\, \lambda^{(j+1)/(2k)}}, \quad \text{as} \quad \lambda \to \infty, \tag{12}$$

where the $\mu_j(c)$ are the moments of $p(c + \sigma x^{2k})$, namely,

$$\mu_j(c) = \mu_j^{\sigma,2k}(c) = \int_{-\infty}^{\infty} x^j p(c + \sigma x^{2k})dx. \tag{13}$$

The expansion of $p(\lambda(c+\sigma x^{2k}))$ follows at once from equation (12) because if p is periodic of period τ then

$$p(\lambda(c + \sigma x^{2k})) = p(c' + \lambda x^{2k}) \text{ if } \lambda c = c' \in \tau\mathbb{Z}$$

and the expansion is given by (12) with c replaced by c'. Furthermore, the moments $\mu_j(c)$ are also periodic of period τ so that $\mu_j(c') = \mu_j(\lambda c)$. Thus,

$$p(\lambda(c+\sigma x^{2k})) \sim \sum_{j=0}^{\infty} \frac{(-1)^j \mu_j^{\sigma,2k}(\lambda c)\delta^j(x)}{j!\, \lambda^{(j+1)/(2k)}}, \quad \text{as} \quad \lambda \to \infty. \tag{14}$$

The expansion of $p(\lambda h(x))$ in the neighborhood of a critical point of h is then derived from (14) by a simple change of variables. Indeed, if x_0 is a critical point of h that satisfies $h^{(j)}(x_0) = 0$, $1 \le j < 2k-1$, $\sigma h^{2k}(x_0) > 0$, where $\sigma = 1$, if x_0 is a minimum and $\sigma = -1$ if x_0 is a maximum. Then if x_0 is the only critical point, we can write

$$h(x) = h(x_0) + \sigma(\psi(x))^{2k}, \tag{15}$$

where ψ is smooth and increasing, $\psi'(x) > 0$ for all x and $\psi(x_0) = 0$. Accordingly,

$$p(\lambda h(x)) = p(\lambda(h(x_0) + \sigma(\psi(x))^{2k})$$

and we have (in view of (14)),

$$p(\lambda h(x)) \sim \sum_{j=0}^{\infty} \frac{(-1)^j \mu_j^{\sigma,2k}(\lambda h(x_0))\delta^{(j)}(\psi(x))}{j!\, \lambda^{(j+1)/2k}} \quad \text{as} \quad \lambda \to \infty. \tag{16}$$

From this formula we derive the first order approximation as

$$p(\lambda h(x)) = \mu_0^{\sigma,2k}(\lambda h(x_0)) \left(\frac{(2k)!}{|\sigma h^{(2k)}(x_0)|\lambda}\right)^{1/2k} \delta(x-x_0) + O\left(\frac{1}{\lambda^{(3/2)k}}\right), \tag{17}$$

where we have used the relations

$$\delta(\psi(x)) = \frac{\delta(x-x_0)}{\psi'(x_0)} \text{ and } \psi'(x_0) = \left(\frac{\sigma h^{2k}(x_0)}{(2k)!}\right)^{1/2k}.$$

When $k = 1$, formula (17) is the distributional version of formula (8) as given by Chapman.

Estrada and Kanwal [94] have also given the following extensions in their paper.

(1) Several variables I: interior critical point,
(2) Several variables II: boundary critical points,
(3) Several variables III: corners.

This paper also contains the references to related studies.

Example 2. The Euler-Maclaurin formula. Recall that if we are interested in evaluating the integal $\int_0^1 \phi(x)dx$, then we consider the numbers $\omega_1, \omega_2, \ldots, \omega_s$, $x_1, x_2, \ldots, x_s$ such that $\omega_1 + \cdots + \omega_s = 1$ so that the quadrature

$$Q(\phi) = \sum_{j=1}^{s} \omega_j \phi(x_j), \tag{18}$$

provides an approximation. An m-copy version $Q^m(\phi)$ of the quadrature is obtained by dividing the interval [0, 1] into m subintervals of equal length and by applying the corresponding quadrature in each of them so that we have

$$Q^m(\phi) = \sum_{k=0}^{m-1} Q_{\frac{k}{m},\frac{k+1}{m}}(\phi)\,. \tag{19}$$

Observe that quadrature (18) can be written as

$$Q(\phi) = \langle g(x), \phi(x)\rangle\,, \tag{20}$$

where g is the generalized function

$$g(x) = \sum_{j=1}^{s} \omega_j \delta(x - x_j)\,. \tag{21}$$

Similarly, (19) takes the form

$$Q^m(\phi) = \sum_{k=0}^{m-1} \langle g(mx - k), \phi(x)\rangle\,. \tag{22}$$

We observe from (1) that the function

$$G(x) = \sum_{k=-\infty}^{\infty} \delta(x - k) \tag{23}$$

is periodic with period 1 and its mean is 1. Thus, the periodic function $G_0(x)$ defined as

$$G_0(x) = G(x) - 1\,, \tag{24}$$

has mean zero. Let $G_n(x)$ denote the nth order primitive of $G_0(x)$ so that

$$G'_{n+1}(x) = G_n(x)\,, \tag{25}$$

and

$$G_n(x+1) = G_n(x)\,. \tag{26}$$

We now follow Estrada [95] and derive the Euler-Maclaurin formula. For this purpose we compute the qth order distributional derivative of the function $(\chi_{[N,M]}(x)G_q(x))$, where $N, M \in \mathbb{Z}$, and $\chi_E(x)$ denotes the characteristic function of a set E. Let us start with the case $q = 1$. When we use formula (5.1.4):

$$\overline{F}'(x) = F'(x) + \sum_{j=1}^{\ell} a_j \delta(x - \xi_j)\,, \tag{27}$$

we find that

$$\frac{\overline{d}}{dx}(\chi_{[N,M]}(x)G_1(x)) = G_1(N)\delta(x - N) - G_1(M)\delta(x - M) + \chi_{[N,M]}G_0(x)\,. \tag{28}$$

When we use the formulas (24) and (25) as well as the fact that $G_1(N) = G_1(M) = \rho_0$ (say), because of the periodicity of $G_1(x)$, the previous formula becomes

$$\frac{\overline{d}}{dx}(\chi_{[N,M]}(x)G_1(x)) = \rho_0\{\delta(x-N) - \delta(x-M)\} + \chi_{[N,M]}(G(x)-1)\,. \tag{29}$$

Next, we evaluate this derivative at a test function and get

$$\sum_{k=N}^{M-1}\phi(k) = \int_N^M \phi(x)dx + \rho_0\{\phi(M) - \phi(N)\} + R_1\,, \tag{30}$$

where $R_1 = R_1(\phi; [N, M])$, is the remainder and is given as

$$R_1 = \langle(\chi_{[N,M]}(x)G_1(x))', \phi(x)\rangle = -\int_N^M G_1(x)\phi'(x)dx\,. \tag{31}$$

The general Euler-Maclaurin formula with remainder is derived by induction, starting with (28) and (29). The first step is

$$\begin{aligned}(\chi_{[N,M]}(x)G_q(x))^{(q)} &= G_q(N)\delta^{(q-1)}(x-N) - G_q(M)\delta^{q-1}(x-M)\\ &\quad + (\chi_{[N,M]}(x)G_{q-1}(x))^{(q-1)}\,,\end{aligned} \tag{32}$$

which leads to

$$(\chi_{[N,M]}(x)G_q(x))^{(q)} = \sum_{j=0}^{q-1} G_{j+1}(0)\{\delta^{(j)}(x-N) - \delta^{(j)}(x-M)\} + \chi_{[N,M]}(x)(G(x)-1)\,. \tag{33}$$

When we evaluate this derivative at a test function $\phi(x)$ we get

$$\sum_{k=N}^{M-1}\phi(k) = \int_N^M \phi(x)dx + \sum_{j=0}^{q-1}(-1)^j\rho_j(\phi^{(j)}(M) - \phi^{(j)}(N)) + R_q\,, \tag{34}$$

where the remainder R_q is given as

$$R_q = (-1)^q \int_N^M G_q(x)\phi^{(q)}(x)dx\,. \tag{35}$$

Finally, we put the formula (34) in the classical form which contains the Bernoulli numbers and Bernoulli polynomials. The Bernoulli polynomials are determined uniquely from the initial value probem [6]

$$B_0(x) = 1\,; \quad B_n'(x) = nB_{n-1}(x) \qquad n \geq 1\,. \tag{36}$$

For instance,

$$B_1(x) = x - \frac{1}{2}, \quad B_2(x) = x^2 - x + \frac{1}{6}, \quad B_3(x) = x^3 - \frac{1}{2}x^2 + \frac{1}{2}x\,. \tag{37}$$

These polynomials satisfy many interesting properties such as

$$\int_0^1 B_n(x)dx = 0\,, \qquad n \geq 1\,, \tag{38}$$

and

$$B_n(1-x) = (-1)^n B_n(x)\,. \tag{39}$$

The Bernoulli numbers are defined as $B_n = B_n(0)$ so that $B_{2k+1} = 0, k = 1, 2, 3, \ldots,$ $B_1 = -\frac{1}{2}$, $B_2 = \frac{1}{6}$, $B_4 = -\frac{1}{30}$, $B_6 = \frac{1}{42}$, etc. Observe that relation (39) yields the values $B_n(1) = B_n$ for $n \geq 2$.

Let the quantity $\{x\} = x - [x]$ denote the fractional part of x, then it follows from Example 2 of Section 5.1 that

$$\frac{\overline{d}}{dx}(B_1\{x\}) = 1 - \sum_{k=-\infty}^{\infty} \delta(x-k)\,, \tag{40}$$

$$G_n(x) = -\frac{B_n(\{x-\beta\})}{n!}\,, \qquad n \geq 1\,, \tag{41}$$

of the function $\sum_{k=-\infty}^{\infty} \delta(x-k-\beta) - 1$. Accordingly, we can write formula (34) as

$$\sum_{k=N}^{M-1} \phi(k+\beta) = \int_N^M \phi(x)dx + \sum_{j=0}^{q-1} \frac{(-1)^{j+1} B_{j+1}(1-\beta)}{(j+1)!} (\phi^{(j)}(M) - \phi^{(j)}(N)) + R_q^\beta\,, \tag{42}$$

where

$$R_q^\beta = \frac{(-1)^{q+1}}{q!} \int_N^M B_q(\{x-\beta\})\phi^{(q)}(x)\,dx\,. \tag{43}$$

Whether we let $\beta \to 0^+$ or $\beta \to 1^-$, in equation (42) we obtain the standard Euler-Maclaurin formula,

$$\sum_{k=N}^{M} \phi(k) = \int_N^M \phi(x)dx + \frac{1}{2}(\phi(N) + \phi(M)) \sum_{j=1}^{q} \frac{B_{2j}}{2j!}(\phi^{(2j-1)}(M) - \phi^{(2j-1)}(N)) + R_q^0\,, \tag{44}$$

where

$$R_q^0 = \frac{1}{(2q+1)!} \int_N^M B_{2q+1}(\{x\})\phi^{(2q+1)}(x)\,dx\,. \tag{45}$$

For extensions, variants and the asymptotic expansion of this formula the reader is referred to Estrada [95].

Example 3. An application in number theory. An arithmetic function $f(n)$ is a function that maps the positive integers to $\mathbb{R}$ or $\mathbb{C}$. The Dirichlet multiplication or convolution of two arithmetic functions $f(n)$ and $g(n)$ is defined as

$$f(n) * g(n) = \sum_{kj=n} f(k)g(j) = \sum_{k/n} f(k)g(n/k)\,. \tag{46}$$

Given a function $f(n)$, an inverse for the Dirichlet multiplication $f^{-1}(n)$ satisfies the relation $f_{(n)} * f_{(n)}^{-1} = I(n)$, where $I(n)$ is the unit element defined as

$$I(n) = \begin{cases} 1, & n = 1 \\ 0, & n > 1. \end{cases} \tag{47}$$

Our aim in this example is to invert the relation

$$\psi(x) = \sum_{n=0}^{\infty} (-1)^n \, \phi\left(\left(n + \frac{1}{2}\right) x\right). \tag{48}$$

For this purpose we consider the periodic distribution

$$m(x) = \sum_{n=-\infty}^{\infty} (-1)^n \delta\left(x - \left(n + \frac{1}{2}\right)\right), \tag{49}$$

which is periodic of period 2 and is an odd function. As such, we can write it as the Fourier sine series

$$m(x) = 2 \sum_{n=1}^{\infty} \sin\left(\frac{n\pi}{2}\right) \sin n\pi x, \tag{50}$$

where the Fourier coefficients are derived by using the sifting property of the delta function.

Next, we observe that

$$\begin{aligned} \int_0^{\infty} m(uv)\phi(u)du &= \int_0^{\infty} (-1)^n \delta\left(uv - \left(n + \frac{1}{2}\right)\right) \phi(u)du \\ &= \frac{1}{v} \sum_0^{\infty} (-1)^n \phi\left(\left(n + \frac{1}{2}\right) \frac{1}{v}\right) = \frac{1}{v} \psi\left(\frac{1}{v}\right). \end{aligned} \tag{51}$$

Accordingly, we have to solve the integral equation

$$\int_0^{\infty} m(uv)\phi(u)du = \frac{1}{v} \psi\left(\frac{1}{v}\right). \tag{52}$$

The solution of this integral equation is

$$\phi(x) = \int_0^{\infty} \left(\sum_{n=1}^{\infty} h(n) \sin\left(\frac{\pi x}{nv}\right)\right) \psi\left(\frac{1}{v}\right) dv,$$

where $h(n)$ is the Dirichlet inverse of the arithmetic function $g(n) = \frac{1}{n} \sin\left(\frac{n\pi}{2}\right)$. Thus we have succeeded in inverting (48), i.e., in obtaining the value of the function $\phi(x)$.

The full details of these results and the references to related work are given by Estrada [96].

15.4. Applications to Microlocal Theory

In Chapter 12 we studied the interplay between the wave propagation and generalized functions. Let us now extend these concepts to the theory of microlocal analysis of wave propagation. The basic idea that singularities have to be classified according to their frequencies goes back to Sato [97, 98] who developed the basic theory of hyperfunctions. Other definitions were given by Iagolnitzer [99], Hörmander, who gave his celebrated theorem on the propagation of singularities [100], and other authors. We follow the analysis as given by Estrada and Kanwal [67].

Let $F(x)$ be a distribution defined in $D'(\mathbb{R}^p)$. Its *singular support* is the complement of the largest open set where $F(x)$ is a smooth function. The *set of singular frequencies* of a distribution of compact support $F(x)$ is the set $S(F)$ of all $\eta \in \mathbb{R}^p\backslash\{0\}$ having no conic neighborhood V such that an estimate of the form

$$|\widehat{F}(\xi)| \leqslant C(1+\|\xi\|)^{-N} \tag{1}$$

for the Fourier transform $\widehat{F}$ of F holds for $\xi \in V$, where C and N are constants. If x belongs to the singular support of $F \in D'(\mathbb{R}^p)$ we introduce the set S_x of singular frequencies of F at x as

$$S_x = \cap S(\phi F), \tag{2}$$

the intersection is taken among all $\phi \in D(\mathbb{R}^p)$ with $\phi(x) \neq 0$. The *wave front set* of the distribution F is the set

$$Wf(F) = \{(x,\xi) \in \mathbb{R}^p \times (\mathbb{R}^p\backslash\{0\}) : \xi \in S_x(F)\}. \tag{3}$$

This set contains the exact information about the location of the singularities of $F(x)$ and the frequencies of such singularities. Because of the invariance of $Wf(F)$ under dilations of the second variable it is also common practice to define the wave front set as a subset of $\mathbb{R}^p \times \Sigma_p$, where Σ_p is the unit sphere in $\mathbb{R}^p$ (that is, by requiring that $\|\xi\| = 1$ in (2) above).

For functions discontinuous across a surface, but otherwise smooth, the wave front set is well known; the singular support of the function $F(x)$ coincides with the surface Σ and if $x \in \Sigma$ then the conic set S_x is invariant under translations in the direction of the normal n to Σ at x. When the jumps of all derivatives of F across Σ are smooth at x then the set S_x coincides with the linear span of n. In the case of a moving surface of discontinuity $\Sigma(t)$ we have the situation that at a point (x, t) of the moving surface the conic set $S_{(x,t)}$ is invariant under translations by $(n, -G)$ and under the smoothness conditions on the jumps it reduces to the linear span of $(n, -G)$. This, of course means that at $x \in \Sigma(t)$ the singularity propagates in the normal direction with speed G. Similar considerations apply to multilayer distributions carried by a surface.

Let us associate a microdistribution or singularity spectrum $\{F\}$ to each distribution F whose singular support is contained in the surface Σ, in the following way. Introduce the equivalence relation $\sim$ given by $f \sim g$ if $f - g$ is smooth in all $\mathbb{R}^p$; the microdistribution or singularity spectrum of F is the equivalence class $\{F\}$ under this equivalence relation. As it is clear, for many problems we need to find only $\{F\}$ and not the whole F. The basic operations on distributions, such as, plus, minus, products by smooth functions, and derivatives, extend immediately to microdistributions.

Let us define the microdistributions $d_K = d_K(\Sigma)$, $K \in \Pi$ by

$$d_K = \{d_n^K \delta(\Sigma)\}, \qquad K \geqslant 0, \tag{4}$$

$$d_{-K-1} = \{F_k\}, \qquad K \geqslant 0, \tag{5}$$

where F_K is a function smooth in the complement of Σ whose jumps across Σ satisfy

$$\left[\frac{d^Q F_K}{dn^Q}\right] = \delta_{QK}, \qquad Q = 0, 1, 2, \dots. \tag{6}$$

Whenever $F(x)$ is a function smooth in the complement of Σ with normal jumps $A^{(Q)} = [d^Q F/dn^Q]$ across Σ, it is natural to expand its microdistribution as

$$\{F\} = A^{(0)}d_{-1} + A^{(1)}d_{-2} + A^{(2)}d_{-3} + \cdots = \sum_{K=-\infty}^{-1} A^{(-K-1)}d_K. \tag{7}$$

The general compatibility formulas given in [13] permit us to find expressions for the microdistributions associated with the derivatives (or, what is the same, for the derivatives of such microdistributions). In fact, if $\{F\}$ has expansion (7) then we have formula (5.5.54),

$$\frac{\overline{\partial} F}{\partial x_1} = \frac{\partial F}{\partial x_i} + n_i[F]d(\Sigma). \tag{8}$$

It follows that

$$\left\{\frac{\overline{\partial} F}{\partial x_i}\right\} = \left\{\frac{\partial F}{\partial x_i}\right\} + n_i[F]d_0. \tag{9}$$

But since $[F] = A^{(0)}$ and

$$\left\{\frac{\partial F}{\partial x_i}\right\} = \left[\frac{\partial F}{\partial x_1}\right]d_{-1} + \left[\frac{d}{dn}\left(\frac{\partial F}{\partial x_i}\right)\right]d_{-2} + \left[\frac{d^2}{dn^2}\left(\frac{\partial F}{\partial x_i}\right)\right]d_{-3} + \cdots,$$

use of formula (12.9.32),

$$\left[\frac{d^P F_{,i}}{dn^P}\right] = \sum_{Q=0}^{P} \frac{P}{Q!}(-1)^{(P-Q)}\mu_{ij}^{(P-Q)}\frac{\delta A^{(0)}}{\delta x_j} + A^{(P+1)}n_i, \tag{10}$$

yields

$$\left\{\frac{\overline{\partial} F}{\partial x_i}\right\} = n_i A^{(0)}d_0 + \left(\frac{\delta A^{(0)}}{\delta x_i} + A^{(1)}n_i\right)d_{-1}$$

$$+ \left(-\mu_{ij}\frac{\delta A^{(0)}}{\delta x_j} + \frac{\delta A^{(1)}}{\delta x_i} + A^{(2)}n_i\right)d_{-2} + \cdots, \tag{11a}$$

and

$$\left\{\frac{\overline{\partial} F}{\partial x_i}\right\} = \sum_{K=-\infty}^{0}\left[\sum_{Q=0}^{-K-1}\frac{(-K-1)!}{Q!}(-1)^{K+Q+1}\mu_{ij}^{(-K-1-Q)} + \frac{\delta A^{(Q)}}{\delta x_j} + A^{(-K)}n_i\right]d_K\,. \tag{11b}$$

These formulas can also be written as follows. If

$$\{F\} = Ad_K\,, \qquad K \leqslant -1\,, \tag{12a}$$

then

$$\begin{aligned}\left\{\frac{\overline{\partial} F}{\partial x_i}\right\} &= An_i d_{K+1} + \frac{\delta A}{\delta x_i}d_K + K\mu_{ij}\frac{\delta A}{\delta x_j}d_{k-1} \\ &\quad + K(K-1)\mu_{ij}^{(2)}\frac{\delta A}{\delta x_j}d_{K-2} + \cdots \\ &= An_i d_{K+1} + \sum_{Q=0}^{\infty}\frac{K!}{(K-Q)!}\mu_{ij}^{(Q)}\frac{\delta A}{\delta x_j}d_{K-Q}\,.\end{aligned} \tag{12b}$$

In other words,

$$\frac{\partial}{\partial x_i}(Ad_K) = A_{n_i}d_{K+1} + \sum_{Q=0}^{\infty}\frac{K!}{(K-Q)!}\mu_{ij}^{(Q)}\frac{\delta A}{\delta x_j}d_{K-Q}\,. \tag{13}$$

The corresponding formulas for $K \geq 0$ follow from equation (12.9.38),

$$\frac{\overline{\partial}}{\partial x_i}(Fd_n^P\delta(\Sigma)) = \sum_{Q=0}^{P}\frac{P!}{Q!}\frac{\delta}{\delta x_k}(\mu_{ik}^{(P-Q)}F)d_n^Q\delta(\Sigma) + Fn_i d_n^{(P+1)}\delta(\Sigma)\,. \tag{14}$$

Indeed, if

$$\{F\} = Ad_K\,, \qquad K \geq 0\,, \tag{15}$$

then

$$\left\{\frac{\overline{\partial} F}{\partial x_i}\right\} = An_i d_{K+1} + \sum_{Q=0}^{K}\frac{K!}{Q!}\frac{\delta}{\delta x_j}(\mu_{ij}^{(K-Q)}A)d_Q\,. \tag{16}$$

In order to find the higher order derivatives of microdistributions we appeal to the general compatibility conditions as given in Chapter 12 and follow the same technique as explained above.

References

[1] Schwartz, L., *"Théorie des Distributions"*, Hermann, Paris, 1966.

[2] Rubinstein, Z., *"A Course in Ordinary and Partial Differential Equations"*, Academic Press, New York, 1969.

[3] Hoskins, R. F., *"Generalized Functions"*, Horwood, Chichester and New York, 1979.

[4] Roach, G. F., *"Green's Functions"*, Van Nostrand, Reinhold, New York, 1970.

[5] Lighthill, M. J., *An Introduction to Fourier Analysis and Generalised Functions"*, Cambridge University Press, London, 1958.

[6] Estrada, R. and Kanwal, R. P., *"Asymptotic Analysis: A Distributional Approach"*, Birkhäuser, Boston, 1994.

[7] Gelfand, I. M. and Shilov, G. E., *"Generalized Functions"*, Vol. 1, Academic Press, New York, 1964.

[8] Estrada, R. and Kanwal, R. P., *Applications of distributional derivatives to wave propagation*, J. Inst. Math. Appl. **26** (1980), 39–63.

[9] Jones, D. S., *"Generalised Functions"*, Cambridge University Press, London, 1982.

[10] DeJager, E. M., *"Mathematics Applied to Physics"*, (E. Roubine, ed.) Springer-Verlag, New York, 1970.

[11] Estrada, R. and Kanwal, R. P., *Distributional biharmonic equation*, J. Math. Phys. Sci. **14** (1980), 285–293.

[12] ———, *Distributional boundary-values of harmonic and analytic functions*, J. Math. Anal. Appl. **89** (1982), 262–289.

[13] ———, *Higher order fundamental forms of a surface and their applications to wave propagation and generalized derivatives*, Circ. Math. Palermo, Series II, **36**, 27–62 (1987).

[14] Farassat, F., *Discontinuities in aerodynamics and aeroacoustics, the concepts and applications of generalized functions*, J. Sound Vib. **55**, 165–193 (1977).

[15] Jain, D. L. and Kanwal, R. P., *Solutions of potential probems in composite media by integral equation techniques*, I, J. Integ. Eqns. **3**, 279–299 (1981), II, **4**, 31–53 (1982), III **4**, 113–143 (1982).

[16] Estrada, R. and Kanwal, R. P., *Regularization and distributional derivatives of* $(x_1^2 + \cdots + x_p^2)^{-n/2}$ *in* $\mathbb{R}^p$, Proc. Roy. Soc. London Ser. A, **401**, 281–297 (1985).

[17] ______, *Regularization, pseudofunction and Hadamard finite part,* J. Math. Anal. Appl. **141**, 195–207 (1989).

[18] Bowen, J. M., *Delta function terms arising from classical point source fields,* Am. J. Phys **62**, 511–515 (1994).

[19] Estrada, R. and Kanwal, R. P., *The appearance of non-classical terms in the analysis of point source fields,* Am. J. Phys. **63**, 278 (1995).

[20] Strichartz, R. S., *"A Guide to Distribution Theory and Fourier Transforms"*, CRC Press, Boca Raton, 1993.

[21] Arsac, J., *"Fourier Transforms and the Theory of Distributions"*, Prentice Hall, Englewood Cliffs, New Jersey, 1966.

[22] Watson, G. N., *"A Treatise on the Theory of Bessel Equations"*, Cambridge University Press, London, 1966.

[23] Beltrami, E. J. and Wohlers, M. R., *"Distributions and the Boundary Values of Analytic Functions"*, Academic Press, 1963.

[24] Bremermann, H., *"Distributions, Complex Variables and Fourier Transforms"*, Addison-Wesley, Reading, Massachusetts, 1965.

[25] Roos, B. W., *"Analytic Functions and Distributions in Physics and Engineering"*, Wiley, New York, 1969.

[26] Kanwal, R. P., *"Linear Integral Equations, Theory and Technique"*, Second Edition, Birkhäuser, Boston, 1997.

[27] Zemanian, A. H., *Distribution Theory and Transform Analysis*, McGraw Hills, New York, 1965.

[28] Wiener, N. and Hopf, E., *Über eine Klausse singulärer Integralgleichungen*, Semester-Ber. Preuss, Akad. Wiss. Berlin, Phys.–Math, Kl., 30/32, 1931.

[29] Estrada, R. and Kanwal, R. P., *Distributional solutions of the Wiener-Hopf integral and integro-differential equations*, J. Integ. Eqns. Applic. **3**, 489–514 (1991).

[30] ______, *Distributional solutions of singular integral equations*, J. Integ. Eqns. **8**, 41–85 (1985).

[31] ______, *Distributional solutions of dual integral equations of Cauchy, Abel, and Titchmarsh types*, J. Integ. Eqns. **9**, 277–305 (1985).

[32] Liverman, T. P. G., *"Generalized Functions and Direct Operational Methods"*, Prentice Hall, Englewood Cliffs, New Jersey, 1964.

[33] Stakgold, I., *"Boundary Value Problems of Mathematical Physics"*, Vol. I–II, Macmillan, New York, 1967, 1968.

[34] Stakgold, I., *"Green's Functions and Boundary Value Problems"*, Wiley, New York, 1979.

[35] Pan, H. H. and Hohenstein, R. M., *A method of solution for an ordinary differential equation containing symbolic functions*, Quart. Appl. Math. **39**, 131–136 (1980).

[36] Littlejohn, L. L. and Kanwal, R. P., *Distributional solutions of the hypergeometric differential equations,* J. Math. Anal. Appl. **122**, 325–345 (1980).

[37] Krall, A. M., Kanwal, R. P. and Littlejohn, L. L., *Distributional solutions of ordinary differential equations,* Canadian Math. Soc. Conference Proc. **8**, 227–246 (1987).

[38] Wiener, J., *Generalized Solutions of Functional Differential Equations*, World Scientific, Singapore (1993).

[39] Vladimirov, V. S., *"Equations of Mathematical Physics"*, Dekker, New York, 1971.

[40] Maria, N. L., *Elementary solutions of partial differential operators with constant coefficients*, Ph.D. Thesis, University of California at Berkley, 1968.

[41] Duff, G. F. and Naylor, D., *"Differential Equations of Applied Mathematics"*, Wiley, New York, 1966.

[42] Farassat, F., *"Class Notes of APSC-215, Analytic Methods in Engineering"*, The George Washington University, 1979.

[43] ______, *Introduction to generalized functions with applications to aerodynamics and aeroacoustics*, NASA Technical paper 3428, Langley Research Center, 1996.

[44] Goldstein, M. R., *"Aeroacoustics"*, McGraw-Hill, New York, 1976.

[45] Tanka, K. and Ishi, S., *Acoustic radiation from a moving line source*, J. Sound. Vib., **77**, 397–401 (1981).

[46] Ortner, N., *Die fundamental losing von produckten hyperbölishes operatoren*, preprint.

[47] Alawneh, A. D. and Kanwal, R. P., *Singuarlity Methods in Mathematical Physics*, SIAM Rev **19**, 437–471 (1977).

[48] Kanwal, R. P. and Sharma, D. L., *Singularity mehtods for elastostatics*, J. Elasticity **6**, 405–418 (1976).

[49] Kanwal, R. P., *Approximations and short cuts based on generalized functions,* Proc. Second Symp. on Innovative Numerical Analysis in Appl. Engng Sciences, University Press of Virginia, 531–542 (1980).

[50] Seodel, W. and Power, D., *A general Dirac delta function method for calculating the vibration response of plates to load along arbitrary curved lines*, J. Sound Vib. **65**, 29–35 (1979).

[51] Chwang, A. T. and Wu, T. Y., *Hydrodynamics of low Reynolds number flow*, J. Fluid Mech., **63**, 607–622 (1974).

[52] Datta, S. and Kanwal, R. P., *Spheroidal rigid inclusion in an elastic medium under torsion*, J. Elasticity **10**, 435–442 (1980).

[53] Alawneh, A. D. and Kanwal, R. P., *Elastodynamic singular points and their applications*, Int. J. Math. and Math. Sci., **4**, 795–803 (1981).

[54] Datta, S. and Kanwal, R. P., *Rectilinear ascillations of a rigid spheroid in an elastic medium*, Quart. Appl. Math., **37**, 86–91 (1979).

[55] ______, *Slow torsional oscillations of a spheroidal rigid inclusion in an elastic media*, Utilitas Math. **16**, 111–122 (1979).

[56] Kanwal, R. P., *Expansion coefficient of wave fronts in continuum mechanics*, Appl. Sci. Res. Section B, **16**, 435–440 (1964).

[57] Ffowcs-Williams, J. E. and Hawkings, D. L., *Sound generation by turbulence and surfaces in arbitrary motion*, Phil. Trans. Roy. Soc. London, ser. A, **264**, 331–342 (1969).

[58] Kanwal, R. P., *"Shock and singular surfaces in three-dimensional rotational gas flows"*, Doctoral dissertation, Indiana University, 1957.

[59] Kanwal, R. P., *On curved shock waves in three-dimensional gas flows*, Quart. Appl. Math. **16**, 361–372 (1958).

[60] ______, *Determination of the vorticity and the gradients of flow parameters behind a three-dimensional unsteady curved shock wave*, Arch. Rat. Mech. Anal., **1**, 225–232 (1958).

[61] Coston, R. C., *Jump conditions for fields that have infinite, integrable singularities at an interface*, J. Math. Phys., **22**, 1377–1385 (1981).

[62] Kanwal, R. P., *A note on jump conditions for fields that have infinite integrable singularities at an interface*, Unpublished Report, Math. Dept., Penn State, 1981.

[63] Estrada, R. and Kanwal, R. P., *Distributional analysis for discontinuous fields*, J. Math. Anal. Appl., **105** 475–490 (1985).

[64] Kanwal, R. P., *Applications of generalized functions to propagating and deforming interfaces*, Proc. First World Cong. Non-Lin. Anal., 3055–3068, Walter de Gruyter, Berlin, New York, 1996.

[65] Drew, D. A., *Evolution of geometric statistics*, SIAM J. Appl. Math., **50**, 649–666 (1990).

[66] Estrada, R. and Kanwal, R. P., *Nonclassical derivation of the transport theorem for wave fronts*, J. Math. Anal. Appl., **159**, 290–297 (1991).

[67] ———, *Some results in discontinuous fields and wave propagation*, J. Math. Anal. Appl., **128**, 389–404 (1987).

[68] ———, *An analysis for the delta function with support on the light cone*, J. Phys. A: Math. Gen., **21**, 2667–2675 (1988).

[69] ———, *A distributional theory for asymptotic expansions*, Proc. Roy. Soc. London A, **428**, 399–430 (1990).

[70] ———, *The asymptotic expansion of some multidimensional generalized functions*, J. Math. Anal. Appl., **163**, 264–283 (1993).

[71] ———, *A distributional approach to the boundary layer theory*, J. Math. Anal. Appl., **192**, 769–812 (1995).

[72] Ramanujan, S., *Notebooks*, 2 volumes, Tata Institute of Fundamental Research, 1957.

[73] Estrada, R., *The asymptotic expansion of certain series considered by Ramanujan*, Appl. Anal., **43**, 191–228 (1992).

[74] Krall, H. L. and Frink, O., *A new class of orthogonal polynomials: the Bessel polynomials*, Trans. Am. Math. Soc., **65**, 100–115 (1949).

[75] Grosswald, E., *Bessel Polynomials*, Lect. Notes Math., 698, Springer-Verlag, 1978.

[76] Morton, R. D. and Krall, A. M., *Distributional weight functions for orthogonal polynomials*, SIAM J. Math. Anal., **9**, 604–626 (1978).

[77] Hernández-Ureña, L. G. and Estrada, R., *Solutions of ordinary differential equations by the series of the delta function*, J. Math. Anal. Appl., **191**, 40–55 (1995).

[78] Hirchman, I. I. and Widder, D. V., *The convolution transform*, Princeton University Press, New Jersey, 1955.

[79] Hohlfeld, R. G., King, J. I. F., Drueding, T. W., and Sandri, G. V. H., *Solution of convolution integral equations by the method of differential inversion*, SIAM J. Appl. Math., **53**, 154–167 (1993).

[80] Murthy, A. S. V., *A note on the differential inversion method of Hohlfeld et al*, SIAM J. Appl. Math., **55**, 719–722 (1995).

[81] Kanwal, R. P., *Interplay between singularities, asymptotics and generalized functions, in Generalized Functions and Their Applications*, Edited by R. S. Pathak, Plenum press, 109–118 (1993).

[82] ———, *Delta series solutions of differential and integral equations*, Integ. Transf. and Special Funct., **6**, 37–50 (1997).

[83] Shanon, C. E., *Communications in the presence of noise*, Proc. IRE, **37**, 10–21 (1949).

[84] Estrada, R., *Fórmulas asintótoticas de reconstrucción*, Proc. Eighth Intern. Cong. Biomath., Costa Rica, 33–40 (1990).

[85] Gelfand, I. M. and Vilenken, M. Ya, *"Generalized Functions"* **4**, *Academic Press, New York, 1964.*

[86] Cristescu, R. and Mariinescu, G., *"Applications of the Theory of Distributions"*, Editura Academici, Bucharest, and Wiley, New York, 1973.

[87] Hida, T., *"Brownian Motion"*, Springer Verlag, New York, 1980.

[88] Lozanov-Grevenković, Z. and Pilipovic, S., *On some classes of generalized random linear functionals*, J. Math. Anal. Appl., **129**, 433–442 (1988).

[89] ______, *Gaussian generalized random processes on $K\{M_p\}$ spaces*, J. Math. Anal. Appl., **181**, 155–161 (1994).

[90] Kythe, P. K., *"Differential Operators and Applications"*, Birkhäuser, Boston (1996).

[91] Estrada, R. and Kanwal, R. P., *Applications of generalized functions in economics*, Nonlinear Analy. **5**, 1379–1387 (1983).

[92] Chapman, C. J., *Time-domain asymptotics and the method of stationary phase*, Proc. Roy. Soc. London A, **437**, 25–40 (1992).

[93] Prentice, P. R., *Time-domain asymptotics I, general theory for double integrals*, Proc. Roy. Soc. London A, **466**, 341–360 (1994).

[94] Estrada, R. and Kanwal, R. P., *Moment expansion analysis for time-domain asymptotics*, Math. Meth. Appl. Sci..

[95] Estrada, R., *On the Euler-Maclaurin Formula*, Bulletin de la Soc. Mat. Mexicana, **3**(3), 117–133 (1995).

[96] ______, *Dirichlet convolution inverses and solution of integral equations*, J. Integ. Eqns. Applic., **7**, 159–166.

[97] Sato, M., *Theory of hyperfunctions I*, J. Fac. Sci. Univ. Tokyo Sect. I 8, 133–193 (1959), II. **8**, 387–437 (1960).

[98] Sato, M., Kawi, T. and Kashiwara, M., *Microfunctions and pseudo-differential equations, in "Hyperfunctions and Pseudo-differential equations"*, Lecture Notes in Math. **287**, 265–529, Springer-Verlag, New York, 1973.

[99] Iagolnitzer,, *Microlocal essential support of a distribution and decomposition theorems – An introduction in "Hyperfunctions and theoretical physics"*, Lecture Notes in Math., **449**, Springer-Verlag, New York, 1975.

[100] Hörmander, L., *The Analysis of Linear Partial Differential Operators*, **I**, Springer-Verlag, Berlin, 1983.

ADDITIONAL READING

M. A. Al-Gwaiz, "Theory of Distributions," Marcel Dekker, New York, 1992.

P. Antosik, J. Mikusińsky, and R. Sikorski, "Generalized Functions, the Sequential Approach" Elesevier, Amsterdam, 1973.

J. Benedetto, "Harmonic Analysis and Applications," CRC Press, Boca Raton, 1996.

F. Constantinescue, "Distributions and Their Applications in Physics, Pergamon, New York, 1980.

A.L. Durán, R. Estrada, and R.P. Kanwal, *Extensions of the Poisson summation formula*, J. Math. Anal. Appl., to appear.

A. Erdelyi, "Operational Calculus and Generalized Functions." Holt, Rienhart and Winston, New York, 1962.

S. Fenyo', and T. Frey, "Modern Mathematical Methods in Technology," Vol. 1. North-Holland, Amsterdam, 1969.

G. B. Folland, "Fourier Analysis and its Applications," Wadsworth & Brooks/Cole, Pacific Grove, California, 1992.

A. Friedman, "Generalized Functions and Partial Differential Equations." Prentice-Hall, Englewood, New Jersey, 1963.

B. Friedman, "Lecture on Application-Oriented Mathematics." Holden-Day, San Francisco, 1969.

J. Horváth, "Topological Vector Spaces and Distributions," Vol 1., Addison-Wesley, Reading, Massachusetts, 1966.

J. Lützen, "The Prehistory of the Theory Distributions," Springer-Verlag, Berlin, 1982.

R. D. Milne, "Applied Functional Analysis, An Introductory Treatment," Pitman, London, 1980.

I. Richards and H.Youn, "Theory of Distributions," Cambridge University Press, Cambridge, 1990.

E. E. Rosinger, "Nonlinear Partial Differential Equations, Sequential and Weak Solutions," North-Holland, New York, 1980.

G. E. Shilov, "Generalized Functions and Partial Differential Equations." Gordon & Breach, New York, 1968.

G. Temple, "100 Years of Mathematics," Springer-Verlag, Berlin and New York, 1981.

F. Treves, "Basic Linear Partial Differential Equations," Academic Press, New York, 1975.

E. Zauderer, "Partial Differential Equations of Applied Mathematics," Second Edition, Wiley, New York, 1989.

Index

SPRINGER NATURE

GPSR Compliance

The European Union's (EU) General Product Safety Regulation (GPSR) is a set of rules that requires consumer products to be safe and our obligations to ensure this.

If you have any concerns about our products, you can contact us on ProductSafety@springernature.com

In case Publisher is established outside the EU, the EU authorized representative is:

Springer Nature Customer Service Center GmbH
Europaplatz 3
69115 Heidelberg, Germany

FSC
www.fsc.org
MIX
Papier | Fördert gute Waldnutzung
FSC® C083411

Kolibri 360 GmbH
Ferdinand-Jühlke-Straße 7
99095 Erfurt, Deutschland
produktsicherheit@kolibri360.de